LOW DAMS

A MANUAL OF DESIGN FOR SMALL
WATER STORAGE PROJECTS

Prepared by the
SUBCOMMITTEE ON SMALL WATER STORAGE PROJECTS

Books for Business
New York-Hong Kong

Low Dams: A Manual of Design for Small Water
Storage Projects

by
Subcommittee on Small Water Storage Projects

ISBN: 0-89499-083-7

Copyright © 2001 by Books for Business

Reprinted from the 1939 edition

Books for Business
New York - Hong Kong
http://www.BusinessBooksInternational.com

September 16, 1938.

Mr. FREDERIC A. DELANO,
  Chairman, Advisory Committee,
    National Resources Committee, Washington, D. C.

DEAR MR. DELANO:

We transmit herewith a report on "Low Dams" and recommend that it be published.

During recent years increasing attention has been given by all levels of Government to the impoundment of water with low dams. The Federal Government has constructed many small projects since the initiation of emergency relief activities during 1933. Experience with some of those structures has been unsatisfactory and has shown the need for improved practices in design.

The report was written by a subcommittee organized during March 1936 at the request of a group of Federal agencies concerned with the design and construction of small water storage projects. It was prepared by, and is for the use of Federal agencies participating in such work. It is intended to replace a number of compilations, no one of which was considered adequate for the purposes of design. It represents, therefore, a pooling of interest and effort on the part of all those Federal agencies concerned with low dams. It is an example of successful correlation of engineering techniques in the field of water conservation.

Sincerely yours

THE WATER RESOURCES COMMITTEE

ABEL WOLMAN, *Chairman*

BUSHROD W. ALLIN	T. B. PARKER
HARLAN H. BARROWS	JULIAN L. SCHLEY
MILTON S. EISENHOWER	represented by
N. C. GROVER	WILLIAM A. SNOW
EDWARD HYATT	THORNDIKE SAVILLE
ROGER B. McWHORTER	R. E. TARBETT
JOHN C. PAGE	

NATIONAL RESOURCES COMMITTEE

HAROLD L. ICKES, *Chairman*
Secretary of the Interior

FREDERIC A. DELANO
Vice Chairman

DANIEL C. ROPER
Secretary of Commerce

HARRY H. WOODRING
Secretary of War

HENRY A. WALLACE
Secretary of Agriculture

FRANCES PERKINS
Secretary of Labor

HARRY L. HOPKINS
Works Progress Administrator

CHARLES E MERRIAM

ADVISORY COMMITTEE

FREDERIC A. DELANO, *Chairman*

CHARLES E. MERRIAM   HENRY S. DENNISON   BEARDSLEY RUML

STAFF

CHARLES W. ELIOT, 2d
Executive Officer

HAROLD MERRILL
Assistant Executive Officer

WATER RESOURCES COMMITTEE

ABEL WOLMAN, *Chairman*

BUSHROD W. ALLIN
HARLAN H. BARROWS
MILTON S. EISENHOWER
N. C GROVER
EDWARD HYATT

ROGER B. McWHORTER
JOHN C. PAGE
JULIAN L. SCHLEY
Represented by
WILLIAM A SNOW

THORNDIKE SAVILLE
R. E. TARBETT
T. B. PARKER
—
GILBERT F. WHITE
Secretary

SUBCOMMITTEE ON SMALL WATER STORAGE PROJECTS

PERRY A. FELLOWS
Works Progress Administration
Chairman

R. W. DAVENPORT
Geological Survey
E. T GILES
Bureau of Reclamation
C. S. JARVIS
Soil Conservation Service
LEWIS A. JONES
Bureau of Agricultural Engineering
T W. NORCROSS
Forest Service

E. F. PREECE
National Park Service
A. L. WATHEN
Office of Indian Affairs
—
With the assistance of
M. B ARTHUR and MERRILL BERNARD
—
Editors
W W. HORNER, *Water Consultant*
and
J. J DOLAND, *Associate Water Consultant*

IV

PREFACE

The need for guidance in the planning of small water storage projects has been demonstrated increasingly during recent years by unprecedented activity in that field and by the reported failure of a number of small dams involved in public works programs in different parts of the United States. A demand has been developed for instructions, standards, and procedures, which will serve as a guide to safe practice for those who are engaged in the design of or are charged with the responsibility for designing such structures. This manual is prepared to meet this requirement.

The committee attempts to distinguish clearly between the project in its broad sense and the dam which may be the most important, and often the only structure involved in the project. The material in this manual is related almost exclusively to the designing of low dams and appurtenant structures. However, it is important that the designer be familiar with the purposes of the project, with the considerations upon which its justification may have been determined, and with the manner of arriving at the size and type of structure to be built. For these reasons, there has been included, as chapter I, an outline discussion of a desirable project investigation. It is expected that this chapter will be helpful to those responsible for project studies.

The concept of low dams, as covered in this work, includes those structures with heights to the spillway crest not exceeding 30 feet above the natural stream channel. This height is necessarily an arbitrary figure, as the same principles of design would apply to structures of slightly greater heights.

Low dams are properly considered as associated with small streams and drainage areas of limited extent. The exception involves larger stream basins with relatively moderate run-off rates due to the aridity of the area, or to the inclusion of unusual storage facilities, such as extensive natural lakes, ponds, or swamps.

This manual is prepared:

(1) To provide engineers fully capable of doing specific work assigned to them, with information and data necessary to the proper accomplishment of such work.

(2) To provide engineers responsible for work done originally by subordinates, with the information and data necessary for checking such work.

(3) To provide opportunity for the subordinate or partially trained or experienced engineer to obtain necessary knowledge in order that

he may improve his own work, add to his own value as an engineer and decrease the amount of review and checking by his superior.

This manual is not intended in any way to encourage the assumption of undue responsibility on the part of unqualified personnel, but rather to point out the importance of specialized training and stimulate the wider use of technically trained and experienced consultants.

This manual should be of service to all concerned with the planning of small water storage projects, but its availability in no way relieves any agency or person using it of the responsibility for safe and adequate design.

PERRY A. FELLOWS, *Chairman.*
R. W. DAVENPORT,
E. T. GILES,
C. S. JARVIS,
LEWIS A. JONES,
T. W. NORCROSS,
E. F. PREECE,
A. L. WATHEN,
The Subcommittee on Small Water Storage Projects.

Contents

INTRODUCTION

A service book cannot cover all details, or even refer to all the alternate methods, processes, or steps applicable to planning a given type of structure or to its design, construction, maintenance, and operation. Space limitations will permit in this volume only a discussion of the basic principles, governing factors and essential elements of design of the various types of low dam structures.

This manual is addressed to the designer of the structure, and does not include in its scope the field of construction practices or methods. However, as the integrity of the design requires adherence to limiting specifications for materials and to the practice of good workmanship in construction, there is included Appendix G, entitled "Construction Methods and Specifications," with references thereto at the end of each chapter. Obviously much more detailed specifications will be required to insure the proper construction of any specific dam.

Preparation of the document involved the collection and synthesis of much material in the files of Federal agencies, and it also required a number of original investigations.

The gathering of data, and the drafting and arranging of textual and illustrative material in preliminary form was largely the work of Mr. M. B. Arthur, Hydraulic Engineer, temporarily assigned from the United States Forest Service.

The preparation of the original text included in Hydrologic Studies was mainly the work of Mr. Merrill Bernard, accomplished shortly before his appointment as Chief of the River and Flood Division, United States Weather Bureau.

The sections on Soil Mechanics and Earthfill Dams were prepared by Mr. E. F. Preece, Assistant Chief Engineer of the National Park Service.

Various other chapters or portions of chapters were written or revised by subcommittee members and their associates.

In April 1937, the second and last preceding draft of the document was confidentially released to a large number of engineers in the Federal service, as preliminary information, but with a specific request for constructive criticism.

In September 1937, on the recommendation of the subcommittee, the Water Resources Committee requested an independent review by W. W. Horner, Water Consultant, as to the scope and general content of the document. Thereafter, all of the notes and criticisms and the

1

last named review were submitted by the subcommittee to the Chief Engineer of the Bureau of Reclamation. The office of the Chief Engineer of that Bureau analyzed all of the comments and prepared its own extensive review and detailed suggestions for further modification. The subcommittee discussed this last document in detail and prepared a rather voluminous set of minutes which contained its conclusions as to the modifications and additions which it approved.

At the request of the subcommittee, the Water Resources Committee then assigned to W. W. Horner, Water Consultant, and to J. J. Doland, Associate Consultant, the task of editing the final draft.

The instructions of the subcommittee gave the editors quite general authority as to rearrangement of material and revision of the language. It also placed upon the editors the responsibility for making those additions and modifications which it had approved. The editors were not authorized to make changes in the document that would substantially involve matters of policy upon which the subcommittee had passed, nor were they requested to make any changes in or assume any responsibility for mathematical presentations, and the use of particular formulae. In several instances, additional material was prepared by the editors. All such new material was reviewed and approved by the subcommittee.

Two sections of the document did not come under the editorial revision:

Appendix B, which was under revision by Mr. Preece at the time the editorial work was done and appendix D, Vegetation and Soil Erosion, which was being revised by a special committee within the Department of Agriculture, were not reviewed by the editors.

Some of the material is new as, for example, figure 4 and the information back of it which was prepared by Dr. C. S. Jarvis, and much of the material in the revised appendix D, which has not heretofore been brought together or put into such usable form. The material in chapter 6:2, relating to loads on conduits, was prepared by the editors with the assistance of the engineers of the Iowa Engineering Experiment Station.

Because of the numerous additions and changes which the Committee had approved, the final draft involved extensive rearrangement and material rewriting of some sections. The reading and rechecking of the finished draft was done by S. W. Jens, Engineering Associate, with W. W. Horner, who also made some suggestions for further improvement in the text.

CHAPTER 1

PROJECT STUDY AND PRELIMINARY INVESTIGATION

PROJECT STUDIES

1. Tests of Project Feasibility.—The first approach in studying a proposed project requires a determination of its feasibility. This involves the necessary project studies which will permit a sound analysis and conclusion with respect to the specific engineering-economic considerations. These are, primarily—

(a) That the service proposed to be performed through the project is justifiable, and that the project is responsive to an urgent social or economic need.

(b) That the project as planned will adequately serve the intended purpose.

The studies should determine that the difficulties inherent in the site which affect economy and safety of construction, and the quality of operation, have been satisfactorily foreseen; and that the structures have been planned to include a design that is technically sound and sufficiently representative of the actual structures that may be expected to be built after more detailed investigation. The soundness of the conclusions with respect to these matters will depend to a considerable degree on the completeness and accuracy of the preliminary investigation.

2. Extent.—There is no simple rule for determining the extent of the investigation which is necessary in any particular case. For example, a dam that is to be founded on a deep and extensively shattered rock and is to impound a depth of water of 15 feet will require much more foundation investigation than will be necessary for a 30-foot high dam to be built on a solid, unfractured rock under a shallow soil mantle. Within the limits of height treated in this manual, it may be said that the size of the physical structure has but little relation to the extent of the investigation necessary. The maximum investigation cost that is justified is, however, limited by the magnitude of the project. The project is generally unjustified if the cost of the necessary investigation would offset a large portion of the expected project value. Cost reduction accomplished by the elimination of a portion of the fundamental investigation is rarely a saving; it generally results in unanticipated construction or functional costs.

3. Primary and Secondary Studies.—The preliminary investigation, if carried to completion, is usually an expensive phase of the development. Moreover, it may indicate that the project is not economically or technically sound. For this reason, it is necessary that the project study be so planned and executed that probable soundness—or unsoundness—will be determined as early and inexpensively as possible. To accomplish this objective the investigation is divided into two phases: First, the primary or project studies; and, second, the detailed studies. The first supplies those data which are necessary to determine whether there is a reasonable justification for the project. The detailed studies should be undertaken after it has been demonstrated that the project is eligible for approval.

4. Aesthetic Value.—The aesthetic value may be of major importance to a project. In the location and design of a dam it should be given appropriate consideration which should begin with the preliminary studies and continue as a part of the design and construction. However, under no conditions should aesthetic values be allowed to outweigh the safety or adequacy of the design.

The applications of the principles involved vary so greatly that the solution with respect to any particular dam requires individual study. This subject cannot be properly covered within the scope of this manual. It is recommended that when aesthetic considerations are important a qualified landscape architect or engineer and, in some instances, an architect should collaborate in the design.

THE PURPOSE OF THE DEVELOPMENT

5. General.—In both the project study and the detailed investigation the purpose of the development determines, to a large extent, what should be investigated. The character of the site also has some bearing on what will be investigated, but for the most part this factor determines the extent rather than the scope of the investigation.

In many cases, the project will be of a dual or multipurpose type. The investigation, for this reason, may embrace a combination of the factors. The listing of such factors must be considered a general one and the investigation should not be limited to these if it is apparent that less common factors are involved in any individual case.

Feasibility studies should always consider possible objection to the project from the viewpoint of mosquito control. Freshly made impounded waters of a constant level are ideal breeding places for malaria-carrying mosquitoes. It should be kept in mind that there is a possibility of malaria in regions which for a long time have considered themselves to be well outside the malaria belt. In 1882 the disease extended into southern Canada and west into the prairie region as far as the outlying fixed military posts.

Many projects for which this manual is intended will be in regions affected by drought and subject to flash floods. Such floods will move large quantities of silt and may soon eliminate the storage capacity of a small reservoir. Loss of capacity due to silting of reservoirs or control of silt-laden streams should be considered in plans for all proposed reservoirs.

6. Irrigation.—(a) The supply of water must be adequate for successful irrigation at all times at a unit cost, in terms of irrigable area, that is economically reasonable.

(b) The quality of the water must be such that it will not be harmful to the crops and soils on which it will be used.

(c) If the distribution system is to depend upon gravity flow, the reservoir must be enough higher than the irrigated area to provide the necessary head for adequate delivery.

7. Recreation.—(a) There must be an adequate supply of water to provide for seepage and evaporation losses and maintain the water level within those limitations assumed as a basis for the recreational development of the shore line.

(b) The water must be free of pollution within the practical limits established for potability.

(c) If bathing is one of the purposes, there must be adequate depth of water in the vicinity of a gently sloping shore.

(d) The shore line should have a relatively steep gradient where possible so that a slight lowering of the water surface will expose a minimum of area. The normal range of the operating level should not include extensive areas of flat shore line which will be unsightly when uncovered. Probable use of marginal lands on shore lines should be considered in proposed plans and in estimates of required reservoir rights-of-way or easements. Easements for maximum and rare floods are cheaper than outright purchase and permit private use of shore line.

8. Wildlife.—(a) There must be sufficient depth and supply of water to maintain livable conditions for wildlife throughout the dry seasons.

(b) Extensive fluctuations in water level are inimical to fish and other wildlife as they prevent or destroy the development of aquatic vegetation required for food.

(c) Satisfactory quality of the water must be definitely assured. Excessively acid or highly alkaline water is harmful to many kinds of wildlife.

(d) The water and the basin must be suited to the production of the right kinds of food supply and must provide adequate shelter.

Projects to impound water for wildlife purposes should never be promoted without the advice of a biologist. An ability to catch fish or shoot game birds is not necessarily a guaranty of knowledge required for raising them.

9. Water Supply.—(a) The supply of water must be adequate to serve the requirements. The immediate demand and a surplus to care for the reasonably predictable increase in requirements are important considerations.

(b) The quality of the water must be such that it can be rendered potable by economical purification methods. The standard for bacterial purity is uniform for all projects. Standards with regard to taste, color, and odor, however, may vary with different sections of the country. The degree to which objectionable characteristics can be corrected will depend upon their concentration in the raw supply and the cost of the remedial measures necessary.

(c) Control and protection of the watershed area is desirable for municipal water supply reservoirs and marginal shore line lands should be purchased to provide control over use and prevent pollution.

10. Flood Control.—In the study and design of flood control projects and structures attention should be directed to the following considerations:

(a) The relation of the cost of control to the benefits derived through the reduction of cumulative damage, should be favorable when considered in the light of public interest.

(b) The temporary storage must be sufficient to lower the major peak flows or to decrease flood frequency.

(c) Insofar as is practicable, the method of control should be automatic rather than manual.

(d) An implied downstream safety that may not exist is more dangerous than no control at all.

11. Power Development.—(a) The capacity of the power generating equipment and the load demand are closely related to the quantity of water available and the amount of storage provided. The height of the dam is usually dictated by these requirements. Its determination is a special study which is outside the scope of this manual.

12. Stock Water.—(a) The quality of the water must be suitable for stock consumption.

(b) The pond should be situated where it is accessible to the stock, either directly or by the economical use of ditches or pipes.

13. Water Storage for Stream Flow Regulation.—There is a definite need for projects of this type in those regions where stream flow either ceases entirely or is reduced to extremely low values during parts of the year. Where such natural stream flow is the principal source of water supply for one or more communities, and where a dependable stream flow is necessary for the dilution of wastes after proper economical treatment, water storage for stream flow regulation may be justified. For such projects it must be determined that—

(a) The dependable annual stream flow after expected losses, including evaporation, have been deducted, is sufficient when properly regulated, to produce the minimum values of regulated flow required for the purpose.

(b) Storage for this purpose will not result in an objectionable alteration of the quality of the water.

14. Miscellaneous Water Conservation Projects.—Under this head will come occasional projects for the regulation of the water level in shallow lakes, swamps or bogs not covered by the purposes heretofore enumerated. This heading will also include projects for the detention or diversion of stream flow, and through the process of infiltration, for transforming surface water to, and conserving it as ground water.

(a) It should be noted that natural shallow lakes, swamps, and bogs generally exist because of an underlying tight subsoil, and that additional storage or surface water under such conditions will rarely be effective in increasing ground water, except where the stored water may be transported to and flooded over areas which are adapted to this type of conservation.

(b) Such projects will generally result in largely increased evaporation and transpiration losses because of the enlargement of water surface or the increased time of exposure of the surface water to evaporation or because of the possible increase in area of semiaquatic plants such as sedge grass and reeds. The net results of such projects after deducting such losses, must be positively beneficial. There is a danger that projects of this type will, in many cases, result in the loss through evaporation of water that might otherwise be beneficially used as stream flow elsewhere.

(c) For projects involving water spreading or the detention of surface water to increase infiltration and percolation opportunity, it must be determined that the soil characteristics are such that infiltration will occur in the quantity which justifies the project economically.

NOTE.—The Geological Survey of the Department of the Interior, the Bureau of Agricultural Engineering of the Department of Agriculture, or appropriate State bureaus should be consulted with regard to projects of this nature.

THE PROJECT INVESTIGATION

15. General.—The preliminary project investigation is a reconnaissance which serves to shed light upon the probable economic feasibility of the undertaking and becomes a basis for an orderly program of more and more detailed investigation as the apparent economic justification becomes more pronounced. The preliminary study program may include some or all the following items: (1) a general plan of development; (2) a review of the need for the project and its relation to other projects; (3) planning for future surveys and investigation programs should they appear to be warranted; (4) the location of and information regarding the dam and reservoir sites; (5) hydrologic investigations; (6) geologic investigations; (7) subsurface explorations; (8) vegetation and erosion studies; (9) sanitary studies; (10) availability of materials; (11) estimate of cost; (12) correlation of studies; and (13) project report.

16. Related Projects and Study of Need for Project.—(a) If a long-range planning program has been adopted for the vicinity, the proposed project should be consistent with this plan.

(b) The entire area to be served by the proposed project should be studied to determine whether there will be a conflict with other projects of a similar nature. This conflict may involve an unnecessary duplication of purpose or dual use of all or part of the same site or water. If the proposed project conflicts with other similar facilities either completed or contemplated, it will usually be found advisable to refer the matter to higher authority for a decision.

(c) The probable existing demand for the services rendered by the project as well as a reasonable estimate of future demand should be carefully determined.

17. General Plan.—It is assumed that the sponsor of the project will outline a general plan of development together with the objectives and purpose of the proposed project. Available maps, data, prior reports, history and other information will be collected for preliminary consideration and study, and a reconnaissance of the locale will probably be necessary to roughly determine possibilities and the scope of the project. A suggested reservoir capacity, height of dam and other requirements will be roughly outlined in the proposed general plan, and the character and scope of required surveys and investigations indicated.

18. Planning Surveys and Investigation Program.—During the project investigation and the reconnaissance, it is advisable to consider plans for further detailed surveys and investigations, particularly if it is known that this work will probably be done. Consideration should be given in a proposed program of surveys and investigations to: (a) The personnel required; (b) housing and subsistence for personnel; (c) location of field office if required; (d) transportation and other equipment, material, and supplies required for this work; (e) character of additional dam foundation investigations and equipment required; (f) availability of local labor and equipment; (g) arrangements with private land owners for entry to site to avoid trespass of lands during surveys; (h) transportation of drilling equipment to isolated and inaccessible locations; (i) consideration of horizontal and vertical control for surveys and available data from other surveys such as railroad and highway surveys, county engineer's surveys, etc.; (j) location of stream gaging stations and available hydrologic data; (k) climatic conditions for work; (l) sanitary conditions, if pollution will be a factor in limiting the utilization of project; and (m) estimate of time and funds required for work.

19. Project Location.—The project location should be established on a topographic map. The quadrangle maps issued by the United States Geological Survey can be used advantageously if available. Photostat enlargements are desirable to furnish maps with a scale of about 2,000 feet to the inch or less. If no map is available, a rough survey and sketch map should be made of the drainage area, showing the plan and governing elevations of watercourses, location and dominant elevations of the watershed, and the situation and character of important cultural and occupational features such as woods, cultivated land, pastures, swamps, roads, railroads, and buildings. Such a map will be adequate to show the proposed location of the dam, the outline of the reservoir that would be created by the dam, the geologic features of the drainage area, existing works, and other pertinent factors. If several dam sites are under consideration, a separate map should be used for the study of each site.

When the site of the proposed dam has been fixed, as the result of a study of the several factors discussed in succeeding sections, a project location map should be prepred and based on an accurate field survey. The

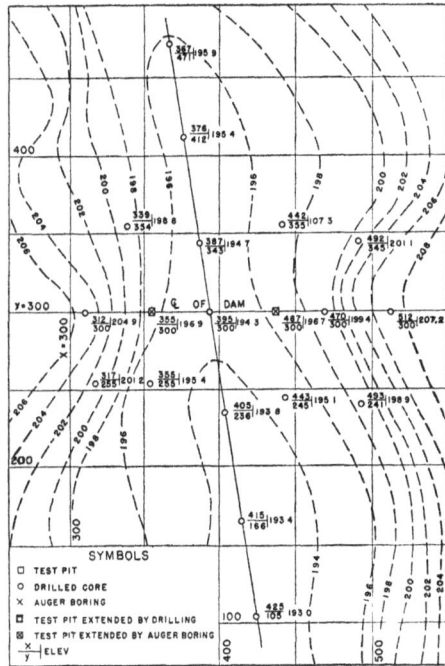

FIGURE 1 —Site plan

scale of the map should be such as to provide a 2-foot contour interval—at least within the construction area. Outside this area, a larger contour interval may be used, particularly in rugged country.

Elevations should be referred, wherever possible, to United States Government bench marks—Geological Survey and Coast and Geodetic Survey monuments—which will give altitudes above mean sea level datum. Property corners and the dam structure should be tied in with public land surveys where these exist.

To facilitate the designation of the location of borings, soil samples, and other investigations, a coordinate grid system can be used. The origin of coordinates should be established so that the entire working area will be within the first quadrant of the grid. This condition will be satisfied generally if the X-axis is approximately parallel to the center line of the dam and the origin so located that the extreme left end of the structure is at the point—X=300, Y=300; see figure 1. If the center line of the dam is a curved or broken line, the grid may be located as shown in figure 2.

Information about land ownership, land values, and attitude of land owners toward the project should be given in as much detail as is practicable.

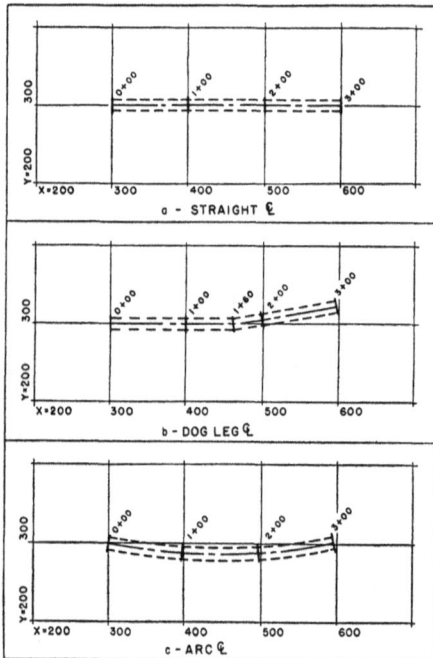

FIGURE 2

20. Hydrologic Investigations.—These will include primarily the collection of data with relation to surface water which must be analyzed for two principal purposes—

(a) In the preliminary or project studies there must be prepared the most accurate estimate possible of the dependable yield of the watershed as a basis of a determination of the justifiable storage. Such a determination is a critical factor in any project where the storage of water is an important purpose. The method of determining the safe yield will depend on the type of hydrologic data available and is outside the scope of this manual. It is suggested however, after all peritnent hydrologic data has been assembled, that the engineer should confer with the District representative of the United States Geological Survey and, if the problem is at all complex, should secure the consulting advice of an experienced hydrologist.

(b) The project studies must also include an estimate of the probable maximum peak flood flow as this is essential to a determination of spillway capacity and indirectly to a determination of the cost of spillway structures as a part of the project. Methods of arriving at a proper estimate of maximum flood flow are discussed in some detail in chapter 2. For the project investigation, the estimate may be prepared through the application of one of the approximate methods suggested. The estimate should be reviewed, revised, and re-checked in connection with the detailed investigations which will precede the actual design.

Ground water.—The primary investigation will also include a ground water study, which may be limited largely to determining its effect on construction methods. However, to the extent that unusual ground water situations are indicated as a result of the examination, this study may have an important bearing on the choice of type of dam to be constructed, and on the estimate of the cost of foundations. Important information with respect to ground water can sometimes be obtained in connection with subsurface investigations of foundation conditions.

21. Geologic Investigations.—Foundation conditions present one of the most important features to be investigated. Consideration must be given to the geology of the drainage area with relation to the watertightness of the proposed reservoir bed and the securing of suitable foundation materials on which to build the base and abutments of the dam. The primary studies should begin with an examination of geological maps, and reports that may be available in publications of the United States and State Geological Surveys. This material must be supplemented by a field examination, which should be made preferably by a geologist, depending on the difficulty and importance of the project.

This examination should indicate: (a) The nature and boundaries of recent deposits by streams, lakes, winds, and ice; (b) the character, structure, dip and strike of beds, shape and magnitude of folds; location, dip, strike and character of fault zones, and flow cleavage; and the direction, extent and width of crevices; (c) the classification of rocks as to age and origin, composition of aggregate and cementing material, and geologic processes which may affect the rock or soil structures; and (d) the relation of these geologic conditions to the permeability of the basin floor, and the future stability and permanence of the dam, spillway and other structures.

22. Subsurface Exploration.—A preliminary investigation is necessary to determine foundation conditions before a decision is reached as to the location of the dam and accessory structures.

Subsoil exploration may be made in various ways, depending on local conditions, as follows:

(a) Open test pits offer the simplest and best method of exploration in shallow soils. Usually pits of depths greater than 6 to 10 feet—depending on soil conditions—require bracing and sheathing, and involve considerable expense and time to excavate. Hence one or more of the following methods are often used, especially where economy of time or expense is required.

(b) A bar or rod may be driven down through light, shallow soils to the rock surface and occasionally give sufficient information.

(c) The earth auger is often used in soils of a cohesive character to bring up subsoil samples, even from a considerable depth.

(d) Wash boring and well drilling is employed to ascertain the nature of substrata at great depths. These methods are unsatisfactory for furnishing information as to structures of the deposit.

(e) Core drilling is the most expensive but the most reliable method of determining the nature, extent, and structure of subsurface materials. The core removed by the drill gives a continuous, accurate, and permanent record of the formations below the ground surface.

Detailed instructions on the use of the above methods of subsurface exploration are given in Appendix C.

(f) Geophysical prospecting employs recent methods and equipment which can be used where the geologic structures are not too complex and a well log or other data of neighboring holes are available as a basis for the interpretation of the geophysical survey results. The apparatus, method of operation, and interpretation of results of such a survey are complex and beyond the scope of this manual.

For small projects, and where the soil mantle is shallow and the geologic structure simple, the methods outlined under (a), (b), and (c) will generally provide sufficient information.

23. Vegetation and Erosion Studies.—(a) For the most part these investigations study the effect of vegetation, or lack of vegetation, on the water yield and run-off characteristics of the catchment basin and the degree of erosion with relation to its effect on both the life of the reservoir and the usableness of the water.

(b) The studies will include classification of the areas as wooded, cultivated, and pasture or barren land. It should determine the portion of each class of cover. The survey should determine the extent of erosion and particularly whether erosion has involved severe gullying which would require major corrective measures.

Where extensive erosion is in progress, a continuing study of the silt load of the stream should be made to determine both the amount of the load and the turbidity.

If the silt load is such that it will limit the life of the reservoir or the resulting turbidity such that it will limit the usefulness of the reservoir in any degree, an estimate must be prepared giving the cost and a description of the necessary corrective measures. See Appendix D.

24. Sanitary Studies.—(a) The necessity for a sanitary study is determined by the degree to which pollution will be a factor in limiting the utilization of the proposed project.

(b) All possible sources of pollution from human, animal, and industrial wastes should be investigated and evaluated.

(c) If municipalities are situated on the catchment basin their sewage disposal systems must be investigated. Water samples, adequately covering the period of the year during which the proposed project will be in use, should be taken from the watercourse below the municipality and analyzed, particularly if there is an outfall emptying into the stream. If the disposal system includes a bypass around the treatment processes directly into the stream, its effect should be evaluated and definite provisions made whereby the authorities in charge of the reservoir project will be notified prior to its use.

(d) Often a municipality has an inadequate disposal system but has plans for eventual improvement. In such cases it may be necessary to plan the proposed project for limited use pending the correction.

25. Availability of Materials.—Availability of natural materials for construction purposes may be an important element in the cost of the project. Such materials usually include: stone for masonry, concrete and riprap; gravel; sand; soil for cores and embankments; and timber.

The preliminary estimate should be based on a careful survey of the nature and extent of all local materials that can be used in the building of the project, and the possible sale of timber cleared from the drainage area. A comparative cost analysis is needed when available materials require processing, as similar materials already processed may be imported at less cost through regular commercial sources of supply.

26. Estimate of Cost.—A rough or approximate estimate of cost may determine the economic soundness and practicability of the project. Such an estimate will include probable costs of lands, water rights, easements and other legal requirements, damages, and the construction costs of dam and accessory structures, of clearing and grubbing the reservoir site and of relocating public highways, railroads, buildings, and other property if required.

The final estimate will be based on the subsequent detailed studies and should be in sufficient detail to serve as a guide for securing bids and awarding a contract for construction.

27. Correlation of Studies and Project Report.—It should be emphasized that incorrect conclusions from the studies are more often arrived at through forced interpretation of the data in favor of the project than through a lack of ability to properly evaluate those data. Under no consideration should a factor be assumed as favorable until that assumption is supported by all available data.

In certain sections of the United States water rights are commonly overlooked. The fact that no difficulties have arisen is not an excuse for omitting investigations of this important factor. Water rights must be investigated in every case.

The Sanitary Study will not be necessary for all projects. It is, on the other hand, often incorrectly omitted in the study of projects where purity of water is a major consideration. It is advisable where such study is required to consider all potential sources of pollution in their most unfavorable light.

Where competent geologists are available, the Geologic Study will offer few difficulties. If such assistance is not available, the greatest caution should be exercised in interpreting geologic characteristics.

When thorough consideration has been given to each of the component studies for a particular project they should each be briefed, listing (a) the favorable and (b) the unfavorable circumstances with regard to the project. Technical honesty should guide in the evaluation, and technical honesty will accept the unsoundness of the project if soundness is not proved beyond a reasonable doubt.

28. Outline of Project Investigation.—The outline given below provides a guide for the field engineer by indicating the items which should be considered for the investigation itself and for the guidance of the designer. The outline necessarily repeats the items which have been discussed in detail and it also contains items which may not be applicable to certain projects.

General Data and Maps.—1. Location and vicinity map. In addition to features outlined in Chapter 1: **18** it may be necessary to show the location of proposed work or features outside the scope of the reservoir and dam site maps such as:

(a) Proposed relocations of highways, railroads, and other public utilities.

(b) Earth dam material and concrete aggregate.

(c) Railroad shipping points.

(d) Stream gaging stations.

(e) Proposed construction camp.

(f) Existing works affected by proposed development.

(g) Existing public utilities, electric power transmission lines, etc.

2. Topographic map of reservoir basin (see also reservoir data required).

3. Land ownership and status map. May be shown on print of reservoir map (see reservoir data.)

4. Detail topography of dam site (see dam data).

5. Climate—seasonal conditions affecting construction and reservoir operation.

6. Hydrologic data (see ch. 1: **20**).

(a) Cross sections of stream, with water surface elevations at dam site, 500 to 1,000 feet upstream and downstream from axis of dam with stream discharge at time of measurement.

(b) All high watermarks, with dates and source of information.

(c) Status of water rights affected by proposed project. This varies with project and is usually considered by sponsor, who should outline under this heading, additional items requiring field investigation.

The Dam.—1. Detail topographic map of dam site using scale of 1 inch = 50 feet or 100 feet for long and large dam sites; with proposed coordinate system. Topography should cover a sufficient area upstream and downstream from the proposed dam and above proposed top of the abutments to provide for the design of auxiliary structures such as spillway outlets and cofferdams. Topographer should map all rock outcrops and apparent geologic features and accurately locate all man-made improvements and existing works on site.

2. Sufficient drill holes and/or test pits to determine character and depth of overburden.

3. Sufficient drill holes or open shafts to determine character of bedrock or impervious foundation stratum.

4. Complete, accurate logs of all drill holes, test pits, shafts and/or drifts, elevation of surface or collar of hole, location coordinates on map (see figs. 1 and 2) and sufficiently detailed remarks for a clear interpretation of records of formations for any subsequent consideration.

5. Location and character of proposed local material to be used in dam. Map borrow area, test material and show location of test holes. Secure representative samples for laboratory tests and analyses if possible.

6. Data on ground water table under abutments and adjacent to site.

7. Local conditions controlling dam design.

(a) Roadway required on crest?

(b) Fishway or fish conservation measures?

(c) Replacement of or provisions for existing works?

(d) Permanent building or quarters for operator?

(e) Spillway and outlet gates? Winter conditions, etc.?

(f) Electric power for construction?

8. Capacities and elevations of required outlets.

9. Local conditions affecting construction.

(a) Additional transportation facilities required for construction?

(b) Location surveys for railroad or highway?

(c) Required improvements to existing transportation facilities?

(d) Estimated cost or sufficient data for preparation of estimates for transportation facilities.

10. Average haul for each class of local material.

11. Average truck haul for shipped material.

(a) Hauling distance from nearest railroad shipping point.

(b) Local trucking rates, or freight rates on branch railroad.

12. Construction camp required?

(a) Estimated population and quarters required for supervisory and construction employees.

(b) Suggested sites shown on map.

(c) Required water supply and sanitation facilities.

(d) Local laws regarding sanitation, stream pollution, etc.

The Reservoir.—1. Topography of reservoir area. (Scale depends upon area, usually not less than 1 inch=1,000 feet; see Chapter 1:**30**). Preferably controlled by a triangulation survey system.

2. Probable life of reservoir; i. e. loss of capacity due to silting.

3. Land classification surveys.

(a) Show cultural classification and status by colors on topographic map.

(b) Show ownership boundaries and owners.

(c) Tabulation of areas and estimated costs for purchase and easements (see item (g)). Will an appraisal survey be required?

(d) Tabulation of areas to be cleared, with estimated cost.

(e) Cost of removing or salvaging buildings, fences, and farm improvements.

(f) Economic or physical limitation to maximum reservoir flowline.

(g) Use of marginal lands—public, private, recreational.

(h) Will easements for submergence during maximum and infrequent floods be satisfactory?

4. Road and public utility surveys.

(a) Relocation and reconstruction of railroads and highways.

(b) Relocation and construction of public utilities.

(c) Make joint reconnaissance with municipality or owner and submit preliminary report with approximate costs of relocation. Necessity for location surveys and who will construct?

5. Geology of reservoir.

(a) Report by qualified geologist desirable (ch. 1:**21**).

(b) Discussion of geologic formations, particularly such as cavernous limestone, exposed lava, exposed gravel and glacial deposits of a permeable nature that might contribute to serious reservoir leakage.

(c) Ground water table observations.

(d) Deleterious mineral and salt deposits.

(e) Photographs showing basin and character of lands.

(f) Geologic cross sections, where necessary.

DETAILED INVESTIGATIONS

29. General.—Under this head comes the collection of all information necessary for the detailed design of the structures and the preparation of construction plans. These investigations will in general only be undertaken after the project has been approved for construction, or when the probability of such approval has been sufficiently determined to justify the further expenditures.

Many of the smaller projects will not require, at this stage, any information in addition to that already obtained in the project investigation and study. The larger and more difficult projects will often require extensive additional surveys and investigation. Project size is not necessarily a criterion with respect to the necessity for further detailed studies. This may rest on a question of complexity of the site, of the foundation conditions, and often of the hydrologic factors.

30. Surveys and Maps.—The character and purpose of the field surveys required for the mapping of the dam site and the reservoir have been outlined under the head of "Project Studies." In many cases, an evaluation of the factors entering into the justification of the project may have required that surveys and investigations be carried out in considerable detail as part of the project studies. For such situations, relatively little additional field work may be required in connection with the design. Where, however, the project studies have been based on reconnaissance surveys and sketch maps, extensive field surveys must be carried out in connection with the detailed studies.

The detailed surveys will include the items listed in Chapter 1:**18** and **28**, and will differ principally from the surveys there set out under "Project Studies" to the extent that they must be more complete and accurate and must provide all of the information needed by the designer of the structure. For projects of importance, these surveys will require—

(a) That a triangulation system should be laid out for the control of topographic surveys for both the dam site and the reservoir. This should include a carefully chained base line centrally located, and all points should be permanently monumented.

(b) The form of coordinate system outlined in Chapter 1·**19** should be maintained as a control for the location of additional drill and test holes.

(c) If not already available, topographic surveys should be made of the dam site, generally by plane-table method, and the resulting map should be on the scale of 1 inch=50 feet. The contour interval should be governed by the topography, but in no case should exceed 5 feet.

(d) The topographic map of the reservoir should be on a scale of 1 inch=500 feet for smaller projects, or of 1 inch=1,000 feet for larger projects. The contour interval should be 5, 10, or 20 feet as indicated by the topography.

(e) Tailwater surveys.—For the purpose of determining the approximate stage-discharge relation below the dam site for various discharges, a profile of the river and cross sections should be taken. The profile should extend approximately 2,000 feet below the dam axis, or farther if local conditions indicate that a greater distance should be covered. Within the distance covered by the profile, cross sections of the river should be taken trom which the hydraulic properties of the channel may be determined. In general, probably four sections will be sufficient, but no definite rule can be made as local conditions change the requirements in this regard. If control sections exist, the sections should be taken at them.

(f) The maps should include all information listed in Chapter 1:**28** that may be pertinent to the particular project.

31. Detailed Hydrologic Investigations.—If the project studies have been relatively complete, the hydrologic investigations included in them may be sufficient for design purposes. If, however, probable maximum flood flow has, for purposes of the project study, been computed by approximate methods or without making full use of all available data, then these studies should be carefully reviewed and extended in detail before the actual design of the structure is undertaken. In general the methods to be used will be the same as those discussed under the heading "Project Studies," and are set out in detail in Chapter 2.

32. Subsurface Investigation and Foundation Studies.—For small projects and where the surface soil mantle is relatively shallow and of obviously satisfactory character, no extensive subsurface exploration may be required. Where the soil structure is more complex, extensive subsurface exploration may have been required as part of the project investigation, and if carried out at that time will be usable as part of the detailed investigation. Various methods of subsurface exploration have been listed in Chapter 1:**22**. The more important methods are described and discussed in detail in Appendix C. This appendix should be carefully examined and the recommendations given there should be followed as closely as possible.

33. Design of Structures.—Technical questions involved in the design of small dams of various types and in the design of the appurtenant structures, such as outlets and spillways, are presented in the various chapters covering outlets, spillways, and the different dam types.

34. Preparation of Estimates.—After determining required capacities and other data controlling the designs, it is necessary to make stress computations and stability analyses to determine detail dimensions of the dam and the auxiliary structures, such as the spillway, outlet, and control works. The general design can then be completed and schedules of construction quantities computed. In preparing estimates for excavation and embankment work, allowances must be made for wasted and unsuitable material, shrinkage of excavation in compacted fills, swell of rock excavation, overbreak in tunnel and channel excavation and in concrete lining and backfill. The contingency percentage usually added to an estimate is for the purpose of covering unforeseen difficulties, changes of plans, detailed items of design that owing to limited funds may be changed or possibly omitted. This contingency percentage does not cover over-excavation or excessive wastes in construction.   In order to be conservative and insure a uniform practice in preparing estimates, correction factors or percentages should be applied to net computed quantities for certain classes of work.   The major items likely to be in error are embankments, excavation, and concrete quantities.   Solid excavation from borrow pits for compacted earth embankments will shrink from 10 to 20 percent when compacted, in general about 15 percent; an additional allowance of 3 to 10 percent, or an average of about 5 percent, should be made for rejected materials.   An allowance of 20 to 40 percent, generally 25 percent, for swell in solid rock excavation for rock-fill and riprap quantities should be made.   From 10 to 25 percent should be allowed for overbreak in tunnel and channel excavation in rock. Adequate allowance for overbreak also must be made for concrete lining quantities of tunnels and channels.   An allowance of approximately 5 percent for cement and aggregate wastes must be included in the concrete quantities.   The estimator must be conservative in estimating quantities and yet avoid unreasonable and excessively high estimates.   Laboratory data, if available, are of considerable importance to the estimator in making the proper allowances for the estimated quantities; this is especially true of embankment materials and concrete mixes.   An important feature of estimating quantities is the general understanding of the definitions for the various items with respect to specifications, or "pay" dimensions.

In preparing a schedule of quantities it is generally advisable to list all items on a standard form and according to standard specifications so that the items may be carefully checked against quantities and omissions.

35. Unit Costs.—Because of widely varying economic and labor conditions in different localities it is not possible to outline safe estimating unit prices for each class of work.   The estimator should familiarize himself with local conditions, probable sources of materials and labor supply, costs of similar work in the locality and probable costs of materials and labor at the time of construction due to economic adjustments.   During periods of economic adjustment consideration should be given to construction cost indices and trends with an intelligent application to the particular project.   Due to improved methods and greater use of machinery and mechanical equipment, the costs of mass operations such as excavation, embankment, and mass concrete work have not risen in proportion to labor costs, in fact, for large-scale operation, costs have been reduced.   On the other hand, the cost of construction operations involving a large percentage of labor (e. g., reinforced concrete, form work, hand excavation, etc.) have greatly increased and if contractors' bids can be relied upon, the percentage of increased cost is somewhat greater than the increase reflected by direct labor costs.

36. Design Report.—A complete report should be prepared covering the investigation, engineering features, and the cost of the proposed dam and reservoir.   To insure a complete description and record of all essential data and calculations and conclusions entering into the design, a uniform procedure is desirable and, as a guide, there is attached hereto as Appendix F an outline of the items which the report should cover.   Obviously, all of the information listed in this outline is not necessary for any particular small dam, but the greater part of it will be required for the larger and more complex structures.

CHAPTER 2

HYDROLOGIC STUDIES [1]

1. Purpose of Hydrologic Studies.—This chapter considers the methods of determining the maximum flood flow to be expected from the drainage area tributary to the reservoir site, for which provision must be made in the design of the dam and the spillway structures, particularly the latter. Also there will be indicated the methods of determining the rate and volume of flood flows less than the maximum together with the probable frequency of their occurrence, primarily for use in connection with the design of supplementary spillways or other outlets for service during the period of construction.

2. Stream Flow Data.—The hydrologic data most directly useful in determining flood flows is an actual stream flow record of considerable length at the location of the dam. Such a record is rarely available. The engineer should determine what stream flow records are available for the general region in which the dam is to be located. He should consult the Water Supply Papers of the United States Geological Survey and, if possible, confer with the Survey's district engineer. He should also make a search of the records of other Federal agencies which may have collected information in the region, the records of State water conservation agencies or State geological surveys, and should determine whether any information may be available from other State departments, county engineer offices, from municipalities in the vicinity, or from utility companies. Where stream flow records are not available, some agencies or inhabitants of the vicinity may have information about high-water marks caused by specific floods.

With respect to the character of the stream flow data available, the expected maximum flood flows at the dam site may be determined under one of the following conditions:

(a) *Stream flow record at or near the dam site.*—If such a record is available and covers a period of 20 years or more, the flood flows shown by the record may be analyzed by one of the accepted methods, and a satisfactory determination for purposes of design may be arrived at both as to the maximum flood flow and as to somewhat smaller floods that may be expected to occur with any particular frequency.

[1] In any project involving water conservation, hydrologic studies for the determination of the safe yield of the drainage basin are essential. Such studies are indicated in chapter I as a necessary part of the project investigation, and their results will in many cases determine the justifiable storage and, thereafter, the height of the dam structure. The method of determining the safe yield is beyond the scope of this document.

If such a record is available, but covers only a few years, it may not include within its limits any flood of great magnitude. It will, however, permit an analysis of the larger floods shown, a determination of their relationship to the precipitation which produced them, and a determination of maximum flood flow through the application of this relationship to maximum expected storm precipitation.

(b) *Stream flow record available on the stream itself, but at a considerable distance from the dam site.*—Such a record may be evaluated as suggested in item (a) with respect to its own location, and an estimate may thereafter be made of the probable maximum flood flow at the dam site by one of the accepted hydrologic methods such as:

1. A transfer of maximum flood to the dam site on the basis of respective drainage areas with proper consideration of variations in topography, soil, and other important factors.

2. Through the application of the unit graph or distribution graph, if sufficient precipitation data also are available.

(c) *No adequate stream flow data available on the specific stream, but a satisfactory record for a drainage basin of similar characteristics in the same region.*—Such a record may be analyzed to determine:

1. The maximum flood flow per square mile for the region or, if satisfactory precipitation information is also available, to determine maximum flood flow characteristics through the application of the unit graph method.

(d) *Stream flow records in the region, but not satisfactorily useful for application and analysis under one of the above methods.*—The records may be assembled and analyzed as reference information in arriving at an estimate of the coefficient of run-off or of the rating percent in the Myers scale (see ch. 2:7).

(e) *Use of high-water marks.*—High-water marks, as pointed out by inhabitants of the valley, should be used with caution in estimating flood magnitudes. However, where there are a number of such high-water marks in the vicinity of the project, and particularly if such marks are obtained from the records of public offices, such as State highway departments or county engineers, they may be used as the basis of a separate supplemental study. These records may be used to determine the cross-sectional area and the water surface slope for the flood to which they refer, and from these data an estimate of that particular flood peak may be prepared using the slope-area method described elsewhere in this chapter. If a record of several great floods can be analyzed by this means, a further estimate of the maximum flood flow can be achieved.

3. Precipitation Data.—In each of the situations outlined in the preceding paragraph, precipitation data are useful in arriving at an estimate of the maximum flood flow. The engineer should assemble the information with respect to precipitation during the greater storms in the region, for all pertinent Weather Bureau stations, and for the full available length of record for each station. Such information can be obtained from the publications of the United States Weather Bureau in the form of daily total amounts. For small drainage basins, it will be desirable that the precipitation be known for shorter periods such as 12, 6, or even 1 hour. Where the source is cooperative Weather Bureau stations using the standard rain gage, such information can sometimes be obtained from the observers' original reports. If a recording gage is included in the group of stations used, more accurate information will be available.

The manner in which precipitation data should be organized will depend on the method or methods chosen for determining maximum flood flows. If used in connection with an application of the Rational Method, it will permit the construction of a rainfall intensity curve of the type shown in figure 5.

4. Collection of Hydrologic Data at the Site.—Whenever it appears that a considerable time may elapse between the selection of the dam site and the beginning of construction of the dam, there should be set up as promptly as possible facilities for securing both a rainfall record and a stream flow record for the project.

An unusual rainfall and flood may occur after the gages are provided and may produce evidence of inadequate capacity which would justify "eleventh hour" revision of the plans, thus adding to the security of the dam.

The rainfall and discharge stations need not be patterned after the more permanent and elaborate installations of the Geological Survey and Weather Bureau. Stream flow can be estimated from the record of water-surface elevation read on the ordinary staff gage after the rating curve for the station has been developed. Rainfall depth can be measured in any receptacle which can be calibrated.

In establishing the project rainfall and stream flow records, the following discussion, which is necessarily limited to important essentials, should be supplemented as far as possible with general reading from authoritative sources, the more important of which are listed at the end of this chapter.

5. The Precipitation Record at the Site.—1. *Type of gage:* The precipitation measurement should record total depth of fall during any convenient time period, usually 24 hours, as observed in some sampling device such as a wide-mouthed can or pail. The standard United States Weather Bureau rain gage consists of a 2-foot deep 8-inch diameter can, topped by an 8-inch diameter intercepter ring which feeds to a small inner tube of such diameter that 1-inch depth of water in the tube equals 0.1-inch depth of rainfall. If it is possible to secure these standard gages with their cedar measuring sticks, they are to be preferred to other forms of nonrecording gages.

Recording gages, such as the tipping bucket type, float type, and weighing type, should be used when available.

The use of a recording gage does not eliminate the need for non-recording gages. Each recorder should be paired with a nonrecording gage to serve as a check on the total catch indicated on the chart. Additional nonrecording gages can be placed between the recording gages.

In the absence of a standard or a recording gage, any wide-mouthed container, deep enough to avoid spattering, can be used. If a non-uniform diameter vessel is used it must be calibrated so that depth readings will be equivalent to average depth in inches per square inch of area of the intercepting section. The gage must be anchored in a vertical position, with the top 30 inches above the ground level.

2. *Location.*—In selecting locations for rain gages the following factors should be kept in mind:

a. If more than one gage is to be used, the total number should be distributed to give, as nearly as possible, equal areal representation to each gage and also to give complete coverage to the watershed. If but one gage is to be used, it should generally be located at about the center of the catchment area. For the larger watersheds, a single gage cannot be expected to give an accurate average depth for the whole of the area.

b. The nearest obstacle which could interfere with normal wind movements should be three times as far away from the gage as the height of the obstacle, and under no conditions nearer than the height of the obstacle itself.

c. In rolling and rugged country an attempt should be made to give equal representation to leeward and windward exposures (with respect to the prevailing wind direction). For slope locations, a site two-thirds of the length of the slope below the crown of the hill is thought to be the least affected by eddies over a sharp hilltop. Insofar as possible, the gage site should be on level ground.

The number of gages to be established will vary with the availability of funds, the size of the watershed in question and the amount of time which the engineer or his assistants can devote to their attention.

3. *Observations.*—Nonrecording gages should be read at least once a day at a fixed time, and preferably twice a day. If possible, notations as to the beginning and ending time of rains should be made, particularly for the intense rains causing sudden and large increases in stream flow.

Intensity of rainfall can be obtained with a nonrecording gage, but only at the expense of considerable time and effort. The observer must note the time when the rain begins and record the depth at regular and frequent intervals, as, for example, every 15 minutes, finally noting the time of the end of the rain.

Snowfall can be collected in the rain gages, or measured directly as depth of fall on bare ground or upon the crust of previous snowfall. Care must be taken to select a site where drifting or blowing has not occurred. Finally, it will be necessary to melt the snowfall to determine the equivalent rainfall depth.

6. The Stream Flow Record.—a. *Gage locations.*—The primary requisite for initiating stream-flow studies is the selection of a gage location close to the proposed dam site, but not so situated that it will be affected by construction operations. Dams are ordinarily located at a constriction of the stream channel and near such constrictions excellent control sections are frequently found. The control is the natural section or reach of the channel below the gage which determines the stage-discharge relation. If the section chosen is permanent, that is, not subject to scouring, filling, variations in vegetal growth, or backwater, then the stage-discharge relation is constant.

One cross section will not always serve as the control throughout all ranges of stage. For low-water conditions a ledge across the channel, a gravel bar, or a riffle caused by large boulders may serve to establish a constant stage-discharge relationship. However, with a rise in stage, these low-water control factors are frequently "drowned out" by the effect of some constriction in the channel farther downstream. Such a change in control factors causes an increase in slope of the curve of plotted discharge against stage measured at the gage. If money and time are available an artificial control of timber, stone, or concrete may be constructed which will stabilize the section and carry its effectiveness to higher stages.

The measurement of stage should be made in the lower end of the pool but sufficiently above the control to be beyond its draw-down effect.

b. *Gage installation.*—If an automatic stage recorder is available, instructions regarding its installation and shelter can be found in "Equipment for Gaging Stations for Measuring River Discharge," prepared by the United States Geological Survey.

Normally, a simple nonrecording gage known as a staff gage will be used. This may consist of a 2-inch by 6-inch, or heavier, timber marked in feet, tenths and half-tenths of elevation, the zero being set by levels to agree with the zero point on the control, the lowest stage of stream flow. The gage markings should be of permanent nature such as saw cuts or staples and should be easily read by the observer at all stages.

If the bank of the stream is sloping, or if the stream occupies a narrow channel during low stages and floods out over its banks during high stages, the gage should be installed in sections so that the section being used will always be accessible to the observer.

A permanent bench mark should be set near the gage for the purpose of periodic checks on its elevation.

3. *Velocity and cross section.*—The method of measuring cross-sectional area and average velocity with the use of a current meter and sounding devices is described in detail in publications of the United States Geological Survey. Frequently in reconnaissance or preliminary investigation the engineer must make these measurements without the current meter and in such cases the velocities may be measured by the "float" method or computed by the slope-area method.

The float method, requires for best results the selection of a fairly straight section of the stream channel from 200 to 1,000 feet in length, with a satisfactory uniform cross section. For low stages specially designed floats are frequently used. For high stages and floods, driftwood or ice cakes can be used.

Range poles are established at the two ends of the course on both banks, normal to the center line of the stream; the distance between them is determined, and the time is observed for the selected floats to travel the known distance. A careful record should be made of the course traversed by the float. Since several such runs must be made with floats in different portions of the channel, it is obvious that a steady flow, i. e., a condition of constant stage, is desirable for satisfactory results.

Depending on the type of float used, a coefficient must be applied to the observed velocity to reduce it to mean velocity for the section of channel represented by each observation. The several mean velocities are averaged to give the average mean velocity, which multiplied by the mean cross-sectional area will give the discharge.

A surface float to be practicable must move with the same velocity as the water surface. Because of its lightness this type of float is very sensitive to air disturbances. Inasmuch as it also measures only the velocity of water at the surface, considerable uncertainty may attend the selection of proper coefficients which must be applied to all such observed velocities in order to obtain the mean values. If this cannot be done by an expert hydraulic engineer, a coefficient of 0.85 may be used for a rough approximation.

Subsurface floats consist of a submerged float and an indicating surface float attached together by an adjustable line. In channel reaches having a uniform longitudinal section, for which the position of the mean velocity in the vertical may be computed, this type of float will measure the mean velocity directly if the submerged float is placed at the computed position, which for average conditions will be about 0.6 of the depth. Allowances, however, must be made when placing the submerged float at the desired position to take into consideration the accelerating or decelerating effect of the connecting line and surface float, which will vary with wind conditions and other local factors.

Tube and rod floats are designed also to measure directly the mean velocity in a vertical plane, but in so doing they must not be permitted to contact the bed of the stream. Consequently, except for longitudinal sections possessing a high degree of uniformity, neither of these types of floats will register the effect of the slower moving water near the bed of the stream, and as a result a coefficient less than unity is necessary to reduce the observed velocity to the mean value.

For rough approximations, the following coefficients may be used:

Ratio of rod submergence to average stream depth.	Coefficients
0.90	1. 00
.75	. 95
.50	. 92

The timing of selected pieces of drift or ice cakes will give results more reliable than the surface float and comparable to those obtained with the floating tube or rod when the latter involves the use of coefficients. Floating ice cakes and heavy drift, comprised of logs and trees which are practically submerged, hold their course well against surface disturbances, and if observed at a sufficient number of points distributed across the width of the stream will give a fair estimate of the surface velocity. This determination also requires a coefficient ranging from 0.85 to 0.90 in order to obtain the mean velocity in the vertical section. As in the case of the slope method the results obtained by the float method are subject to considerable inaccuracy or error except under the most favorable conditions and should be used with appropriate caution.

The slope-area method of measuring the discharge of a stream consists in determining (a) the mean area of the channel cross section, (b) the mean hydraulic radius, (c) the slope of the water surface, and (d) the character of the channel lining, in order to choose a suitable roughness factor. With these data, the mean velocity of the stream may be found by the Chezy formula, $v = C\sqrt{rs}$. The discharge will be the product of this velocity and the mean area of the cross section. C is a coefficient depending on the roughness of the stream channel, the hydraulic radius, and, to a slight extent, the surface slope. r is the hydraulic radius and is equal to the quotient of the mean area of the cross section in square feet divided by the mean wetted perimeter in feet. s is the slope of the water surface expressed as a decimal.

In order to determine the slope of the surface of the stream, it is necessary that a course be chosen similar to that required in the float method—i. e., of satisfactory uniform cross section and fairly straight. The length of the course should be as great as possible—up to 1,000 feet and never less than 200 feet—in order that the percent error in the slope determination shall be as small as possible. This is

particularly important for streams of very flat slopes. On the other
hand, since the slope to be used in the Chezy formula should be fairly
uniform, it may be necessary to shorten the length of the course in
order not to include rapids or other abrupt falls.

The slope will be determined by means of gages placed at the ends
of the course and read simultaneously. It is desirable that there be
at least two gages at each end, one at each bank, and the average of the
two readings be used. All gages should be set with reference to some
bench mark and tied together by means of levels. Inasmuch as the
gages should be read to hundredths of feet, they should be protected
from all wave action. This is most easily accomplished by surrounding each gage with a stilling box.

Where gages are not installed, or when flood stages overtop the
gages, the slope may be determined by means of reference marks on
posts, trees, bridge piers, etc., in as great a number as practicable,
these marks being referred to a common elevation.

Unless the channel cross section is satisfactorily uniform it will be
necessary to correct for changes in velocity head or make other hydraulic adjustments which are of such complexity that they can be successfully performed only by the most experienced engineers. In artificial
channels, the section is practically constant so that only one section
may have to be measured, but in the case of natural streams the cross
sections may be different at every point in the stream so that there is
no single section which can safely be used as a basis for the area and
wetted perimeter in applying the Chezy formula. In such instances
the number of sections to be measured will depend on the length of
the course and the configuration of the channel of the stream. Enough
sections should be taken to furnish a reasonably accurate value of the
average area and wetted perimeter.

The remaining factor C can be evaluated by means of formulas, of
which Manning's $C=\frac{1.486}{n}r^{1/6}$ is the simplest in practical application.
In this formula n is the roughness factor [2] and r the hydraulic radius.
When substituted in the Chezy formula there results,

$$v=\frac{1.486}{n}r^{2/3}s^{1/2} \text{ (commonly called Manning's formula)}.$$

Kutter's formula for C is also well known and widely used. It
likewise contains a roughness factor, n, which for a wide variety of
channel conditions in small streams is equivalent to that in the
Manning formula.

A general discussion of the Kutter formula will be found in the
section on Open Channels, Chapter 5—Spillway Structures, and in
Appendix A.

d. *Use of Records.*—It is expected that the rainfall stations have

[2] See Appendix A for values of n, also ch. 5, table 5B.

been so located as to best represent the rainfall distribution over the whole of the area, in which case the arithmetical average of the records will give an acceptable average depth for the watershed. However, if the distribution of stations does not provide for areas of equal size, those stations representing the larger areas should be accorded greater significance.

The record of rainfall should be closely correlated with that of stream flow, particularly on the smaller watersheds.

The first step in the use of the record of stream stages is the development of the rating curve for the station which is drawn by plotting stage against discharge as computed from the velocity-area measurements. As time passes, failure of the plotted points to conform to the curve as reliably defined, indicates an unstable, shifting control. If the differences are not great, one average curve may serve the needs of the project, but if the deviations from the average curve are consistently greater than 10 percent it will be better to compute the discharges for the various periods using rating curves developed within each period.

The record of stage is converted to discharge by means of the rating curve or a table prepared from the curve.

Hydrographs of the larger flood flows as measured at the site should be plotted and studied to indicate the relation between rainfall and stream flow for the various storm periods.

7. Methods of Estimating Flood Peak Discharge.—The method to be used in estimating flood peak discharge will depend—

(a) On the character and applicability of the stream-flow data available;

(b) On the size of the project and the extent to which a determination of spillway capacity may affect the cost of the spillway structure, and consequently of the total project, in an amount that may be critical to the project justification; and

(c) On the size of the drainage basin above the dam site.

For important projects, and particularly under the conditions outlined in (b) above, the best possible use of stream-flow data should be made in the manner suggested in Chapter 2:2-a, b, and c. Except for the situation outlined under 2:2-a, that is, for a long time stream-flow record at or near the dam site, the hydrologic studies involved are extremely complex, and should only be undertaken by a hydraulic engineer or hydrologist experienced in this class of work. For all important projects, a determination of maximum flood peak discharge should be referred to an expert in this field.

For small projects, or for those projects in which spillway capacity can include an ample factor of safety at relatively low cost, a sufficient approximation of probable flood peak discharge may be arrived at through an application of the Rational Method or through the use of a flood flow formula as outlined in the succeeding section. Frequently it may be desirable to apply both methods as an aid to final judgment.

8. Use of Flood Flow Formulas.—Discussion of several long-established flood-flow formulas can be found in texts on hydrology and also in the recent Water-Supply Paper No. 771 of the United States Geological Survey entitled "Floods in the United States." The hydraulic engineer, in the course of his work, may have established confidence in one or more of these formulas as the means of reflecting his own experience and judgment. The older empirical formulas, however, disregard many important factors that are known to influence the magnitude of flood peaks. There is also obvious danger in using the coefficients of such empirical formulas on watersheds having characteristics differing from those upon which they were developed.

A major distinction in flood type—namely, run-off predominantly from melting snow and ice—sets apart the rugged western regions

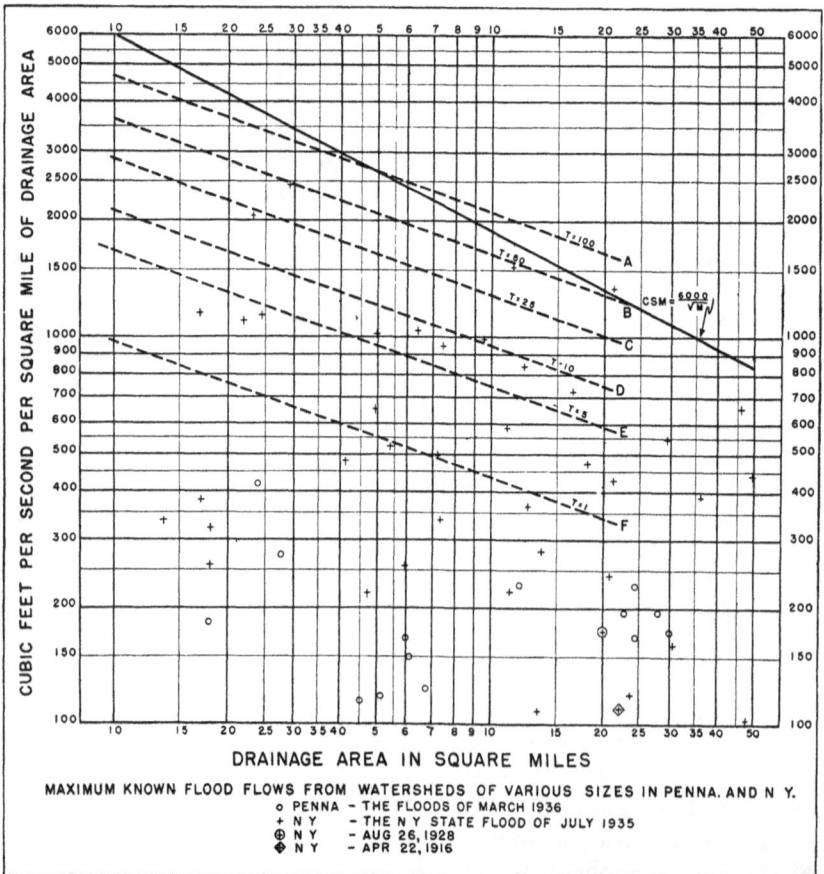

MAXIMUM KNOWN FLOOD FLOWS FROM WATERSHEDS OF VARIOUS SIZES IN PENNA. AND N Y.

o PENNA – THE FLOODS OF MARCH 1936
+ N Y – THE N Y STATE FLOOD OF JULY 1935
⊕ N Y – AUG 26, 1928
◆ N Y – APR 22, 1916

FIGURE 3

from the central and eastern portions of the country. The west is also typified by a paucity of basic hydrologic data which definitely precludes the possibility of estimating floods from determinable controlling factors such as excessive rainfall rate.

West of the one hundred and third meridian, flexibility of method and the accompanying economy in design may have to be sacrificed in all except the isolated and infrequent case where adequate local records make possible an evaluation of the hydrologic factors for the locality. For drainage areas not thus served, the engineer must resort to a limiting flood formula, in which maximum peak flow is expressed as a function of area.

East of the one hundred and third meridian, limiting flood-flow formulas may also be used to advantage, and factors making them applicable to different parts of the country are generally available. However, in this area precipitation data are also available to an extent that makes their use, together with a consideration of physiographic factors, often the more satisfactory basis of estimating flood flow.

The two methods above referred to are compared with observed data in figure 3. The curves A, B, C, D, and E shown represent average conditions for southern New York and western Pennsylvania and have been computed by the Rational Method. (See ch. 2:**10**.) The points shown are observed floods maxima for the same region.

9. The Limiting Flood Method.—When flood maxima q in cubic feet per second per square mile are plotted against area of drainage basin, it is found that the enveloping curve can generally be expressed by the equation

$$q = \frac{C_m}{\sqrt{M}}$$

in which C_m is a coefficient varying with locality and M the area of the drainage basin in square miles.

The curve

$$q = \frac{6,000}{\sqrt{M}}$$

reasonably defines the upper limit of flood peaks for eastern United States, expressed in terms of watershed area and is exceeded only by rare occurrences which are in general confined to the Gulf States.

Wherever flood peak discharge has to be determined by the use of this formula, stream-flow data assembled as suggested in chapter 2:2 should be reduced to flood flow per square mile and entered on a diagram similar to figure 3 which will permit an evaluation of C_m from the best local information. Obviously, stream-flow records used for this purpose must be of considerable length, or if only short records are available a generous factor of safety must be used in fixing the enveloping line.

9a. Flood Ratings.—If the formula given above is multiplied by the total drainage area M, it will represent total peak flow instead of flow per square mile, and will take the form

$$Q = C_m \sqrt{M}.$$

In either form, Q or q is in cubic feet per second, and area M is expressed in square miles. In the Myers formula, the base factor C_m is taken as 10,000 but for any particular area is multiplied by a percentage which is referred to as the Myers rating. Thus the Myers percentage rating of 30 is equivalent to a C_m factor of 3,000.

To assist in visualizing the flood potentialities of the various regions within the United States, as experienced at widely scattered stream-gaging stations listed in Appendix I (though possibly exceeded in many locations for which no records are available), figure 4 has been prepared showing the maximum ratings in percentage on Myers scale. These are really one one-hundredth of respective numerical values assigned to C_m in the foregoing equation, or $\dfrac{C_m}{100} = p$. Thus, the flood peak in cubic feet per second,

$$Q = qM = C_m \sqrt{M} = 100p \sqrt{M}.$$

These percentages are not strictly comparable with each other due to the wide variation in (1) length of periods covered by respective records; (2) the influence of departures from the mean, either plus or minus, in average intensities, durations, and maximum rates for both rainfall and run-off; and (3) departures from periodicity corresponding to length of record, manifested by the maximum observed run-off; for example, the 100-year flood occurring during a very short period of record.

It will be noted that wide variations of these percentage ratings occur along a given stream, as for instance the Arkansas, the Ohio, or any other large drainage system, with the preponderance of numbers considerably below the maximum. The enveloping curve of maximum ratings may possibly give undue weight to outstanding values, as for example, in southern Texas, where several stations have recorded 100 to 200 percent on the Myers scale, while other stations interspersed in the same area have not exceeded 50 percent.

It must be understood that not only each stream but each gaging station of a drainage area presents an individual problem peculiar to itself with major modifying factors represented by rainfall and run-off habits and all physical features which affect them. Thus are included both rainfall and run-off intensities and durations, facilities

for stream-flow concentration, regulation by storage or detention, infiltration, surface slope, maturity of drainage systems, diversions and losses enroute, and opportunities for smoothing of flood crests during their progress downstream.

For the western portion of the United States, the upper limiting values of the coefficient C_m might vary from 2,000 or less for the Great Basin region to 7,000 (equivalent to 70 percent on Myers scale) in the coastal and mountainous regions of highest rainfall intensities. Frequently, the appropriate coefficient for a given locality would be considerably less than the limiting values.

These coefficients are applicable to areas of more than 4 square miles. For lesser areas, the first power of M as used in the Rational Method seems to be more nearly applicable and the formula becomes $Q = C_m M$ for 4 square miles or less.

The propriety of the use of the limiting flood values represented by these formulas becomes more questionable as the area reduces in size. For example, small areas in any region may include a basin of extremely rugged topography and one of flat or rolling character. The limiting flood as expressed by this formula will be that applicable to the steeper and rugged conditions and may be several times the appropriate value for the flatter basin. Because of this situation, the limiting flood-flow formulas should be used with caution for areas less than 25 square miles, and where possible should be checked against a separate determination—possibly through the use of the Rational Method.

Where circumstances appear to warrant a reduction from the regional maximum, fullest possible use should be made of all pertinent local and neighboring hydrologic data.

To illustrate the use of hydrologic records from neighboring drainage basins to determine the best value of the factor C_m, let us assume that reservoirs are to be established at two sites on Red Willow Creek, Nebr., a headwater tributary of the Republican River, with drainage areas of 100 and 500 square miles, respectively, for which no hydrologic data are available. It so happens that most of the rainfall and run-off records in neighboring areas are either brief or intermittent in character, perhaps reflecting the highly erratic occurrence of precipitation and resulting streamflow.

From a broad viewpoint, we may consider a large area of the Great Plains, perhaps within a 100-mile radius, centering about the proposed projects as essentially similar and comparable as to rainfall potentialities even though the run-off may vary more widely because of differences in soil, slopes, vegetation, shape, and other physical characteristics of the drainage basins concerned.

Maximum stream-flow records and flood ratings for the locality are given in table 1. From this tabulation, it is apparent that ratings according to the Myers scale of 10 to 15 percent may occur frequently, and under unusual circumstances may be as high as 30 percent. Neglecting the moderating influence of storage at the sites under consideration, elongated shape of the drainage basin, or favorable detention factors, it would be advisable to provide spillway capacity for $3,000 \times \sqrt{100} = 30,000$ c. f. s. and $3,000 \times \sqrt{500} = 67,000$ c. f. s., respectively.

10. The Rational Method.—For situations where no useful stream-flow data are available it is necessary to estimate peak flood flow entirely from precipitation information and a consideration of the physical characteristics of the drainage basin. An estimate of this character can generally best be made by an application of the Rational Method. For very small drainage basins, generally less than 10 square miles, the Rational Method is often the only means by which maximum probable flood flow can be determined. This method is outlined in principle in this chapter and is discussed and illustrated in detail in Appendix A.

The rate at which water is discharged from a drainage basin depends directly upon the rate at which that basin receives, in the form of precipitation, that water which it is to discharge. It also depends in varying degrees upon other things which serve to modify that rate by diverting or storing a part of this flow or augmenting it in some manner.

TABLE 1

Stream and station	Years of record	Drain-age area, square miles	Date	Maximum stream flow, c f s		Myers scale rating
				Total	Per square mile	
						Percent
Republican River, Newton, Colo	----------	1, 270	May 1935	103, 000	81 1	28 9
Republican River, Kansas-Nebraska line	----------	2, 550	----do	150, 000	58 8	29 7
Republican River, McCook, Nebr	----------	12, 000	----do	245, 000	20 4	22 3
Republican River, Wakefield, Kans	1917–34	24, 700	June 1915	70, 000	2 83	4 4
Republican River, Junction City, Kans	1895–1905	25, 000	May 1903	71, 000	2 84	4 5
Kansas River, Ogden, Kans	1917–34	45, 200	June 1935	170. 000	3 76	8 0
Kansas River, Topeka, Kans	1903–34	56, 700	May 1903	220. 000	3 88	9 2
Arikaree River, Cope, Colo	1935	690	May 1935	25, 000	36 2	9 5
Arikaree River, Idelia, Colo	1935	1, 190	----do	54, 00C	45 4	15 7
Arikaree River, Haigler, Colo	1935	--------	----do	50, 000	31 2	12 5
Solomon River, Niles, Kans	{ 1897–03 1917–34	} 6, 770	June 1903	41, 000	6 06	5 0
Smoky Hill River Near Abilene, Kans	{1904–21	}18, 700	{August 1915	13, 600	73	1 0
At Solomon, Kans	{1922–34		{September 1928	18, 400	98	1 34

If 1 inch of rain falls on 1 acre of area, the total quantity thus to be discharged is $43,560 \times \frac{1}{12} = 3,630$ cubic feet. If the rate at which it flows away were the same as the rate at which it falls, then 1 inch of rainfall per hour on 1 acre would produce a discharge of 3,630 cubic feet per hour or 1.008 cubic feet per second. Thus it may be said that, subject to some modification, rates expressed in cubic feet per second per acre and in inches per hour, may be considered as equivalent.

The intensity of rainfall varies considerably during any one storm, but the average intensity for any period will always be more than the average intensity for a longer period. For any specific location, this difference in intensity for varying observed periods of duration may be shown by a curve (see fig. 5).

RAINFALL INTENSITY-DURATION-FREQUENCY CURVES FOR SOUTH CENTRAL OHIO

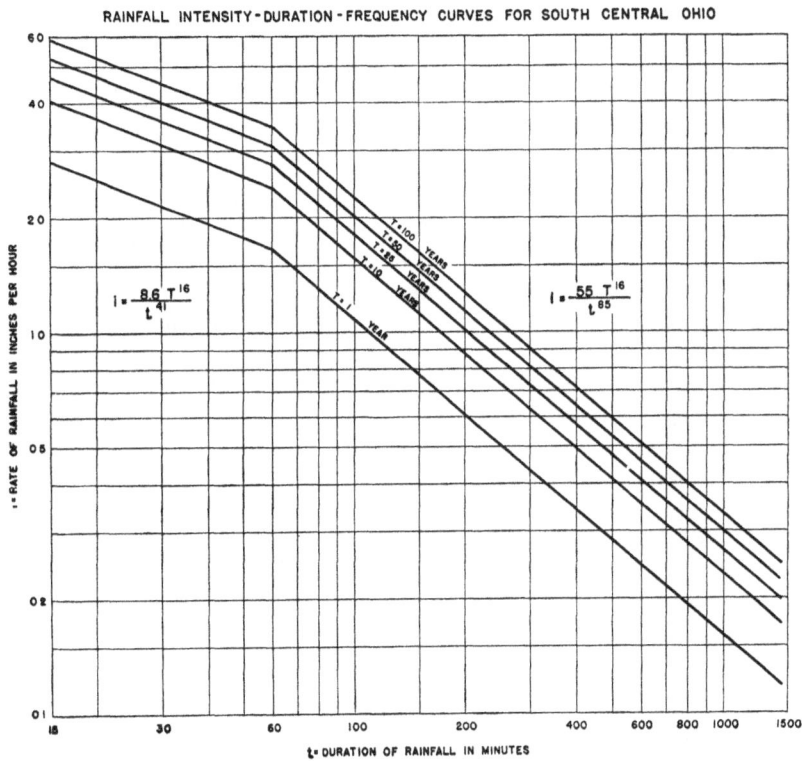

$$I = \frac{8.6 \ T^{16}}{t^{41}}$$

$$I = \frac{55 \ T^{16}}{t^{85}}$$

T = 100 YEARS
T = 50 YEARS
T = 10 YEARS
T = 1 YEAR

I = RATE OF RAINFALL IN INCHES PER HOUR

t = DURATION OF RAINFALL IN MINUTES

FIGURE 5

This difference in intensity for observed periods of duration varies throughout the country and has been the subject of several extensive studies, those of the Miami Conservancy District and David L. Yarnell being the more outstanding.

Figures 114 to 118 of Appendix A have been prepared from the Yarnell rainfall depth-frequency charts. These charts and their nomenclature are described in Appendix A. From them the rainfall equation for any point in eastern United States may be developed. To illustrate, if a project is located in south central Ohio, the rainfall intensity equation for durations of 60 minutes or less is

$$i = \frac{8.6 T^{0.16}}{t^{0.41}}$$

and for durations of from 60 minutes to 1,440 minutes the equation is

$$i = \frac{55 T^{0.16}}{t^{0.85}}$$

Figure 5 presents the intensity-duration-frequency curves for the locality.

It is assumed that water will be discharged from a drainage basin at a higher rate if the entire basin is contributing than if only part of the basin is contributing; therefore the least time that should be taken into consideration in computing maximum flood flows is the time required for the flow to arrive from all parts of the tributary area. This is known as the "time of concentration" for the drainage area. If a greater time were to be considered, the average intensity of the rainfall would be less; therefore, the greatest period that should be considered should not extend beyond that needed for all of the drainage area to contribute its run-off to the stream flow at the point being studied.

One important exception to the above assumption is related to those drainage basins which have a headwaters section of a tongue-like shape involving a relatively long time of flow but a comparatively small portion of the basin area.   Often it will be found for such basins maximum flood flow will occur when the small extended area is not contributing to the flood.   For such areas, the rational method should be tested also by eliminating the narrow headwaters section from the computation.

The rational formula is as follows: $Q = CiA$, in which

Q = the peak discharge in cubic feet per second.

i = a quantity equivalent to the average rainfall rate in inches per hour for a duration equal to the concentration time of the drainage area.

A = The area of the drainage basin in acres.

C = The ratio of the maximum peak flow per acre given in cubic feet per second to the average rate of rainfall in inches per hour throughout the period of concentration.

The accuracy with which peak discharge, Q, can be determined under the assumptions of the rational method depends to a considerable extent upon the accuracy with which the duration of the selected rainfall intensity conforms to the concentration time of the particular watershed.   The intensity-duration curve will show, for the short duration periods typical of small watersheds, how rapidly the value of i increases with comparatively small reductions in duration. Much then depends upon the accuracy with which concentration time is determined.   This may vary between wide limits on small watersheds of the same area, but of variable characteristics in other respects.

A preliminary determination of the time of concentration should be made by estimating the average velocities for the principal reaches of the stream, and applying these to the respective lengths of channel to determine time of flow. The summation of this application gives the actual probable time of flow from the headwaters to the outlet. However, rapid lateral inflow into a stream in flood generally creates a flood wave which is in a sense superimposed on the normal stream flow. This wave will travel at a more rapid rate than the normal stream velocity, increasing apparent velocities by from 30 to 50 percent, and correspondingly shortening the time at which the peak flow occurs at the outlet.

It is generally necessary to make preliminary approximations of peak flow in which the following steps may be taken:

1. Determine the average velocity for the principal channel under an assumed condition of flood peak flow by the ordinary methods of flow computation for open channels. (This must be revised and velocities corrected, after an approximate value of flood peak flow has been calculated.)

2. This value, being average velocity, is increased 30 percent to give the approximate average velocity of the flood wave.

3. Average wave velocity is converted into time in transit through the principal channel and expressed in minutes. This is taken as the approximate concentration time for the watershed.

4. The average rainfall intensity, i, is taken from the selected intensity-duration curve for the watershed, for a duration equal to the concentration time previously determined.

5. The proper value of the coefficient, C, is selected to meet the conditions of the problem.

6. The estimated peak discharge is obtained by multiplying the factors C, i, and A.

Hydrologists of broad practical expereince may make acceptable first approximations by the above method, drawing upon first-hand knowledge of run-off coefficients and the influence of watershed physiography on concentration time. To the less experienced engineer the accumulated experience of several leaders in this field is made available in Appendix A under the heading of "The Modified Rational Method." The procedure, supplemented by working charts and an illustrative example, makes it possible to consider and evaluate the more important factors.

11. Determination of Spillway Capacity from the Maximum Flood Flow.—The maximum flood flow at the dam site as determined by one of the methods discussed in this chapter is the estimated value for the natural stream in the absence of the reservoir under consideration.

In many cases this value may be taken as the maximum flood flow into the reservoir after construction. However, in some cases the area occupied by the reservoir may have been subject to deep flooding in a state of nature. Inasmuch as the construction of the reservoir will maintain the area in a permanently flooded condition, any such natural storage will no longer be effective, and the peak inflow into the reservoir will, in such cases, be somewhat greater than the maximum peak flood flow at the dam site under natural conditions. For important projects, and particularly for those projects creating relatively shallow storage in a valley where flood flows have normally involved the submergence of wide over-bank areas, this condition should be carefully evaluated.

The required capacity of the spillway structure must be equal to the maximum peak flood flow or to such peak flood flow as may justifiably be reduced through storage in the reservoir above spillway level. Where a consideration of such reduction may be justified because of over-spillway storage, not only the peak flood flow rate must be known but also the volume of such maximum floods.

In this chapter, discussion of flood flows has been related primarily to peak flood flow rate, and the total volume of flood flow has not been discussed. Where the peak rate may have been determined from hydrographs of stream flow, or arrived at through the application of the unit hydrograph or similar methods, an estimate of volume of flood flow may be easily made.

Where the spillway design proposed will carry the outflow at relatively low depths above the spillway crest, the total storage above spillway level will not be great unless the reservoir is of an unusually large area. For many projects having spillways of this type, considerations of storage above spillway crest will not justify any considerable reduction of spillway capacity below the maximum peak flood flow. If, on the other hand, the spillway is relatively short and the outflow will be carried at considerable depths, storage above spillway level may be sufficient to justify a spillway capacity very much less than the peak rate of the maximum flood flow. For all important projects, and particularly where cost of the spillway section of the structure will be a considerable part of the project cost, the storage item should be carefully evaluated by an experienced hydraulic engineer. For small projects which may not justify such an analysis in detail, or where spillway capacity can be secured at relatively low cost, this capacity should probably be taken as equal to the maximum rate of flood flow. An indication as to the possible propriety of reducing spillway capacity may be arrived at in the following manner, even where no flood flow hydrograph is available:

1. Make a preliminary determination of spillway characteristics by

selecting a length of spillway and determining the depth on the crest necessary to carry the maximum flood flow.

2. Determine the reservoir storage above crest level which would result from this increased depth.

3. Determine the approximate total volume of maximum flood flow. This may be approximated from an analysis of assembled precipitation records, a determination of the volume of precipitation on the watershed which might be expected in the type of storm producing maximum floods, and by an application to this precipitation volume of a percentage factor which will give an indication of the volume of flood run-off. During maximum floods, the percentage of the precipitation which appears as run-off is generally high. An indication of the probable range of this factor may be arrived at through a review of the material in United States Geological Survey Water Supply Papers 771 and 772. For some areas, maximum flood flow may involve importantly the run-off from melting snow.

4. Determine the ratio of reservoir storage above spillway level to the total volume of storm run-off.

5. Table 2 will give an indication under average conditions of the extent to which spillway capacity may be reduced. If this table indicates possible reductions in capacity of more than 10 percent, then further and more accurate computations will be justified. The percentage reductions shown in this table should never be used for the actual design of important projects, as they are only indicative of average conditions, and may be far from representative of the justifiable reduction in a particular case.

TABLE 2 [1]

Percentage of total storm run-off volume, represented by reservoir capacity above spillway level:

	Percentage reduction of peak flow due to storage
5	1
10	3
20	7
30	14
40	23
50	35
60	47
70	60

[1] Waterworks Practice Manual, American Waterworks Association, 1929, p. 72.

The necessary spillway capacity can be calculated accurately by the use of mass curves, if the inflow is in the form of a hydrograph, through an application of the storage equation, in which the outflow for any period of time will be equal to the maximum flood inflow for that time less the storage resulting from a rise in reservoir level. The storage equation requires a preliminary design of the spillway, in order that a relation between reservoir stage and spillway outflow may be introduced into the computations.[3] In cases where the watershed above the dam is relatively small, as less than 400 square miles, the use of mass diagrams of the inflow to the pool above the spillway has limitations which impair the definition of the rapid change of the rates of flow of the flood hydrograph. A more sensitive method employing the hydrograph of flood rates directly is that developed by the Los Angeles County Flood Control District of California and reported in the Transactions of the American Geophysical Union, April 1937, pp. 435–437, by R. S. Goodridge: "A Graphic Method of Routing Floods through Reservoirs." The method is rapid, flexible, and readily adaptable to all spillway pondage problems and is recommended for use. A few such trial spillway designs may be necessary to achieve a desirable balance, and in these initial design phases the propriety of reducing spillway capacity because of reservoir storage above the spillway crest will be determined. The actual design of the spillway structures is covered by Chapter 5.

12. Spillway Adequate for Maximum Flood Flows.—In this chapter it has been assumed that spillway capacity will be provided for the estimated maximum flood flow, or for this amount as it may properly be reduced by reservoir storage above spillway level. While such maximum flood flows will be of rare occurrence and will actually be approached in value only once in a hundred years or more, yet it must be recognized that they are possible of occurrence in any year and have a 10 percent chance of occurring in the first decade. If the spillway structure is not adequate to care for such flows, the result would normally be an over-topping of the structure and the destruction or partial destruction of the dam. Failures of this type, even for small dams, may involve a heavy loss of life and good practice must recognize safe provision for the expected ultimate condition.

Note.—It is recognized that there are rare situations which may justify spillway capacity for only the more frequent and therefore somewhat smaller floods. This will never be justified where it can be visualized that life could be endangered as the result of failure of the structure. Consequently, it would never be justified where the valley below the dam site is used for permanent habitation. Neither would it be advisable where such lands are subject to temporary occupancy, unless they can be so administered and policed that no such hazard will exist.

Where such smaller capacity may be justified and is in accordance with the policy of the agency under which the work is being carried out, the actual capacity will be predicated upon economic considerations, that is, a balance between the cost of replacement or repair of the structure and assumption of other damages at certain estimated intervals as compared with the cost of providing a spillway for the maximum flood.

---

[3] See "Reservoir Storage above Spillway Level," *Civil Engineering*, vol. 3, p. 233.

BIBLIOGRAPHY

"Floods in the United States," U. S. Geological Survey, *Water Supply Paper* No. 771, 1936.

"Rainfall and Run-off in the United States," U. S. Geological Survey, *Water Supply Paper* No. 772, 1936.

C. S. Jarvis, "Flood Flow Characteristics," *Trans. Am. Soc. C. E.*, Vol. 89, 1926.

R. L. Gregory and C. E. Arnold, "Run-off—Rational Run-off Formulas," *Trans. Am. Soc. C. E.*, Vol. 96, 1932.

C. E. Ramser, "Run-off from Small Agricultural Areas," *Journal of Agricultural Research*, Vol. 34, No. 9, 1927.

David L. Yarnell, "Rainfall Intensity-Frequency Data," U. S. Department of Agriculture, *Misc. Pub.* No. 204, 1935.

William A. Liddell, *Stream Gaging*, McGraw-Hill Book Co., Inc., 1927.

Merrill Bernard, "Formulas for Rainfall Intensities of Long Duration," Trans. Am. Soc. C. E., Vol. 96, 1932.

Don M. Corbett, "Gaging the Flow of Streams," an unpublished manuscript, U. S. Geological Survey.

C. E. Ramser, "Flow of Water in Drainage Channels," U. S. Department of Agriculture, Technical Bulletin No. 129, 1929.

"Instructions for Cooperative Observers," Weather Bureau, *Circulars B and C,* 8th Ed. Revised, U. S. Government Printing Office, 1935.

"Measurement of Precipitation," Weather Bureau, *Circular E, Revised,* U. S. Government Printing Office, 1922.

Horace W. King, *"Handbook of Hydraulics."*

U. S. Geological Survey Water Supply Paper 773–E. New York State Flood of July 1935.

U. S. Geological Survey Water Supply Paper 796–B. Flood on Republican and Kansas Rivers, May and June, 1935.

U. S. Geological Survey Water Supply Paper 796–C. Flood in La Canada Valley, January 1, 1934.

U. S. Geological Survey Water Supply Paper 798. Floods of March 1936 in New England Rivers.

U. S. Geological Survey Water Supply Paper 799. Floods of March 1936, Hudson River to Susquehanna River Region.

U. S. Geological Survey Water Supply Paper 800. Floods of March 1936, Potomac, James, and Upper Ohio Rivers.

U. S. Geological Survey Water Supply Paper 816. Major Texas Floods of 1936. C. S. Jarvis.

"Rainfall Characteristics and Their Relation to Soils and Run-off," *Trans. Am. Soc. C. E.*, Vol. 95, p. 379, 1931.

CHAPTER 3

SELECTION OF TYPE

1. General.—It is only in exceptional circumstances that an experienced designer can say that only one type of dam is suitable or most economical for a given dam site. Except in cases where the selection of type is entirely obvious, it will be found that preliminary designs and estimates may be required for several types of dams before one can be shown to be most economical. It is, therefore, important to emphasize that the project is apt to be unduly expensive unless decisions regarding selection of type are based upon adequate study and after consultation with competent engineers.

In the selection of type for important structures it is also usually wise to secure the advice of an experienced engineering geologist in connection with the relative applicability of possible types to the foundations available.

In numerous cases excessive cost of spillway protection, limitations of outlet works, and the problem of diverting the stream during construction have an important bearing on the selection of type.

In certain cases the selection of type may also depend upon the availability of labor and equipment. These may be particularly important considerations when the element of time is involved. Inaccessibility of site may also have an important bearing on selection.

In general it may be said that type selection will be dictated by: The physical characteristics of the site, by economic features, and by practical considerations.

The following types are considered, in this manual, as applicable to dam structures up to 30 feet in height.

1. Earth embankment.
2. Rock-fill.
3. Solid gravity.
4. Arched masonry.
5. Hollow gravity or buttress.
6. Timber.

PHYSICAL FEATURES OF THE SITE

2. Topography.—Topography in large measure dictates the first choice of type of dam. A narrow stream flowing between high, rocky walls would naturally suggest a concrete overflow dam. A low, rolling plains country would, with equal fitness, suggest an earth dam with a separate spillway. For intermediate conditions, other considerations take on more importance, but the general principle of satisfactory conformity to natural conditions is a safe primary guide.

The location of the spillway is an important item that will be governed very largely by the local topography, and will in turn have a material bearing on the final selection of type of dam to be used.

3. Geology and Foundation Conditions.—Foundation conditions depend upon the geological character and thickness of the stratum which is to carry the weight of the dam, its inclination and permeability, its relation to underlying strata and existing faults and fissures. The foundation will limit the choice of type to a certain extent, although such limitation will frequently be modified considering the height of the proposed dam. The different foundations commonly encountered include:

(a) *Silt or fine sand* foundations can be safely used for the support of low masonry and earth embankment dams. The main problems are the prevention of piping and excessive percolation losses, and protection of the downstream toe from erosive action of the overflow.

(b) *Clay* foundations require the same treatment and restrictions as for those on silt or fine sand. There may be considerable settlement of the dam if the clay is unconsolidated and the moisture content is high. Tests should be made of bearing value as well as of consolidation characteristics (see ch. 7 and Appendix B).

(c) *Gravel foundations*, if well-compacted, are suitable for earth and rock embankments and for gravity masonry dams. As gravel foundations are frequently subject to water percolation at high rates, special precautions should be taken to provide effective water cutoffs or seals.

(d) *Solid rock foundations*, because of relatively high bearing power and resistance to erosion and percolation, offer few restrictions as to the type of dam that can be built upon them; economy of materials, or over-all cost will be the ruling factor. The sealing of seams and fractures in the rock by grouting will frequently be necessary.

If an arch dam is under consideration, it is essential that the abutments be adequate to carry the horizontal thrust of the arch. This condition is explained in more detail in Chapter 10. Often the rock will be badly fractured, faulted, or weathered, and while it may have adequate bearing capacity, it may at the same time be as permeable as many sand and gravel foundations. Under such conditions, a type of dam which is suitable for pervious foundations must be selected.

(e) Occasional situations may occur where reasonably uniform foundations of any of the foregoing descriptions cannot be found, and where a nonuniform foundation of rock or soft material or of both must be used if a dam is to be built. Such unsatisfactory conditions can often be overcome by special design features. Each site presents, however, a problem for appropriate treatment by experienced engineers, and no attempt will be made in this manual to cover such unusual problems.

4. Spillway Location.—The spillway is the most vital element of a reservoir, and frequently the natural restrictions in its location will be a controlling factor in the choice of the type of dam. One of the common spillway arrangements is that in which a channel is excavated through one or both of the abutments, outside of the limits of the dam. As a general rule, it is desirable to have the spillway apart from the dam to reduce the danger of erosion of the dam or its foundation. Where such a location is adopted, the dam can be of the nonoverflow type, which extends the choice to include earth and rockfill structures. Conversely, failure to locate a site for a spillway away from the dam requires the selection of a type of dam which can include an overflow spillway.

On some projects the only available location for the spillway will be in the main river channel. If the length of the dam is short and the required spillway capacity is relatively large, the length of crest needed for the spillway may often be nearly as great as the length of the dam. In this case, there are two possible solutions; one is the use of a masonry overflow dam with sufficient capacity to pass the flood waters, and the other is the use of a spillway that is more or less independent of the dam, such as the side channel, horseshoe, or vertical shaft types, with which earth, rockfill, or masonry dams may be used, the final choice being influenced by other requirements and restrictions.

Where it is possible to find a suitable spillway location which is apart from the dam site, the condition will be favorable for an earth or rockfill dam.

5. Materials Available.—*Materials* for dams of various types which may sometimes be available at or near the site are:

(a) Soils for embankments.

(b) Rock for embankments or riprap.

(c) Concrete aggregates:

    Sand.

    Gravel.

    Crushed stone.

(d) Masonry:

    Sand.

    Cut stone.

(e) Timber.

The elimination or reduction of transportation expense for construction materials, particularly those which are used in great quantity, will effect a considerable reduction in the total cost of the project. The most economical type of dam will often be the one for which materials are to be found in sufficient quantity within a workable distance of the site.

The availability of adequate deposits of suitable sand and gravel for concrete at a reasonable cost locally, and perhaps even on property which is to be acquired for the project, is a factor favorable to the use of a concrete structure. On the other hand, if suitable soil for an earth embankment can be found in conveniently located borrow pits, an earth dam will usually prove to be the most economical. Advantage should be taken of every local resource to reduce the cost of the project without sacrificing the efficiency and quality of the final structure.

6. Earthquake and Other Hazards.—If the dam lies in an area that is subject to earthquake shocks, the design must include provision for the added loading and increased stresses. The types of structure best suited to resist earthquake shocks without damage are earth embankment, solid gravity, and single arch dams. For earthquake areas neither the selection of type nor the design of dams should be undertaken by anyone who is not experienced in this class of work.

The vulnerability of a dam to malicious deliberate destruction may in some situations become an important factor in the choice of type. The loss of storage capacity in a water supply system, or the damage that might be caused by the sudden release of a large volume of water stored above an urban area, makes this factor worthy of consideration for some projects. For such cases the choice of the concrete gravity dam offers the greatest protection.

LEGAL, ECONOMIC AND AESTHETIC CONSIDERATIONS

7. Statutory Restrictions.—Statutory restrictions exist with respect to control of the waters of navigable streams. Plans for diversion or control of waters in such streams are subject to approval by the Corps of Engineers, United States Army. There are numerous other Federal and State regulations relating to dam construction and operation which may be determining factors in the choice of type. Before proceeding with detailed design, reference should be made to the proper authorities. See Appendix H for a summary of State legislation.

8. Purpose and Benefit Cost Relation.—Consideration of the purpose which the dam is to serve will often suggest the type most suitable, as for example, whether its principal function is to furnish continuous and dependable storage of the water supply for irrigation, power or domestic use; to furnish temporary detention for the control of floods; or to regulate the flow of the stream; or to be a diversion dam or weir without storage features.

Few sites exist where it would be impossible to build a dam that would be safe and serviceable, but in many instances conditions inherent in the site will result in a project cost in excess of the justifiable expenditure. The results of a search for desirable dam sites often determine whether a project can be built at a cost which will be consistent with the benefits to be derived from it. These benefits are easily evaluated for water power, irrigation, and water supply uses, are less well defined for flood control, and there is no satisfactory measure of value of recreational projects. Justification for this last type of development must be evaluated on the basis of a comparison of the population that will be benefited, the location of other projects of the same kind, the trend of development in the district (appreciative and depreciative) all as related to the cost of the project and the money available. In a case where the need is great but the number of people to be served is limited, the development of an expensive site may not be justified. In another case, the present need may be great but a tendency toward decline of population and property values must be considered. In both such instances, the development should be made as inexpensive as possible, probably with a low dam of small storage capacity. The use of timber dams, under such circumstances, may be justified in the first case because it would meet the low cost requirement, and in the second case because of low cost, and also because the initial structure might outlast the need for the development.

Similar relations of benefit to cost will dictate the use of simple and in some cases temporary structures for stock-water use. Of the various types of dam described in this manual, the timber dam is the only one which is classed as a short-lived or temporary structure. Other types if properly built and maintained and provided with adequate spillway capacity, will be of indefinitely long life.

In time of emergency, the volume of employment that the construction of a project provides may be one of the considerations in the choice of type. For example, at a site adaptable to either type a rock-fill dam with an earth apron may in some cases be justifiably selected as more desirable than a concrete structure because of the greater social and community benefit derived in employment of the manpower needed to move the greater volume of material.

9. Cost.—Usually the final choice of one type of dam over another or of one site over another, should be made on the basis of lowest cost. Where there will be a marked difference in maintenance charges, or in the life of the structures, the comparison should be on the basis of annual cost, but otherwise, only the construction cost figures need be considered.

The first step in preparation of comparative cost estimates consists in making tentative preliminary designs from which quantity estimates can be made for each type considered. As detailed refinement is not necessary at this stage, it is common practice to use standard curves or tables giving the quantity of materials for standard sections of varying heights. Information useful for this purpose will be found in several of the subsequent chapters devoted to dam types. Because different types of dams will require different layouts for spillway and outlet structures, these structures must be included in the preliminary layout for estimating purposes.

10. Appearance.—In general, the completed structure, regardless of type, should have a finished, workmanlike appearance. The alinement and texture of finished surfaces should be true to the design requirements and free from unsightly irregularities. Aesthetic considerations may have an important bearing on the selection of type (see ch. 1: **4**).

Highway Across Dam.—Where the dam may carry a highway across the stream, a nonoverflow type of structure will be necessary unless it is possible to design a safe and economical highway as a bridge carried on piers projecting above the crest of the dam. Where the highway is laid directly on the dam, consideration must be given to the live loads which may be transmitted to the foundation and to any conduit through the dam.

CLASSIFICATION OF TYPES

11. Earth Embankment.—The earth embankment is the most common type of dam for small projects, mainly because the embankment material can often be obtained cheaply in the reservoir area or at other convenient locations nearby. This situation is also an advantage in projects where a high ratio of labor cost to material cost is important. The type includes rolled fill, semihydraulic, and hydraulic fill dams as described in Chapter 7. An earth dam requires a supplementary structure to serve as spillway. The principal disadvantage to the use of an earth dam is that it deteriorates or may be destroyed under the erosive action of water flowing over it. It is also subject to serious damage or even failure due to burrowing of animals unless special precautions are taken.

The earth fill dam is frequently the most suitable type where the foundation is of loam, sand, gravel, or other uncemented and more or less pervious material. The relatively wide base introduces a long path of resistance to seepage losses, and pervious foundations are often susceptible of inexpensive treatment in connection with this type. Earth dams may also be built economically and safely on various types of rock foundations with cores or diaphragms to reduce seepage.

Economical construction of earth dams requires a proper balance between skilled and unskilled labor, the use of equipment suited to the job and a coordinated construction program. The availability of the necessary labor and equipment may be an important factor in the selection of this type.

As regards safety and satisfactory performance, an important requisite is care in selection, mixing, placing, and compacting of the fill materials. Provision must be made for diversion of the stream during construction through the dam by means of a culvert or other type of conduit. Flood flows may precipitate an occasional and serious hazard and hence earth fill and rock fill dams are not generally desirable on drainage areas subject to flashy flood flows greatly in excess of normal.

FIGURE 6 —Tensleep earth dam, showing riprap on upstream face, Bighorn National Forest, Wyo

12. Rock Fill.—The rock fill is a modified form of the earth dam, using rock of all sizes to provide stability, and an impervious membrane to provide watertightness. The membrane may be a blanket of impervious soil, a concrete slab, a steel plate facing, a timber deck or other similar device.

Like the earth embankment, the rock fill dam is subject to deterioration or destruction by the overflow of water, and so must be supplemented with a spillway of adequate capacity to prevent overtopping. It is preferably built on a rock foundation, because of its weight and the fact that any seepage will flow over the base through the voids in the rock.

This type is adapted to remote locations where suitable soil for an earth fill dam is not available, where transportation of cement for a concrete dam would be costly, or where the site or climate is not suitable for thin concrete structures.

The construction requirements are about the same as for earth embankments, except that rock quarrying equipment and personnel will be necessary. The construction program will be in two major operations: first, the placing of the fill, and second, the placing of the flow retarding membrane.

13. Solid Gravity Masonry.—The solid gravity masonry type of dam is desirable where a reasonably sound rock foundation is available. It is well adapted for use as an overflow crest for spillways, and be-

FIGURE 7—Swift Creek Recreational Demonstration Project, Va , SP-24   Swift Creek Dam Ogee type gravity spillway designed for 8 foot maximum head with protected walkway on wing walls   Operating pedestals for sluice gates concealed below parapet in walkway of far end

cause of this advantage, is often used for the spillway feature of other than masonry dams and as an overflow dam constructed on gravel or sandy foundations. In the latter situation adequate cutoff devices must be provided.

The masonry may be of concrete or of stone, depending somewhat on the material and the kind of labor available. The construction of a concrete dam requires ordinary form carpenters and concreting labor. For a masonry dam, a few skilled stone masons will be required, and difficulty may be encountered in getting qualified men.

Gravity dams may be either straight or curved in plan. Where the abutments are capable of resisting arch thrust and where the radius of curvature is not too great, the curved plan may offer some advantage in matters of both cost and safety.   The upstream offset may occasionally locate that part of the dam on higher bedrock foundations.   The curved plan may also afford some advantage in providing additional length of spillway.   Certain structural disadvantages arising from wide span and relatively long radius may also be present, but detailed discussion of them is outside the scope of this manual.

14. Masonry Arch.—An arched masonry dam of thin section is adaptable to sites where the width between abutments is not great, and where the foundation at the abutments is solid rock capable of resisting arch thrusts.   Other requirements are: first, the span must

FIGURE 8 — Concrete arch dam on Quanah Creek, Wichita National Forest, Okla

not be too great to produce arch action with a reasonable ring thickness; second, the arch ring must join the abutments at a safe angle, without being too flat for economy of material; and third, the ratio of the length of the arch ring to its thickness at any elevation must come within the accepted limit.   It will be seen from these requirements that comparatively few sites will be fully suitable.   Because of its low yardage content and the limited foundation area to be prepared, the construction cost will usually be lower than for any other type.   By using conservative design standards, it can have a very high factor of safety, without great extravagance in the use of materials.

Arch dams may be built of concrete or stone masonry. Due to the small volume of material needed, the time required for construction is relatively short. For the structures designed on the constant angle theory, the variations in thickness are often quite complicated, and the layout and construction of the formwork will require skilled supervision. For small structures, such as may be designed by approximate methods complicated formwork may be avoided and the requirement of skilled labor will be about the same as it is for gravity masonry dams. Somewhat closer engineering supervision of the work is desirable because of the relatively greater effect of inaccuracies on thin sections.

15. Buttress Dams.—Buttress dams comprise the flat-deck and multiple-arch structures. They require less concrete than solid gravity dams (usually about 60 percent less) but the increased formwork and reinforcing steel required may offset any saving in concrete. The relative cost of materials and skilled labor together with suitability for the site will determine whether the solid gravity or buttress type will be the less expensive.

The flat-deck type consists of a flat structural slab which transmits the water load to a series of piers or buttresses. In the multiple-arch dam the flat slabs are replaced by concrete arches, which permit the use of wider spacing of the buttresses. If the dam is of the nonoverflow type, the deck covers only the upstream side of the buttress, whereas for the overflow or spillway dam, both upstream and downstream faces are enclosed by the deck. Hence the term "hollow dam" is sometimes applied to this type.

In buttress dams, the load on the foundation may be concentrated in the buttresses, and therefore give higher unit bearing stresses, or the buttresses may rest on a slab which distributes the load over the foundation. The stresses are not usually excessively high for moderate dam heights. On pervious foundations where the amount of uplift pressure might be an important factor in the design loads of a solid dam, the buttress type has the advantage because the uplift pressure is relieved between the buttresses (except where slab construction is used) and is therefore effective over a much smaller area. A relatively impervious foundation is ordinarily required to prevent excessive seepage losses because of the short path of water travel under the narrow foundation.

16. Timber.—The annual expense of a timber dam will usually exceed the annual cost of a dam made of more durable materials. There are however many places where a timber dam may be used economically; e. g., where the need for a structure is only temporary and will not extend beyond the life of the timber, or where the timber may be had for the cutting and labor is available as an incidental part of other programs.

Timber dams can be built almost entirely of material cut in the forests, their life being limited to the period for which the timber will resist decay. After the first 5 years or more of service, it is probable that continuous maintenance will be required to keep the dam in good condition. If loss of life or damaging of property is apt to result through the failure of a dam and adequate maintenance cannot be provided for timber dams, a more permanent type should be selected.

17. Other Types.—Dams of other types than those mentioned above have been built, but in most cases they meet some unusual local requirement or are of an experimental nature. In a few instances structural steel has been used both for the deck and for the supporting framework of dams. This and other unusual types are not treated in this manual

18. Summary.—The selection of the best type of dam for a particular site calls for thorough consideration of the characteristics of each type as related to the physical features of the site, the adaptation of the purposes the dam is proposed to serve, as well as economy, safety, and other pertinent limitations. The final choice of type of dam will generally be made after consideration of these factors. It is possible, however, that restrictions in the use of the appropriations from which the work is to be financed will influence the choice. If there is a low limitation on the proportionate amount of the total cost which may be used for the purchase of construction materials, the earth embankment, with its high labor and low equipment and material purchase requirements, will possibly be the most suitable.

BIBLIOGRAPHY

H. K. Barrows, *Water Power Engineering*, McGraw-Hill Book Co.
F. C. Hanna and R. C. Kennedy, *Design of Dams*, McGraw-Hill Book Co.
W. P. Creagor and J. D. Justin, *Hydroelectric Handbook*, John Wiley & Sons.
American Civil Engineers Handbook, Fifth Edition, John Wiley & Sons.

CHAPTER 4

FOUNDATIONS

1. Definition.—A study of the dam foundation should include not only the material as it occurs in a natural state below the bottom level of the proposed structure, but also the probable alteration in its condition and structure under additional loading or a change in moisture content. The study should include the natural banks at the ends or abutments of the dam as well as the supporting materials.

2. Function.—The purposes of a foundation under a dam are: (1) to insure stable support for the structure under all conditions of loading and (2) to provide the necessary resistance to the passage of water so that the purposes of the dam may be fully attained. A failure of a dam which is produced by a failure in the foundation may be due to one or more causes such as: crushing, plastic flow, sliding, piping and scouring, and excessive uplift. If the bearing power of the material on which the dam is founded is exceeded, the foundation is displaced or crushed and the dam fails for lack of support. If the frictional resistance between the layers or strata of a foundation, or between the base of the dam and the top of the foundation is exceeded, one portion of the foundation may slide upon another or the dam may slide over the foundation and cause failure. Where water passes through a foundation in a concentrated stream or streams with an erosive velocity, piping results and the foundation may be gradually worn away and leave the dam unsupported. The flow of water over an overflow dam or a spillway may discharge near the base of the dam and erode the foundation to such an extent as to weaken it and cause it to give way. Water entering through the permeable strata under a dam may cause hydrostatic uplift sufficient to critically increase the overturning moment; or it may so decrease the frictional resistance so as to cause failure of the structure.

3. Classification of Foundation Materials.—Foundation materials under small dams may be classified in two general types, (1) those subject to negligible consolidation under load, and (2) those subject to considerable consolidation or displacement due to plastic flow. In the consideration of foundations it must be remembered that an examination of the material in immediate contact with the base of the dam is not sufficient to insure the reliability of the underlying materials which will be affected by the structure.

Materials of the first class are usually ideal for foundation purposes in the matter of load bearing possibilities. In exceptional cases, they may, however, be of such a nature that they offer little resistance to sliding. Some such materials have very different characteristics in the dry condition from those in a wet condition.

The information with respect to foundation conditions which will be secured from superficial examination, geological investigations, and subsurface exploration, must be such as to permit a determination of the character of the foundation strata with regard to the following:

(a) Its ability to sustain the weight of the structure without appreciable settlement or serious consolidation. Where the preliminary examination has determined that only a shallow soil mantle exists over a massive rock strata, samples or cores of the rock may be observed to determine its probable capacity to provide ample factor of safety against the maximum pressure that can prevail under the proposed dam. Many rock formations will obviously provide far greater capacity than is required for dams under 30 feet in height; others may have doubtful capacity, and will require that crushing tests be made of representative specimens obtained from core drilling or excavation. Sufficient core or other drilling should be done to determine the stratification, the inclination of the strata, and the occurrence of seams or fissures.

Where the overlying soil mantle is of considerable depth and it seems at all probable that the dam structure may be founded on the natural soil, the subsurface exploration should provide soil samples from which its characteristics may be determined in the laboratory. These samples should be studied to determine the probable consolidation of the soil under the expected loads and its shearing strength as a measure of its resistance to flow or shear failure under load.

(b) Its ability to sustain the lateral forces on the structure without sliding. This will be affected by the character of the material, including the changes that may be produced by submergence of a material which may be fairly dry in the natural state, and by the thickness and inclination of the strata. Soil and rock samples secured in preliminary examinations should be subjected to laboratory analysis if doubt exists as to any of these qualitites.

(c) The permeability of the subsurface structure. Where this structure consists of rock at relatively small depths, but where it appears that the dam may be founded on such rock, this investigation will relate largely to the examination of the rock as to fissures and seams through which water may pass.

If it is probable that the dam is to be founded on undisturbed soil, the ability of the soil to resist percolation and piping should be investigated.

Soil Samples.—Undisturbed samples of the full range of soil structure may be required for laboratory study. However, unless the preliminary examination or the borings indicate a soil of high plasticity, relatively low bearing capacity, or of high permeability, and unless this material extends to a considerable depth, the laboratory analysis will generally be unnecesssary for small projects.

For those situations where the soil samples indicate a high permeability, or relatively low shearing values, their analysis may have an important bearing on the type of structure and of the foundation.

Foundation Examination.—An outline of essential subsurface exploration is contained in Chapter 1. Methods of making such examinations are described in detail in Appendix C. The preliminary examination carried out as part of the project will indicate the extent of further examination to be undertaken as part of the detailed investigation.

For small projects auger borings may sufficiently confirm an assumption of firm soils sufficiently impermeable and of ample bearing capacity. However, one or more test pits are desirable.

4. Resistance to Flow of Water.—The resistance of the foundation to the flow of water from the upstream toward the downstream side of the dam can be examined from two standpoints; first, the permissible total flow, and second, the permissible maximum rate of flow. The permissible total *volume* of flow through the foundation depends on the relative value of the water which is lost and the cost of measures necessary to conserve the loss. The maximum *rate* at which water may be safely permitted to pass through the foundation is that rate above which the stability or permanence of any part of the dam or bank or foundation would be endangered. See Flow of Water Through Soils, in Soil Mechanics, Appendix B.

Water seeping through the structure or at the junction of the foundation and structure is flowing under pressure. It is important to dissipate the head before the downstream side of the dam is reached in order to reduce the velocity below any possibility of dangerous erosion. This may be accomplished by construction of a core or cut-off wall at the upstream toe, or, in earth dams, at about one-third the total cross-sectional width of the base from the upstream toe, extending the core or wall vertically, well into the dam. Sometimes, due to the character of the foundation or the materials of which the dam is composed, it is advisable to have multiple core walls. The principle governing the situation is that the flow of water through the embankment and the foundation should be reduced to the lowest practicable volume so far as storage loss is concerned and, in every case, the pressure must be dissipated well below that at which the resulting velocity at emergency tends to cause a condition of flotation. See Appendix B, sections 24 and 25.

Drains in the foundation of the downstream section of a dam may be beneficial under some circumstances and harmful under others, and should be used only after a careful analysis of the several ways in which they may influence the effectiveness of the embankment. See Chapter 7:18 and Appendix B, section 25 for more detailed discussion.

Dams built on foundations of high permeability should provide for the restriction of the velocity of flow so that the fine material will not be entrained and carried out in suspension from under the dam. This action, commonly known as piping, may be directly through the foundation, or along the surfaces of its contact with the dam. The path of flow along this contact surface is termed the line of creep and its length is the total length of path from headwater to tailwater. If either the length of the path or the resistance at the upper end of the path is increased, the seepage flow will be reduced. Flow-retarding devices, such as cores or cut-off walls previously mentioned, and clay blankets on the bottom of the stream above the dam, make effective use of both principles; they add resistance to flow in the path across which they are placed and increase the length of possible alternate paths.

An empirical method involving a term known as the safe percolation factor is sometimes used in analyzing structures on pervious foundations to determine whether the design is safe against excessive or destructive percolation. The percolation factor may be defined in either of two ways depending upon whether the Bligh values or Lane values are used in the design. (See table below.) (1) If the Bligh values are used, the percolation factor is determined by dividing the total length of the path of creep by the head causing the flow. (2) If the Lane values are used, the percolation factor equals the total length of the vertical paths of creep plus one-third of the horizontal length (weighted length), all divided by the head causing the flow. In either case, the path of creep is measured along the lines of contact between the structure and the foundation. The following table gives suggested safe values for different kinds of materials as recommended by the respective investigators. If for any structure the computed factor for a critical cross section is less than that given in the table, the insertion of additional flow-retarding elements into the design is indicated. The Lane values represent results of more recent research.

	Lane safe-weighted [1]	Bligh
Very fine sand or silt	8.5	18
Fine sand	7.0	15
Medium sand	6.0	------
Coarse sand	5.0	12
Fine gravel	4.0	------
Medium gravel	3.5	------
Gravel and sand		9
Coarse gravel and cobbles	3.0	------
Boulders, cobbles, and gravel	2.5	------
Boulders, gravel, and sand		4–6
Soft clay	3.0	------
Medium clay	2.0	------
Hard clay	1.8	------
Very hard clay—Hardpan	1.6	------

[1] Weighted creep=vertical creep plus ⅓ horizontal creep

Too much dependence should not be placed upon this method of design. It should never be used in place of a detailed analysis (see Appendix B) for important structures which involve human safety or considerable expense.

5. Preparation of Foundations.—The design must provide for the effective sealing of the foundation, which may be accomplished in various ways dependent upon the type of the structure and the character of the foundation.

A rock foundation may be made watertight by the application of cement grout, clay grout, liquid asphalt, or other materials to the seams and crevices of the rock. (See Appendix G, sec. 19.)

Cut-off walls extending into the stratum on which the dam rests and upward into the structure of the dam to protect the junction may be employed for an impervious stratum other than rock.

The floor of the reservoir should also be carefully examined to detect pervious spots, such as shale strata that may disintegrate or cause slippery places, and crevices in exposed ledges of rocks, and places where the soil has been eroded, leaving sand or gravel exposed in the area to be flooded. Such features may lead to possible water channels passing underneath the dam and should be adequately sealed before the dam is closed. The ledge crevices may be grouted in the same manner as in the foundation rock, and the gravel areas may be sealed by clay blankets.

Consideration should be given to the dip or slope of the strata—especially in porous rock—below the floor of the proposed reservoir. For example, if the slope of the porous strata is inclined away from the reservoir under the dam, water may be carried out of the reservoir through or under the foundation of the dam or under the natural walls, but if the downward slope of the strata is toward the reservoir or vertical, the water flow may also be toward the reservoir.

Foundation conditions have been discussed in Chapter 3:**3** in their relation to the selection of the proper type of dam.

Foundation requirements with respect to design criteria are discussed in each of the chapters devoted to specific dam types. Certain other important considerations are presented here in the following:

(a) *Earth and Rockfill Dams.*—In Chapter 7:**9** are set out basic principles of foundation stability and the devices that are effective in eliminating seepage through the dam. For an earth dam on a permeable foundation, the dam and foundation may be considered as a unit with regard to seepage. Flow-retarding elements are discussed in detail in Chapter 7:**9**. For the higher dams covered by Chapter 7:**10** unless the foundation is unusually uniform and simple in structure, reference to Soil Mechanics, Appendix B, is advisable.

(b) *Masonry Dams.*—The borings which have been made prior to the selection of the site will have given general information concerning the rock or other consolidated strata upon which the dam is to rest, and the geological study will have indicated whether faults or fissures are likely to traverse the site. The area should be subjected to a further detailed examination before the design is prepared, particularly to determine the existence of seams, faults, or fissures. If such weaknesses are found the information should be promptly reported and expert advice should be secured. Such foundation conditions may be of sufficient importance to affect the final decision as to the site of the dam, the depth of excavation, the provision of cut-offs, or the necessity for grouting.

(c) In *buttress dams*, excavation into the underlying stratum is required only for the footings of the buttresses and for the extensions of the deck slabs to form cut-off walls. Otherwise the design should follow the same principles as for mass masonry structures.

Where the foundation is not solid rock, but is of thick beds of consolidated clays or gravel conglomerate, the excavation should be adequate to afford the necessary resistence to sliding and the whole area should be covered with a bed of concrete to form the base for the dam structure.

(d) *Arched Masonry Dams.*—The remarks on sealing the foundations of mass masonry dams apply with equal force to arched dams where special attention should be given to the abutment walls receiving the thrust of the arch. Where the bedding of the strata is steeply inclined toward the vertical, provisions should be made for grouting a greater distance along the line of thrust and on each side of the area of pressure so as to obtain a resistance more nearly approaching that of a monolith.

(e) *Timber Dams.*—Requirements for foundations for timber dams are discussed in Chapter 12.

BIBLIOGRAPHY

Stanley M. Dore, "Permeability Determinations, Quabbin Dams," *A. S. C. E Proceedings,* March 1936.

Glennon Gilboy, Soil Mechanics Research, *Trans. A. S. C. E.,* 98–218, 1933.

Charles Terzaghi, "The Science of Foundations—Its Present and Future," *Trans. A. S. C. E.,* 93–270, 1929.

Grouting the Foundation of the O'Shaughnessy Dam, *Trans A. S. C. E ,* 93:1473, 1929

E. W. Lane, "Security From Under-Seepage Masonry Dams on Earth Foundations," *Trans. A. S. C. E ,* 100:1234, 1935.

D. W. Cole, "Stabilizing Constructed Masonry Dams by Means of Cement Injections," *Trans. A. S. C. E.,* 101, 1937.

W. P. Creager, *Masonry Dam,* John Wiley & Sons.

Joel D. Justin, *Earth Dam Projects,* John Wiley & Sons.

Warren J. Mead, *Geology of Damsites in Hard Rock,* Civil Engineering, 1937.

Geology and Engineering for Dams and Reservoirs, Technical Publication No. 215, Am. Inst. Min. & Met. Eng., 1929.

CHAPTER 5

SPILLWAY STRUCTURES

The determination of the maximum flow to be provided for has been treated in Chapter 2—Hydrologic Studies. In the design of the spillway, the problem is to provide a structure capable of passing this flow without exceeding the permissible maximum flow line and without damage to the dam.[1] The common types of spillways may be grouped as follows:

1. Overflow dams and weirs.
2. Open channels.
3. Side inlet channels.
4. Drop inlet shafts.

The hydraulic characteristics of each of these types will be considered first, to be followed by specific examples of design.

1. Overflow Dams and Weirs.—This type of spillway is usually made of stone masonry or concrete. The shape of the crest may take one of several forms, but the general equation for computing the overflow capacity is

$$Q = CL_e H^{3/2} \qquad (1)$$

in which Q=total discharge capacity in c. f. s.

   C=a constant depending on the shape of the crest and the depth of overflow.

   L_e=effective length of crest in feet.

   H=head, measured from top of crest to reservoir level.

If the sides of the approach channel coincide with the ends of the weir, the effective length, L_e, is the same as the over-all length, L, of the crest. As a rule the crest of the dam is narrower than the approach channel, with the result that the flow is contracted to a narrower channel as it flows over the crest. In addition to the two ends of the dam, there may be other contractions introduced by piers along the crest, each such pier introducing two additional contractions. To correct for the hydraulic losses due to these contractions, the net length,

$$L_e = L - 0.1\, nH \qquad (2)$$

in which L=the over-all or gross length of the crest,

   n=the number of complete end contractions (with square corners).

---

[1] See ch 2 11, 12.

Well-rounded abutments and sharp-nosed, round-shouldered intermediate piers cause less disturbance to the water than square corners, so that the coefficient 0.1 for square corners may be reduced to as low as 0.05 for rounded corners which will give smooth flow conditions. (See fig. 9.)

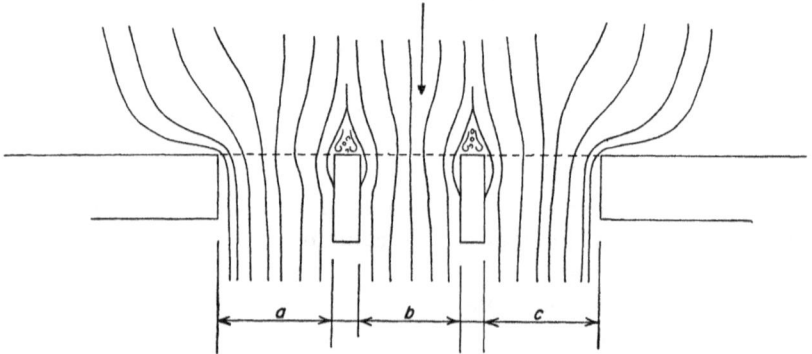

$$L = a + b + c$$
$$L_e = (a+b+c) - 0.1(6)H$$

COMPLETE CONTRACTIONS CAUSED BY ABUTMENTS

AND SQUARE NOSED PIERS

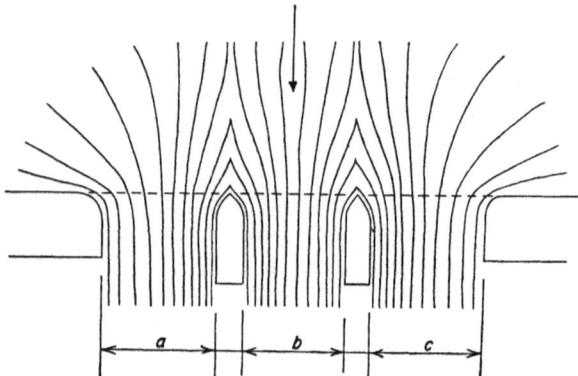

$$L = a + b + c$$
$$L_e = (a+b+c) - 0.06(6)H$$

PARTIAL CONTRACTIONS CAUSED BY ROUNDED ABUTMENTS

AND SHARP NOSED PIERS

FIGURE 9.—Complete and partial contractions caused by abutments

For greater accuracy there may be a correction for velocity of approach [2] in the channel above the crest, because the initial velocity increases the discharge as though the equivalent velocity head were added to the measured head. With a velocity V_o in the approach channel, the value of the velocity head $\left(h_o = \dfrac{V_o^2}{2g}\right)$ should be added to the head H at the crest, and equation (1) becomes

$$Q = CL_e \, (H + h_o)^{3/2} \qquad (3)$$

For large reservoirs, the effect of velocity of approach is usually slight, but in small ponds and for diversion dams which provide little storage capacity it may be considerable. It is preferable not to use any value for the velocity of approach rather than an incorrect one, as the omission of this factor in the computation of discharge (Q) will give a value less than that which the spillway will actually be capable of handling.

The value of the constant, C, is a function of the shape of the dam and is the result of experimental investigation with models or completed structures. For the ordinary working heads encountered on small projects, the range in values of C is given in the following table:

Type of weir:	Range of values of C
Broad-crested	2.63 to 3.33
Sharp-crested	3.33
Round-crested	3.30 to 3.98

Control sections (see sec. 4 below) 3.09 for all heads.

Numerous patterns for broad-crested dams have been used experimentally, some of which justify the use of coefficients even higher than 3.33 for low heads. Most of these patterns give cross sections which are not adapted to general use, and the curved crest and downstream profile commonly known as the ogee spillway has been generally accepted as the most satisfactory. Fig. 12 indicates the outline of the downstream face of such a spillway, which it will be noted takes the form of a parabola for its upper portion; this is designed to conform to the shape of the under side of the nappe corresponding to the maximum overflow. The lower part of the downstream face of an ogee spillway is concave upward to discharge the overflow in a horizontal direction at the toe of the structure.

Having designed a section for a certain head, the coefficient C will be greatest when the actual head equals the designed head. Where

---

[2] The velocity of approach can be evaluated roughly as follows Measure the cross section of the approach channel just above the probable spillway location; then divide the spillway discharge being investigated by the cross-sectional area below the reservoir level to obtain a rough value of V_o. This can then be converted to h_o and introduced into formula (3) to ascertain the effect of correcting the computed spillway discharge for velocity of approach

the designed head is exceeded, the under side of the nappe leaves the
face of the dam and creates a vacuum unless some means is provided
for admitting air to relieve it.   Figure 10 [3] gives values of coefficient
C for standard overflow sections, showing the relation between C
and the ratio of effective to designed head.   It will be noted that
there is an abrupt increase in the value of C when the actual head

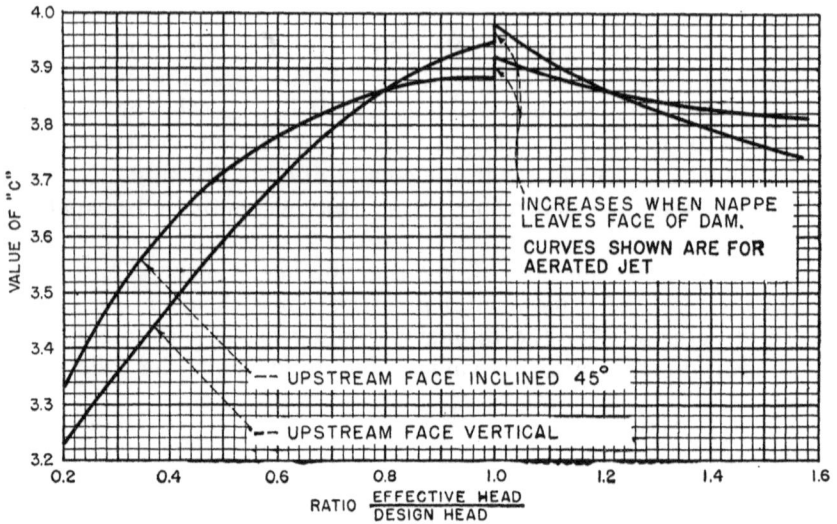

INCREASES WHEN NAPPE
LEAVES FACE OF DAM.

CURVES SHOWN ARE FOR
AERATED JET

-- UPSTREAM FACE INCLINED 45°

-- UPSTREAM FACE VERTICAL

VALUE OF "C"

RATIO EFFECTIVE HEAD / DESIGN HEAD

FIGURE 10

and breaking of the contact of the nappe with the face of the dam
is an undesirable and sometimes dangerous condition.   It is therefore
advisable that the head used in the design be conservatively large, so
that the point of maximum required overflow can be reached with a
value C of not over 3.9.   Figure 11 gives the relation of the head to
the discharge per foot of length for five different values of C ranging
from 2.65 to 3.9.   Multiply the discharge obtained from these curves
by the net effective length of crest to obtain the total discharge
capacity.

The equations given for overflow capacity have been for the condition of free flow below the structure.   For low dams, such as might be
used for diversion, and in some spillway arrangements, the tailwater
may build up until it interferes with the overflow.   According to
experiments by the United States Board of Engineers in 1899, the
reduction in overflow capacity on a model dam with rounded profile,
reasonably comparable with the modern ogee profile, is reduced in
the ratio given in table 1.   This table gives the relative effect of
submergence up to 90 percent; to use it, multiply the normal discharge

---

[3] From "Hydro-electric Handbook", Creager and Justin, p 133.

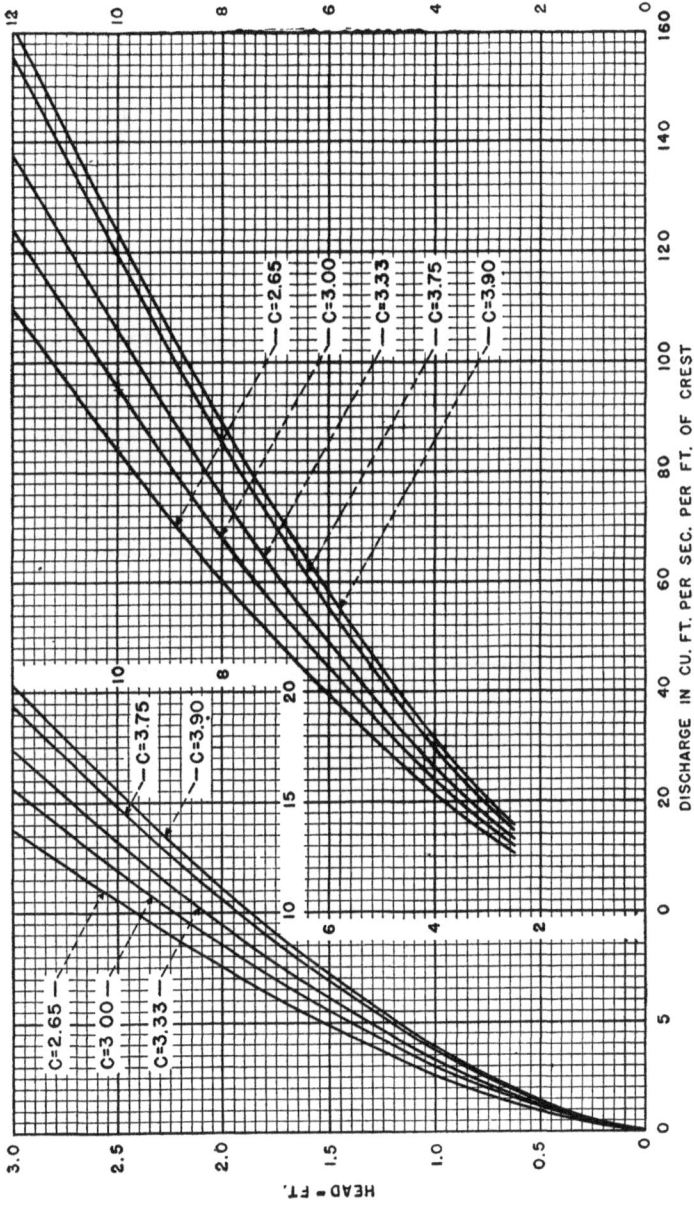

FIGURE 11.—Discharge capacity of overflow dams

by these factors to determine the discharge under submerged conditions. In the model tests, the depth of water above the weir was 6.6 feet.

Similar coefficients for sharp crested weirs were determined experimentally by Clemens Herschel, and are also given in table 1.

The Soil Conservation Service has conducted tests on V-notched weirs with a trapezoidal crest, with heads up to 3 feet over the weir. The results of these tests are given in table 1. Tests of small scale models indicate that similar results may be expected on such weirs with side slopes varying from 2:1 to 5:1 with heads over the weir up to 6 feet.

In selecting factors for use in computations, choose the factor which was determined under conditions which give the closest approximation to the actual conditions. The most common condition for diversion or other submerged dams will approximate the conditions surrounding the United States Board of Waterways model tests. The use of these factors will not give highly accurate results and should be used in full realization of this fact, but the results will be sufficiently accurate for the ordinary design purposes.

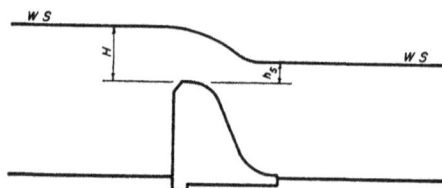

SUBMERGED DAM

FIGURE 12

TABLE 1.—*Discharge coefficients for partially submerged dams and weirs*

$\frac{h_s}{H}$	U S Board Waterways [1]	Herschel	Soil Conservation Service				
			Condition A	Condition B	Condition C	Condition D	Condition E
0 0	1 00	1 000	1 000	1 000	1 000	1 000	1 000
.1	991	1 005	1 000	1 000	1 000	1 000	1 000
.2	983	985	1 000	1 000	1 000	1 000	.998
.3	972	959	.999	996	998	998	997
.4	956	929	995	992	996	997	995
.5	937	892	989	989	992	992	993
.6	907	846	978	976	982	982	988
7	856	787	952	970	960	970	971
8	778	703	902	910	919	937	931
9	621	574	800	831	843	856	832

[1] Water Supply and Irrigation Paper No. 12.

H=Upstream head; h_s=downstream head; see fig. 12
Condition A V-notch weir slopes 2 to 1, width of weir 30 inches; weir notch 2 feet above channel bottom, upstream and downstream, channel 16 feet wide.
Condition B: V-notch weir slopes 2 to 1, width of weir 30 inches; weir notch 2 feet above downstream channel and 0 5 feet above upstream channel; channel 16 feet wide
Condition C Same as B, but channel width upstream was 8 0 feet with 1 to 1 side slopes on sides
Condition D: Same as C, but channel silted up to bottom of notch upstream from weir.
Condition E: V-notch weir slopes 3 to 1; width of weir 30 inches; weir notch 2 feet above downstream channel and 0.4 feet above upstream channel; channel 16 feet wide upstream and downstream

2. Crest Elevation.—The determination of the elevation of the spillway crest is ordinarily the result of consideration of several factors. The weight which each of these factors will have in the determination will vary with the purpose of the project, but they may be summarized as follows:

1. Capacity and water level of reservoir.
2. Cost of storage and diversion.
3. Effect of backwater.
4. Shore line topography.
5. Surface area.
6. Physical restrictions.

Each of these factors will be discussed individually. Spillway level, as used in this discussion is assumed to determine the normal operating level, or the maximum level to which usable storage is permitted except as this level may be modified through the use of flashboards or crest gates.

1. *Reservoir capacity.*—The need of a predetermined amount of storage will have been determined in the project study; the design problem is therefor reduced to a determination of the relation of capacity to the surface elevation.

The spillway crest will be set at the elevation that will give the required storage capacity but in the design of the spillway, the crest level may have to be adjusted so there will be sufficient head to pass the maximum flood discharge without exceeding the economical maximum flood level. For diversion dams the spillway crest may be fixed at the height necessary for diversion.

It is advisable on silt bearing streams to make allowance, if feasible, for the loss of live storage [4] space when considering storage requirements and capacities. Any dead storage space available may for a considerable length of time care for material carried in suspension. But, in some reservoirs a considerable amount of live or useful storage space may be filled in with coarse sand and gravel deposits brought down by tributary streams. The extent to which allowance for silting and basin erosion should be made in fixing the crest elevation is finally a matter of judgment which should be based upon an analysis of all pertinent factors, particularly of the character of the vegetal cover, of the soil and the extent of erosion within the drainage basin.

2. *Cost of storage.*—The usual problem in connection with storage capacity is to develop as much as is economically feasible, providing it does not exceed the amount that can be beneficially used. A reservoir capacity curve such as fig. 13 should be prepared, showing the storage available below different elevations. This should be

---

[4] "Live" storage is that above the outlet level which can be withdrawn for the purposes for which the project is built, "dead" storage is that below the outlet level which cannot be used for the project purposes.

followed by estimates of construction costs of dams of several different heights, which can be plotted against storage capacity, as in figure 13. The curves shown in this chart are for a hypothetical case and will necessarily be different for various designs and locations. Further analysis to show the cost of storage per acre-foot at several heights of dam will indicate the economic limit of the development. If the unit cost increases near the upper limit, as in the following tabulation, it may be that storage in excess of the most economic capacity as measured by unit costs (15 or 16 thousand acre-feet, in the example) can be developed at a lower cost at some other location. This analysis will also show whether the cost of added storage capacity exceeds the value of the water.

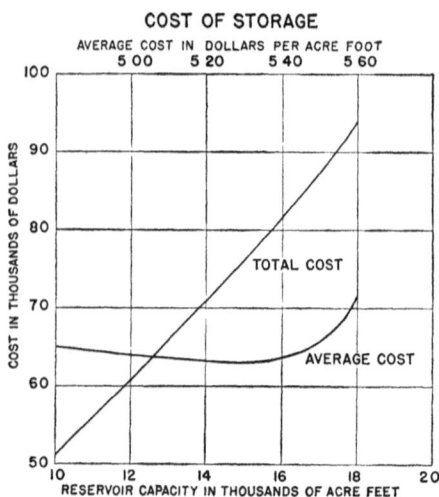

COST OF STORAGE

AVERAGE COST IN DOLLARS PER ACRE FOOT

COST IN THOUSANDS OF DOLLARS

RESERVOIR CAPACITY IN THOUSANDS OF ACRE FEET

TOTAL COST

AVERAGE COST

FIGURE 13

Cost of storage (hypothetical case)

Spillway elevation	Reservoir capacity	Total cost	Average cost per acre foot	Increase in capacity	Increase in cost	Increment cost per acre-foot
	Acre-feet			*Acre-feet*		
4,652	10,000	$51,000	$5 10			
				3,300	$16,500	$5 10 / 5 00
4,660	13,300	67,500	5 07			
				2,100	10,395	4 95
4,665	15,400	77,895	5 06			
				2,200	13,200	6 00
4,670	17,600	91,095	5 18			
				400	3,000	7 50
4,672	18,000	94,095	5 23			

For diversion projects, the determination of the crest elevation may be dependent upon studies of the economical relation between heights of dam and consequent benefits that would accrue from the resulting water levels. The cost of flowage rights may be a controlling factor, particularly if highway or railway relocations are involved.

3. *Effect of backwater.*—In reaching a decision as to the proper elevation of spillway crest, there must be determined for each possible crest elevation the maximum water level in the reservoir at the time of maximum spillway discharge. There must also be determined what, if any, increase in stream stage will occur upstream from the reservoir due to backwater effects. When these elevations have been determined, an estimate must be made of the value of lands and improve-

ments that would be damaged by the operation, and the resulting expected expense should be treated as a part of the project cost. The amount of said submergence and flowage damage may be determined from an appraisal, but if possible should be substantiated and confirmed by negotiations, options or agreement with the property owners affected.

In many instances, the cost of such damages will be a considerable part of the total project cost and may have an important bearing on a determination of the economic spillway elevation as analyzed in the above table.

4. *Shoreline topography.*—For practically all types of storage projects, excepting those related to wildlife conservation, the chosen crest elevation should be that which will avoid large areas of shallow water. For recreational projects, it is important that the shoreline at normal pool level be sharply defined and so placed that wide areas of mud buttom will not become exposed for small drops in water surface. The avoidance of normal pool levels that will involve large areas of shallow marsh is particularly important to any program of mosquito control. The effect of diversion projects upon shoreline topography should be taken into account in connection with a study of benefits and damages which would be created by dams of different heights.

5. *Surface area and evaporation.*—In many parts of the country, the rate of evaporation from water surfaces is so great that evaporation constitutes one of the serious sources of loss of stored water. In the Southwest, evaporation losses may average as high as 72 inches per year equivalent to 6 acre-feet of water for each acre of pond area and generally about half of this loss will occur in the four summer months of May to August, inclusive. For such regions, it is evident that reservoirs having the greatest possible mean depth and least surface area will be the most efficient for storage purposes. The effect of evaporation on the determination of pool level may be analyzed from the typical area-capacity curve for the reservoir such as that shown in figure 14. This particular curve is representative of a most undesirable type of small reservoir, involving relatively small maximum

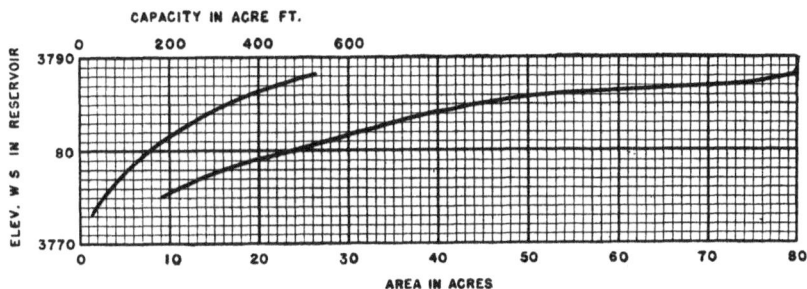

FIGURE 14

depth and a comparatively large surface area. A comparison of this project for pool elevations at 3785 and 3786 shows the choice of the higher level would result in an increase in capacity of about 50 acrefeet and an increase in area of 7 acres. If the inflow were such as to permit this reservoir to be filled to the higher elevation practically all of the time, the increase in area would result in the loss of 42 acrefeet through evaporation, as compared to an increase of 50 acre-feet in storage. It is obvious that a consideration of any higher pool level would reduce the reservoir to an absurdity, and its operation would be more nearly that of a large evaporation pan than of a water conservation project.

6. *Physical restrictions.*—There will be projects where certain definite limitations such as railways and highways not susceptible of relocation will fix the spillway level. For diversion dams, the elevation of the water surface at the outlet channel will be a controlling factor. In other cases, the amount of head required to carry the water to a definite point of use will determine the normal operating level.

The choice of location of a desirable site for the spillway may be so restricted as to offer only one plan of development, and that within a narrow range of spillway elevation which leaves no choice. Likewise, there may be conditions at the damsite which limit its development, such as changes in the foundation conditions above a particular elevation.

3. Control of Reservoir Level.—1. *Flashboards.*—Closely related to the problem of determining the most economic spillway level is the question of propriety of using flashboards on the spillway crest. Flashboards provide a means by which the normal pool level (spillway crest level) can be raised during those periods when the spillway is not in use for discharging floods. They can obviously only be used with safety under conditions where their removal in advance of the flood wave can be absolutely assured. Several different types of flashboards have been developed, some of which must be placed and removed manually, some of which are designed to fail under certain heads, and others which drop out of position automatically when the level is increased a certain amount. The common installation of wooden panels supported behind vertical pins on the spillway crest is ordinarily unsightly and involves large losses through leakage. As small projects will rarely justify mechanical equipment for placing and removing these boards, and as they cannot be removed by hand unless the water level is below them, they can practically never be justified from the viewpoint of safe spillway operation. They should never be utilized on small dams where the reservoir area is small and the stream subject to flash flows. For small projects intended to control storage in very large lakes, and where the rise in pool stage for flood conditions will be at a very slow rate, use of flashboards may be permissible provided that an operating crew for their removal is always available. If they are used, the profile of the overflow crest shown in figure 88 may be modified by the introduction of a short tangent of 6 to 12 inches at the point of compound curvature to provide a level base of concrete for them.

The simpler types of flashboards that are designed to fail when the water reaches certain stages are uncertain in their operation, are generally unsightly, and, if conditions are such that they will function frequently, their repeated replacement will involve a large expense item.

Use of flashboards for spillways of earth dams, or other structures where failure of the mechanism to operate would result in overtopping and destruction of the dam, is never justified.

2. *Crest Gates.*—Where the value of the additional storage above spillway crest during the low-water season justifies the expense, stop logs or mechanically operated gates can be used to partially close the spillway opening and removed or opened to make the full spillway capacity available during high water. When correctly built they are expensive features, which cannot be justified in the ordinary small project. Where used, consideration should be given to the flood crest that may be created downstream by the automatic opening of the full spillway. Gates used for this service include stop logs, sliding gates, radial gates, and roller gates.

Stop logs.—The simplest kind of crest gate is the stop-log arrangement, consisting of grooved piers, usually of concrete, with wooden timbers spanning the space between them. The timbers will vary in thickness from 2 inches up to 12 inches or more, and the depth of an individual piece may range between 4 and 12 inches. The timbers are placed and removed individually, either by hand with hooks or by use of small hand-operated hoists. A typical installation is shown in figure 15.

Stop logs have several objectionable features in common with flashboards. The bottom log should have a rubber gasket (such as a rubber hose) along its lower edge to prevent leakage, but there usually will be leakage between the logs, and considerable time may be required for the removal of the logs from the opening, particularly if they become jammed in the slot; also the length of the spillway crest must be increased to allow for the reduction in discharge caused by the concrete piers.

The timber used for stop logs may be of any structural species conveniently obtainable. It may be cut on the job or obtained from small nearby mills. It should be free from defects that will markedly lower its strength, such as large knots or groups of knots near the center, sharply inclined.

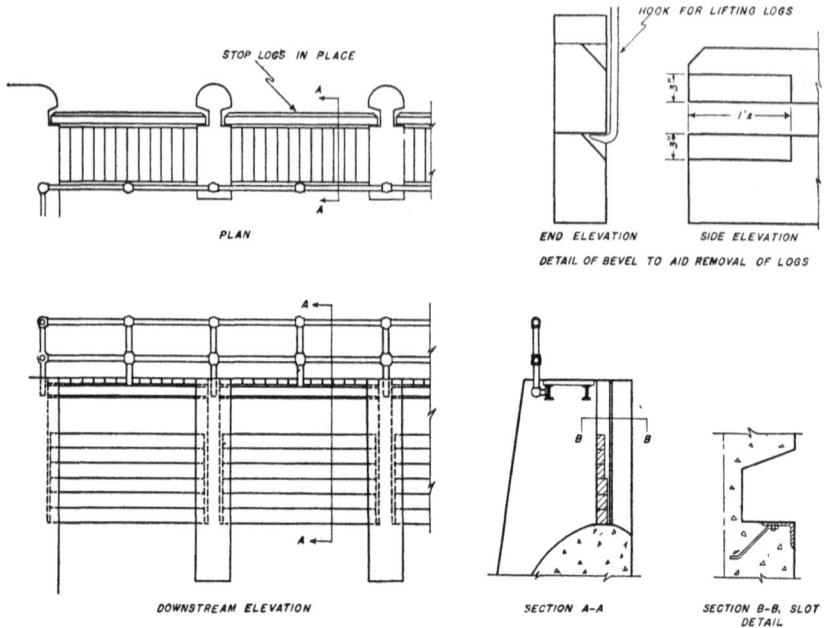

FIGURE 15 —Typical stop logs installation

TABLE 2.—*Design stresses for roughly graded structural timber*

[Pounds per square inch]

Species	Extreme fiber stress in bending	Compression perpendicular to grain	Horizontal shear
Ash, white	1,100	500	100
Chestnut	800	300	70
Cypress, southern	1,000	350	80
Douglas fir, west coast	1,200	325	70
Douglas fir, Rocky Mountains	900	275	70
Fir, commercial white	900	300	60
Hemlock, eastern	900	300	60
Hemlock, western	1,000	300	60
Maple, hard	1,200	500	100
Oak	1,100	500	80
Pine, western or ponderosa	700	250	70
Pine, Norway	900	300	70
Pine, southern yellow	1,200	325	100
Redwood	1,000	250	60
Spruce, Englemann	600	175	60
Spruce, red, white, Sitka	900	250	70
Tamarack	1,000	300	80

If commercially graded material is used, the stresses should conform to those of the association under whose rules it was graded. The table in Appendix E lists the commercial stress grades and species, the equivalent commercial grade names, the rules under which the species are graded, and the allowable stresses. The grade selected as suitable for stop logs is generally an intermediate grade.

The stresses in table 2 and Appendix E are for untreated timber used under conditions unfavorable to decay and continuously dry or for pressure-treated timber used under any conditions. For service where the untreated timber is subject to decay hazards, such as alternate wetting and drying, the bending and compression stresses should be reduced by from 20 percent to 40 percent of the values given, depending upon conditions.

Because of the severe service to which most stop logs are subjected, the use of pressure-treated material will usually prove economical.

FIGURE 16 — Downstream face of Echo earth fill dam and spillway, Salt Lake Basin Project, Utah

Stresses in the stop logs from the water pressure are computed from the simple beam formula for uniform loading. The design stresses should be conservative, because excessive deflection of the logs will cause a concentration of bearing pressure on the inside corner of the pier and make the logs hard to remove under load.

For the determination of the horizontal shearing stress in rectangular beams, use the following equation:

$$q = \frac{3V}{2bd}$$

in which q is the unit horizontal shear, V is the external shear or maximum end reaction, and b and d are the width and thickness, respectively, of the rectangular beam.[5]　Figure 17 gives the required

---

[5] See "Wood Handbook", prepared by the Forest Products Laboratory, Forest Service, Department of Agriculture.

thickness for different spans and various heads where the allowable fiber stress is 1,000 pounds per square inch.

Sliding gates.—This type of gate may be made of wood, cast iron, or structural steel.

Wooden gates should be made of oak, hard pine, or other durable species. The individual timbers are held together by through bolts from top to bottom, and watertightness is increased by the use of

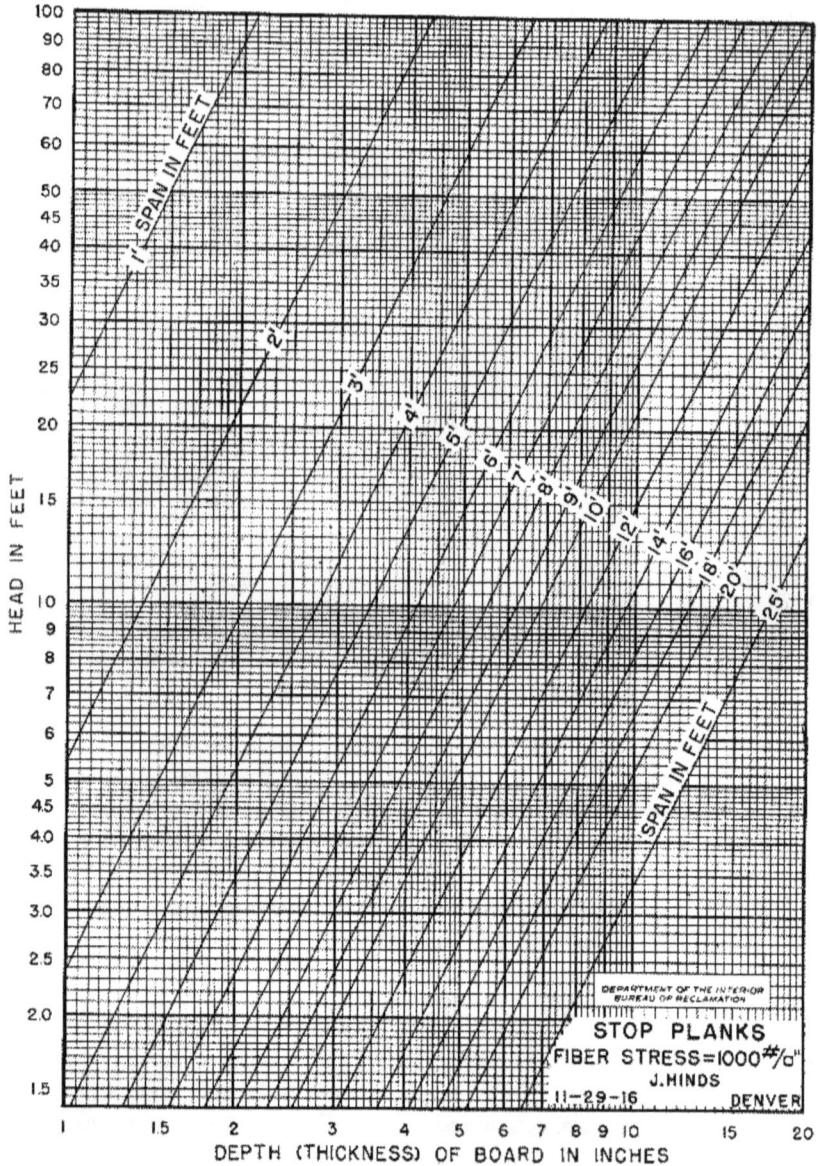

FIGURE 17

splined joints. The thickness of the timber used is computed from the pressure on the gate and the allowable unit stresses. (See table 2.) For small gates where the compressive stresses will not be exceeded, the timber can bear directly on the frames. If frequent manipulation of the gate will be necessary, as in daily operation, the timber should be protected against wear by steel faceplates.

Timber gates are usually not heavy enough to lower into place of their own weight when under pressure and therefore the stems must be rigid so that closing pressure can be exerted. For gates exceeding 5 feet in width, two stems are recommended, as they will tend to reduce vibration during operation and will make it easier to keep the gate in line.

The load on the gate stems when operating is a function of the total pressure on the gate and of the coefficient of friction between the materials of which the bearing surfaces of the gate and the frame are made. The opening or closing force due to water pressure is

$$W = fP$$

in which P is the total horizontal pressure load on the gate, and f is the coefficient of friction between the two bearing surfaces. The starting friction is usually about twice as great as the moving friction, and the operating mechanism must be designed for the more severe condition. Values of f for commonly used materials are given in table 3.

TABLE 3.—*Coefficients of friction [1] and bearing values for gate seats*

Material	Starting friction	Sliding friction	Bearing pressures
Timber on steel	0 62	0 31	
Timber on bronze	60	30	
Bronze on bronze	70	35	500 pounds per square inch.
Steel on bronze (no rust)	70	35	400 pounds per square inch.
Cast iron on steel	75	38	Do

[1] These values are for gates which will remain closed for long periods

The total load on the gate stems when the gate is being raised is the sum of the force W and the dead load of the gate.

Figure 18 shows typical details of a timber sliding gate.

Iron or steel gates cost much more than wooden gates, but usually have longer life. Ribbed cast-iron gates are made in stock patterns in sizes up to 10 feet square or 9 feet by 12 feet rectangular shapes. If the bearing surfaces of the gate or the frame are made of iron or steel, they will eventually become rusty and the friction factor will be greatly increased for both the starting and moving conditions. Bronze contact surfaces for the gate and frame will give much smoother operating conditions and should always be used for gates which will remain closed for long periods.

The design of a structural-steel gate is similar to that for the wooden type. The skin plate acts as a continuous beam in transmitting the pressure to the horizontal members, which in turn are simple beams which transmit their loads to the frame. The load from the frame is transmitted through the gate seat to the supporting piers and the base structure.

Structural steel gates are usually assembled inside a channel or I-beam frame, with horizontal I-beam cross members, and a steel skin plate on the upstream face. If the dead weight of the gate is insufficient to give satisfactory closure under pressure, the space between the cross beams is frequently filled with concrete to give additional weight. Concrete filling makes a gate very rigid, and great care must be taken before and during concreting operations to keep the bearing surfaces in a true plane, so that it will fit properly against the frame. A rigid gate which has become warped cannot be made watertight.

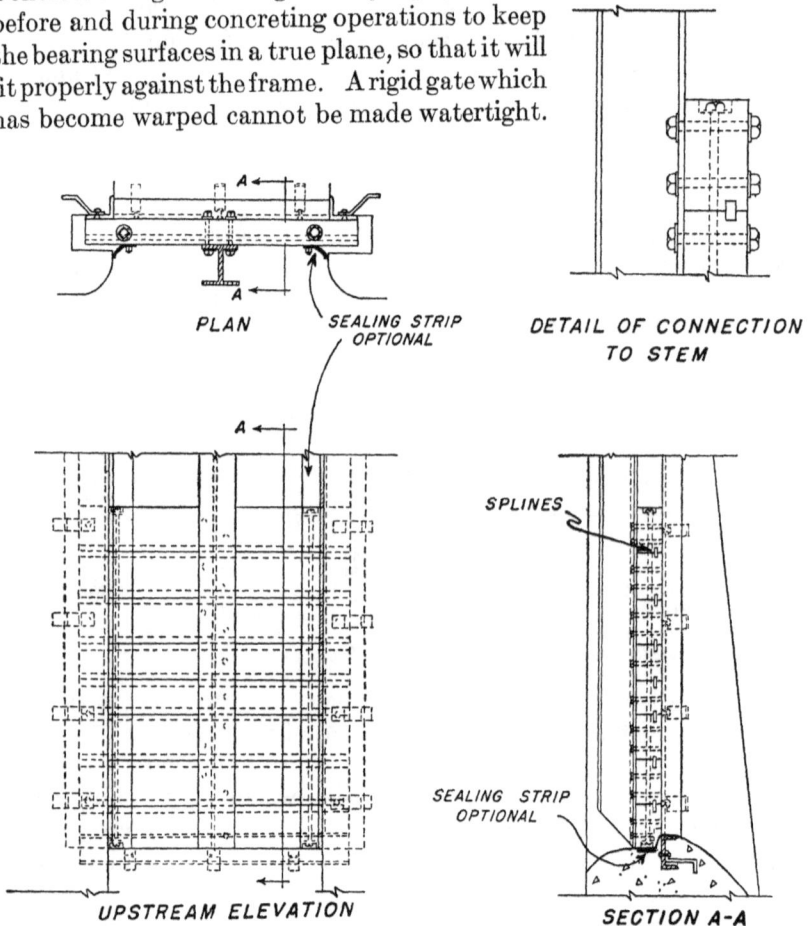

PLAN

SEALING STRIP OPTIONAL

DETAIL OF CONNECTION TO STEM

SPLINES

SEALING STRIP OPTIONAL

UPSTREAM ELEVATION

SECTION A-A

TYPICAL TIMBER SLIDING GATE

Figure 18

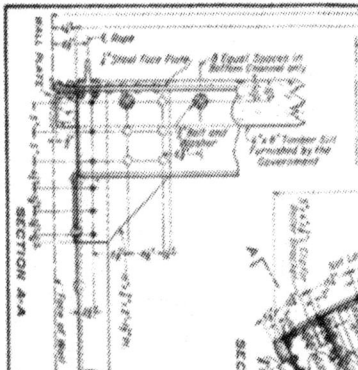

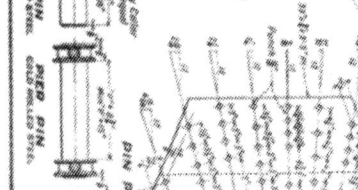

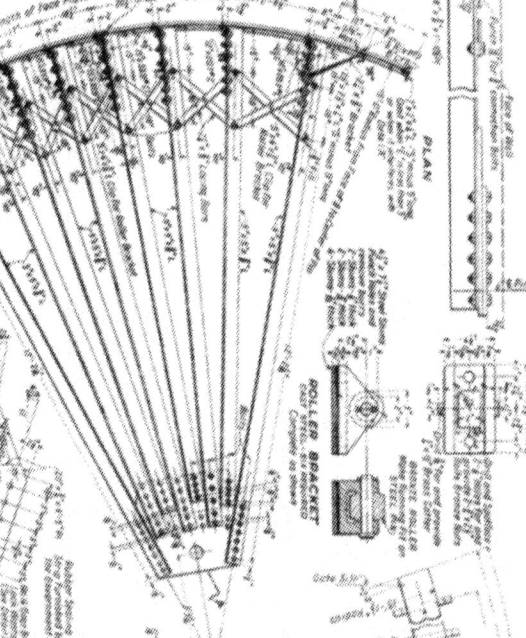

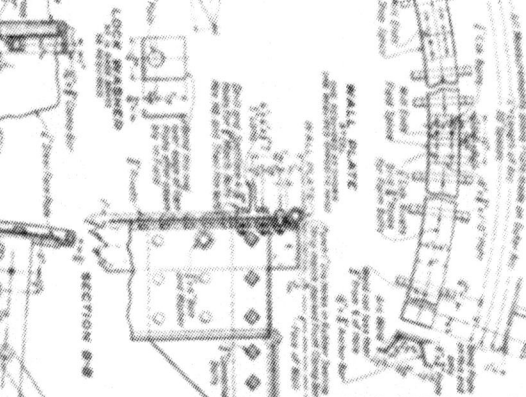

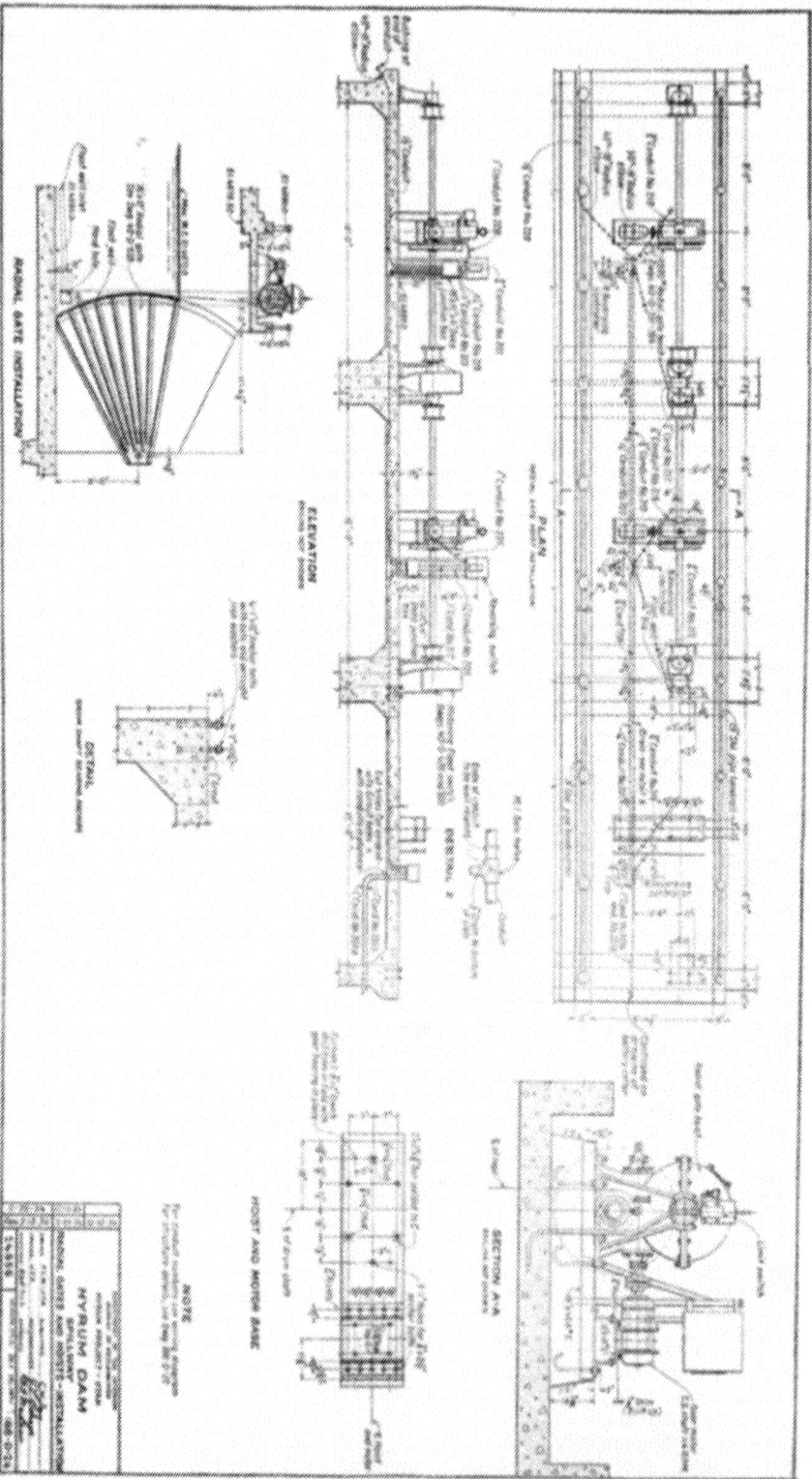

The design of gates of this kind should not be attempted without experienced supervision. The information given here is intended only to be sufficient to give an understanding of the general requirements that would be useful in early studies of a project and to indicate the different kinds of crest gates that are available.

Radial Gates.—The radial gate, which is also commonly known as The Taintor gate, is particularly well adapted to crest control. The thrust from the water pressure is carried by the trunnion bearing where it offers little resistance to the operation of the gate. The operating load is the dead weight of the gate with a liberal allowance for ice in some cases, plus a small frictional resistance of the flexible sealing strips between the ends of the gate and the pier faces, plus the torsional resistance of a trunnion bearing.

$$L = \frac{Wx + k_1 F r_1 + k_2 P r_2}{r}$$

in which, as shown in figure 20.

L = lifting load.

W = the weight of the gate.

F = the total water pressure transmitted by the sealing strips at the ends of the gate to the bearing area of the strips.

P = the total water pressure on the gate.

x = the horizontal distance from center of bearing to center of gravity of the gate.

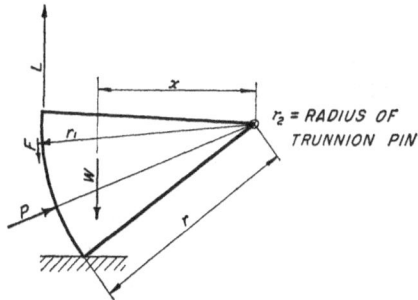

RADIAL GATE FORCE DIAGRAM

FIGURE 19

r = the radius of the gate from center of pin to face of gate.

r_1 = the radius from center of bearing to sealing strips.

r_2 = the radius of the bearing pin.

k_1 = the coefficient of starting friction of the sealing strips. (See table 4.)

k_2 = the coefficient of friction of the trunnion bearing. (See table 3.)

TABLE 4 —*Coefficient of starting friction of gate seal strips on wet surfaces* [1]

Bearing surface	Coefficient
New steel plates	0.64 to 0.73
Rusty steel plates	.82 to .88
Rough concrete	.56 to .60
Smooth concrete	.60 to .70

[1] From 1924 report of the hydraulic committee of the National Electric Light Association.

Radial gates have been built within a wide range of sizes upward from 3 feet high by 6 feet wide. The gates are operated by a hoist-

FIGURE 20 —Taintor gates installed at diversion dam on North Platte River

ing device on a bridge overhead, the hoist being connected by cable
or chain to the two lower corners of the gate.

The sill on which the gate rests may be of wood embedded in the
concrete crest, or in cases where the wooden sill might be damaged
by ice, logs, and such in the water the timber is replaced by a steel
channel or H–beam.   The bottom of the gate which bears on the sill
may be of steel, but is usually faced with a timber nosing bolted to
the bottom gate member.   At the ends, a steel plate is anchored into
the concrete piers, forming an arc corresponding to the path of the
gate face.   The water seal between the gates and the piers at the ends
is formed by flexible fabric belting or a special rubber sealing strip
made for the purpose.   This sealing strip bears against the steel plate
in the face of the pier, which gives a smooth surface for the contact
and prevents excessive wear of the sealing strip.   The face plate also
provides a bearing surface for guide rollers at the ends of the gate to
prevent it from binding against the piers.   Figure 21 shows a typical
design of a radial gate by the Bureau of Reclamation, and figure 22
shows the arrangement of the hoisting mechanism.

In the design of the radial gate, the water pressure is borne by the
skin plate and transmitted to horizontal members, usually channels,
which in turn are supported by the radial arms.   The arm thrust is
transmitted through a hub to the pin or trunnion bearing and thence
to the pier.   The trunnion carries the entire thrust of the gate and is
usually bolted to the downstream face of the pier by long anchor rods
which transmit the load into the concrete.

Roller Gates.—In order to avoid the high lifting pressures due to
friction in sliding gates, several arrangements have been devised
to provide roller contacts.   In the common type, rollers or wheels

are attached to the gate and bear on a track which is part of the frame. The Stoney gate uses roller trains that are independent of both gate and frame, but roll between them. In the Broome gate, there is a roller-train arrangement similar to the treads of a caterpillar tractor which bears on a track on the gate frame. The gate seat is built at a slight angle with the roller track, and the frame is parallel to the gate seat. As a result of the inclination of the path of travel to the plane of seating, the movement of the gate at the time of contact has a normal component and the sliding friction is eliminated. All of these arrangements permit the use of larger gates such as are not usually required for small projects. They should be designed only under supervision of engineers experienced in their use.

4. Open channels.—The two features which control the design of open channels are the inlet capacity and the channel capacity.

Referring to figure 23, the following nomenclature will be used in developing capacity formula:

D=depth of water above channel bottom at entrance.

H=depth of water above channel bottom at outlet just above any drawdown influence.

h_0=velocity head.

y_0=mean velocity at upper end of channel.

b=width of channel bottom.

z=slope of sides of channel, horizontal to vertical.

s=slope of water surface in channel.

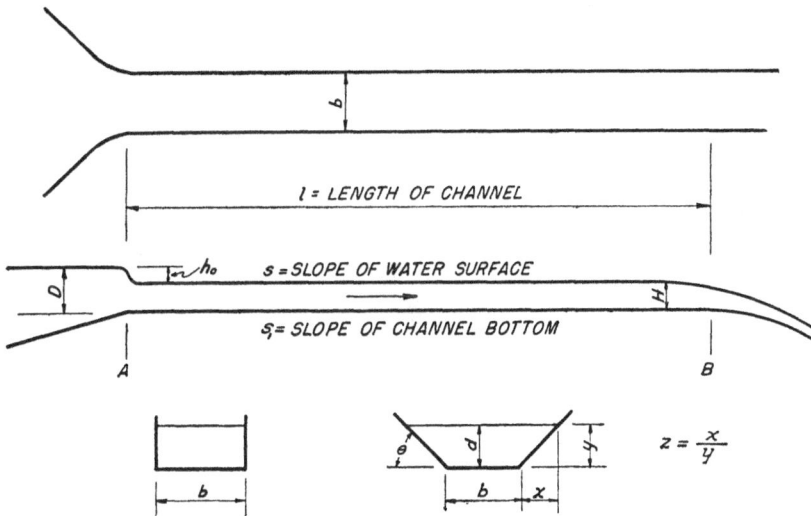

FIGURE 23

v = the average velocity in feet per second.

n = the roughness coefficient.

r = hydraulic radius, or $\dfrac{\text{area of water prism}}{\text{wetted perimeter}}$

C = coefficient of discharge.

a = cross-sectional area of the water.

As the water flows from the comparatively still pond, through the entrance to the channel, there will be a marked drop in the water surface, represented by h_0. This drop in head will produce a velocity that may be expressed

$$v_0 = C\sqrt{2gh_0} \tag{4}$$

and the quantity of water flowing in the channel will be $Q = a\,v_0$. Since

$$a = b(D - h_0)$$
$$Q = b(D - h_0)C\sqrt{2gh_0} \tag{5}$$

To determine the condition which will give the **maximum capacity**, equation (5) is differentiated with respect to h_0, equated to zero, and solved for h_0. The value of h_0 is found to be $1/3\,D$, and substituting this value for h_0 in equation (5), the maximum capacity is

$$Q = 3.087 C b D^{3/2} \tag{6}$$

The constant C represents the entrance losses, or the hydraulic efficiency, and will vary from 1.0 for the perfect entrance of smooth curves and gradual transitions to 0.82 for the more abrupt type, such as a rectangular shaped concrete structure with square corners.

The foregoing solution is for inlets and channels with vertical sides. The method of solution is the same for trapezoidal sections, but the

FIGURE 21—Damage from flood water over an open spillway channel protected only by vegetation in
North Dakota

resulting equation can not be reduced to such simple terms. The value of h_0 which gives maximum discharge for a trapezoidal section is

$$h_0 = \frac{3(2zD+b) - \sqrt{16z^2D^2 + 16zDb + 9b^2}}{10z} \tag{7}$$

and

$$Q = 8.03Ch_0^{1/2}(D-h_0)[b + z(D-h_0)] \tag{8}$$

Before solving (7) it will be necessary to assume a value for z, the side slope ratio.

The capacity of a channel depends upon the cross-section area, the slope of the channel bottom and the water surface, the hydraulic radius, and the roughness of the channel lining. For channels of this type, the velocity as expressed by the Manning formula is

$$v = \frac{1.486}{n} r^{2/3} s^{1/2} \tag{9}$$

Values of n are determined experimentally, and have been established for nearly all of the materials and conditions encountered.

In using the Manning formula, the roughness coefficient will be selected from table 5, dependent upon the channel lining that is to be used. Two of the other three unknowns must be given or assumed for the solution. For ease of calculation there is given in figure 25, a standard chart for the solution of the Manning formula.

For example, with a channel cross section having a roughness coefficient (n) of .019, a hydraulic radius (r) of 2, and a slope (s) of .00257, we enter the chart at the left-hand side marked "s=slope," follow the dash line to the right until it intersects the vertical dash line above 2 at the bottom of the chart marked "r=hydraulic radius—feet," then follow the curved dash line downward and to the right until it intersects the vertical line under .019 for "n=coefficient of roughness"; directly opposite at the right-hand edge of the chart we read 6.3 as the "v=velocity—feet per second," which gives us the velocity of flow in the channel.

In selecting the best proportions of a channel section it will be noted that for a given slope and roughness coefficient, the velocity will be greatest where r is greatest. From the relation $Q = a\,v$, it is seen that for a given water area, the rate of flow varies directly with the velocity. For greatest hydraulic efficiency, therefore, choose the section which gives the greatest value for r for a given area. For a trapezoidal section as shown in figure 23, the hydraulic radius is greatest where

TABLE 5 *—*Horton's values of n.*   *To be used with Kutter's and Manning's formulas*

CONDITION OF SURFACE AS AFFECTING FLOW

Surface	Best	Good	Fair	Bad
Uncoated cast-iron pipe	0 012	0 013	0 014	0 015
Coated cast-iron pipe	011	¹ 012	¹ 013	
Commercial wrought-iron pipe, black	012	.013	014	015
Commercial wrought-iron pipe, galvanized	013	.014	015	017
Smooth brass and glass pipe	009	.010	011	013
Smooth lockbar and welded "OD" pipe	010	¹ 011	¹ 013	
Riveted and spiral steel pipe	013	¹ 015	¹ 017	
Vitrified sewer pipe	010 / 011	¹ 013	015	.017
Common clay drainage tile	011	¹ 012	¹ 014	017
Glazed brickwork	011	012	¹ 013	015
Brick in cement mortar, brick sewers	012	013	¹ 015	017
Neat cement surfaces	010	.011	012	013
Cement mortar surfaces	011	012	¹ 013	015
Concrete pipe	012	013	¹ 015	016
Wood stave pipe	010	.011	012	013
Plank flumes.				
Planed	010	¹ 012	013	014
Unplaned	011	¹ 013	014	015
With battens	012	¹ 015	016	
Concrete-lined channels	012	¹ 014	¹ 016	018
Cement-rubble surface	017	020	025	030
Dry rubble surface	025	030	033	035
Dressed ashlar surface	013	014	015	017
Semicircular metal flumes, smooth	011	012	013	015
Semicircular metal flumes, corrugated	0225	025	0275	030
Canals and ditches				
Earth, straight and uniform	017	020	¹ 0225	025
Rock cuts, smooth and uniform	025	030	¹ 033	035
Rock cuts, jagged and irregular	035	040	045	
Winding sluggish canals	0225	¹ 025	0275	030
Dredged earth channels	025	¹ 0275	030	033
Canals with rough stony beds, weeds on earth banks	025	030	¹ 035	040
Earth bottom, rubble sides	028	¹ 030	¹ 033	035
Natural stream channels				
(1) Clean, straight bank, full stage, no rifts or deep pools	025	0275	030	033
(2) Same as (1), but some weeds and stones	030	033	035	040
(3) Winding, some pools and shoals, clean	033	035	040	045
(4) Same as (3), lower stages, more ineffective slope and sections	040	045	050	055
(5) Same as (3), some weeds and stones	035	040	045	050
(6) Same as (4), stony sections	045	050	055	060
(7) Sluggish river reaches, rather weedy or with very deep pools	050	060	070	080
(8) Very weedy reaches	075	100	125	150

* From Handbook of Hydraulics, by H. W King
¹ Values commonly used in designing.

$$b = 2d \left(\sqrt{z^2 + 1} - z \right) \tag{10}$$

or

$$b = 2d \, \tan \frac{\theta}{2} \tag{11}$$

and

$$r = \frac{d}{2} \tag{12}$$

For rectangular sections θ is 90°, and the bottom width, b, equals two times the depth, d. While this relation holds true for best hydraulic efficiency, and structurally for channels that are to be built above ground, for excavated channels it will be subject to modification.

For example, the two excavated channels shown in section in figure 26 have the same capacity, and the same slope. Section A has the ideal proportions below the water line, the depth being one-half of the width, but the reduced amount of excavation required above the water line in section B makes B the more economical section from the viewpoint of excavation yardage. A comparison can be made from the following tabulation of values:

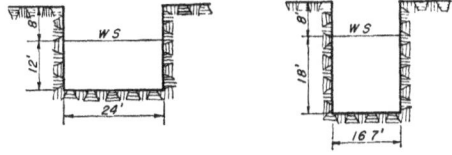

SECTION-A   SECTION-B

FIGURE 26 —Excavated channels

Item	Section A	Section B
Slope	0 005	0 005
Capacity	2,880 cubic feet per second	2,880 cubic feet per second.
Roughness coefficient	0 035	0 035
Water area	288 square feet	300 6 square feet
Hydraulic radius	6	5 70
Velocity	10 feet per second	9 58 feet per second.
Excavation area	480 square feet	434 2 square feet.
Lining to water level	48 linear feet	52 7 linear feet

The comparison shows that the excavation required for section B is 9.5 percent less than for section A, even though the section B water area is 4.5 percent greater. The difference comes in the portion above the water level, where section B is 30.5 percent smaller than A. It should be pointed out that if the ground surface were only 1 foot above the water surface, the advantage would rest with section A, for its area would then be nearly 2 percent less than B.

With $n=0.035$, these sections obviously were not intended to be lined. However, the relative values are the same for any value of n, so the comparison can be made. In this case it is evident that the 4.7 linear feet of extra lining required for section B would not cost as much as the rock excavation represented by the additional 45.8 square feet area of section A, but the method of comparison is illustrated.

The use of the Manning formula for the determination of channel capacity is for a condition of steady and uniform flow. That is, the slope of the channel is just sufficient to overcome the friction losses without appreciable change in the average velocity. For steeper or flatter slopes, there will be either acceleration or deceleration of the water velocity, in which case rational values of v and s by the Manning formula cannot be computed until the increasing velocity is high enough to cause a friction head equivalent to the slope of the channel.

Nonuniform flow.—The condition of nonuniform flow is represented in figure 27 in which longitudinal section (a) represents a channel with either a uniform section, or a section which is changing at a uniform rate, with increasing velocity; and longitudinal section (b) represents

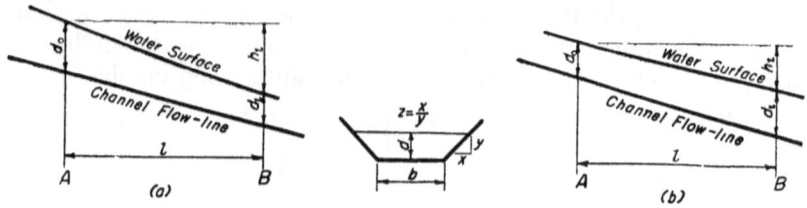

ACCELERATED AND DECELERATED FLOW

FIGURE 27

like conditions but with decelerating velocity caused by backwater. The method of computation of velocity or depth at any point along the channel for a known flow is the same for either condition.

The following nomenclature in addition to that given above is used in subsequent formulas:[7]

l=length of channel considered.
s_l=slope of channel bottom.
h_l=fall of water surface in length l.
d_0=depth of water at upstream end.
d_l=depth of water at downstream end.
b_0=width of channel bottom at upstream end.
b_l=width of channel bottom at downstream end.
v_0=mean water velocity at upstream end.
v_l=mean water velocity at downstream end.
v=effective velocity in channel section being considered.

b, d, and v are corresponding values for the midsection of channel length l.

In the channel from A to B, it is assumed that the effective velocity, v, is the average of v_0 and v_l or $v=1/2\ (v_0+v_l)$ and that it occurs midway between A and B. If the sections such as $A\ B$ are not too long, or in other words, if the difference between v_l and v_0 is not too great, the error can be kept within reasonable limits. As a general rule, select values of l such that v_l will not vary more than 25 percent from the value of v_0. The depth at any point for a trapezoidal channel is

$$d=\sqrt{\frac{b^2}{4z^2}+\frac{Q}{zv}}-\frac{b}{2z} \qquad (13)$$

in which the values are for an average cross section midway between A and B, in an isolated length of channel, l, determined by

---

[7] "Handbook of Hydraulics," King.

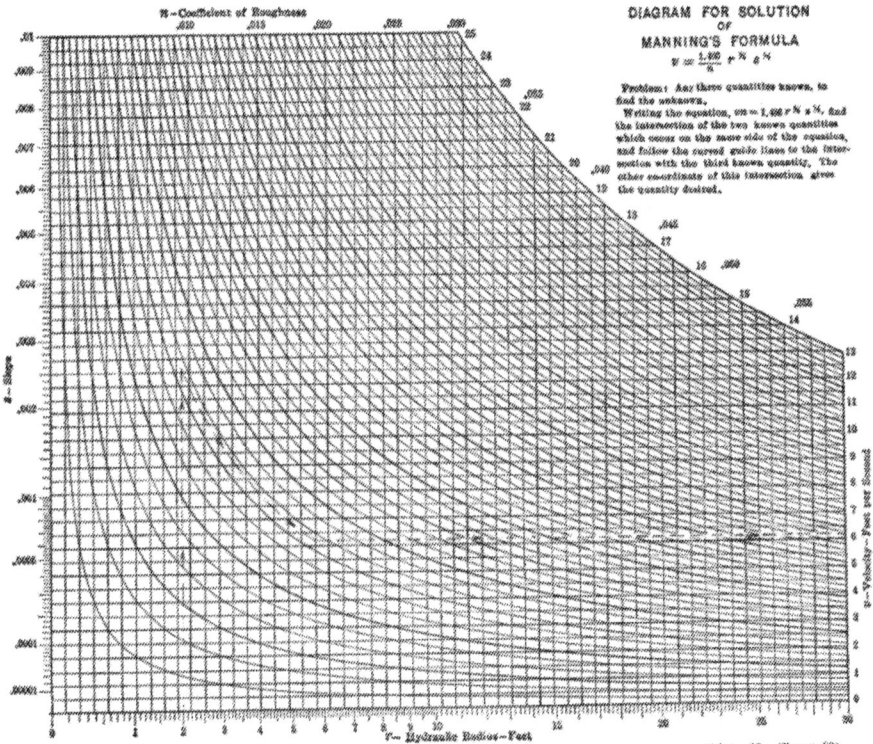

n — Coefficient of Roughness

.010 .015 .020 .025 .030

DIAGRAM FOR SOLUTION
OF
MANNING'S FORMULA

$$V = \frac{1.486}{n} r^{\frac{2}{3}} s^{\frac{1}{2}}$$

Problem: Any three quantities known, to find the unknown.

Writing the equation, $vn = 1.486 \, r^{\frac{2}{3}} \, s^{\frac{1}{2}}$, find the intersection of the two known quantities which occur on the same side of the equation, and follow the curved guide lines to the intersection with the third known quantity. The other coordinates of this intersection gives the quantity desired.

s — Slope

V — Velocity — Feet per Second

r — Hydraulic Radius — Feet

Figure 25

78003°—39 (Face p. 80)

$$l=\frac{\dfrac{v_i^2}{2g}-\dfrac{v_o^2}{2g}+d_i-d_o}{s_i-\dfrac{(nv)^2}{2.208r^{4/3}}} \qquad (14)$$

In using formulas (13) and (14), s_e and n are constant for any section. At A, the quantity, velocity, and channel dimensions will be known. By assuming a value for v_i at section B (not more than 25 percent greater than v_0), the average velocity and channel dimensions for the section midway between A and B can be computed. Use these mid-section values to determine r in (14), so that the value of l can be found.

For rectangular channels, formula (13) becomes

$$d=\frac{Q}{bv} \qquad (15)$$

1. *Design of outlet channel with steep slope.*[8]—Where a spillway outlet discharges into an open channel with a steep slope, the inlet velocity will be that which will give the maximum discharge, or the velocity which will occur at critical depth. This is called a control section. In figure 28, it is assumed that the outlet is from a still pool in which the velocity of approach is negligible. The drop from D to d_0, represented by h_0, is the head required to produce the velocity corresponding to critical depth. For rectangular openings, $h_0=1/3D$, and for trapezoidal openings it is given by equation (7). Knowing the amount of outflow required, and the depth which it will assume, the required width for a rectangular outlet is

$$b_0=\frac{0.324Q}{D^{3/2}} \qquad (16)$$

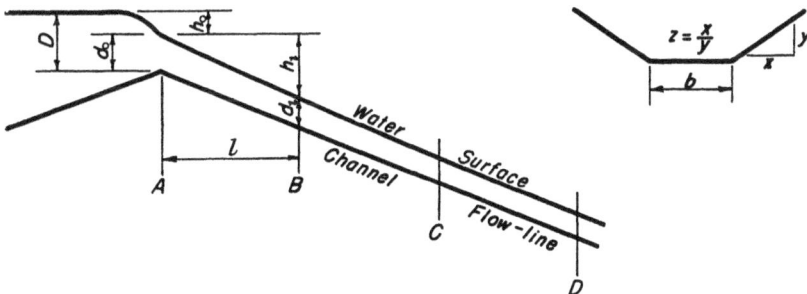

SHORT CHANNEL WITH STEEP SLOPE

FIGURE 28

8 "Handbook of Hydraulics," King.

For trapezoidal outlets, an approximate value of the bottom width is

$$b_0 = \frac{0.324Q}{D^{3/2}} - 0.7zD \qquad (17)$$

The mean velocity at the point of entrance is, for trapezoidal channels

$$v = \frac{Q}{a_0} - \frac{Q}{d_0(b_0 + zd_0)} \qquad (18)$$

and for rectangular channels

$$v_0 = \frac{Q}{d_0 b_0} \qquad (19)$$

We can now find the water velocity and the dimensions of its cross section as it enters the channel. From this initial section compute successive reaches downstream, reducing the size of the channel where necessary to maintain economic proportions. Each successive section should be similar in shape, and the transitions should be gradual and uniform. On long slopes, the flow will become uniform when the velocity has increased to the point where the loss of head due to friction is offset by the slope of the bottom of the channel. This condition is represented by the following adaptation of the Manning formula for trapezoidal sections:

$$Q = \frac{1.486s^{1/2}[d(b+zd)]^{5/3}}{n[b + 2d(1+z^2)^{1/2}]^{2/3}} \qquad (20)$$

and for rectangular sections

$$Q = \frac{1.486s^{1/2}(db)^{5/3}}{n(b+2d)^{2/3}} \qquad (21)$$

From these equations, if the quantity of flow and the channel dimensions are known, solve for s, the slope at which the flow will become uniform. Conversely, knowing the quantity and the slope, the equations can be solved to give the dimensions of the channel in which uniform flow will be maintained, but in this case the solution will have to be by trial of assumed values for b and d.

2. *Transitions.*—For spillway designs as described in the previous pages, it is evident that there will be a sharp increase in the average velocity below the inlet when the effect of the steep slope of the channel bottom is realized. The area must be reduced in inverse proportion to the velocity change if an economical section is to be maintained. Changes in shape or cross sectional area of a stream of water produce disturbances in the flow and loss of head. The required increase in velocity head to overcome these losses must be provided for in the design of the structure by allowing a drop in the water surface

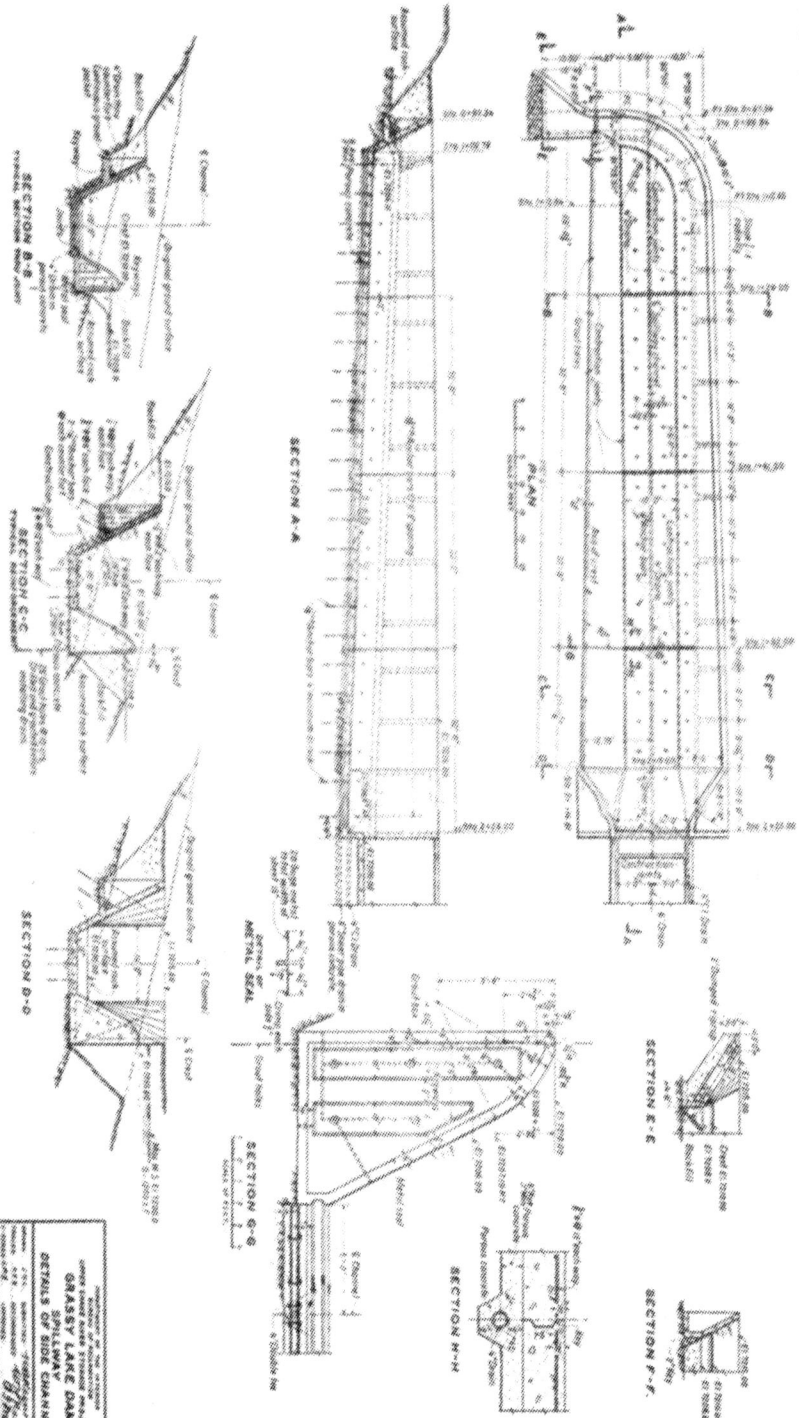

GRASSY LAKE DAM — SPILLWAY — DETAILS OF SIDE CHANNEL

curve. Similarly in channel outlets where the change is from high to low velocities, the water surface will rise an amount equivalent to the reduction in velocity head less the energy losses.

Theoretical computation of transition losses is too complicated to be of practical use, so designs are usually based on empirical rules developed from practice. After much experience with structures of this kind, the United States Bureau of Reclamation has adopted the following criterion:[9] "that the computed water surface profile through the transition shall be a smooth, continuous curve, approximately tangent to the water surface curves in the channels above and below." Figure 29 illustrates this point.

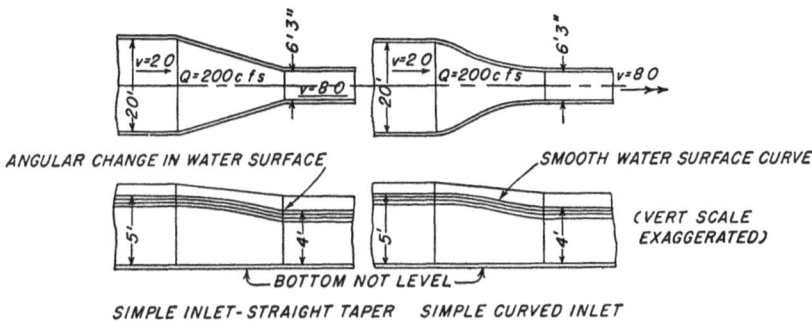

FIGURE 29.—Design of flume and siphon transitions

An arbitrary rule commonly used in determining the length of transition structures is illustrated in figure 30. A straight line joining the flow line at the two ends of the transition makes an angle of approximately $12\frac{1}{2}°$ with the center line of the channel. With this length as a starting point, assume a transition section and compute the water surface profile. If it does not fulfill the criterion requirements, make the indicated changes and recompute the profile, and continue the revisions until the results are satisfactory. Several operations will usually be required before the final layout is accomplished.

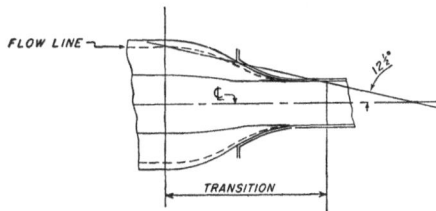

FIGURE 30

5. Side Channel Spillways.—These devices, an example of which is shown in figures 32 and 33, consist of an overflow weir or dam discharging into a narrow channel in which the direction of flow is approximately parallel to the axis of the weir. The abrupt change in the direction of flow and the constant addition of more water along the course of the channel introduce a high degree of turbulence and impact loss which is hydraulically inefficient. Nevertheless, the physical

9 "Hydraulic Design of Flume and Siphon Transitions", by Julian Hinds, Trans. A. S C. E , October 1927, p. 1423.

FIGURE 51.—Beford State Fair, Iowa, 51-24   Paved spillway with denuated sill well back from the
end of the paving

advantages sometimes are so great as to make it the most economical arrangement that can be found.

The design of a side channel layout involves two major factors, namely, the water capacity, and the economic proportions from a construction cost viewpoint. These factors are interrelated and must be considered together in the design of the structure.

The basic hydraulic equations are derived from the application o the law of the conservation of linear momentum,[10] and are as follows

$$Q=bx \qquad\qquad (22)$$

$$V=ax^n \qquad\qquad (23)$$

$$y=\frac{n+1}{n}(h_v) \qquad\qquad (24)$$

The nomenclature of the terms of these equations and of those which follow is:

Q = Discharge in cubic feet per second.

Q_1 = Discharge at upper end of two adjacent sections.

Q_2 = Discharge at lower end of two adjacent sections.

A = Cross-sectional area of water prism.

a = An arbitrary coefficient of x in the velocity equation.

b = Inflow per foot length of weir crest.

d = Depth of water in channel.

hv = Velocity head.

---

[10] See "Side Channel Spillways Hydraulic Theory, Economic Factors, and Experimental Determination of Losses", by Julian Hinds, Trans of Am. Soc. of C. E , vol 89, 1926, p 881.

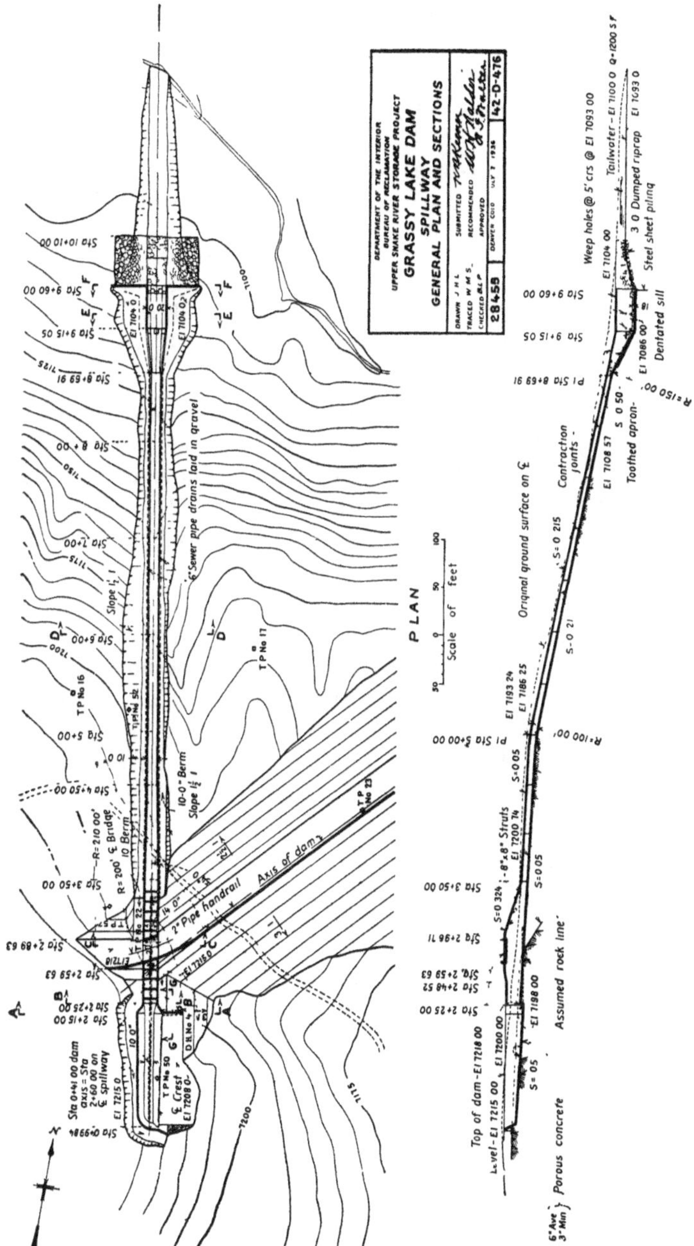

FIGURE 33.

H = Head on weir crest.

n = An arbitrary exponent of x in the velocity equation.

T = Width of channel at water surface.

V = Velocity in feet per second.

V_1 = Velocity at upper end of two adjacent sections.

V_2 = Velocity at lower end of two adjacent sections.

$\Delta V = V_2 - V_1$ change in velocity.

   x = Distance along axis of the channel.

Δx = Distance between consecutive cross sections of the channel.

   y = Vertical ordinate of the water surface curve in the channel.

The spillway channel is completely developed by equations (22), (23), and (24) after the choice of a shape and size of the cross section, and of values of a and n. Choice of the proper values, controlled by economic considerations, is the essential feature in the design of a new structure.

The usual layout will consist of a channel excavated from a steep side-hill location, for which a trapezoidal cross section is suitable. The following discussion on the effect of the shape of the channel on the amount of excavation is quoted from Mr. Hinds' paper. "Safety usually demands that the channel be set well into the original formation. It may be required that the waterway be entirely in rock. If the water-surface elevation, channel side slopes, area of water prism, and location of point of outcrop, A, are fixed, it is evident from figure 34 that the excavation is reduced by a narrow bottom width of channel. It is similarly evident from figure 35 that, other things being constant, the side slopes should be made as steep as feasible. The minimum practical width of bottom will depend on the equipment to be used for removing the material from the trench. If the excavation is to be done by machinery, a width of 15 or 20 feet may be required. For team work a somewhat narrower base may be used. The reduction in excavation for extremely narrow widths is not great. The side slopes should be trimmed to the steepest angle at which the materials will safely stand.

"In many cases it will be necessary to line the spillway channel with concrete. Other things being constant, the cost of lining, which

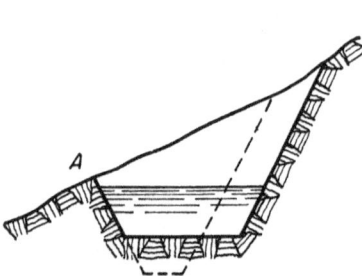

FIGURE 34 —Economical bottom width   FIGURE 35 —Economical side slope

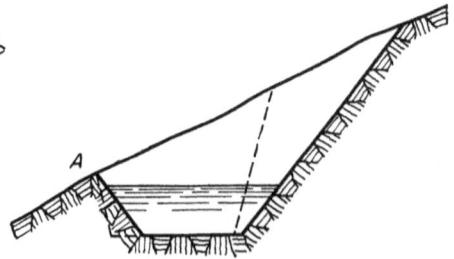

is an important item, is least when the bottom width is such that the wetted perimeter is a minimum. With steep side slopes this will require an average width of water prism equal approximately to twice the depth of the water. The bottom of the channel may be made somewhat narrower without greatly increasing the amount of lining, but the cost of lining should be considered in the final selection of channel width.

"Figures 34 and 35 are drawn to represent that part of the channel downstream from the crest structure, but the same principles apply to the part of the channel opposite the crest."

Having selected the shape of the cross section, the next consideration is the longitudinal profile. The constants a and n control the form that it will take. Figure 36 shows the water surface and channel bottom profiles made during the design of the channel shown in figure 31, using three values for n, and three values for a for each value of n. Each of these profiles can be fitted to the topography, and estimates made of the cost of each. That combination of a and n which gives the lowest cost will determine the best theoretical profile. This is sometimes modified to simplify the structural features, as was done in the Grassy Lake spillway which is used here as an illustration.

In this example, the crest elevation was established at 7210 and a discharge capacity of 1,200 cubic feet per second under a head of 2 feet was required. Using a weir coefficient of 3.7, it was found that a crest 115 feet long and discharging 10.4 cubic feet per second per foot of length would be adequate. A study of the rock conditions, construction methods, and probable depth of cut led to the adoption of a trapezoidal section 10 feet wide at the base with $\frac{1}{2}$ to 1 side slopes.

The effect of submergence on an overflow crest has been discussed in section 1 of this chapter. With a submergence of 67 percent, the net flow will be about 81 percent of the flow over a dam which has no submergence. It has been found from experience that a submergence as great as this at the upstream end of a side inlet channel will cause a reduction in total capacity which is negligible in amount. This is due to the fact that the maximum submergence occurs only at the upper end and is reduced further downstream by the drop in the water surface. It is therefore effective over a comparatively short portion of the whole crest. The initial point on the water surface profile is at the upper end at an elevation that will cause a maximum submergence of 67 percent. In this example it starts at elevation 7211.33.

Results of computations to determine water surface and channel bottom profiles using equation (22), (23), and (24) are given in the following tabulations. At the start, three or four values of n, varying from 0.5 to 1.0 are assumed. For each of these, three values of a are assumed such that the drop in the water surface will be within reason-

DESIGN OF SIDE CHANNEL SPILLWAY
FOR GRASSY LAKE DAM

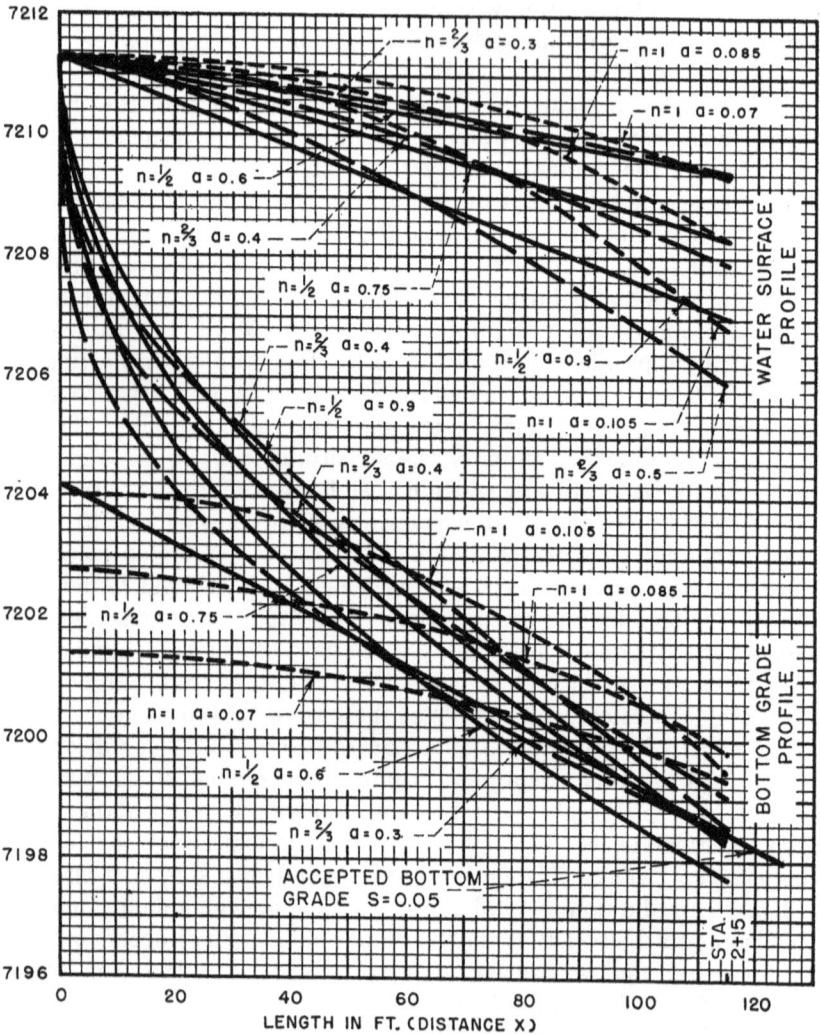

FIGURE 36.

preliminary values of y at the lower end of the weir were assumed to be 2.0, 3.0, and 4.5 for each value of n.   From (24) the velocity head is computed, and is converted to the corresponding velocity.   Since values of V, x, and n are known, the value of a is found from equation (23).

A study of these values from the economic and practical viewpoints led to the adoption of a plane-surfaced channel bottom sloping upward at the rate of 5 feet per hundred from elevation 7198.00 at station 2+25. In addition, a control weir 2 feet high is placed across the channel at this station. The channel cross section downstream from this point is changed from trapezoidal to rectangular shape to shorten

Results of computations to determine water surface and channel bottom profiles for the Grassy Lake spillway using three values of (n) and three values of (a) for each value of (n)

$Q = bx$ (22)

$V = ax^n$ (23)

$y = \dfrac{n+1}{n} h_v$ (24)

$n = \frac{1}{2}$

$y = 3 h_v$

			Approximate values for a		
	y	h_v	V	$x^{1/2}$	a
	2	0 67	6 56	10 7	0 61
	3	1 0	8 02	10 7	75
	4 5	1 5	9 8	10 7	92

	$n = \frac{1}{2}$					$a = 0\ 6$				
x	Q	x^n	V	h_v	y	Elevation W S	A	d	$d+y$	Elevation bottom
0	0	00 0	0 00	0 000	0 00	7, 211 33	0 0	0 0	0 00	7, 211 33
2	21	1 41	85	.011	033	11 30	24 6	2 25	2 28	9 05
5	52	2 24	1 34	02ʻ	084	11 25	39 0	3 35	3 43	7 90
10	104	3 17	1 90	056	168	11 16	55 0	4 50	4 67	6 66
25	261	5 00	3 00	.140	.420	10 91	87 0	6 55	6 97	4 36
50	522	7 07	4 25	281	843	10 49	123 0	8 60	9 44	1 88
75	782	8 66	5 20	.420	1 260	10 07	150 0	10 00	11 26	7, 200 02
100	1, 044	10 00	6 00	.560	1 680	9 65	174 0	11 20	12 88	7, 198 45
115	1, 200	10 73	6 44	.645	1 935	9 39	186 5	11 75	13 69	7, 197 64

	$n = \frac{1}{2}$					$a = 0\ 75$				
x	Q	x^n	V	h_v	y	Elevation W S	A	d	$d+y$	Elevation bottom
0	0	0	0 00	0 00	0 00	7, 211 33	00 0	0 00	0 00	7, 211 33
2	21	1 41	1 06	017	05	11 28	19 7	1 90	1 95	9 38
5	52	2 24	1 68	044	.13	11 20	31 1	2 75	2 88	8 45
10	104	3 17	2 38	088	.26	11 07	43 9	3 70	3 9ʻ	7 37
25	261	5 00	3 75	218	65	10 68	69 6	5 50	6 15	5 18
50	522	7 07	5 31	438	1 31	10 02	98 4	7 20	8 51	2 82
75	782	8 66	6 51	.659	1 98	9 35	120 0	8 45	10 43	7, 200 90
100	1, 044	10 00	7 50	873	2 62	8 71	139 2	9 45	12 07	7, 199 26
115	1, 200	10 73	8 05	1 008	3 02	8 31	149 0	9 95	12 97	98 36

	$n = \frac{1}{2}$					$a = 0\ 90$				
x	Q	x^n	V	h_v	y	Elevation W S	A	d	$d+y$	Elevation bottom
0	0	0 00	0 00	0 00	0 00	7, 211 33	00 0	0 00	0 00	7, 211 33
2	21	1 41	1 27	025	08	11 25	16 5	1 55	1 63	9 70
5	52	2 24	2 02	063	19	11 14	25 8	2 35	2 54	8 79
10	104	3 17	2 85	126	38	10 95	36 6	3 10	3 48	7 85
25	261	5 00	4 50	315	95	10 38	58 0	4 70	5 65	5 68
50	522	7 07	6 37	631	1 89	9 44	81 8	6 20	8 09	3 24
75	782	8 66	7 78	940	2 82	8 51	100 0	7 30	10 12	7, 201 21
100	1, 044	10 00	9 00	1 260	3 78	7 55	116 0	8 20	11 98	7, 199 35
115	1, 200	10 73	9 66	1 450	4 35	6 98	124 2	8 70	13 05	98 28

the span for the bridge which crosses the channel a few feet further down. Having adopted a layout which varies somewhat from the theoretical, it is necessary to determine its hydraulic properties by another set of computations.

Results of computations to determine water surface and channel bottom profiles for the Grassy Lake spillway using three values of (n) and three values of (a) for each value of (n)—Continued

$$y = \frac{n+1}{n} h_v \qquad (24)$$

$$n = \tfrac{2}{3}$$

$$y = 2\ 5\ h_v$$

			Approximate values for a		
	y	h_v	V	$r^{2/3}$	a
	2	0 8	7 18	23 62	0 30
	3	1 2	8 79	23 62	37
	4 5	1 8	10 75	23 62	45

	$n = \tfrac{2}{3}$						$a = 0\ 5$			
x	Q	x^n	V	h_v	y	Elevation W. S.	A	d	$d+y$	Elevation bottom
0	0	0 00	0 00	0 00	0 00	7,211 33	0 0	0 00	0 00	7,211 33
2	21	1 59	0 80	010	02	11 31	26 1	2 35	2 37	8 96
5	52	2 92	1 46	033	08	11 25	35 8	3 10	3 18	8 15
10	104	4 65	2 32	084	21	11 12	45 0	3 80	4 01	7 32
25	261	8 55	4 28	285	71	10 62	61 0	4 90	5 61	5 72
50	522	13 58	6 79	72	1 80	9 53	77 0	5 95	7 75	3 58
75	782	17 80	8 90	1 23	3 08	8 25	87 9	6 60	9 68	7,201 65
100	1,044	21 56	10 78	1 81	4 53	6 80	97 0	7 10	11 63	7,199 70
115	1,200	23 62	11 81	2 17	5 43	7,205 90	101 7	7 40	12 83	98 50

	$n = \tfrac{2}{3}$						$a = 0\ 4$			
x	Q	x^n	V	h_v	y	Elevation W S	A	d	$d+y$	Elevation bottom
0	0	0 00	0 00	0 00	0 00	7,211 33	0 0	0 00	0 00	7,211 33
2	21	1 59	0 64	006	02	11 31	32 6	2 85	2 87	9 46
5	52	2 92	1 17	021	05	11 28	44 6	3 75	3 80	7 53
10	104	4 65	1 86	054	13	11 20	56 2	4 60	4 73	6 60
25	261	8 55	3 42	182	46	10 87	76 4	5 90	6 36	4 97
50	522	13 58	5 43	46	1 15	10 18	96 2	7 10	8 25	3 08
75	782	17 80	7 12	79	1 97	9 36	110 0	7 90	9 87	7,201 46
100	1,044	21 56	8 62	155	2 89	8 44	121 0	8 50	11 39	7,199 94
115	1,200	23 62	9 45	1,39	3 48	7 85	127	8 85	12 33	99 00

	$n = \tfrac{2}{3}$						$a = 0\ 3$			
x	Q	x^n	V	h_v	y	Elevation W S	A	d	$d+y$	Elevation bottom
0	0	0 00	0 00	0 00	0 00	7,211 33	0 0	0 00	0 00	7,211 33
2	21	1 59	48	004	01	11 32	43 5	3 65	3 66	7 67
5	52	2 92	88	012	03	11 30	59 4	4 80	4 83	6 50
10	104	4 65	1 40	030	08	11 25	74 7	5 80	5 88	5 45
25	261	8 55	2 57	103	26	11 07	101 5	7 40	7 66	3 67
50	522	13 58	4 07	26	65	10 68	128 2	8 90	9 55	1 78
75	782	17 80	5 34	.44	1 10	10 23	146 5	9 85	10 95	7,200 38
100	1,044	21 56	6 47	.65	1 63	9 70	161 3	10 60	12 23	7,199 10
115	1,200	23 62	7 09	78	1 95	9 38	169 5	10 95	12 90	98 43

Results of computations to determine water surface and channel bottom profiles for the Grassy Lake spillway using three values of (n) and three values of (a) for each value of (n)—Continued

$$y = \frac{n+1}{n} h, \quad (24)$$

$n = 1$

$y = 2h,$

	Approximate values for a			
y	$h,$	V	x^1	a
2	1 00	8 02	115	0 07
3	1 50	9 82	115	085
4 5	2 25	12 03	115	105

	$n = 1$					$a = 0\ 07$				
x	Q	x^n	V	$h,$	y	Elevation W S	A	d	$d+y$	Elevation bottom
0	0	0	0 00	0 00	0 00	7,211 33	0 0	0 0	0 00	7,211 33
2	21	2	14	000	00	11 33	149 1	9 95	95	1 38
5	52	5	35	002	00	11 33	149 1	9 95	9 95	1 38
10	104	10	70	008	02	11 31	149 1	9 95	9 97	1 36
25	261	25	1 75	048	10	11 23	149 1	9 95	10 05	1 28
50	522	50	3 50	191	38	10 95	149 1	9 95	10 33	1 00
75	782	75	5 25	43	86	10 47	149 1	9 95	10 81	7,200 52
100	1,044	100	7 00	76	1 52	9 81	149 1	9 95	11 47	7,199 86
115	1,200	115	8 05	1 01	2 02	9 31	149 1	9 95	11 97	99 36

	$n = 1$					$a = 0\ 085$				
x	Q	x^n	V	$h,$	y	Elevation W S	A	d	$d+y$	Elevation bottom
0	0	0	0 00	0 00	0 00	7,211 33	0 0	0 00	0 00	7,211 33
2	21	2	17	000	00	11 33	122 8	8 60	8 60	2 73
5	52	5	42	003	01	11 32	122 8	8 60	8 61	2 72
10	104	10	85	011	02	11 31	122 8	8 60	8 62	2 71
25	261	25	2 12	070	14	11 19	122 8	8 60	8 74	2 59
50	522	50	4 25	28	56	10 77	122 8	8 60	9 16	2 17
75	782	75	6 38	63	1 26	10 07	122 8	8 60	9 86	1 47
100	1,044	100	8 50	1 12	2 24	9 09	122 8	8 60	10 84	7,200 49
115	1,200	115	9 77	1 48	2 96	8 37	122 8	8 60	11 56	7,199 77

	$n = 1$					$a = 0\ 105$				
x	Q	x^n	V	$h,$	y	Elevation W S	A	d	$d+y$	Elevation bottom
0	0	0	0 00	0 00	0 00	7,211 33	0 0	0 00	0 00	7,211 33
2	21	2	21	001	00	11 33	99 4	7 25	7 25	4 08
5	52	5	52	.004	01	11 32	99 4	7 25	7 26	4 07
10	104	10	1 05	017	03	11 30	99 4	7 25	7 28	4 05
25	261	25	2 62	107	21	11 12	99 4	7 25	7 46	3 87
50	522	50	5 25	.43	86	10 47	99 4	7 25	8 11	3 22
75	782	75	7 87	96	1 92	9 41	99 4	7 25	9 17	2 16
100	1,044	100	10 50	1 715	3 43	7 90	99 4	7 25	10 68	7,200 65
115	1,200	115	12 07	2 265	4 53	6 80	99 4	7 25	11 78	7,199 55

It is evident that the control section is at station $2+25$, so the computations start there and work upstream to determine the water surface profile to the end of the channel. The depth at the control section will be the critical depth, which is

$$d_c = \sqrt[3]{Q_1^2/g} = \sqrt[3]{\left(\frac{1,200}{14}\right)^2 \times \frac{1}{g}} \tag{25}$$

$$= 6.11 \, foot.$$

$$h_{vc} = \frac{d_c}{2} = 3.06 \, foot.$$

There will be a warped transition channel from the trapezoidal shape at station 2+15 to the rectangular shape at station 2+25. (See figure 37.) By Bernoulli's theorem of the conservation of energy

$$d_1 + h_{v1} = h_{dc} + d_c + h_{vc} + h_i \tag{26}$$

in which h_i represents the transition loss, or the loss in the velocity head between 2+15 and 2+25. This is assumed to be two-tenths of the velocity head increase between the two sections. The other factors are as shown in figure 37.

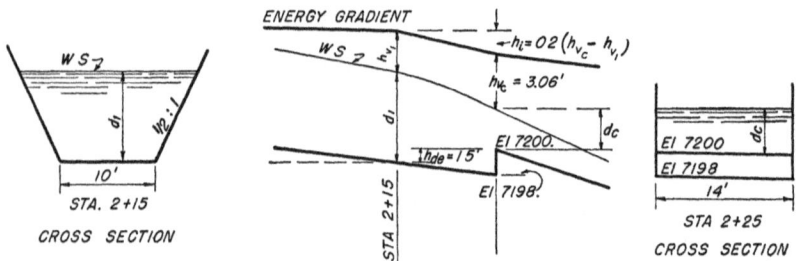

FIGURE 37.

$$d_1 + h_{v1} = 1.50 + 6.11 + 3.06 + 0.2 (h_{vc} - h_{v1})$$

$$d_1 + 1.20 h_{v1} = 11.28$$

This equation is solved by trial, as follows:

Assume $d_1 = 11.0$; $A = 170.5$; $V = 1,200/170.5 = 7.04$; $h_{v1} = 0.77$

$$d_1 + 1.20 \, h_{v1} = 11.00 + 0.92 = 11.92 \text{ (too large)}.$$

Assume $d_1 = 10.20$; $A = 154.0$; $V = 1,200/154.0 = 7.79$; $h_{v1} = 0.94$

$$d_1 + 1.20 \, h_{v1} = 10.20 + 1.13 = 11.33 \text{ (too large)}.$$

Assume $d_1 = 10.12$; $A = 152.4$; $V = 1,200/152.4 = 7.88$; $h_{v1} = 0.97$

$$d_1 + 1.20 \, h_{v1} = 10.12 + 1.16 = 11.28.$$

Therefore at station 2+15, which is the end of the weir, the water surface elevation is 7,198.50+10.12=7,208.62, the velocity is 7.88 feet per second; and the velocity head is 0.97 foot. These data provide the necessary starting point for computing the backwater curve to the upper end of the spillway.

Knowing these data for the control section, the channel upstream is divided into short sections so that similar data may be computed for other points. The following equation gives the value of Δy, the rise in the water surface from the lower end of the section to the upper end, in terms of the velocity, quantity, and distance from the end of the channel:

$$\Delta y = \frac{Q_1}{g}\left(\frac{V_1+V_2}{Q_1+Q_2}\right)\left[\Delta V + \frac{bV_2\Delta x}{(Q_1)}\right] \tag{27}$$

It must be solved by trial, first assuming a value for Δy, from which the depth of the water is determined. From the relation of the depth to the shape of the channel cross section the water area can be computed. The relation between the quantity of water and the area at any point gives the velocity, from which ΔV is computed. The other terms in equations (27) are self-explanatory. In addition to the losses from impact and turbulence, there is the usual loss from friction, which is computed for each isolated stretch of channel, using the average velocity and depth. Knowing the velocity, hydraulic radius, and Manning's n coefficient (0.014 for concrete), use the chart of figure 25 to determine the slope necessary to overcome the friction loss. The product of this slope and the distance Δx will give the value of h_f, which is added to the assumed Δy to give the actual rise in the water surface.

From the results of the computations which appear in table 6, it is seen that the actual design causes less submergence at the upper end than was originally assumed, which is of course a favorable condition as far as discharge capacity is concerned.

Another factor which has not yet been mentioned is the swell in volume of water in the channel due to entrapped air. Accurate determination of the amount is not practicable, but it may amount to as much as 10 percent in extreme cases. With this in mind, the walls or pavements surrounding the channel should be made high enough so that there is little danger of having the water slop over the edges.

6. Drop Inlet Spillways.—The drop inlet, as the name implies, is a closed conduit structure which the water enters over an overflow lip, drops through a vertical or sloping shaft, and discharges downstream in a more or less horizontal direction. The structure may be considered as being made up of four elements (see figure 38): 1, Inlet; 2, riser; 3, transition; and 4, outlet barrel.

TABLE 6 —Backwater curve computations (spillway for Grassy Lake Dam)

Station (x)	Δx	Elevation bottom	Trial Δy	Water surface elevation	d	A	Q₂	V	Q₁+Q₂	$\frac{Q_1}{g(Q_1+Q_2)}$	V₁+V₂	ΔV	$\frac{bV_2\Delta x}{Q_1}$	$\Delta v+\frac{bv_2\Delta x}{Q_1}$	Δy	h_f*	Δy+h_f	Error	Remarks
(1)	(2)	(3)	(4)	(5)	(6)	(7)	(8)	(9)	(10)	(11)	(12)	(13)	(14)	(15)	(16)	(17)	(18)	(19)	(20)
115		7,198 50	—	7,208 62	10 12	152 4	120 0	7 88	2,087 0	0 01323	14 41	1 35	2 78	4 13	0 79	0 02	0 81	0 16	
85	30	7,200 00	0 65	7,209 27	9 27	135 7	887 0	6 53	2,087 0	01323	14 24	1 52	2 78	4 30	81	02	83	02	O K
			85	7,209 47	9 47	139 5	887 0	6 36	2,087 0	01323	14 26	1 50	2 78	4 28	81	02	83	00	O K
55	30	7,201 50	83	7,209 45	9 45	139 2	887 0	6 38	1,460 9	01222	10 98	1 78	3 48	5 26	71	02	73	01	
			74	7,210 19	8 69	124 7	573 9	4 60	1,460 9	01222	10 99	1 77	3 48	5 25	71	02	73	00	O K.
30	25	7,202 75	73	7,210 18	8 68	124 5	573 9	4 61	886 9	01097	7 41	1 81	3 84	5 65	46	01	47	10	
			57	7,210 75	8 00	112 0	313 0	2 80	886 9	01097	7 46	1 77	3 84	5 60	46	01	47	02	O K.
			45	7,210 63	7 88	109 8	313 0	2 85	886 9	01097	7 45	1 76	3 84	5 61	46	01	47	00	
10	20	7,203 75	47	7,210 65	7 90	110 2	313 0	2 84	417 4	00777	3 92	1 76	5 68	7 44	23		23	02	O. K.
			25	7,210 90	7 15	97 1	104 4	1 08	417 4	00777	3 92	1 76	5 68	7 44	23	0	23	00	
1	9	7,204 20	23	7,210 88	7 13	96 7	104 4	1 08	114 8	00283	1 20	1 96	9 72	10 68	04	0	04	01	O. K.
			03	7,210 91	6 71	89 6	10 43	0 12	114 8	00283	1 20	1 96	9 72	10 68	04	0	04	00	
			04	7,210 92	6 72	89 8	10 43	0 12											

*Computations for friction not shown

Elevation of weir crest=7,210 00
Elevation water surface above crest=7,212,00

Elevation water surface in channel=7,210 92.
Maximum submergence=46 percent

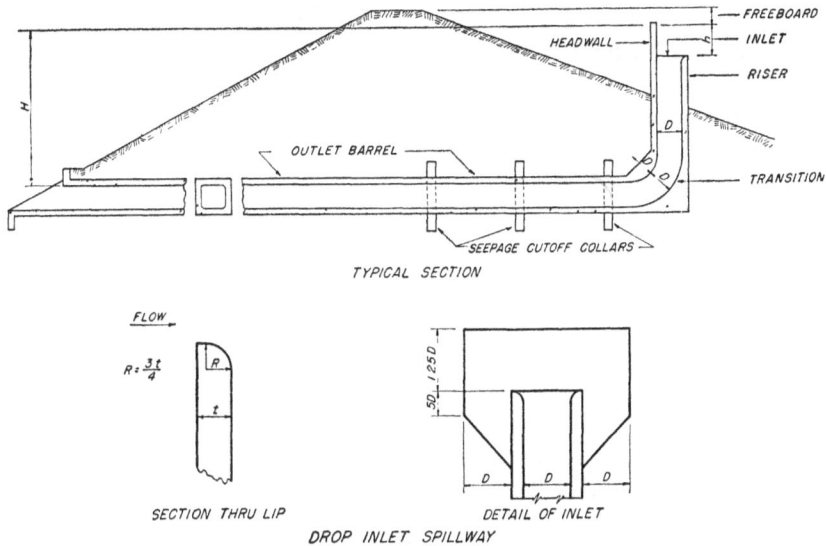

TYPICAL SECTION

SECTION THRU LIP

DETAIL OF INLET

DROP INLET SPILLWAY

FIGURE 38

The problem of hydraulic design is to get the correct size to give the desired capacity and to proportion the various parts for best hydraulic efficiency. The best hydraulic efficiency and the maximum flow capacity will be reached when all the water channels are flowing full from the outlet back to the inlet lip. This condition causes a mild vacuum in the riser, which tends to raise the value of the inflow coefficient at the inlet.

The best available determination of the hydraulic coefficients and characteristics of drop inlets is from the results of model experiments made at the University of Wisconsin, under the direction of Asst. Prof. Lewis H. Kessler during 1933. The capacity curves for a wide range of sizes and heads as given in figure 39 have been taken from the published report of this work.[11]

In a drop inlet structure there are three sources of hydraulic loss, (1) at the inlet, (2) at the transition, and (3) the frictional losses for the entire length of conduit.

The inlet conditions have two distinct phases. At low heads, the flow is the usual weir flow, which continues up to the point where the overflow streams from the three sides begin to merge together. Up to this point the overflow is represented by the equation $Q=CLh^{3/2}$ in which $C=3.8$ for the lip design shown in figure 38. When the nappes from the three sides merge together the under sides adhere to the wall, giving evidence of negative pressures against the wall, with a tendency

[11] "Experimental Investigation of the Hydraulics of Drop Inlets and Spillways for Erosion Control Structures," by Lewis H. Kessler. Bulletin No. 80, Engineering Experiment Station Series, University of Wisconsin

to increase the discharge. With the riser flowing full, the discharge capacity will be further increased by this vacuum, the magnitude of which increases with the overall head H up to the atmospheric limit. The velocity coefficient in the equation for entrance losses increases as the head on the lip h increases, but becomes practically constant when the inlet throat becomes full so that air cannot be sucked into it. This full condition will be reached in all cases when the head h is at

CAPACITY OF DROP INLETS WITH SQUARE CROSS-SECTION AND TYPE A ELBOW
(FROM EXPERIMENTS BY L.H KESSLER, UNI. OF WISCONSIN)

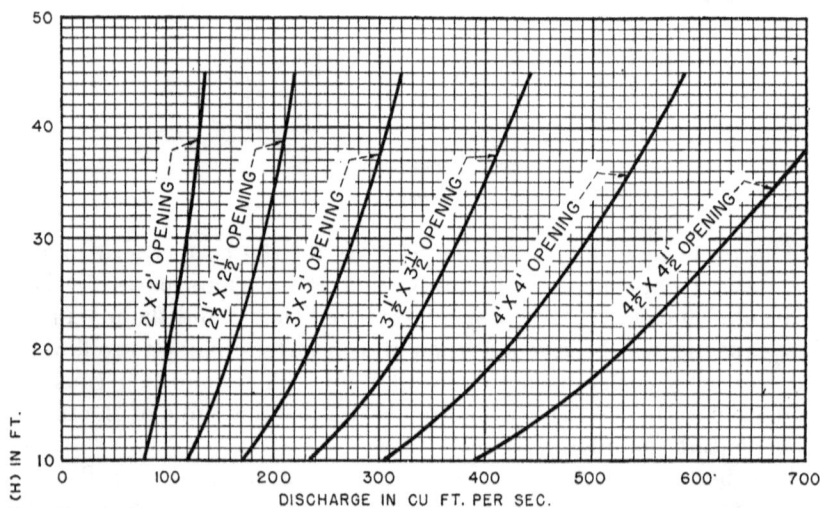

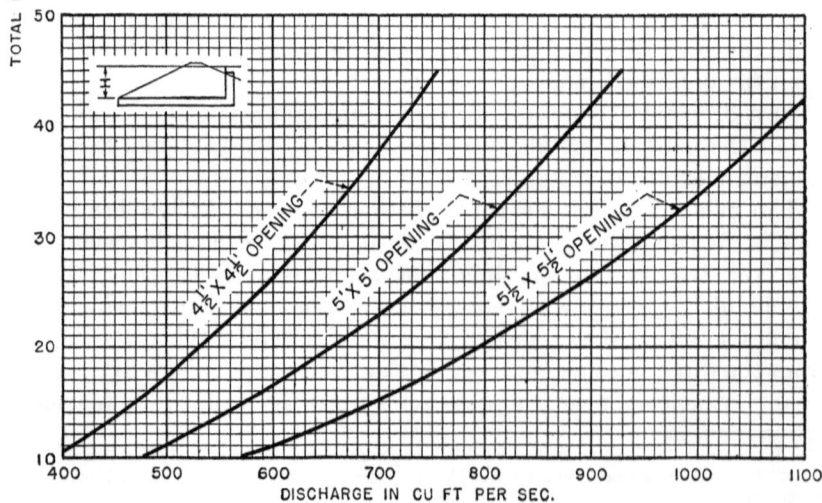

FIGURE 39.

least 1.2 times the width D of the inlet. This head is known as the design head, and represents the head over the lip which will produce the full capacity of the structure. Greater values of h do not increase the flow materially.

The friction loss in the transition section, or elbow, is determined from the equation

$$H_t = K_t \frac{v^2}{2g} \qquad (28)$$

in which K_t is a coefficient which shows a wide variation with different forms of transition. Figure 40 shows two designs which were tested by

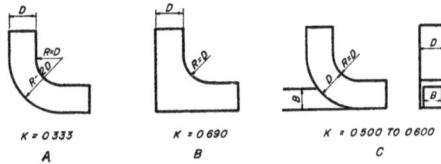

K = 0.333   K = 0.690   K = 0.500 TO 0.600
A   B   C

ELBOW TYPES FOR TRANSITION

FIGURE 40.

models at the University of Wisconsin. Type A has a low coefficient, is not complicated to build, and is the type used for the discharge charts presented here. Several designs that were easier to build were tested, but type A was by far the most satisfactory, its hydraulic advantages greatly outweighing the somewhat greater cost of form work during construction, as compared, for example, to type B. Type B shows a design which has a coefficient that is over twice as great as type A, and in addition it is subject to severe surges which prevent smooth flow and cause rapidly changing pressure variations.

In certain locations it may be advantageous to combine an outlet gate with a drop-inlet structure, in which case the transition section might be modified as shown in figure 40, type C. The opening on the outside of the elbow will obviously be a source of eddy disturbance and will increase the coefficient of 0.333 for type A. The amount of increase will be dependent on the relative size of the outlet opening. It is estimated that where dimension B is less than one-half of D, a value of 0.500 would be safe for K_t, but where $B = D$, the value of K_t should be at least 0.600.

Corrections to the discharge curves of figure 39 should be made if the elbow design differs from type A. This can be done by a comparison of the elbow loss in each case. For example, if we assume the case of a 4 foot by 4 foot section having an overall head of 30 feet, the capacity from figure 39 is 490 cubic feet per second, and the velocity will be $490/16 = 30.6$ feet per second. Using equation (28)

$$H_t = \frac{K_t v^2}{2g}$$

$$II_t = 0.333 \frac{30.6 \times 30.6}{64.4}$$

$H_t = 4.8$ feet

This loss of head in the transition is included in the standard curves. With a *different* design, which we will assume has a coefficient of 0.600, the capacity and therefore the velocity will be slightly lower, and is assumed to be 30.0.

$$H'_t = 0.600 \frac{30 \times 30}{64.4}$$

$H'_t = 8.4$ feet

The difference between 8.4 and 4.8 indicates that an additional head of 3.6 feet is required for the new design to give the same capacity, or, for the same head, the actual capacity will be that of the standard design with a head of $30 - 3.6 = 26.4$ feet, which is 470 cubic feet per second. The actual velocity would be $470/16 = 29.4$ feet per second, which in turn gives a corrected value of 8.1 feet for the head lost. The new difference is 3.3 feet and the net head is therefore $30 - 3.3 = 26.7$ feet, giving a capacity of 472 cubic feet per second. Such refinement is unnecessary and is not justified by the accuracy of the basic data.

A third source of loss of head is the friction loss throughout the rest of the conduit. This factor will vary with the velocity and with the roughness of the conduit interior. If L is the length of the conduit, the loss of head is

$$H_f = f \frac{L}{D} \frac{v^2}{2g} \tag{29}$$

in which f, the friction coefficient, will be 0.14 for the ordinary range of heads for concrete structures. The discharge curves in figure 39 have been computed on this basis.

If the outlet barrel is built with a downward slope which is greater than that required to maintain the flow through it, acceleration will take place, and the water area will decrease so that it will not fill the barrel. When the barrel is not full, air can enter between the water surface and the crown of the conduit and will travel upstream to the elbow. If it passes around the elbow and enters the riser it will break the vacuum and interfere with the hydraulic behavior of the inlet. When this happens, the reduced capacity of the inlet will cause a rapid rise in the headwater, which might easily overtop the dam and cause failure. To avoid this condition, a safe rule to follow is to find the maximum slope of the barrel by the equation:

$$S_m = 0.00016 v^2 / D \tag{30}$$

Referring to figure 41 section A–A, it will be seen that the downstream wall of the riser has been continued to a height of $1.25D$ above the lip. This headwall is a result of model tests, and was found to be indispensable to prevent the formation of a vortex at the inlet. Water flowing over three sides of the riser is sufficient to maintain capacity flow in the structure at the design head of $1.2D$. However, it is possible to modify the inlet by flaring the sides, and so reduce the head above the lip. In cases where the maximum height of the dam is limited and the desired pond level does not leave ample freeboard, the flared inlet offers a means of lowering the flood level. Referring to figure 41, B–B shows an inlet in which two sides have been flared outward so that the length of the crest is increased 33⅓ percent. Model tests indicate that a reduction of 10 percent can be made in the design head of the straight inlet to give the same capacity. If all three sides are flared as in C–C, a 20 percent reduction in head is accomplished.

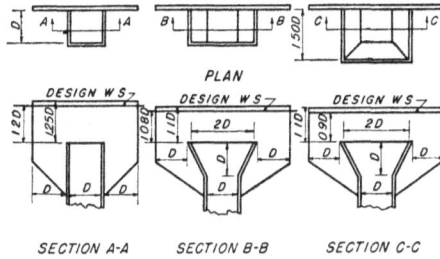

PLAN

SECTION A-A   SECTION B-B   SECTION C-C

INLET DESIGNS

FIGURE 41

. Table 7 gives the discharge capacity of drop inlets made of smooth round pipe, and coefficients to be applied to these values where corrugated pipe is used, i. e., multiply the tabular values by the coefficient, k, to obtain the capacity of corrugated pipe drop inlets.

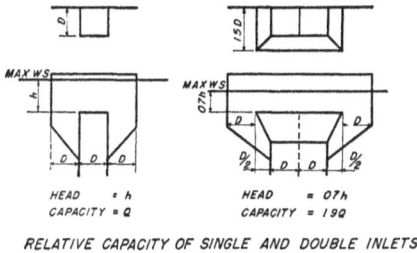

HEAD   = h
CAPACITY = Q

HEAD   = 07h
CAPACITY = 19Q

RELATIVE CAPACITY OF SINGLE AND DOUBLE INLETS

FIGURE 42

Under some conditions, it is possible that one large structure might be replaced with two smaller structures placed side by side with a common wall. In this case, the twin structures with three sides flared can safely be assumed to carry 1.9 times the quantity that would be carried by a single straight inlet of the same throat dimensions, and the design head will be 30 percent lower than for the single structure (see fig. 42). Equal capacity is reached in an inlet with two sides flared with a design head that is 15 percent lower than for the single structure.

Various other patterns of inlet design were investigated by Professor Kessler, including the round "morning glory," but for small

TABLE 7.—*Discharge capacity of drop inlets made of round pipe*

[For smooth pipe]

Pipe diameter in inches	Head in feet					Corru-gated pipe factor K
	5	10	15	20	25	
12	5	10	16	18	20	0 70
15	18	20	23	27	28	70
18	24	29	34	40	45	70
24	40	55	70	80	83	73
30	70	90	110	130	140	75
36	100	135	165	185	205	77
42	135	185	230	255	280	78
48	180	250	300	335	370	80

For corrugated pipe multiply the tabular value by the factor given in the last column.

structures the hydraulic advantages are not great enough to offset the added expense involved in the more complicated construction problems that they present.

For small projects of minor importance, where the value of the improvement will not justify the cost of a reinforced concrete drop-inlet structure, and where the useful life of the project is limited to comparatively few years, metal pipe may be used. It is evident that when the pipe rusts and fails, the enbankment through which it passes will be washed away unless other provisions are made to dispose of the run-off. However, pipe which is protected by heavy galvanizing, or bituminous or other protective coating may give from 15 to 25 years or more of service.

Metal pipe should never be used where the failure of the structure may involve a hazard to life or serious damage to property. Note particularly the limitations on the construction of conduits through earth dams discussed in chapter 7: 17.

Requirements for the safe design of outlet conduits through embankments are discussed in detail in chapter 6: 2–3. These requirements relate to provision against cracking, settlement, and leakage; provisions for the prevention of the movement of the water along the outside of the conduit; considerations of foundation conditions and a proper bedding of pipe and a discussion of the loads on conduits under various conditions of construction.

When pipe is used for this kind of service, only the heavier weights should be used, following the more conservative recommendations of culvert pipe manufacturers. All joints should be caulked or otherwise sealed securely against leakage, and collars should be used to prevent seepage along the pipe. Sheet metal collars or diaphragms as furnished commercially by culvert pipe manufacturers are equipped with bands which have a watertight connection to the outside of the pipe. This type of collar is superior to concrete for metal pipe because distortion of the pipe from loading will usually crack the concrete collar, whereas the metal diaphragm will remain intact.

Trash racks for drop inlet structures are important because there is considerable danger of logs or debris lodging in the elbow and creating serious obstruction to flow. Since small debris will readily pass through the structures, it is necessary to prevent only larger pieces from passing the inlet. Some form of floating boom is usually the more desirable, but screens are occasionally used. Under conditions where there is much debris in the water some arrangement must be made to keep the screen clean. Whichever arrangement is used, it should be placed at a considerable distance from the inlet.

7. Spillway and Outlet Protection.—Spillway and outlet works usually develop high velocities and it is therefore necessary to provide devices which will dissipate the energy and reduce the velocity to a value which will avoid erosion in the flowing away channel.

The protective measures more commonly used in dam and spillway design are:

(a) For free falling water the construction of a water cushion or pool to absorb the impact. This may be accomplished by the construction of a low dam immediately downstream from the main structure to create a pool of the required depth. This second low dam may also require the introduction of some device to protect the channel below.

(b) For water flowing on a steep slope such as in the case of an ogee crest dam or spillway channel or issuing from a sluiceway or outlet works, energy dissipation in such cases may be accomplished by adjusting the flow channel so that a hydraulic jump will occur.

As stated in section 4, the flow on a long steep slope becomes uniform when the velocity reaches a point when the loss of head due to friction is offset by the slope of the bottom of the channel. When the stream reaches the base of an ogee spillway, a long spillway channel, or is issuing from a sluice way channel, it is usually flowing at a depth less than critical. The relatively flat slope of the tailrace channel will not support flow at the shallow depth and hence the depth will be increased and the velocity diminished. The phenomenon is called the hydraulic jump. Figure 43 shows this condition at the foot of a dam. The water is here shown as passing from the high depth stage above the dam through the critical depth at the crest into low stage depth as it reaches the foot of the dam and back to a high stage depth after passing through the jump.

For a given discharge, if the depth D_2 and the velocity V_2 of the high stage are known, the depth D_1, of the low stage jet which will cause the hydraulic jump can be determined from this equation:

$$D_1 = -\frac{D_2}{2} + \sqrt{\frac{D_2^2}{4} + \frac{2V_2^2 D_2}{g}} \qquad (31)$$

If the depth and velocity of the low stage are known, the depth of the high stage which will cause the jump is

$$D_2 = -\frac{D_1}{2} + \sqrt{\frac{D_1^2}{4} + \frac{2V_1^2 D_1}{g}} \qquad (32)$$

for which

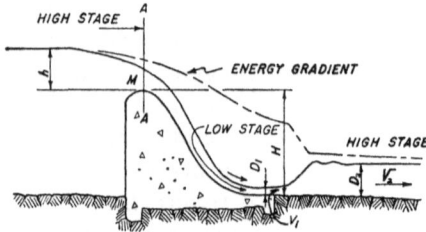

HYDRAULIC JUMP AT TOE OF SPILLWAY

FIGURE 43

D_1 = the depth of the low stage in feet, above the jump.

D_2 = the depth of the high stage in feet, below the jump.

V_1 = the velocity of the low stage in feet per second.

V_2 = the velocity of the high stage in feet per second.

g = 32.2 feet per second per second.

(See fig. 43.)

When for a given flow, the tailwater depth is greater than D_2 the water will back up to the face of the dam and there will be no jump. On the other hand, if the low-stage depth is less than D_1 the jump will occur at some distance downstream where the friction has increased the depth to D_2.

Determination of the tailwater depth D_2 is made from discharge computations which take into consideration the channel area, the roughness coefficient, and the slope for different discharges ranging from the minimum flow. Using these values in formulas (31) and (32), with values of velocity and depth for corresponding discharges over the dam will determine the position of the jump. The elevation and length of the apron are selected so that it will protect the toe of the dam from damage.

Since the solution of the hydraulic jump problem involves the depth and velocity of the water at the toe, some method of determining these values is required. At the crest line A–A on figure 43 the depth is approximately 0.67 H and the mean velocity occurs at M, which is $H/3$ above the crest. From this point the velocity increases in proportion to the drop minus the friction loss on the surface of the dam. For smooth concrete dams this loss is not great, and for lack of better knowledge as to its value, it may be neglected. The net velocity, V_1, at the toe of the dam is therefore the initial velocity at M plus that due to the drop below M, and $D_1 = Q/V_1$, at the same point. Figure 44 gives the relations between variables in the hydraulic jump formulas for rectangular channels.

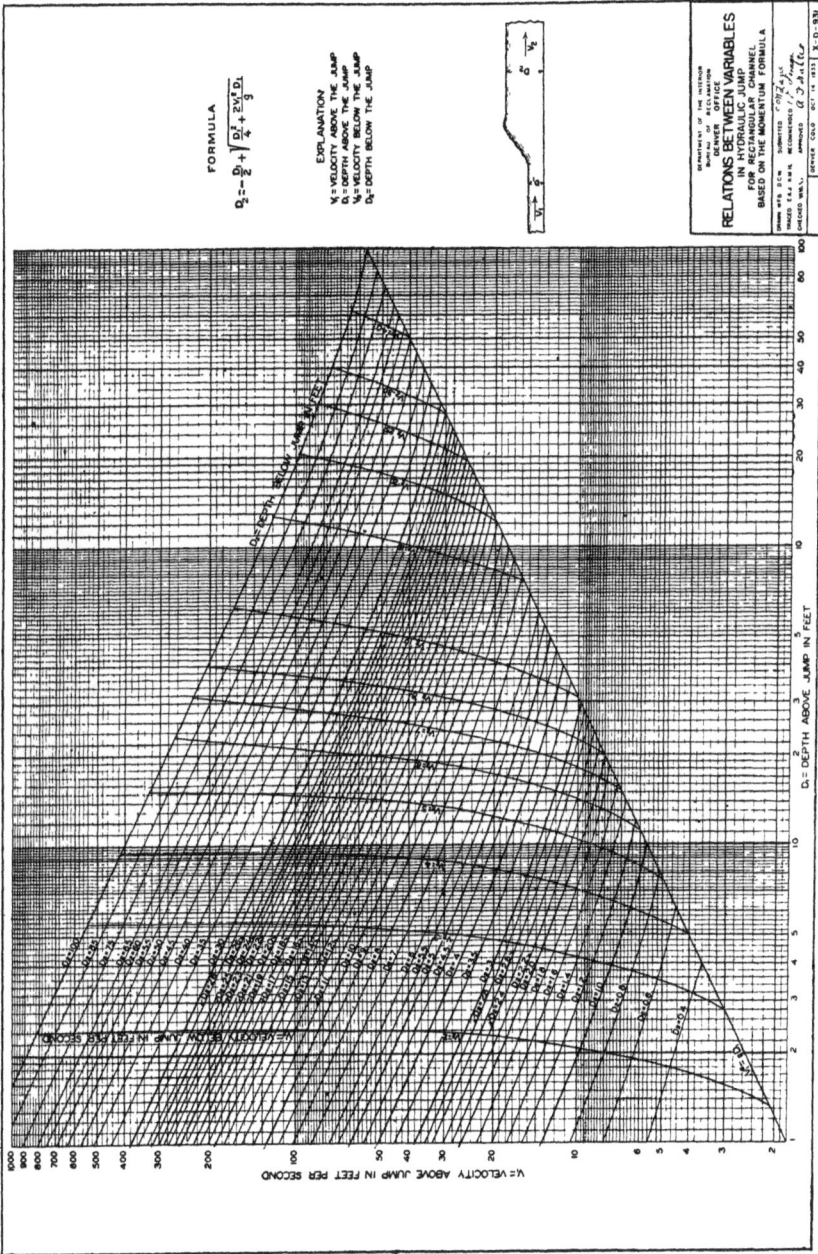

FIGURE 44.

Because of the high velocity above the jump and the turbulence in the jump itself destructive erosion will occur unless properly designed aprons or stilling pools are provided.

In the design of stilling pools below overfall dams or spillway channels, it is important that a jump be formed on the apron or in the pool for all stages of flow. The length of the pool should be from 4 to 5 times the jump height unless model experiments indicate otherwise. Pools with sloping sides should be avoided if possible. Where they are used, the slopes should not be greater than one horizontal to two vertical.

A sill at the end of the pool serves to divert high bottom velocities away from the stream bed and to allow a roller to form which will keep material banked against the floor and cut-off wall at the lower end. Wing walls are usually placed at the end of the pool to prevent scouring behind the walls. Riprap may be added to reduce erosion below the pool.

The floor and walls of the pool should be designed to resist uplift pressure equal to the hydrostatic head produced by the tailwater. The weight of the floor slab must be sufficient to resist the unbalanced pressure due to the difference in elevation between the depth of the tailwater and the depth of flow above the jump. If the foundation material is porous, cut-off walls and drains may be used to reduce the pressure but neither drains nor cut-off walls should be depended upon to decrease the uplift to a value less than that developed by low tailwater. A cut-off wall at the lower end is required to prevent undermining.

In the design of stilling pools for outlet works or sluiceways, other complications are involved and much depends upon the judgment and experience of the designer. The problems that arise may differ from the problems which are encountered below an ogee crest or a long spillway chute. These may be listed as follows:

(a) The flow is horizontal rather than curving.

(b) The arrangement of gates may produce unsymmetrical flow.

(c) Desirability of maintaining open flow conditions in the outlet portal and throughout the conduit.

(d) Prevention of freezing.

(e) Methods of stabilizing the jump.

(f) Smaller quantities of water are involved.

Expert advice should be obtained in connection with designs of stilling pools for outlet works. Figure 45 shows an example of the details of this type of construction.

The foregoing discussion is intended to point out the methods by which erosion below spillway and outlet works must be controlled under most conditions. For certain low structures where the maximum velocities are not high, it may be sufficient to provide a simple

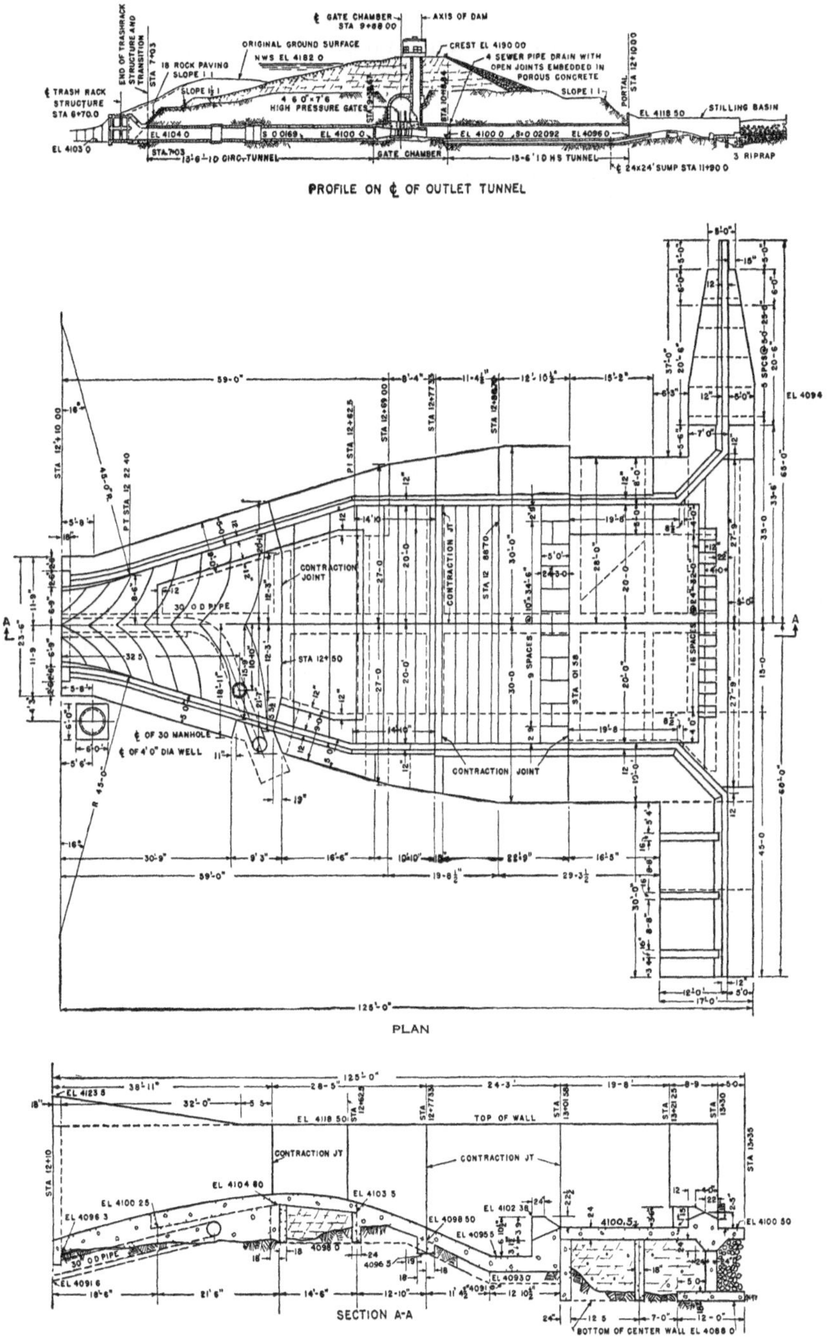

PROFILE ON ℄ OF OUTLET TUNNEL
℄ GATE CHAMBER STA 9+88 00
AXIS OF DAM
ORIGINAL GROUND SURFACE
NWS EL 4182 0
CREST EL 4190 00
4 SEWER PIPE DRAIN WITH OPEN JOINTS EMBEDDED IN POROUS CONCRETE
18 ROCK PAVING SLOPE 1 1
END OF TRASHRACK STRUCTURE AND TRANSITION
℄ TRASH RACK STRUCTURE STA 6+70.0
4 6 0 x 7 6 HIGH PRESSURE GATES
SLOPE 1 1
PORTAL
STA 12+10 00
EL 4118 50
STILLING BASIN
EL 4103 0
GATE CHAMBER
13-6 I D H S TUNNEL
℄ 24x24 SUMP STA 11+90 0
3 RIPRAP
CONTRACTION JOINT
CONTRACTION JT
30 O D PIPE
℄ OF 30 MANHOLE
℄ OF 4'0" DIA WELL
EL 4094
PLAN
SECTION A-A
TOP OF WALL
EL 4118 50
EL 4104 60
EL 4103 5
EL 4100 25
EL 4096 3
EL 4091 6
EL 4098 50
EL 4095 5
EL 4093 0
EL 4102 38
EL 4100 50
BOTTOM OF CENTER WALL EL 4088 0

FIGURE 45.

apron of concrete, or riprap, which will give satisfactory results. It is quite important, however, under all conditions that the net result of energy dissipating devices be to reduce the velocity and effect a velocity distribution that will insure the safety of the tailwater channel. Table 8 gives empirical values recommended by Etcheverry for safe velocities in different materials. If the energy dissipating devices produce a mean velocity at or below the values given, the result should be satisfactory inasmuch as the bottom velocities will not exceed about 75 percent of the mean.

TABLE 8 —*Maximum mean velocities safe against erosion*

	Velocity in feet per second
Very fine sandy soil or loose silt	0. 50
Pure sand	1. 00
Light sandy soil, 15 percent clay	1. 20
Light sandy loam, 40 percent clay	1. 80– 2. 00
Coarse sand	1. 50– 2. 00
Loose gravelly soil	2. 50
Ordinary loam	2. 50
Ordinary firm soil or loam, 65 percent clay	3. 00
Still clay loam	4. 00
Firm gravelly clay soil	5. 00– 7. 00
Stiff clay	6. 00
Conglomerates, soft slate	6. 50
Stratified rocks	8. 00
Small boulders	8. 00–15. 00
Hard rock	13. 33
Concrete	15. 00–20. 00

BIBLIOGRAPHY

Design of Spillways of O'Shaughnessy Dam, *Trans. Am. Soc. C. E.*, 93:1451, 1929.

E. W. Lane "Determining Design of Dam Spillways Through Use of Hydraulic Models," *Western Const. News*, April 1935.

H. W. King, *Handbook of Hydraulics*, McGraw-Hill Book Co.

A. F. Meyer "High Velocity Discharge of Overfall Dams and Forms of Spillway Profile", Eng. N., 94:597–9, April 9, 1925.

Julian Hinds "Side Channel Spillways: Hydraulic Theory, Economic Factors, and Experimental Determination of Losses," *Trans. Am. Soc. C. E.*, 89:881, 1926.

E. M. Burd "Spillway Design on Glacial Drift," *Trans. Am. Soc. C. E.*, 99:815, 1934.

Spillway Gates on or Adjacent to Dams, N. E. L. A., *Serial Report No. 289–84:1–29. July 1929.*

CHAPTER 6

OUTLET STRUCTURES

1. Function.—The relative importance of the outlet structure varies with the purpose of the development. If water is being stored for such uses as irrigation, water supply, and flood control, it is important that the appurtenant structures be so designed that the necessary releases may be made as required. Other projects, such as those serving recreational and wildlife interests, have different requirements, in that normal operation of the project requires the release of little or no water. The use of an outlet in the latter case may be limited to the release of water required by water rights or needs downstream or to intermittent discharge for sanitary control of the reservoir level, provision for fish life in the stream below the dam, or emergency drawdown for repair work or other contingencies.

For flood-control projects, the outlet arrangement should be such that dependence for effective operation will not rest on the manipulation of gates, as the projects which will come within the scope of this manual will seldom be large enough to justify the employment of full-time operators who would regulate the gates to meet hydrologic conditions. Unless continuous supervision can be assured the arrangement of outlet and spillway should be such that they will function without manipulation. This is particularly important for small watersheds, where the interval between the storm occurrence and the flood wave is relatively short.

2. Design considerations.—The essential features of outlet works to be considered include:

1, location; 2, capacity; and 3, safety and permanence.

1. *Location.*—If the purpose of the outlet is to drain the reservoir, it must of course be located at or near the bottom of the reservoir. For earth and rock-fill dams, the outlet conduit may be through the base of the dam or it may be a tunnel in an undisturbed formation at one end.[1] The latter is a desirable and convenient arrangement from the construction viewpoint, because the streamflow can be diverted through the tunnel while work on the dam progresses, but it is seldom economical for dams of 30 feet or less in height. Outlets for masonry or concrete dams are usually built into the base of the structure.

The intake end of the conduit should be so placed that there will be the least possible tendency for silt and submerged debris to collect

---

[1] See requirement for conduits through embankment, ch 7 17.

in front of it.   Locations in deep pockets should be avoided in favor of open level areas when possible.

For many irrigation, power, or municipal water supply projects, and occasionally for other developments, the reservoir will feed directly into an open canal or pipe line, in which case the location and elevation of the canal or pipe will be a determining factor in the placing of the intake.   Since floating debris during flood periods may often obstruct intake structures, a layout plan which will direct the main flood currents away from the intake is desirable.   Such a plan will be effective whenever the spillway discharge exceeds the discharge at the outlet.

2. *Capacity*.—The capacity which an outlet must have will be determined either from the requirement of the purpose which it serves, or for some projects will depend upon the time that can be permitted for drawdown of the reservoir for repairs or maintenance.   The assumption can ordinarily be made that any repair work required can be delayed until service demands are light and can generally be done at times of low inflow and at a season favorable to such construction.   Under these circumstances, several days' time can be allowed for complete drawdown.   A determination of the capacity and size required for drainage involves consideration of the inflow to the reservoir during the discharge period and the discharge capacity available under the diminishing head of water.   The capacity at low reservoir level should be at least equal to the average inflow expected during draining operations.

The proposed outlets works may often be used for diversion during construction, thus avoiding the construction of supplementary or temporary works for that purpose, and this use, rather than the function they are to perform after the dam is completed, may determine the capacity.

The capacity of open channel outlets is computed by the method shown for open spillways in chapter 5.

The determination of the capacity of outlet structures with submerged entrances is based on the general formula

$$H = \frac{V^2}{2_g} + H_0 + H_1 + H_2 + \text{etc.} \tag{1}$$

in which H = the gross or overall head.

$\dfrac{V^2}{2_g}$ = the velocity head.

H_0 = the loss of head at the entrance.

H_1 = the loss of head due to conduit friction.

H_2, etc. = other losses of head due to bends, contraction, enlargements. etc.

The problem for the usual small project includes only the velocity head, the entrance loss, and the friction loss.

Entrance loss.—The head lost at the entrance to a conduit is determined from the equation

$$H_0 = K_0 \frac{V^2}{2g} \tag{2}$$

which expresses the loss as a percentage of the velocity head, the factor K_0 varying with the shape of the entrance. Figure 46 shows

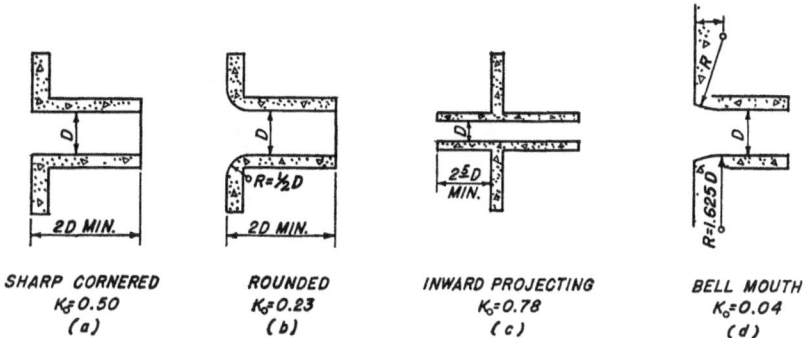

SHARP CORNERED	ROUNDED	INWARD PROJECTING	BELL MOUTH
$K_0 = 0.50$	$K_0 = 0.23$	$K_0 = 0.78$	$K_0 = 0.04$
(a)	(b)	(c)	(d)

TYPES OF CONDUIT ENTRANCES
FIGURE 46.

four types of entrance which cover the conditions ordinarily encountered. Values of K_0, which have been determined experimentally, are shown for each type.

A diagram giving the loss of head at entrance to conduits of various sizes from 10 inches in diameter up to 6 feet by 6 feet is shown in figure 47. This diagram gives entrance losses for a sharp-cornered inlet as in (a) of figure 46. Conversion factors for finding the loss due to other types of inlet are also given.

For example, assume that a flow of 100 cubic feet per second must be provided for. Entering the diagram at a flow of 100 cubic feet per second, it is found that a 2 foot by 3 foot conduit will discharge this amount with a velocity of 16.6 feet per second and will have an entrance loss of 2.2 feet of head. A round conduit 30 inches in diameter will discharge the same amount with a velocity of approximately 20 feet per second and an entrance loss of 3.2 feet of head. And so on for other sizes. The velocity for the type chosen is converted to the equivalent head and added to the head lost at entrance and by friction throughout the conduit to find the gross head required.

In another form of this problem frequently encountered, it is required to determine the size of conduit to give a known discharge under the existing head the procedure being as follows: Deduct from the total head an approximate value for the entrance loss, and convert the remaining head to velocity. Enter the diagram from the right with this velocity, to find the nearest conduit size that will give the

LOSS OF HEAD AT ENTRANCE OF SQUARE
CORNERED CONDUITS

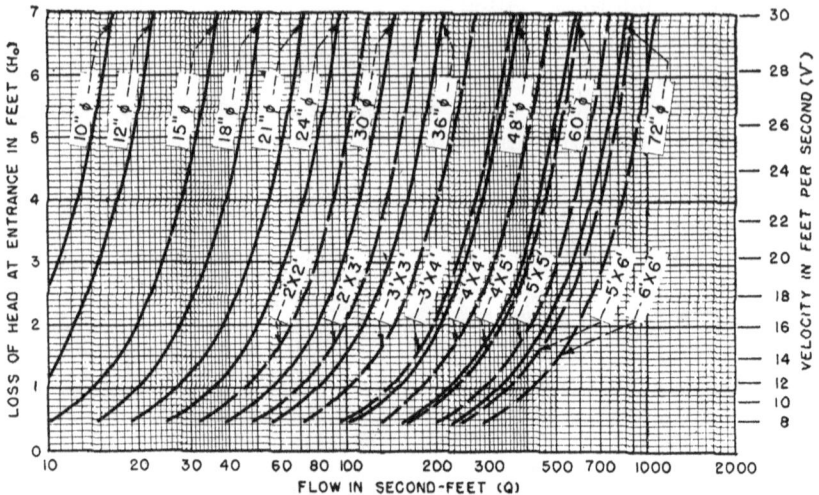

CONVERSION FACTORS

ROUNDED	— — — —	MULTIPLY BY	0 46
BELL MOUTH	— — —.	" "	0 08
INWARD PROJECTING		" "	1 56

$$Q=\frac{0\ 463}{n}d^{8/3}s^{1/2} \qquad S=\frac{h_f}{100}$$

Diagram reads direct for concrete conduits in perfect condition and for arc welded steel conduit or old cast-iron pipe $(N=0\ 012)$ For other kinds of conduit material, multiply the given friction loss by the following factors

Concrete in fair condition	$(N=0\ 015)$	1.25
Concrete in poor condition	$(N=\ 018)$	1.5
Cast iron—new	$(N=\ 010)$	.8
Vitrified clay	$(N=\ 014)$	1.2
Riveted steel	$(N=\ .016)$	1.3
Corrugated metal	$(N=\ .036)$	3 0

FIGURE 47

required capacity.   Also find the entrance head loss, and compare it with the original assumption.   If the two values are not reasonably close, make a new assumption guided by the first result, and repeat the calculation.   The conduit size which has the required capacity and for which the sum of the velocity head, the entrance head, and the friction head loss does not exceed the gross head available gives the desired answer.

Friction loss.—The head lost due to friction in the conduit is comparable to similar losses in open channels and can be determined by an adaptation of the Manning formula.   For round pipes

$$Q=\frac{0.463}{n}d^{8/3}s^{1/2} \tag{3}$$

and for square pipes

$$Q=\frac{0.59}{n}d^{8/3}s^{1/2}$$

in which $n=$ roughness coefficient.

$\quad d=$ diameter or side dimension of the conduit.

$\quad s=$ slope or hydraulic gradient which will maintain uniform flow.

The diagram of figure 48 gives the loss of head per 100 feet of length of conduit for various capacities and sizes of round, square, and nearly square conduits. This has been made for $n=0.012$. Conversion factors are given to permit evaluation of lost head for other values of n as may be required for conduits made of various materials.

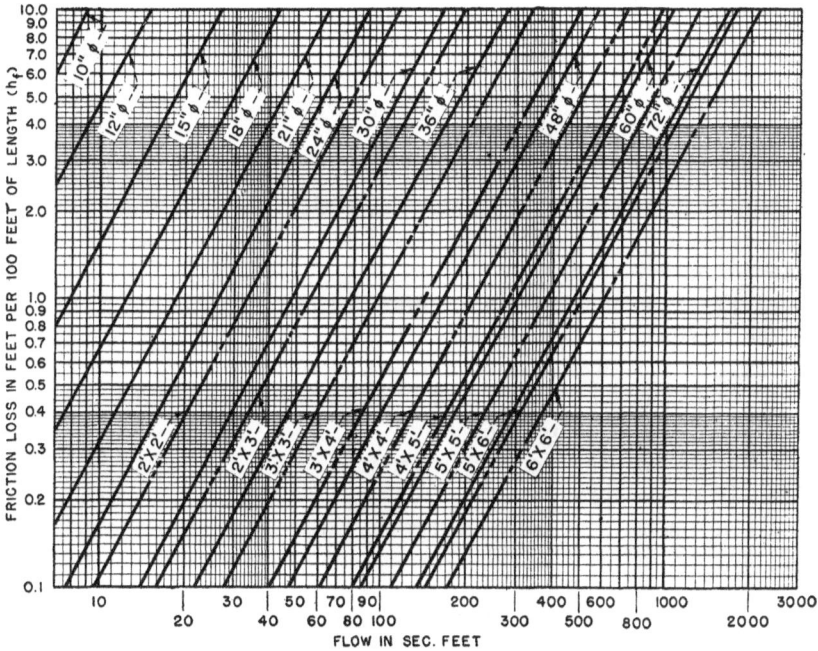

FIGURE 48.—Friction loss for conduits.

3. *Safety and permanence.*—For earth and rockfill dams the security of the dam depends to a large degree on the safety of the outlet structure. In some cases all or part of the conduit is under the pressure of the maximum depth of the water, so that any leakage or failure of the conduit may open up water passageways which will gradually be enlarged until partial or complete failure results. There is also the danger of leakage developing along the contact surfaces between the outlet structures and the earth fills, which may cause serious damage. These facts emphasize the importance of using durable materials, conservative designs, and construction methods and details that will insure safe structures.

Replacement of a conduit through an earth or rockfill dam is usually a difficult and expensive operation which can be avoided by the use of permanent material, such as cast-iron pipe for the small sizes and reinforced concrete for the larger sizes. For small projects with comparatively low heads exceptions may be made where there is a limited amount of money available for the purchase of material, or where the value of the service rendered will not justify a greater expenditure and where the possible damage from failure is of little or no consequence. In such cases, the use of iron or steel pipe protected by galvanizing, bituminous coating, or other rust-resisting treatments may be justifiable. The boundary line between the two types of construction cannot be determined by any general formula, and each case must be decided on its merits. For a further discussion of the use of steel or light metal conduits see chapter 5: **6**.

Provisions for safe design construction of outlet conduits must cover the following conditions:

(a) Prevention of seepage along the outside conduit surface.

(b) Provision against cracking which might result in leakage either of water into the fill around the conduit or, in extreme cases, of soil into the conduit itself.

(c) Provision of satisfactory foundation conditions both for the sake of the integrity of the conduit structure and for the prevention of cracking and leakage as set out in (b).

(d) Safe structural design of the conduit structure to carry the loads to which it will be subjected.

These factors are discussed in the following paragraphs:

(a) If proper provision is made against settlement which might result in a separation between the conduit surface and the adjacent soil, further provision against seepage along the outside conduit surface may be accomplished through the liberal use of projecting fins or collars. The National Park Service regulations require that the fin projection shall be not less than 3 feet, that it should encircle the conduit on all sides, and that the spacing of fins shall be such that the increased length of the path around the fins shall not be less than 20 percent of the length of the conduit.

Where such fins are used, it is important that they shall not introduce additional stresses in the conduit barrel. This may be accomplished by constructing the fins after the conduit barrel has been completed—separating them from the barrel structure possibly by a double layer of graphite-coated paper which will permit a slight lateral slipping of the fin on the conduit. If there is danger that torsional stress may be introduced, the fin may be separated from the conduit barrel by a space of a half inch or more, and this space may be filled with a watertight bituminous compound.

It is important to recognize that seepage along the conduit surface
is most likely to occur along the sides and lower quarters of the structure, and the fins should be particularly designed to meet this danger.
Figure 49 shows a typical construction drawing of fin construction

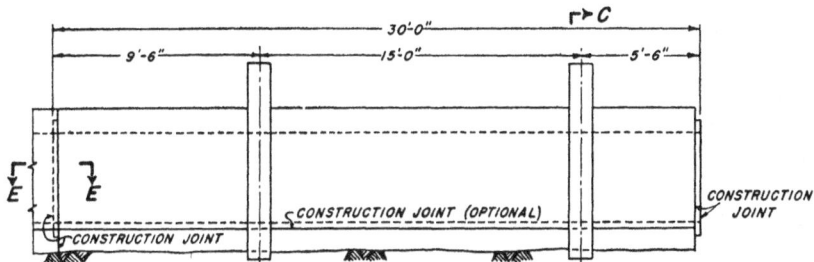

ELEVATION – TYPICAL 30'-0" UNIT
WHERE ROCK LINE IS BELOW INVERT

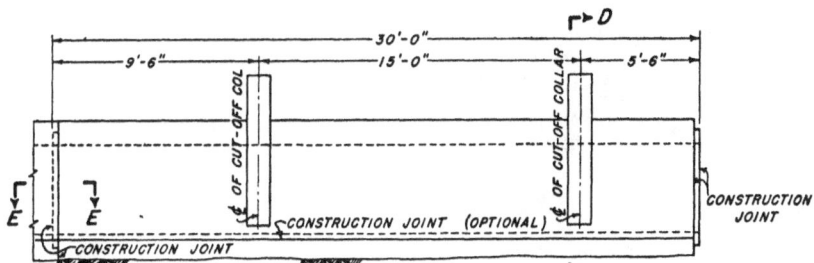

ELEVATION – TYPICAL 30'-0" UNIT
WHERE ROCK LINE IS ABOVE INVERT

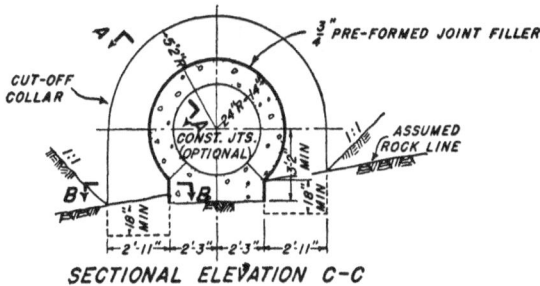

SECTIONAL ELEVATION C-C

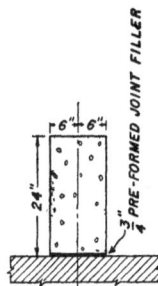

SECTION A-A
SCALE OF FEET

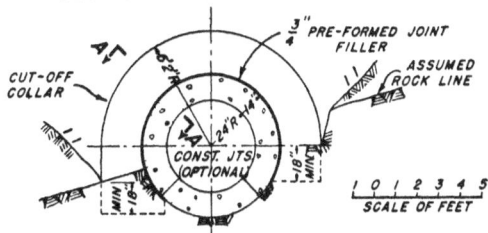

SECTIONAL ELEVATION D-D

FIGURE 49.

for a single-barrel 4-foot conduit.   Figure 50 shows the actual construction of such fins on a double-barrel conduit.

(b) Cracking which might occur on account of contraction of the conduit barrel may, for short concrete conduits, be prevented by liberal use of longitudinal reinforcement.   For long conduits, contraction may be concentrated in this manner at a predetermined point and provision made at this point for its occurrence without permitting leakage.   The contraction of cast-iron pipe will commonly be taken care of at the joints without an appreciable opening of the joint filling.

More serious danger may be present due to the cracking that may result from unequal settlement of the foundation.   Such settlement may often be insufficient to cause structural failure, but may open up

cracks which will permit water under pressure to enter the surrounding fill.   It can best be prevented by a careful analysis of foundation conditions and by a design loading which will produce approximately the same consolidation and settlement on the foundation throughout. If the plan includes a gate tower through which the conduit passes, the junction of the structure and the tower is a particularly vulnerable point which requires careful analysis to assure that unequal settlement will not occur.

Where the outlet structure consists of pipe, whether of reinforced concrete or cast iron, the method of bedding the pipe and the backfilling around it should be such as to insure, insofar as possible, against unequal settlement and also to secure the best possible distribution of the load on the foundation material.

When filling around these structures, great care should be taken to secure tight contact between the fill and the conduit surface; this is important not only for the prevention of seepage along the conduit, but also to insure that the fill develops a satisfactory lateral restraint on the structure in order to prevent excessive internal stresses from the load on the conduit. Optimum moisture content of the soil is of vital importance, as is also adequate tamping of material in thin layers. In this connection, the engineer is cautioned against allowing an appreciable reduction in the moisture content of elastic soils after the tamping has been done. Certain soils with high shrinkage coefficients, particularly very plastic clays, will pull away from the conduit surface if allowed to dry out. To prevent this action, the use of selected soils and the maintenance of a uniform moisture content during and after the construction operations is important.

(c) The requirement for approximately uniform distribution of the load on the foundation has been discussed in the preceding paragraphs. The safe foundation condition is that in which the conduit is supported throughout by concrete or masonry on a rock foundation. Where this is not possible, it should have continuous support from undisturbed material carefully trimmed to the exact subgrade and no lower. Whether laid in a trench or at the ground surface, the fill at the sides should be placed to secure the greatest possible compaction, as this will not only improve the foundation support and develop lateral support to the structure but will have an important bearing on the vertical load which the structure will carry. The supporting of conduits on piers, as has sometimes been done, is an unsatisfactory and generally dangerous practice. Even where the conduit is sufficiently strong to carry the load between supports, settlement may occur underneath the conduit which will leave openings and facilitate dangerous seepage conditions.

(d) The load which a conduit through an embankment is required to carry will vary between wide limits, depending on the condition of the conduit construction and character of compaction of the embankment. If the conduit is laid in a trench in natural soil under the embankment, it will practically never receive the full weight of the fill immediately above it. On the other hand, if the conduit is laid under conditions where it is exposed in whole or in part above the natural ground surface before the embankment is placed, the load which may come upon it can in some cases be as much as 50 percent in excess of the weight of the backfill directly above it.

It is important, therefore, before the conduit is designed, that the conditions under which it is to be constructed be carefully determined and specified, and the load calculations must give full consideration to these conditions. The amount of load which the conduit must carry has been determined experimentally by some of the Federal agencies and several of the State highway departments for a variety of conditions, but the most complete set of experiments are those which have been carried out and analyzed at the Iowa State Agricultural College and which are described in various of its bulletins. If a highway is laid directly on top of the fill, consideration must be given to any live load that may be transmitted to the conduit.

In designing an outlet conduit, it is suggested that the following procedure be used:

1. Make a careful examination of the ground profile along the line of the conduit, and note from point to point the extent to which the top of the conduit will extend above the natural ground surface. If the top of the conduit is not appreciably above the natural ground surface, but if it is to be constructed in a trench with rather flat side slopes, it should for safety be considered as projecting up to the point where the side slopes will meet the side of the conduit. From these notes, determine the greatest projection of the conduit top above the ground surface that will be projecting for any appreciable distance. The height of this projection divided by the width of the conduit will be taken as the projection ratio p.

2. An important factor in determining the vertical load on the conduit is the settlement ratio r_{sd}, which indicates the relative expected settlement and deflection in a vertical plane through the conduit as compared to the similar settlement immediately adjacent to the conduit. It is not practical to determine this ratio in the field, and it is suggested for safety that it be assumed as follows:

(a) For rigid conduits, including reinforced concrete and cast-iron pipe, where laid with a concrete cradle on a rock foundation, use r_{sd} as 1.00.

(b) For rigid conduits, including reinforced concrete and cast-iron pipe laid on rock or other hard foundation, when bedded on an earth cushion of appreciable thickness, use r_{sd} as 0.70.

(c) For flexible conduits, such as corrugated metal pipe for which concrete cradle is not recommended, use r_{sd} as -0.10.

3. The values of the projection and settlement ratios are combined into a single ratio $r_{sd}p$ for use in load calculations.

4. Divide the height of the fill H above the top of the conduit by the outside width of the conduit B_c and note the value of the $\frac{H}{B_c}$ for use in the load calculation chart figure 51.[2]

5. In using figure 51, enter the diagram at the left-hand side with the value of $\frac{H}{B_c}$ and follow horizontally to an intersection with the line representing the proper value of $r_{sd}p$. Vertically below this intersection may be found the value of the coefficient C_c. If a safe value of $r_{ds}p$ has been chosen, no consideration need be given to the

2 This diagram has been prepared from material contained in the bulletins of the Iowa engineering experiment station, with the advice and assistance of Mr W. J. Schlick of Iowa State College, Ames, Iowa.

VERTICAL LOADS ON CONDUITS THROUGH EMBANKMENTS
VALUES OF COEFFICIENT C_0

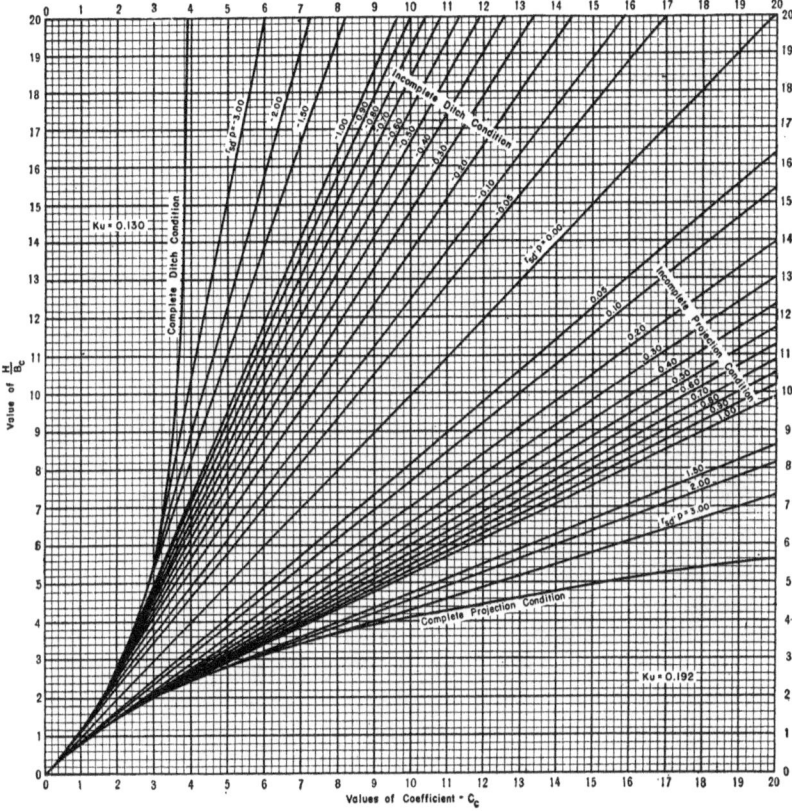

terms "projection condition," "incomplete projection condition," etc. These terms cannot be interpreted simply without a consideration of the type of conduit being used. It may be stated generally, however, that the "complete ditch condition" may refer to any conduit completely buried in ditches excavated in comparatively solid material, and also to very flexible conduits which may project somewhat above the subgrade of low embankments. The "complete projection condition" may refer to a rigid or semirigid conduit projecting in part above the subgrade under a low embankment, while the "incomplete projection condition" may differ from it only in being under an embankment of greater height. For a detailed interpretation of these terms, a study of Bulletin 112 of the Iowa Engineering Experiment Station is necessary.

6. The vertical load on the conduit may then be determined by the formula $W_c = C_c \times w \times B_c^2$, in which w is equal to the weight of the filling material per cubic foot, and B_c is the width of the conduit.

For conduits having a flat top, the load per lineal foot given by the above formula may be assumed to be uniformly distributed across the width.   For circular conduits, it may, with safety, be taken as being distributed on the upper quarter of the circumference.

For reinforced concrete conduits constructed in place a conservative design will be based on the vertical loading as determined above without consideration of horizontal pressures.   For projects in which the cost of the outlet conduit is an important part of the total cost, and where construction regulations as to compacting of the material at the sides of the conduits may be expected to be rigidly adhered to, a determination of stress relieving horizontal pressures may be justified. The intensity of horizontal pressure may be determined by the Rankine formula, in which the intensity of horizontal pressure is equal to $K \times$ the intensity of vertical pressure and

$$K=\frac{\sqrt{\mu^2+1}-\mu}{\sqrt{\mu^2+1}+\mu}$$

in which μ is equal to the coefficient of internal friction of the fill material.   Unless this coefficient shall have been determined in connection with studies of foundation and embankment stability, the horizontal pressure might be taken as equal to one-third of the vertical pressure as determined from the formula above.[3]

Where the outlet conduit consists of precast reinforced concrete pipe, it is assumed that it will be carefully set on a good foundation and well bedded in concrete.[5a]   For such conditions, the strength of the pipe may be safely taken at two and one-half times the test load [4] specified for the particular pipe in standard A. S. T. M. specifications. A proper pipe to be selected for any particular case should have a load carrying capacity under these conditions sufficiently in excess of the actual load, as calculated above, to give a reasonable factor of safety. If horizontal pressures are neglected, a considerable factor of safety will result on that account.

For small conduits through low structures, complete calculations of load as above set out may not be necessary, but where this is not done the pipe selected should have sufficient strength to carry at least twice the weight of the fill which will be superimposed above the width of the conduit.

---

[3] The above conditions refer to circular conduits and represent the loadings that will produce the maximum expected stress in the structure   It must be recognized that after the dam is filled a part of the conduit length will in any case be below the seepage line through the dam   Under these conditions, the coefficient of internal friction will be very low, the vertical pressures will be reduced, and the horizontal pressure increased   Where reinforced concrete conduits are of the box type section, it is therefore necessary also to design the vertical side slabs to resist bending from horizontal pressures resulting from the full height of the saturated fill and equal to the vertical pressures that will occur under this condition.

[5a] The use of precast concrete pipe through earth dams is not permitted by some Government bureaus, including the Forest Service and the National Park Service   It should not be used unless approval is obtained from responsible officials.

[4] Three-edge bearing test.

For important projects where the outlet conduit may represent a considerable part of the cost of the structure, it is suggested that full use be made of the material contained in Bulletin 112 of the Iowa Engineering Experiment Station.

3. Gates.—Control of the discharge through outlet structures is an important function for which gates of several different types are used. For the open channel outlet, the gate installations are similar to those already discussed under spillway crest control in Chapter 5.

For submerged outlets there are several types of gate suitable, including the sliding gate, the gate valve, needle valve, and the butterfly valve. The first two are most applicable to low heads, and are the only ones which will be described here.

For conduits through earth or rockfill dams, the most desirable location for the gate is at the upstream end, because with this arrangement the conduit is under water pressure only while the gate is open, and the danger from leakage is limited to the same period. However, if the conduit is well built and properly sealed against leakage, an arrangement as shown in figure 53 with the valve chamber near the middle of the dam has some advantages. Figures 54, 55, and 56 show typical arrangements of outlet valves and conduits for different conditions. Figure 54 shows a simple and inexpensive layout in which the operating stem follows the slope of the fill, suitable only for small projects with a maximum depth of approximately 20 feet. Disadvantages are that settlement of the fill may throw the stem out of line and also that ice or floating debris may injure the stem at the water surface or hinder operation of the gate. The stem of the gate should be in a pipe with an inside diameter at least 2 inches larger than the diameter of the torque rod. The pipe should be packed with grease and have bushings to maintain rod alinement.

A more satisfactory arrangement is shown in figure 55, in which the sliding gate is operated from a tower by a manually operated gear stand. This particular installation was designed for a recreational reservoir where infrequent operation of the gate is expected. The usual convenience of a bridge from the dam to the tower was omitted to reduce the hazard of interference by unauthorized persons. In any case the operating stand should be provided with a substantial lock.

An important feature in most projects is that the outlet gate operating mechanism be accessible at all times. Although the safety of the structure should not depend on the release of water through the outlet during flood flows, the gate control should be accessible from the shore as a special measure of precaution. This condition is not met with in the arch dam design of Chapter 10. However, in that example, capacity of the outlet is very low in comparison with

Bureau of Reclamation

FIGURE 52.—Cast-iron outlet gates and vent and water gage pipes at Clear Creek Dam, Yakima project, Washington.

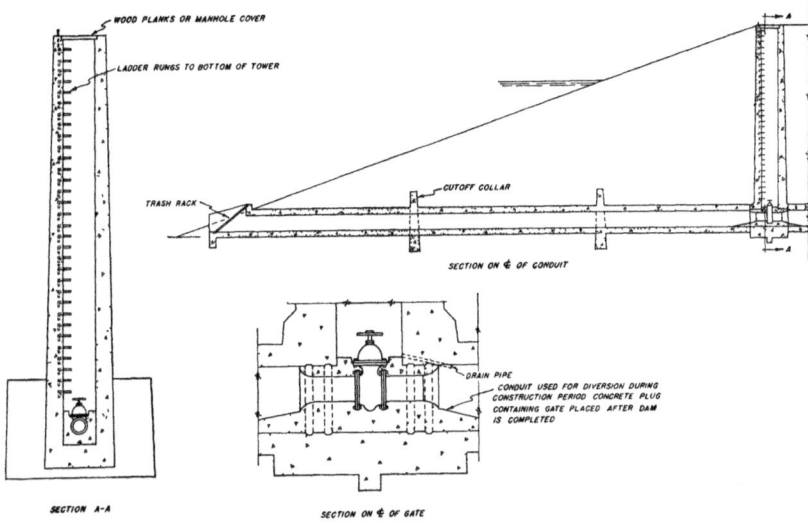

WOOD PLANKS OR MANHOLE COVER

LADDER RUNGS TO BOTTOM OF TOWER

TRASH RACK

CUTOFF COLLAR

SECTION ON ℄ OF CONDUIT

DRAIN PIPE
CONDUIT USED FOR DIVERSION DURING
CONSTRUCTION PERIOD CONCRETE PLUG
CONTAINING GATE PLACED AFTER DAM
IS COMPLETED

SECTION A-A

SECTION ON ℄ OF GATE

GATE VALVE INSTALLATION WITHIN THE DAM

FIGURE 53.

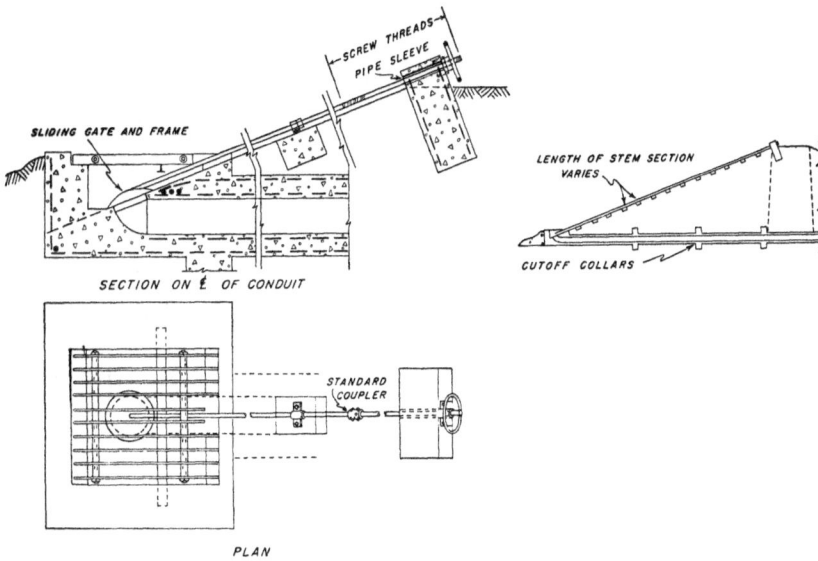

SCREW THREADS

PIPE SLEEVE

SLIDING GATE AND FRAME

LENGTH OF STEM SECTION
VARIES

SECTION ON ℄ OF CONDUIT

CUTOFF COLLARS

STANDARD
COUPLER

PLAN

SMALL SLIDING GATE INSTALLATION ON THE UPSTREAM SLOPE

FIGURE 54.

LADDER RUNGS
TO BOTTOM OF
RESERVOIR

STEM GUIDE

VENT PIPE

CARRY FOOTINGS TO SOLID
FOUNDATION

SECTION ON ℄ OF TOWER

MAXIMUM WATER SURFACE

NORMAL WATER SURFACE

TRASH
RACKS

CUTOFF COLLAR

SECTION ON ℄ OF CONDUIT

CARRY WALL TO
SOLID FOUNDATION

SECTION A-A

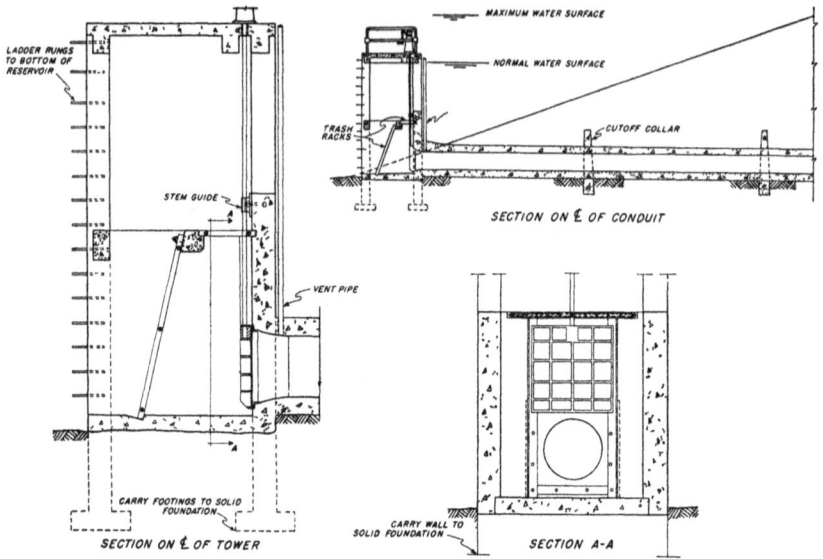

SLIDING GATE OPERATED FROM TOWER AT UPSTREAM END OF CONDUIT

FIGURE 55

PIPE SLEEVE

SECTION ON ℄ OF GATE

STEM GUIDE

UPSTREAM ELEVATION

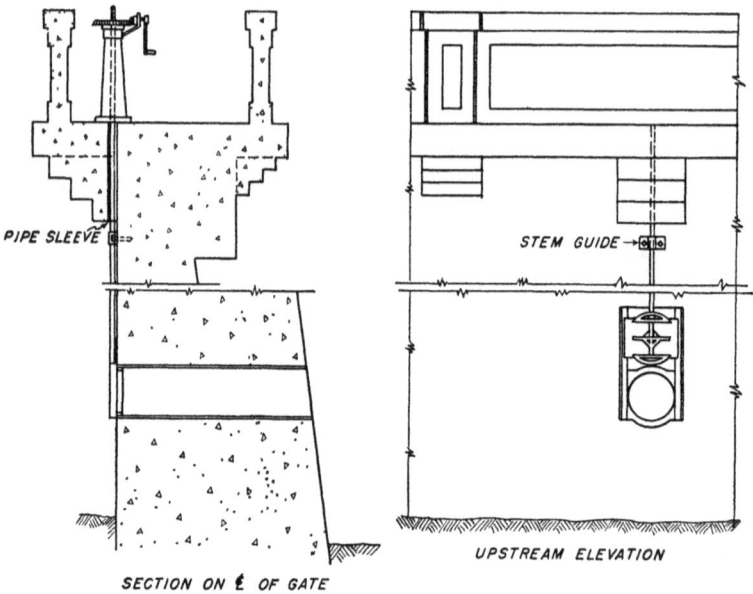

SLIDING GATE INSTALLATION FOR CONCRETE NON-OVERFLOW DAM

FIGURE 56.

the spillway capacity, and the entire dam can be overtopped to a considerable depth without damage. To locate the operating stand above high water would have placed an obstruction near the middle of the crest which would materially reduce the overflow capacity. The arrangement shown in the drawing however is not recommended for frequent use. A more satisfactory layout from the operating viewpoint would place the gate operating stand above flood level on the nonoverflow section, as in figure 56, or on the downstream side of the dam as in figure 57.

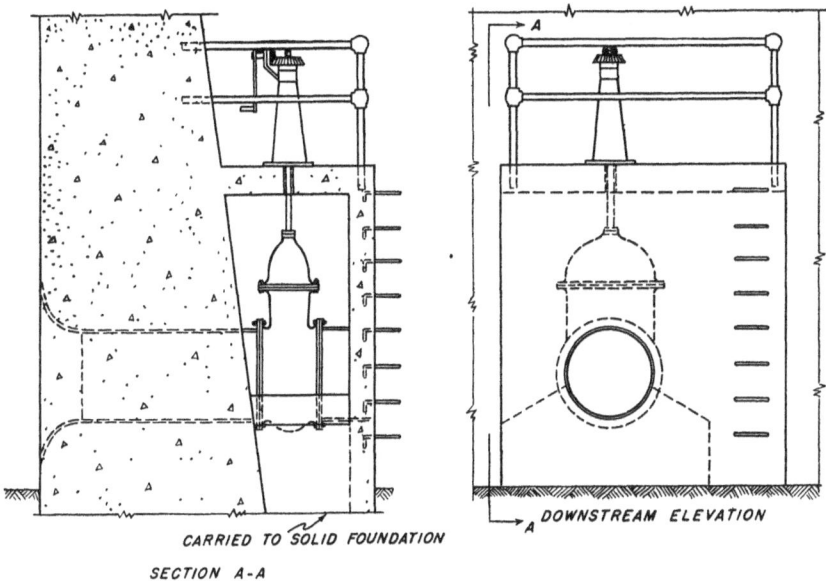

CARRIED TO SOLID FOUNDATION

SECTION A-A

DOWNSTREAM ELEVATION

OUTLET USING GATE VALVE AT DOWNSTREAM END

FIGURE 57

Because many small storage projects for stock water include an arrangement for utilization of the water outside the reservoir there is shown in Figure 58 a layout for automatic regulation from the reservoir to the watering trough. A gate valve for manual control is included.

Figure 59 shows details of a rectangular cast iron sliding gate designed by the Bureau of Reclamation. This type of gate and the determination of operating loads has been explained in Chapter 5 in the section on crest gates.

Operation of gates for most small projects is done by hand, using either a rack-and-pinion or a screw-stem hoist. All gate valves operate under the latter principle, which is also preferable for sliding gates up to twenty or more square feet in area, depending on the head.

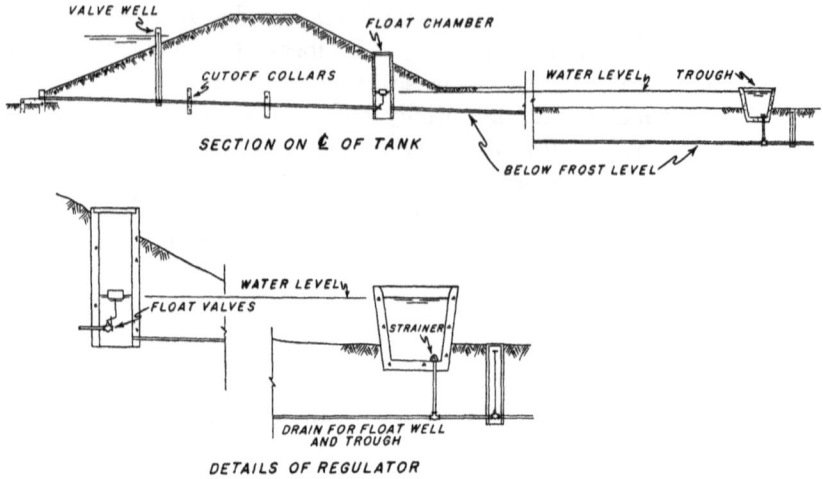

AUTOMATIC REGULATION FOR STOCK WATER TANK (NON-FREEZING)

FIGURE 58.

Most requirements for gates and joists can be met from standard designs of gate manufacturers. Before accepting stock designs however, the designer should satisfy himself as to the strength of the various parts.

Stresses [5] in screw stems are caused by direct compression and by torsion, the combined stress of closing being represented by the formula

$$t = \frac{1}{2}S_c + \sqrt{S_s^2 + \frac{1}{4}S_c^2} \tag{4}$$

$$s = \sqrt{S_s^2 + \frac{1}{4}S_c^2} \tag{5}$$

in which t is the maximum compressive stress and s is the shear due to the combined torsional and compressive stresses, S_s and S_c, all in pounds per square inch.

$$S_c = \frac{P}{A} \tag{6}$$

$$S_s = 103 \left(\frac{H}{nd^3}\right) \tag{7}$$

in which

P = vertical load to move the gate in pounds.

A = net area of the stem in square inches.

H = foot-pounds per minute applied to the gate stem.

---

[5] "The Design of Dams," by Hanna and Kennedy   McGraw-Hill Book Co p 381

$n=$ the number of revolutions of the stem per minute.
$d=$ the diameter of the stem in inches.

The allowable compressive stress is determined by

$$p=\frac{S_c}{1+\dfrac{L^2}{25,000R^2}}$$

where

$p=$ working stress in pounds per square inch.
$S_c=$ allowable unit compressive stress for short columns.

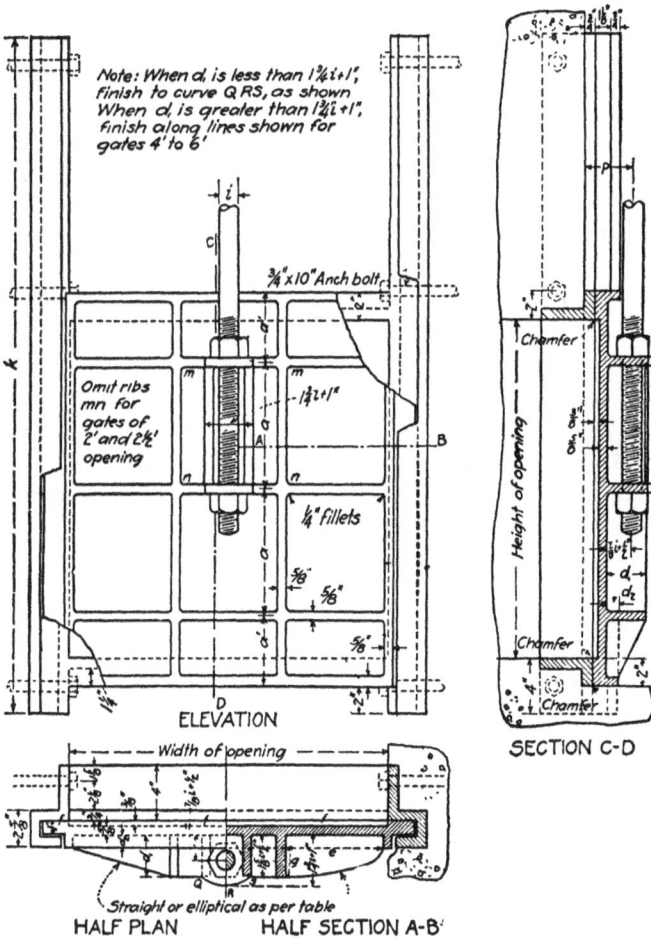

Note: When d is less than $1\frac{3}{4}i+1''$, finish to curve QRS, as shown.
When d is greater than $1\frac{3}{4}i+1''$, finish along lines shown for gates 4' to 6'

$\frac{3}{4}'' \times 10''$ Anch bolt

Omit ribs mn for gates of 2' and 2½' opening

$1\frac{3}{4}i+1''$

$\frac{1}{4}''$ fillets

$\frac{5}{8}$

$\frac{5}{8}''$

$\frac{5}{8}''$

ELEVATION

Chamfer

Height of opening

Chamfer

Chamfer

SECTION C-D

Width of opening

Straight or elliptical as per table
HALF PLAN　　　　HALF SECTION A-B

STANDARD CAST-IRON GATES
OPENING 2 FEET TO 3½ FEET

Figure 59.—Slide gate, lap-closure type

Bureau of Reclamation

L=the unsupported length of the stem in inches.

R=least radius of gyration of the stem in inches.

The horsepower required to operate a gate is

$$H=\frac{Fv}{33,000E}\qquad(8)$$

when

F=the pull or push exerted on the gate.

v=the speed of gate travel in feet per minute.

E=the overall hoist efficiency, expressed as a decimal.

4. Trash Racks.—All submerged outlets should be protected from possible clogging with debris by the use of steel trash racks. The usual rack is made of flat steel bars assembled into sections that can be handled with the equipment available. The bars are placed from 3 to 6 inches apart with the long axis of the bar in the direction of the current. The gross area of the screened opening should be such that the velocity past the racks will not exceed 10 feet per second. The loss of head through trash racks may be determined from the diagrams shown in figure 60.

Unless racks are cleaned frequently, the water passage through them will be obstructed by debris so that a considerable pressure may be exerted on the rack surface. Rack bars of mild steel designed to be stressed to 38,000 pounds per square inch under a head of the maximum depth of water above them will usually be safe. The rack supports should be designed for normal working stresses under the same conditions.

Figures 61 and 62 show two kinds of rack assembly—61 being the conventional pipe spacer and tie rod type, and 62 being a more recent development in which welding is used. The latter style is usually lighter in weight and will be as durable as the other type if the welding is properly done.

5. Fish Screens.—In some localities where water is distributed in open channels for irrigation use, preservation of fish life in the reservoir requires that the outlet be provided with fish screens. The openings in the screen, being limited by the size of the young fish, are so small that if installed as stationary barriers they would be continually clogged with debris and interfere seriously with the discharge. Completely satisfactory arrangements have not been developed, but the rotary drum type shown in figure 63 shows the most recent development made under the direction of the Bureau of Fisheries of the Department of Commerce. One of the most important features is the small bypass conduit from the upstream side of the screens to the river channel below. When fish travel downstream to a barrier they will seldom turn upstream again, but instead will charge the barrier until they are carried over it or else destroy themselves. The bypass, through which water flows continuously, provides a means of escape

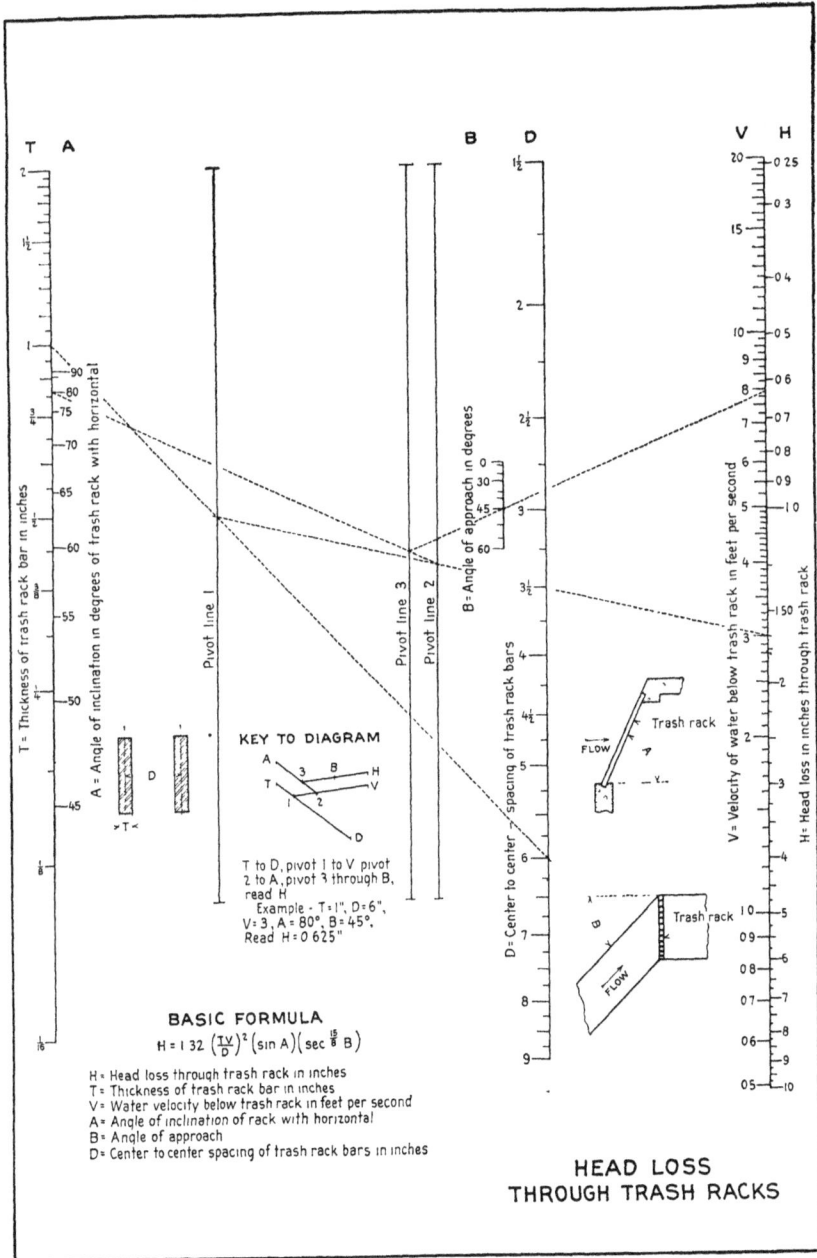

KEY TO DIAGRAM

T to D, pivot 1 to V pivot
2 to A, pivot 3 through B,
read H
    Example - T = 1", D = 6",
V = 3, A = 80°, B = 45°.
Read H = 0 625"

BASIC FORMULA

$$H = 1\,32 \left(\frac{TV}{D}\right)^2 (\sin A)\left(\sec^{\frac{15}{8}} B\right)$$

H = Head loss through trash rack in inches
T = Thickness of trash rack bar in inches
V = Water velocity below trash rack in feet per second
A = Angle of inclination of rack with horizontal
B = Angle of approach
D = Center to center spacing of trash rack bars in inches

**HEAD LOSS
THROUGH TRASH RACKS**

FIGURE 60

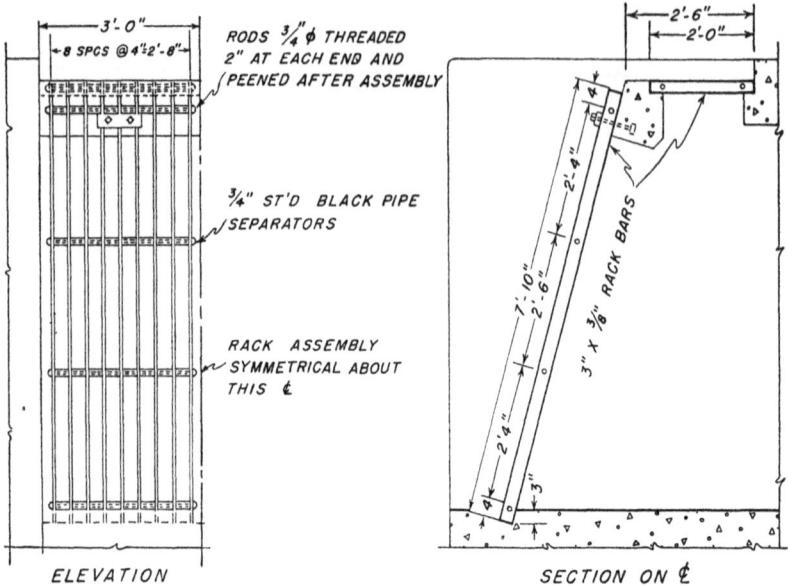

TYPICAL TRASH RACK INSTALLATION

(BOLTED RACK)

FIGURE 61.

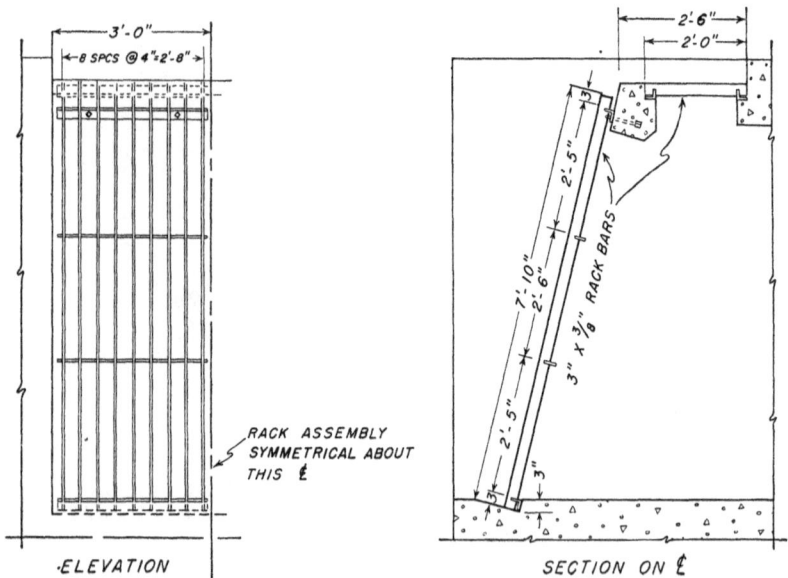

TYPICAL TRASH RACK INSTALLATION

(WELDED RACK)

FIGURE 62.

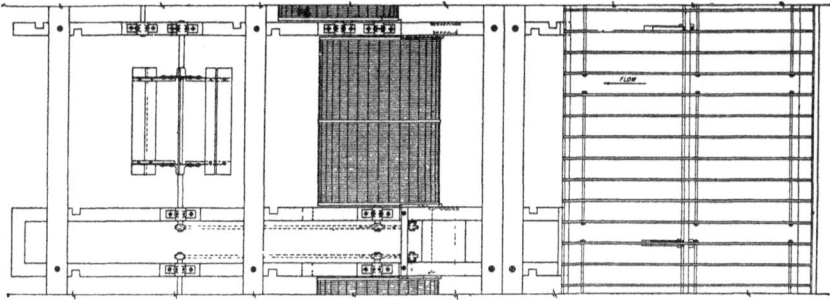

PLAN SHOWING ARRANGEMENT OF UNITS

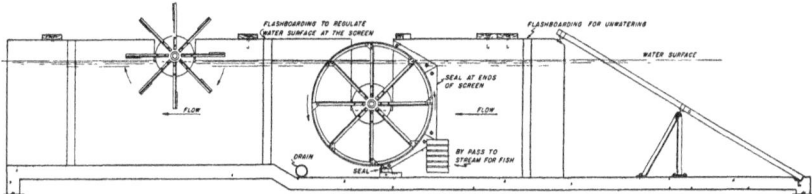

ELEVATION SHOWING SECTION OF SCREEN

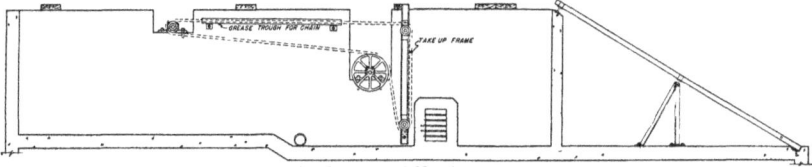

ELEVATION SHOWING DRIVING MECHANISM

ROTARY FISH SCREEN
DESIGNED BY BUREAU OF FISHERIES
DEPARTMENT OF COMMERCE
FIGURE 63.

FIGURE 64.—Rotary fish screens, Yakima irrigation district, Washington.

to livable water in the stream below. Where a series of drums are placed across a canal or stream they should be arranged in echelon with the outlet near the downstream drum.

6. Fish Ladders.—Federal and State regulations require the use of fish ladders in dams on certain streams which are frequented by migratory game fish, and they may also be desirable in other places to meet local conditions. Figure 65 shows the essential features of these structures. The pools are stepped to give an average slope of 3 or 4 to 1 and their size is determined by the probable peak re-

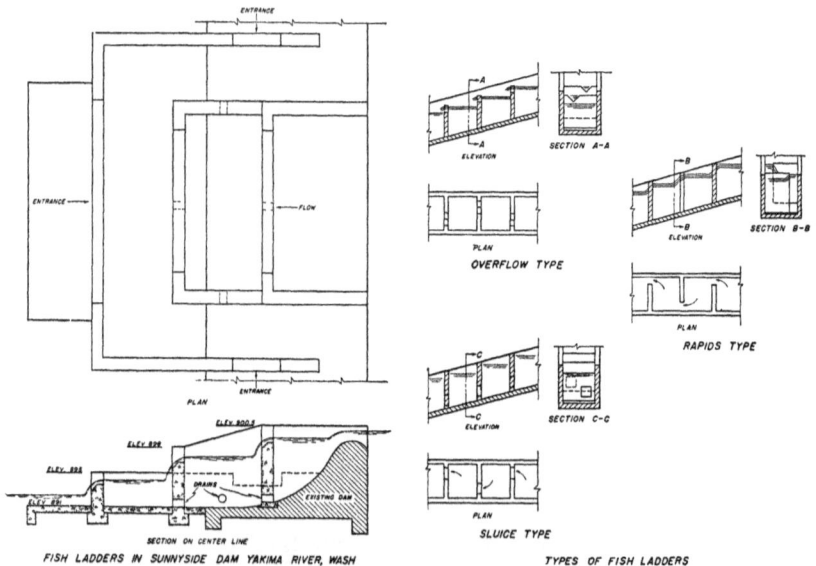

FISH LADDERS IN SUNNYSIDE DAM YAKIMA RIVER, WASH

TYPES OF FISH LADDERS

COMMON TYPES OF FISH LADDERS

FIGURE 65.

quirements during a fish run. Federal, State, or local agencies concerned with fish life should be consulted before the design or placement of fish ladders is attempted.

7. Log Sluices.—On streams where log driving operations may exist, it will be necessary either to provide a logway in the dam or make provisions for carrying the logs around it. The logway or logsluice is usually the least expensive in the long run.

A logway consists of a chute or flume built from the top of the dam and descending to the tailwater on a slope of 3 or 4 to 1. The entrance is usually provided with stop logs to prevent loss of water when the sluice is not in use. The size will depend on the size of the logs to be run, but should not be less than 4 feet deep and 6 feet wide. The lower end should discharge into a pool deep enough so the logs will not be damaged by hitting the bottom.

FIGURE 66 —Easton concrete gravity diversion dam and fish ladder (at right), Yakima project, Washington.

8. Construction of Outlet Works and Appurtenances.—Information regarding the building of structures for outlet works will be found in Appendix G which contains specifications for concrete work and other pertinent material.

CHAPTER 7

EARTH DAMS

CLASSIFICATION

1. General.—Earth dams may be classified in two ways; first, according to the method of construction, and second, according to the method of design. The terms "rolled fill," "hydraulic fill," and "semihydraulic fill" indicate types based upon distinctive methods of construction; the terms "simple embankment," "core," and "diaphragm," on the other hand, designate types based upon distinctive design. Cores, diaphragms, and upstream blankets are often referred to as dense or flow retarding elements which terms may be defined as the part or parts of the structure designed to develop the principal resistance to the passage of water.

2. Rolled Fill.—This designation is applied to that type in which the embankment material is spread and compacted by means of mechanical equipment. Some compaction may be attained by simple methods such as the movement of the trucks hauling the fill, but this is not a satisfactory method to use on structures of any importance even when carefully controlled. The best compaction is usually secured by the sheepsfoot roller; road rollers are commonly used but the results are generally less satisfactory. The tamper type of compactor, more recently introduced in American practice gives good results in competent hands. While it does not involve rolling operations, structures built by its use may properly be included in the "rolled fill" classification.

3. Hydraulic and Semi-hydraulic Fills.—In both types, the emplacement of the fill is accomplished by means of water. While the control of deposit varies somewhat in the two types, the essential difference is that with the "hydraulic" method the material is excavated, transported and placed hydraulically, while in the "semihydraulic" methods the material is excavated mechanically and hauled to the site where it is dumped. It is then transported into place hydraulically. These types are not recommended for dams within the limitations of this manual and for this reason, further consideration is not given them.

4. Simple embankment type consists of reasonably uniform material throughout. It has neither diaphragm nor core though it may have an upstream blanket, which for this type consists of a layer of

134

highly impervious material placed on the floor of the reservoir and extended up the upstream slope of the dam.

5. Core type dams are used more than any of the other types. The core is of substantial dimensions and is ordinarily constructed of selected soil. In addition to the core this type may also have an upstream blanket which is connected to the core. In design the core type of structure is considered pervious throughout though the core is constructed to be much less so than the rest of the embankment.

6. Diaphragm type dams include a relatively thin section of concrete, steel, or wood which forms a barrier to the flow of percolating water. With favorable site conditions and carefully controlled construction, this type can be made reasonably watertight. The "full diaphragm" type is that in which the diaphragm extends from the level of the impounded water down to a competent seal in an impervious foundation. The "partial diaphragm" type is that in which the diaphragm does not meet the conditions for the "full type," that is, it may not extend to the level of the impounded water, or it may not seat in an impervious foundation. Usually the diaphragm extends from some depth in a pervious foundation to a height in the dam well below the impounded water level; such construction is usually called a cut-off wall.

DESIGN

7. General.—It is important to emphasize that the design and construction of earth dams are closely interrelated, and it is frequently true that good construction methods are even more important than many features of the design.

The design is based to a considerable degree upon the result of investigations, and the refinements to which such investigations need be carried depends largely upon the height and importance of the structure. The manner of satisfying the design criteria makes it desirable to separate dams into two arbitrary groups, first, those having a normal operating depth of 15 feet or less, and second, those having a normal operating depth greater than 15 feet.

When the depth of water to be impounded does not exceed 15 feet, a relatively small increase of construction quantities may provide so ample a margin of safety that an expensive detailed examination of subsurface conditions may be omitted. There are exceptions to this statement, particularly when there is any suspicion that the structure might be founded upon plastic or semiplastic material. Above the arbitrary figure of 15 feet, the need for detailed investigations becomes increasingly important as the height increases.

It is important to emphasize that while complete detailed investigation of the properties of the soils is ordinarily not justified for the lower dams, the principles of soil behavior as disclosed by the technique of soil mechanics should be borne in mind as a guide to the most effective construction regardless of height.

8. Design Criteria are stated in the following paragraphs as they apply to all earth dams; these criteria are discussed in section 9 with regard to their application to dams 15 feet high or less; and in section 10 to dams higher than 15 feet. All earth dams must satisfy the following criteria regardless of their height:

Criterion 1.—The dam must be so designed that destruction through erosion of the embankments is prevented. This implies—

(a) That the spillway capacity is sufficient to pass the peak flow for which it is designed;

(b) That over-topping by wave action at maximum high water is prevented;

(c) That the original height of the structure is sufficient to maintain the minimum safe freeboard after settlement has occurred; and

(d) That erosion of the embankments due to surface run-off will not occur.

Criterion 2.—The foundation must be structurally competent to support the load of the structure.

Criterion 3.—The flow of water through the structure must not be sufficiently large in quantity to defeat the purpose of the structure, nor at a pressure sufficiently high to cause flotation (see Appendix B, sec. **24** and **25**).

Criterion 4.—The embankment must be stable under all conditions.

The various requirements under Criterion 1 are treated in other sections of the manual as follows: Spillways in chapter 5, Freeboard under that title in this chapter. The other criteria are discussed in greater detail under Soil Mechanics (see Appendix B), as follows: Settlement in section **29**, Criterion 2 in section **28**, **29** and **31**; and Criterion 4 in sections **39** to **44**, inclusive.

9. Design of Dams Impounding a Normal Depth of Water at Dam Not Greater Than 15 Feet.—*Criterion 1.*—Spillway capacity and freeboard should conform to the instructions contained in the references given above. The allowance that must be made for settlement will vary with the degree of compaction and consolidation [1] obtained in the construction. The weight of such low dams (below 15 feet) will cause a certain consolidation of the foundation but this will generally be much less than the consolidation of the structure itself, particularly where the depth to the underlying rock is relatively small. With reasonably satisfactory compaction during construction it is probably sufficiently safe to allow for a total settlement of both the structure and

---

[1] Compaction refers to increase in density due to dynamic action such as rolling, tamping, etc. Consolidation refers to the gradual increase in density due to static loads.

the foundation equal to 5 percent of the height of the structure alone.

Criterion 2.—Unless the foundation material is highly plastic the loads from these low structures will not over-stress the foundation and this criterion may be assumed to be adequately satisfied. If the structure crosses swampy or similar area where the foundation material will be of a plastic nature, it will be advisable in the general case to excavate to more suitable material. If the depth does not permit such excavation, the problem will usually require laboratory analysis. The difficulty of placing and maintaining even very low fills on a semi-plastic foundation may be observed in highway construction so widely that additional emphasis on the necessity of laboratory analysis in such cases should not be necessary.

Criterion 3.—The resistance of an earth dam to harmful percolation is developed by proper composition and the compacting of the flow retarding elements and embankments. Proper composition is secured by selection of materials and proper compacting is controlled by good construction. (See Appendix B, sec. **32**.) For many very low dams and occasionally for higher dams, sufficient suitable soil will be available so that all of the cross section, except possibly the outer layer of the downstream slope, may act to retard the flow. In other cases, particularly for higher dams, insufficient material having a low coefficient of permeability will be available for the entire cross section. In such cases, flow retarding or dense elements such as cores and blankets are introduced and the remainder of the cross section is constructed of whatever suitable material is available in sufficient quantity.

Material which will have a sufficiently low coefficient of permeability, when properly compacted, will contain a wide range of particle sizes from the coarsest to the finest. A soil which is graded from coarse sand particles to clay particles and contains not less than 20 percent by weight of the latter will usually be suitable. Soil which is made up of very uniform sand particles, whether they be coarse or fine, will ordinarily be unsuitable. Material such as clay will not be suitable for exposed surfaces because of its tendency to shrink when dried (Appendix B, sec. **11** and **15**), opening up cracks which may permit free flow through it when again submerged.

The suitability of material for flow retarding elements may be determined approximately by a simple sedimentation test on a small sample of soil. Remove all particles that will not pass through a No. 10 sieve. If a No. 10 sieve is not available a reasonably satisfactory sieve may be made of ordinary fly screen (about 14 mesh). Place the material passing through the sieve in a bowl, adding an amount of water approximately twice the volume of soil. Water-glass (sodium silicate) should be added to the water in the approximate proportion of one-half teaspoonful to a quart of water. Stir the soil and water vigorously with an ordinary eggbeater for 5 minutes. After beating, pour into a bottle having about four times the volume of the soil sample and add sufficient additional water to fill the bottle. The bottle should be straight sided preferably, and long in proportion to its diameter. After filling the bottle invert back and forth a few times to mix thoroughly and then permit the soil to settle. Not more than 75 to 80 percent of the sample should settle in the first 10 minutes. If the soil contains approximately a suitable proportion of fines the remaining 25 to 20 percent will settle during the succeeding 24 hours.

Even where the entire dam may be built of material of low permeability it is usually advisable to make the outer 2 feet of both the upstream and downstream embankments of coarser material to give greater stability. Unless a relatively impervious stratum in the foundation is at a very shallow depth it is not economical to carry the full width of a dam below the original surface except as required under Criterion 2.

A dam built of material of low permeability throughout includes the characteristics of both a dam with a core and a dam with a blanket on the upstream embankment. Therefore, in the following discussion where either a core or a blanket is recommended it must be understood that for the portion of the dam above ground the uniform material dam may be considered as equivalent to either type of construction.

There are two general classes of foundation which determine the type of flow-retarding element to use. If the depth to a relatively impervious stratum in the foundation is such that it may be reached by either a core or a diaphragm this is the most suitable element. If the depth is so great that it cannot be reached by these means, then the blanket will be the most effective except that if the foundation includes an extremely porous stratum, that porous stratum should be cut off by a core or diaphragm.

Where the core type of construction is used, the core must be carried deep enough into the relatively impervious stratum to afford an effective seal (see fig. 67). The actual dimensions of the core cut-off will depend largely on the permeability of the material. The dimen-

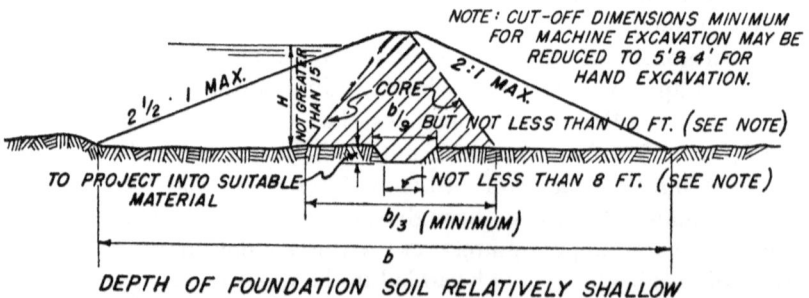

NOTE: CUT-OFF DIMENSIONS MINIMUM FOR MACHINE EXCAVATION MAY BE REDUCED TO 5' & 4' FOR HAND EXCAVATION.

DEPTH OF FOUNDATION SOIL RELATIVELY SHALLOW

FIGURE 67.

sions given in figure 68 are on the assumption that the compacted material of the core is very dense and will reduce the flow to a negligible quantity. A tough, workable clay will answer the requirement. The width dimensions should be increased as the permeability increases.

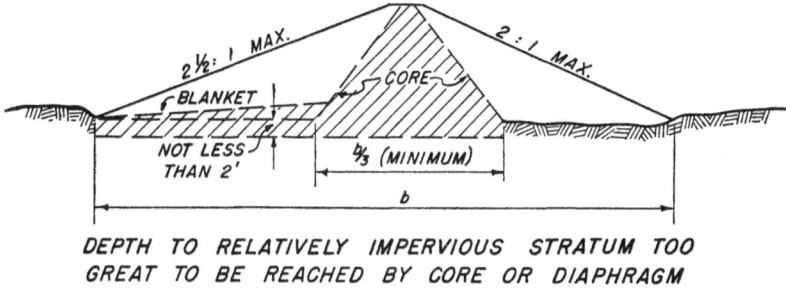

DEPTH TO RELATIVELY IMPERVIOUS STRATUM TOO GREAT TO BE REACHED BY CORE OR DIAPHRAGM

FIGURE 68

It must always be borne in mind that it is much easier to construct a core large enough in the first place than it is to repair or improve it later.

The approximate thickness of the blanket shown in figure 69 is given by the formula

$$t = 2 + 0.02d$$

where

d = distance, in feet, to any point of blanket from upstream end.

t = thickness, in feet, of blanket for very tight material.

This formula is based on material having a very high resistance to flow through it and if more permeable material is used the thickness should be increased.

NOTE.—Usually either type A or type B, figure 69, is used but not both. Type B is preferred.

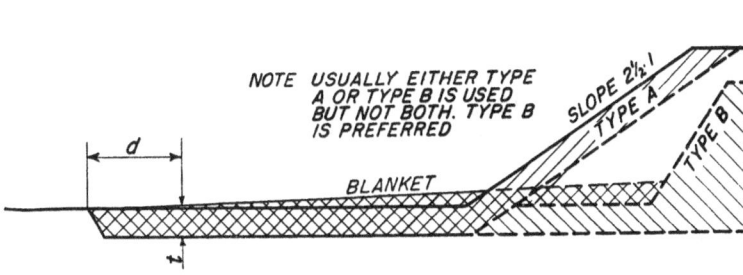

NOTE USUALLY EITHER TYPE A OR TYPE B IS USED BUT NOT BOTH. TYPE B IS PREFERRED

SLOPE 2½:1 TYPE A

TYPE B

BLANKET

FIGURE 69.

If the dam is exposed to burrowing animals it will be safer to use the core design and to protect this by a layer of stone of sufficient size to discourage burrowing. While this protective layer of stones may abut the upstream face of the core it should not be placed adjacent to the downstream face. Since the percolation velocity upstream will be determined by the core, the addition of a highly porous layer on the upstream face will not affect it. On the downstream side, however, the velocity would be increased and there would be a tendency for the core material to wash into the stone layer.

If, after the normal head has been impounded, it is found that the downstream embankment shows signs of saturation it will usually be advisable to place a rock fill at the toe as in figure 164, Appendix B. This fill need not extend as high up the slope as the saturation appears. If the lower one-third to one fourth is weighted down with rock, sloughing will usually be prevented. If this condition develops, the flow retarding blanket should also be thickened and extended upstream at the first opportunity.

Many dams, particularly those for stock watering, impound waters for only a part of the year and at other times are dry. Because of the difficulties introduced by the shrinkage of fine grained materials the blanket design is not well suited to these conditions.

Ordinarily drains in the downstream embankment should be avoided. While the drain discussed in connection with figure 167, of Appendix B, served a useful purpose without increasing the flow, it must be borne in mind that this condition is entirely dependent upon the competency of the dense elements. If the blanket or core is penetrated so as to permit flow through it the drain concurrently becomes a positive disadvantage. Drains near the surface tend to dry the outer surface. When it is considered that the surface sod is a most effective preventive of erosion, it is apparent that overdrainage may cause a more serious problem than that which it attempts to solve.

Criterion 4.—Stable embankments may be secured with most materials suitable for dam construction, if: the upstream slope is built not steeper than 2½ (horizontal) to 1 (vertical); downstream slope not steeper than 2 to 1; and the top width is made preferably 10 feet but not less than 7 feet. These specifications are applicable to dams which impound water to a normal depth of not more than 15 feet under extreme flood flow conditions.

10. Design of Dams Impounding a Normal Depth of Water at Dam Greater Than 15 Feet.—In the design of dams exceeding this height every possible advantage should be taken of the technique and principles of "soil mechanics." This does not mean that a complete laboratory analysis is either required or even justified in every case, but it does mean that the design will be based upon an understanding of the principles of soil science. The difference in application as between projects will be chiefly in the refinement of method rather than principle.

Criterion 1 —Spillway capacity is discussed in Chapter 5 and free-board under that title in section **13** of this chapter. Reference should also be made to the discussion of settlement due to the consolidation of the foundation and the structure in Appendix B, section **29**, "Soil Mechanics." The scope of the necessary investigation will vary between rather wide limits depending chiefly on the depth and character of the unconsolidated soil in the foundation. If the base of the dam rests on a suitable consolidated material such settlement as will occur must be confined mainly to the structure itself. In such cases, an allowance of 2 percent of the height for a properly compacted material will undoubtedly prove sufficiently safe. On the other hand, if the foundation soil is deep and poorly consolidated it should be investigated thoroughly. Between these two limits good judgment, based upon an adequate understanding of the principles involved, must determine the extent to which the investigation must be carried.

As an indication of the necessity of proper investigation, it may be pointed out that dam failures due to the reduction of essential freeboard through settlement are far from infrequent (see sec. **19**).

The erosion caused by surface run-off may become a serious problem, particularly in areas too deficient in rainfall during parts of the year to maintain a proper sod cover. Berms and other erosion control measures are described in many easily obtainable references which should be consulted.

Criterion 2.—The solutions to the problem of the distribution of stresses in foundations are beyond the scope of this manual.[2] It is recommended that experienced assistance be obtained, particularly in those instances in which the foundation investigation discloses a plastic stratum. In many cases it will be obvious that the empirical bearing value for the material will afford an adequate margin on the side of safety for the particular maximum load. In such cases detailed investigation is, of course, not required. The degree of refinement of investigation must be determined for each case by good judgment based upon an intelligent understanding of the principles of soil mechanics and their applicability to the problem at hand.

Criterion 3.—The problem of the rate and effect of percolation is unquestionably the most difficult and at the same time the most important problem in the design of a dam. There is a wide variation in the rate of flow for relatively simple variations in structure (Appendix B, sec. **22** to **25**, incl.). When horizontal and vertical variations in permeability of the various strata are added to variations in structure, it is obvious that only a very approximate solution may be made even with the most careful methods of investigation and analysis. The value and importance of careful design methods is

---

[2] Appendix B, sec. 44.

strikingly apparent in a tabulation of dam failures for the period extending from 1799 to 1931 inclusive.[3]   The failures due to percolation are 22 percent of the total.   Insufficient spillway capacity which caused 28 percent of the failures is the only factor which exceeds or approaches the number of failures due to faulty percolation conditions.   The latter failures, grouped according to height, were—

Height of dam:	Percent of total failures
Height not given	25
0–25 feet	25
25–50 feet	28
50–75 feet	19
75–100 feet	3
Over 100 feet	0
	100

A study of this record indicates that consideration of the percolation factor has been inadequate or entirely lacking in the design of many dams less than 75 feet high.   Rule-of-thumb and empirical methods have been, too frequently, considered adequate because of low height.

A discussion of coefficients of permeability of the more commonly used formulas of transmission of water through soils and tables on constants are given in section **22**, Appendix B, Soil Mechanics, together with illustrations of their application.   With this information an estimate of storage loss through percolation may be obtained.   The accuracy of the solution will depend upon two factors; first, the correctness of the determination of the coefficient k, and second, the validity of the assumption that the sample is representative of a certain stratum.

A large volume of percolation is not necessarily an indication of structural or economic incompetency.   Some dams, particularly those which serve to retard all or part of a flood flow, do not impound water for any extended period.   In other cases, the water supply is adequate to afford the loss without detriment to the purpose.   Under such circumstances a heavy percolation loss is of no concern provided it is at pressures which will not affect the structural competency of the dam.   In other cases, such as recreational lakes, where the maintenance of a constant shore line is an important consideration, a low rate of percolation well below any possible adverse effect on the competency of the structure may be a vital factor if the water supply to the lake is deficient during certain periods.

As is pointed out in sections **22** and **23**), Appendix B, Soil Mechanics, there is no mathematical solution for the problem of flow pressures in a complex structure.   The study of flow nets developed with the aid of a model flume may serve as a guide to the judgment.   Knowing the volume of percolation and the flow pattern, it is possible to gain some idea of the velocity of flow at critical points.   A roughly approximate estimate as an indication of whether a model test is advisable will be possible through the use of a series of flow patterns, such as those illustrated in figures 145 to 156 inclusive, Soil Mechanics, Appendix B. . Since the pattern of flow in the model will be similar to that of the prototype where the elements are proportional in linear and permeability scales, it follows that an approximate proportionality will establish at least a possibility to be either investigated further or guarded against by allowing a wide margin of safety.

The treatment of the foundation and the design of dense or flow-retarding elements will vary with the peculiarities of the site, the materials available, and the purpose of the dam.   Probably the simplest set of conditions is that in which the overburden above bed-rock is shallow and the rock surface rises above the proposed water levels at both abutments, and where material for concrete is available.   A most difficult situation is that in which the foundation is a deep, highly permeable glacial deposit containing relatively thick lenses of plastic materials, where there is a deficiency of available material suitable for the construction of dense elements, and concrete construction is out of the question.   A concrete diaphragm is clearly indicated for the first case.   This is usually the most satisfactory type of design, particularly if the diaphragm can be seated in sound rock.   In such a case, the diaphragm alone is depended upon to limit percolation.   In the example involving the deep deposit, the flow-retarding elements may project either downward or horizontally upstream, or a combination of both.   Where the deposit is so deep that a core or diaphragm can be extended economically less than half of the depth of the deposit, it is probable that an upstream blanket will be not only more effective than a core or diaphragm but will be simpler to extend if the need develops.   If, however, the underlying deposit contains a stratum of extremely high permeability such that an increase in the length of path will have only a slight effect upon the rate of flow, the upstream blanket would be ineffective and it will probably be necessary to cut off the stratum by the use of piling or deep grouting.

Criterion 4.—A stable embankment up to 30 feet in height can usually be secured, with the ordinary run of embankment material, with an upstream slope of 2.75 to 1 and a downstream slope of 2 to 1, provided there is no possibility of submergence by tail water.   A minimum slope of 2½ to 1 should be used for any part of the embankment subject to submergence.   By ordinary run-of-material is meant that having at least 50 percent by weight of its particles larger than 0.2 mm in diameter and having no unusual peculiarities.   Complete laboratory data should be obtained for all questionable materials and for all material to be used in embankments exceeding 30 feet in height.   The methods of analysis given in Appendix B, sections **41** to **43**, will give sufficient information for a preliminary design.   Final designs should be based on laboratory determinations of values of c and tan ϕ used in formulas (**73**) and (**74**) of Appendix B, section **27**.   Even where the expense of such determinations cannot be absorbed through the facilities of Federal bureaus or other public agencies, the cost should not exceed $100.

11. Design Procedure (15 Feet or Less).—The following brief outline may be used as a guide to the preparation of a schedule of design procedure for a particular job.   It must not be taken as fixed either in content or arrangement.   It is assumed that the type of dam, the maximum flow to be accommodated, the spillway elevation, and the spillway capacity have been determined, and they are not, therefore, included in the outline.

(1) Plot the cross section, making allowance for settlement, and determine maximum unit load on foundation (height of embankment $\times$ weight per cubic foot).

(2) Determine empirical or actual bearing capacity of weakest stratum and compare with actual load.   If factor of safety is less than 1.5, obtain consulting advice.

(3) Decide whether probable foundation permeability requires more detailed study.

(4) Lay out *dense elements:* core, diaphragm, or blanket.

(5) Obtain representative samples of material for dense elements and check by method given under section 9.

(6) Decide necessity of toe rock-blanket, drains, etc.

(7) Determine the necessity of adding water to the embankment materials as described in Appendix B, section **37**.

12. Design Procedure (Greater Than 15 Feet).—Since it may involve the assistance of a laboratory and other consulting advice, no specific procedure will be suggested for the design of dams in this category. A scheduled procedure should be adopted and, in general, the order of consideration followed in the treatment of soil mechanics (see Appendix B) may serve as the skeleton for such a schedule.

MISCELLANEOUS DESIGN DETAILS

13. Freeboard is defined as the vertical distance from the top of the embankment to the reservoir surface during maximum flood conditions.   This distance should be sufficient to prevent waves from overtopping the dam or from reaching portions of the crest which may be weakened by erosion or by frost disturbance.

The required allowance for waves is based on the effect of a wind of maximum velocity blowing down the reservoir toward the dam and is expressed by Stephenson in the equation

$$h = 1.5\ (D)^{\frac{1}{2}} + 2.5 - (D)^{\frac{1}{4}}$$

in which h is the height of the wave in feet from trough to crest, and D is the length of the reservoir, or the exposure, in miles. Although only one-half of this height is above the mean water level, the full height is ordinarily used to allow for the run of the waves up the slope of the dam. For reservoirs less than one-fourth of a mile long, a minimum freeboard of 2 feet is recommended.

The minimum freeboard for safety against frost disturbance depends on the depth to which frost action is noticeable. It will vary from nothing in the south to 6 feet or more in the most northerly parts of the country; 3 feet is a common value for most of the northern half of the United States.

14. Top Width.—The minimum top width may be determined by the empirical formula $W = 2(H)^{\frac{1}{2}} + 3$

in which $W =$ top width in feet.

$H =$ maximum height of embankment in feet.

Ten feet is a safe and conservative minimum top width for the lowest dams. If the top of a dam is to be used for a roadway, the minimum requirement should allow for at least a 2-foot shoulder on each side of the traveled way to prevent raveling.

15. Slope Protection.—Both the upstream and downstream slopes may require protection against damage from wave action, erosion and in some cases burrowing animals; where it is required one of the methods described in the following paragraphs may be used.

1. *Stone riprap.*—If upstream slopes are to be protected by stone riprap, it should extend from the maximum water level, including allowance for wave action, down to a few feet below the lowest expected level. For severe conditions in important dams, a shoulder or berm should be provided in the embankment at the base of the riprap.

There are various acceptable methods of laying stone riprap, the choice depending largely on the anticipated severity of the wave action and the material available. The riprap should be placed on a gravel cushion of 6 to 12 inches depth graded to a uniform surface. This type of construction gives excellent protection and a course 18 inches in thickness will withstand very severe conditions. If the stone is of a flat, stratified nature, it should be laid with the principal bedding planes normal to the slope. Rounded or irregular shaped rock is less satisfactory than squared rock and should therefore have a greater depth for the same conditions. Hand placing of the rock into a compact layer tends to reduce the amount required. Direct openings to the underlying fill should be avoided by careful placing of rock of various sizes and by closing the interstices with spalls to obtain a smooth layer of stone pavement. Rock used for this purpose must be hard and durable and not subject to slaking, solubility, or rapid weathering.

2. *Concrete.*—In locations where suitable rock is not available and where safe protection measures are indicated, reinforced concrete slabs or blocks may be used. If these are to be poured in place they should be divided into small sections, usually not larger than 8 or 10 feet square, and locked together with flexible joints that will permit some movement but prevent major displacement. Precast blocks of small dimensions can also be used, and they also should be tied together. The entire mat should be anchored into the face of the

FIGURE 70.—Riprap on upstream face of Crystola earth dam, Pike National Forest.

embankment at frequent intervals—e. g., at every third or fourth row of blocks.

Sacked concrete riprap may be used to advantage where rock is not available. While it is less expensive than reinforced concrete, it has a far less attractive appearance. A lean mixture of cement and gravel, in the proportion of approximately 1 to 8, is placed in cloth cement sacks and laid up shingle fashion on the face of the dam, the long dimension of the sack usually being normal to the axis of the dam.

3. *Willow mattress.*—There will be many small dams where rock is not available for riprap and where the cost of concrete paving is prohibitive. Protection can be effectively accomplished through the use of a mattress of willow saplings, but provision should be made for frequent replacement.

Saplings from 1 to 2 inches in diameter in lengths as great as 20 feet are assembled into bundles from 12 to 18 inches in diameter and tied with light wire. The bundles are laid on the face of the dam with the long dimension running up and down the slope, butts downhill, and woven together with a heavy wire or cable, which is anchored occasionally to a stout post set deep into the embankment, or preferably to a concrete anchor block. The tie cables should be not over 3 feet apart, and should be closer if a single-strand wire is used. The mat should have a minimum thickness of 1 foot for the smaller dams and 1½ or 2 feet for larger structures.

Figure 71 —Riprap on upstream face of small earth dam   Black's Dam, on Burney Creek, Calif

Under ordinary climatic conditions, the downstream slope can be surfaced with topsoil and planted to native vines, shrubs, or grass which form dense root growth as a protection against erosion. Trees should not be used because their heavy roots penetrate too far into the dam. In arid localities where vegetation cannot be maintained coarse gravel or loose cobblestone surfacing may be used.

In certain parts of the country there is danger that burrowing animals will open up passageways which start leakage that may ultimately result in failure of the dam. A protective layer of rock on all exposed surfaces will often help prevent this damage, but in most cases a diaphragm barrier is also advisable. (See below.)

Where trouble may be anticipated in maintaining a good protective cover on the downstream slope, the surface drainage may be improved by the use of a berm or shoulder midway down the slope and wide enough to hold a paved gutter which will intercept the drainage and dispose of it away from the face of the dam. This will seldom be required for dams of less than 30 feet in height.

4. *Fencing.*—Where cattle may graze, the growth of vegetative protection may be hindered and trails worn in which erosive action may start. Such situations will require the installation of fencing, but no fence should be built that will obstruct the passage of water over the spillway or through the spillway channel.

16. Diaphragms.—Where an earth dam is founded on rock, there is no known method of bonding the soil to the rock, and there is a tenddency for water to seep along the contact surfaces of the two materials. The danger is particularly acute if the rock surface is smooth. In cases of this kind, where the embankment is to contain an impervious core, seepage along the rock surface may be intercepted by a partial diaphragm, or low concrete wall around which the core material can be tightly compacted.

The partial diaphragm or core wall, as it is usually termed, should be located in the middle of the base of the soil core, and extend from 3 to 5 feet above the general level of the rock. It should be set deeply enough into the rock foundation to have a firm base and to prevent appreciable passage of water underneath. The top width is usually 12 inches, and the sides are built with a batter of 1 to 10 or 12 so that any settlement of the fill will tend to pack it more tightly againt the core wall. Construction joints are permitted as required and are provided with the customary keyways for interlocking the adjacent sections. Copper membrane water-sealing strips are not usually required under these conditions. The use of reinforcing steel is not required from any design loads that can be determined, but is usually used to take care of stresses that may be introduced by unequal placing or settlement of the embankment and to reduce the effect of cracking due to shrinkage of concrete. Horizontal rods one-half inch in diameter, spaced 12 inches apart, are commonly placed in both faces.

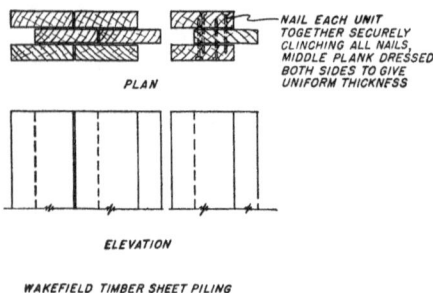

NAIL EACH UNIT TOGETHER SECURELY CLINCHING ALL NAILS, MIDDLE PLANK DRESSED BOTH SIDES TO GIVE UNIFORM THICKNESS

PLAN

ELEVATION

WAKEFIELD TIMBER SHEET PILING

FIGURE 72

The full diaphragm is used to seal the flow of water not only along the foundation but through the embankment as well. It is used with embankment soils which are not otherwise sufficiently impervious. It should be founded in undisturbed material of sufficient imperviousness to protect the reservoir against excessive water loss.

A top width of 1 foot is common for these diaphragms, the thickness being increased at the base by a batter of 1 in 20 on each face. Reinforcing steel is commonly used in both faces and in both directions, using one-half inch diameter rods on 24-inch centers in each direction. Because the diaphragm is the principal barrier to the passage of water, all construction joints should be sealed against leakage through the use of 16-ounce soft rolled copper membrane sealing strips. The 16-ounce weight is heavy enough to stand a reasonable amount of handling, but is pliable enough to facilitate handling and shaping. Splices in the copper should be soldered. The top of the diaphragm should extend at least to the maximum high-water level, and is usually stopped about 2 feet below the crest of the earth fill.

FIGURE 73 —An illustration of what might happen to Wakefield type sheet piling   This diaphragm was driven and later excavated

Partial or full diaphragms may also be made of sheet piling of either steel or wood. If either of these materials is used it should be of the interlocking type, and driven before the embankment is placed. In cases where it is to extend to the top of the dam, considerable care must be taken during the placing of the fill to avoid throwing the piling out of line by unbalanced pressures. Wood piling of the laminated type which provides a tongue-and groove joint may be satisfactory for low structures if the wood is creosoted or otherwise treated with preservatives.

If no diaphragm is required for water-sealing purposes, but there is danger from burrowing animals such as gophers, ground hogs, or mice, a treated wood barrier in the upper part of the embankment to a depth of 6 or 8 feet below the top will be adequate. When the fill has reached the proper elevation, the barrier, usually of 3-inch tongue-and-groove plank, is set up along the axis of the dam and the fill placed around it, taking particular care to compact the soil tightly against the barrier.

dense, downstream embankment porous

17. Sluiceways.—In many structures the normal location of the outlet conduit is through or underneath the embankment. When so located, any water which escapes from the conduit enters the surrounding embankment and may be the cause of failure.

Because of this danger, it is the rule in some agencies that conduits under pressure through earth dams will not be permitted, whereas others will permit them under certain restrictions. In questions of this kind, the policy of the responsible agency must be followed. The security of outlet conduits is of paramount importance and when located in the embankment, the following precautions should be taken:

1. The outlet should be founded on solid, undisturbed material which will not settle under the load of the dam and of the outlet structure.

2. The conduit must be adequately designed to resist all external and internal pressures, and be free from leakage.

3. The conduit must be built of materials which will last as long as the dam will be in use.

4. Adequate provision must be made to prevent seepage of water along the outer surfaces of the conduit into or through the dam.

These features are covered in Chapter 6, Outlet Structures, and in part in Chapter: **5-6.**

In preventing seepage along the conduit, the use of collars which

project into the fill is generally resorted to. The usual rule is to
increase the length of the path of travel along the conduit by 25 percent by adding these collars. If a conduit is 120 feet long, collars
sufficient to increase this distance by 30 feet are required. If each
collar projects 3 feet and is one foot thick, each collar increases the
distance by 6 feet, and five of them are required.

   18. Drains.—Surface drains are often required to carry off surface
run-off from the downstream face of the dam and from the banks
adjacent to it. The intersection of the face of the dam with the
abutments forms a natural gutter which often requires paving to
prevent erosion.

   Internal drains within the embankment should rarely be needed
but when necessary they should be specified with great care and placed
only within the downstream third of the embankment. If the embankment is made of suitable soil which has been well compacted
when placed, the presence of water in it is expected and will not

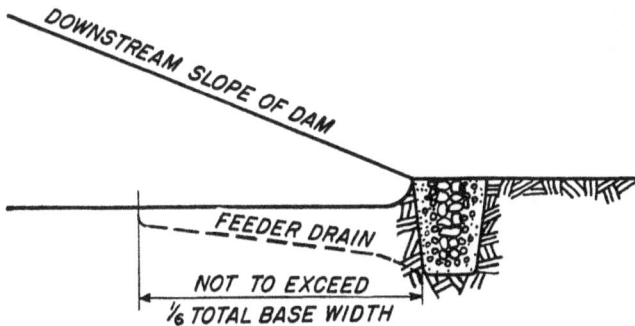

SECTION THRU TOE DRAIN
FIGURE 75.

be harmful if the dam has been designed with proper consideration
of the percolation factors. However, under the following conditions,
drains would be beneficial:

   (1) Where the percolation through and under the dam, although
safe as far as the dam is concerned, raises the level of groundwater
below the dam to a point where it may cause boggy areas. If the
percolation is collected by internal drains just inside the limits of the
dam and conducted to the old stream channel, the trouble will be
eliminated

   (2) If the embankment is founded on rock, or other very impervious
base, there may be a tendency for the water which percolates through
the core to concentrate in flow along the contact plane. A few lateral
drains leading to a header that will dispose of the flow at a safe outlet
will relieve this condition.

   The design details of the collecting drain requires special attention.
It should be built so as to act as a reverse filter so that the seepage

water may escape without erosive or piping action on the embankment
soil.   To accomplish this, the outer layers of the drain are composed
of fine gravel, graduated to coarser sizes toward the middle of the
drain.   This is illustrated in figure 75 which shows a section of a toe
drain.   A similar reverse filter should be constructed at the inlet of
interior drain pipes.

19. Causes of Failures.—The most common cause of failure of
earth dams is overtopping, caused by inadequate spillway capacity.
Other frequent causes are from piping through the dam or the foun-

FIGURE 76 —Sherrando earth dam, Va   Placing and rolling earth embankment

dation, and from weaknesses in the outlet conduit arrangement.
Table 1[4] summarizes the causes of 80 earth dam failures, most of
which occurred in the United States.   These figures emphasize the
importance of adequate spillway capacity, and of careful design and
construction of pressure sluiceways through the embankment, where
they are permitted

20. Construction.—Suggestions for specifications for earth dams
will be found in Appendix G.   Information in regard to construction
methods is also given in that appendix.   See sections G–1 to G–8
inclusive and section G–9–1   See also App. B sec. **37.**

TABLE 1.—*Causes of failure of earth dams*

Cause of failure	Number of failures	Percent of total
Insufficient spillway	31	40
Outlet pipe arrangement	14	17
Piping through dam or foundation	14	17
Miscellaneous	21	26
Total	80	100

[4] Barrows, H. K —"Water Power Engineering," McGraw-Hill Co

CHAPTER 8

ROCKFILL DAMS

1. General.—A rockfill dam consists of an embankment of loose rock provided with a dense or flow retarding element. Its similarity to the earthfill type suggests at once that similar criteria for design apply to both. In some cases a combination of earthfill and rockfill types may be used. The design of the flow retarding elements is practically the same for both types except for certain details. In general, the rockfill type is characterized by the use of large and small rock to form the greater portion of the embankment. This material

FIGURE 77.—Placing rock and articulated concrete slab on upstream face of rockfill dam. Salt Springs Dam in California

serves as a support for the dense elements which provide resistance to the passage of water through the structure. Proper construction methods are of great importance to produce a structure that will fulfill the functions which are anticipated in design.

2. Design.—Design criteria for rockfill dams are in general the same as those for earthfill dams (see ch. 7). The following paragraphs are therefore intended to point out and emphasize certain factors particularly pertinent to the rockfill dam.

1. *Foundation.*—A foundation which will allow a minimum of settlement is highly essential for a rockfill dam. It is therefore important that the structure be founded upon rock or upon well con-

153

solidated gravels or clays. Materials which tend to flow under load should be avoided as foundations.

2. *Materials.*—Soils used for any part of a rockfill dam should meet the specifications, for the purpose which they serve, as indicated in Appendix B. Rock must be hard, durable, and have the necessary composition to withstand the effects of exposure to air and water. Large rock is usually used for the lower courses and in general, the rock is dumped into the fill as it comes from the quarry except in rather rare cases where hand placement is used. Dumped rock fills cannot be placed in as dense and compact a condition as can finer materials and consequently a great deal of settlement must be anticipated both during and after construction. The after-settlement presents a dangerous situation as it is apt to cause rupture or disintegration of the flow retarding element. Provision for such settlement should be made in the design and also in connection with the construction operations. The porous nature of a rockfill should also be considered when the flow-retarding element is composed of soil. An inverted or reverse filter may often be used to advantage between the earth fill or blanket and the rockfill to prevent piping of the impervious material into and through the rockfill. Sand and gravel or rock, graded from fine to coarse, may be interposed between the two elements for this purpose.

3. *Cross section.*—The top width of the dam is dependent upon the height and may be determined in the same manner as for earth dams. Structures within the scope of this manual (30 feet or less of normal water depth) should have a top width of 10 to 14 feet unless a roadway requirement demands greater width. Side slopes for rock fills on the downstream side should have a minimum steepness of from 1¼ (horizontal) to 1 (vertical) to 1½ to 1. Side slopes of ¾ to 1 on upstream slopes are not uncommon where provision is made to prevent raveling and in some cases where a timber or concrete deck is used as a flow retarding element. When the upstream portion of the embankment is an earthfill (fig. 78) the limitations for slope given in Chapter 7 and Appendix B should prevail.

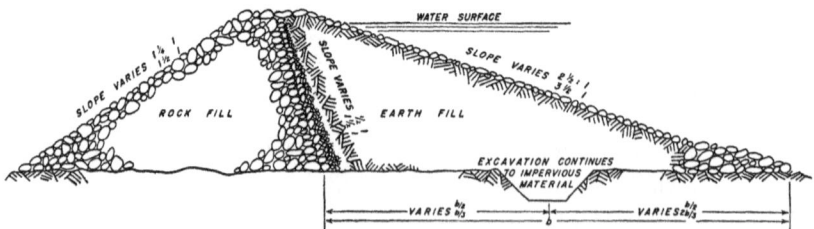

ROCKFILL DAM WITH EARTH BLANKET
FIGURE 78.

4. *Flow-Retarding Elements.*—The two principal types of retarding elements in common use are (a) the earth blanket and (b) the articulated concrete slab. A third type (c) the timber deck is occasionally used.

(a) *The earth blanket.*—Where suitable soil is available, a blanket or toe-fill against the upstream slope of the rockfill will usually prove to be the most economical (fig. 78). The soil must be of a grade suitable for selected core material for earth dams as described in chapter 7, and should be subjected to the same tests (see also Appendix B, Soil Mechanics).

The minimum thickness of the earth blanket will depend on the percolation, which must be low enough to prevent loss of soil particles from the fill and excessive loss of water from the reservoir. The usual slope of 2½ to 1 or 3½ to 1 is used for the upstream face, which must be protected by riprap above the low water line, as in earth dams.

The placing of the earthfill should be given the same careful attention as the impervious portions of earth dams, using the same moisture control and rolling methods to obtain density and compaction.

The foundation under the earthfill should be prepared in the same manner as for earth dams under similar conditions, using even greater care, because the base width is usually less in proportion to the water depth. If the foundation is smooth rock, a concrete core wall is usually necessary to prevent seepage along the rock surface. A wall 3 or 4 feet high, located near the middle of the base of the earth blanket should be sufficient. The sides of the wall should be battered moderately, and special care must be taken to tamp the fill against both sides.

If the foundation is clay or gravel, a core trench similar to that used in earth dams is required. It is usually located near the middle of the earthfill base, and is deep enough to secure an adequate seal. These details are illustrated in figure 78.

As previously stated, special provision should be made to prevent piping of the materials of the earth blanket into and through the rockfill.

(b) *The articulated concrete slab.*—If suitable soil is not available, the upstream slope may be covered with a thin slab of concrete which is sealed into the foundation at the base and sides. The slab is divided into rectangular sections with flexible joints between, so that settlement of the rockfill will not cause rupture and leakage.

Figure 79 shows a typical design of this type; it also shows a concrete parapet wall at the top which is sometimes used to give additional freeboard.

This type of watertight membrane is most effective when it can be sealed into a rock foundation. A cut-off trench is excavated into the rock to a depth of 2 to 6 feet or more, as may be necessary to give an adequate seal against leakage.   Any seams or fissures below the trench are sealed by pressure grouting after the concrete has been poured.   If a concrete deck is to be used above a pervious gravel or other uncemented soil of considerable depth, an effective water cut-off can often be secured by the use of interlocking or Wakefield type of sheet piling, in which case the concrete sill at the base of the deck is poured around the top of the piling to make the seal complete.

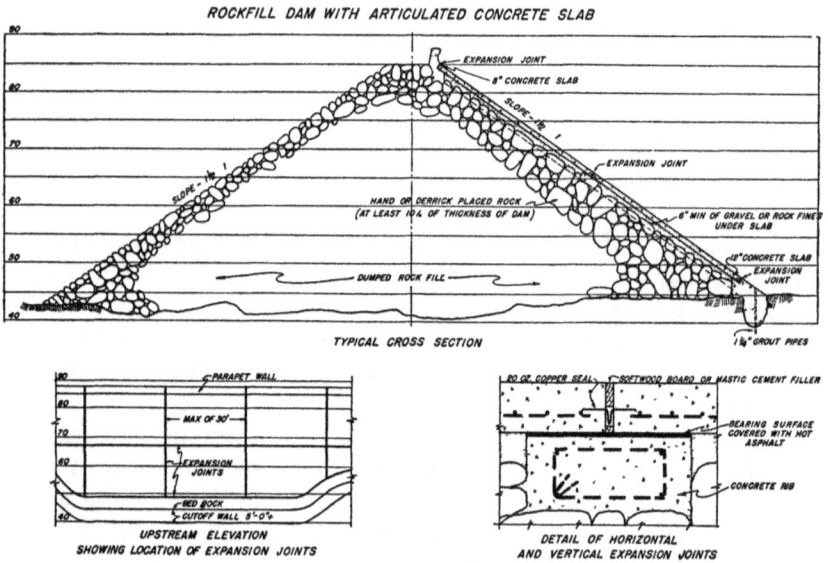

ROCKFILL DAM WITH ARTICULATED CONCRETE SLAB

FIGURE 79 —Rock-fill dam with earth blanket.

(c) *The timber deck.*—A timber deck for sealing rockfill dams against leakage is not recommended for general use.   However, it is often the cheapest type of construction, and may have occasional advantageous application.   The use of treated timber will increase the material cost but can often be justified by longer life and less maintenance expense.   Pressure treatment should be required, using creosote or other approved preservative.[1]

The deck plank, either in single or double layers, is spiked to heavy timber sills, which are buried in the upstream face of the rockfill. Floating of the deck is prevented by fastening the sills to posts buried deep in the fill.   The spacing of the sills depends upon the strength of the deck planks, which should be designed to carry the full water pressure on the span between the sills without support from the rock-fill.   At the base of the slope, the deck rests on a heavy longitudinal

---

[1] Consult the Forest Products Laboratory, Forest Service, Dept of Agriculture for latest developments in wood preservatives and treatments.   See also "Manual on preservative treatment of wood by pressure" U. S. Dept. of Agr Pub. No 224 by the same bureau.

sill, to which is secured vertical sheathing which has been driven into the foundation.   If the foundation is rock, a concrete sill may be used. These features are illustrated in figure 80.

5. *Ample freeboard and spillway* capacity are necessary to prevent overtopping of the dam, from floods and wave action (see ch. 7).

6. *Outlet conduit.*—The design and location of the outlet conduit requires special consideration, because of possible damage from the

ROCKFILL DAM WITH TIMBER DECK

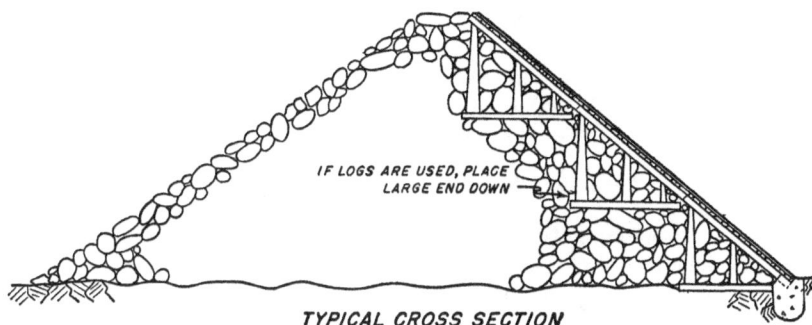

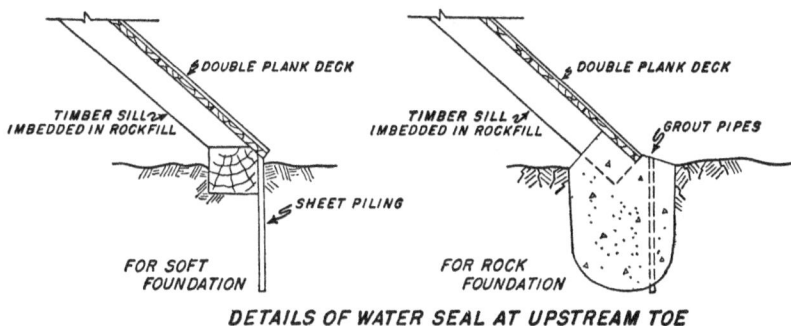

IF LOGS ARE USED, PLACE
LARGE END DOWN

TYPICAL CROSS SECTION

DOUBLE PLANK DECK

DOUBLE PLANK DECK

TIMBER SILL
IMBEDDED IN ROCKFILL

TIMBER SILL
IMBEDDED IN ROCKFILL

GROUT PIPES

SHEET PILING

FOR SOFT
FOUNDATION

FOR ROCK
FOUNDATION

DETAILS OF WATER SEAL AT UPSTREAM TOE

FIGURE 80.

impact of the rock during the filling operation.   For the larger dams, a tunnel through rock around one end of the dam will be the more satisfactory outlet structure from this viewpoint, although it will also usually be the more expensive solution.   A thick and well reinforced concrete conduit built on a solid foundation will be satisfactory if the fill is placed carefully until the conduit is well covered.

3. Construction.—Suggestions for specifications for rockfill dams will be found in appendix G.   Information about construction methods is given in the same appendix.   See sections G–1 to G–8 inclusive and section G–9–2.

BIBLIOGRAPHY

I. C. Steele, "High Rockfill dam designed with jointed concrete face; Salt Springs Dam in California 328 feet high," *Engineering News*, Jan. 16, 1930.

H. K. Barrows, *Water Power Engineering*, McGraw-Hill Book Co.

Wm. P. Creager, *Hydroelectric Handbook*, John Wiley & Sons.

Edw. Wegmann, *The Design and Construction of Dams*, John Wiley & Sons.

"Bonita Rockfill Dam, New Mexico," *Western Construction News and Highway Builder*, Sept. 10, 1932.

"Failure of Castlewood Rockfill Dam," *Western Construction News and Highway Builder*, Sept. 1933.

"Preliminary Work on the World's Largest Rockfill Dam," *Western Construction News*, May 1934.

"Data on Castlewood Dam Failure and Flood," *Engineering News-Record*, Sept 7, 1933.

"Construction, Subsidence, and Repair of San Gabriel Dam No. 2," *Engineering News-Record*, Mar. 7, 1935.

"Corewall in Rockfill Dam Tilts When Reservoir Fills," *Engineering News-Record*, Nov. 3, 1932.

"San Gabriel No. 2—A Rockfill Dam of Unusual Interest," *Excavating Engineer*, May 1933.

"Flexible Timber Facing Applied to San Gabriel Dam No. 2," *Engineering News-Record*, June 13, 1935.

"Build Dams With Material at Hand," *Engineering News-Record*, Mar. 28, 1935.

"Not a Rockfill Dam," *Engineering News-Record*, Oct. 19, 1933.

"The Design of Rockfill Dams," by J. D. Galloway, *Proceedings, Am. Soc. C. E.*, Oct. 1937 and subsequent discussion.

CHAPTER 9

CONCRETE AND MASONRY GRAVITY DAMS

1. Definition.—The gravity dam utilizes the weight of the masonry composing it to resist the forces exerted upon it. The term masonry includes rubble or ashlar stone laid in mortar, though in most instances concrete is now used. There is no type of dam more nearly permanent and none that requires less maintenance than the solid gravity dam.

2. Forces Acting on the Dam.—In the design of dams, it is first necessary to determine the forces acting on them. The forces which must be considered for gravity dams within the range of this manual are those due to (1) water pressure, (a) external, (b) internal or uplift; (2) silt pressure; (3) ice pressure; (4) earthquakes; (5) weight of the structure; and (6) reaction of the foundation.

Other forces include impact, wind, waves, and unbalanced atmospheric pressures; the first three are negligible for small dams, and the last can be avoided by the use of proper design provisions.

The nomenclature used in this chapter is as follows:

w=unit weight of water; assumed at 62.5 pounds per cubic foot.

w_1=unit weight of masonry.

W=total weight of masonry structure.

P=horizontal force due to water pressure.

P_i=horizontal force due to ice pressure.

P_e=horizontal force due to earth pressure.

R=resultant of vertical and horizontal forces.

R_v=vertical component of resultant.

V=vertical force due to water pressure.

h=head or depth of water below the reservoir surface.

H=head or depth of water above the crest of an overflow dam.

a=distance from maximum water surface to top of dam (freeboard).

g=32.2 feet per second per second—the acceleration due to gravity.

f=coefficient of static friction.

e=eccentricity of loading.

w_2=unit weight of silt or other soil.

w'_2=unit weight of submerged silt or other soil with a dry weight w_2.

h_2=total depth of silt deposit.

a=angle of repose of soil.

k=percentage of voids in soil.

θ=angle of inclination of the resultant with the vertical.

P_g=horizontal inertia force of the dam.

$$\lambda = \frac{\text{Maximum horizontal acceleration of the foundation}}{g}; \text{ usually}$$

assumed to be 0.1.

b=width of base of dam (see fig. 81).

e=distance from midpoint of base to point of application of resultant of horizontal and vertical forces (see fig. 81).

n=one-third of width of that portion of base of dam subject to compressive forces.

f=coefficient of static friction.

t=the effective top width of a trapezoidal dam section.

s=specific gravity of the masonry.

Q=discharge in cubic feet per second.

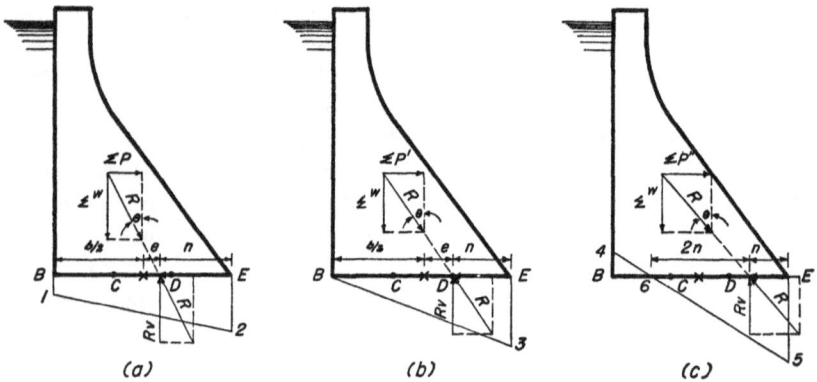

FIGURE 81.

1. (a) *Water pressure.*—The *external water pressure* is due to the weight of the water and the depth. The weight of water is generally accepted as 62.5 pounds per cubic foot, so that the unit pressure at any point is $62.5h$ pounds per square foot. For a nonoverflow dam, the total pressure on a vertical section of dam that is 1 foot long, is $P=1/2 wh^2$, which acts at a depth of $2/3h$. If the shape of a dam is such that water is vertically above part of the dam, there will be a vertical pressure acting downward, as in figure 82 (a). The total load V is equal to the weight of the water vertically above surface 1–2, acting through the center of gravity of the figure 1–2–3–4.

The horizontal pressure is represented by the pressure diagram 5–6–8–7 in figure 82 (b), in which the unit pressure at the top is $62.5H$, and at the bottom is $62.5h$. The total pressure is represented by the area of the trapezoid 5–6–8–7, and its line of action passes through its center of gravity.

On overflow dams without control features, the vertical component of the water which is flowing over the top of the spillway is not used in the analysis because the water approaches spouting velocity, which greatly reduces the vertical pressure on the dam. Likewise, because of its high velocity the stream of water on the downstream face does not exert enough pressure on the dam to warrant consideration.

The question of backwater pressure exerted by tail water on the downstream side of a dam is determined in the same manner as for water pressure on the upstream face. In the case of the spillway section shown in figure 82 (b), the depth of tailwater in contact with the dam is greatly reduced by the impact of the water jet, and the backwater pressure on the structure is negligible.

Where control features consisting of flashboards or crest gates are used, they are treated as part of the dam as far as the application of water pressure is concerned.

1. (b) *Internal or uplift pressure.*—All rock or earth foundation materials are porous, although the degree of porosity varies with different materials. Some foundations contain additional void space due to seams and fissures.

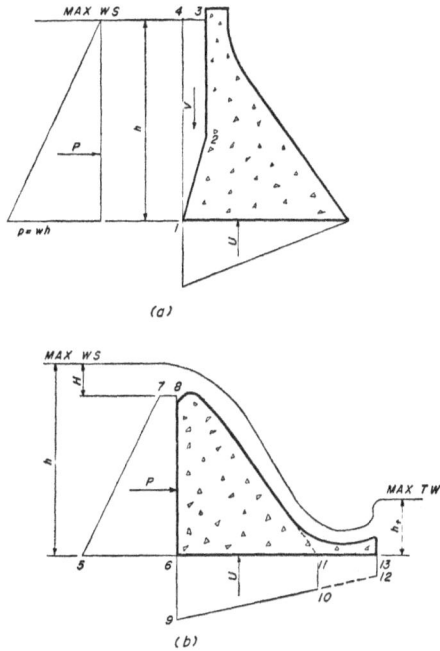

(a)

(b)

FIGURE 82.

It is evident that these spaces in a dam foundation will be filled with water from the reservoir, and that it will exert its normal pressure in all directions. The component which acts vertically upward will have an important effect on the stability of a dam, and so must be included in the analysis.

As the water travels from the upstream face of the dam to the downstream toe, the pressure is usually assumed to reduce uniformly from that due to the head in the reservoir to that due to the head at the downstream toe of the dam. In the case of overflow dams, the tailwater pressure may be considerable, but in non-overflow dams there usually is no tailwater, in which case the pressure is assumed to be zero.

The area over which the uplift pressure is effective will depend on the porosity of the foundation material, and the existence of seams, fractures, and fissures. Pressure can be exerted only on the portion of the base that is occupied by water. This is generally treated as a percentage of the total area, the factor ranging from 100 percent to zero. For earth foundations, 100 percent should be used. For rock foundations the usual allowance is between 50 and 67 percent of the total base area, and these values are recommended here for the ordinary run of small projects. In practice this percentage is considered applicable to each square foot of the whole base area. The effect of different percentages of uplift pressures upon the cross section of the dam may be determined from a study of figure 81.

For foundation materials that have a high degree of porosity, such as loose grained sandstone or for any rock which is seamy or fractured, the uplift pressure allowance at the upstream face of the dam should be $0.67wh$ per square foot. At the downstream toe the pressure will be fully or partially relieved, depending on the condition of tailwater.

In figure 82 (a) there is no backwater, and the uplift pressure at the downstream toe is zero. The uplift pressure diagram is a triangle, and the maximum pressure is uwh, in which u represents the percentage factor. When tailwater backs up to the downstream toe, as in figure 82 (b), the uplift pressure at the edge of the dam cannot be less than the pressure due to the water immediately above it. When a dam has an apron or bucket as shown in the drawing, the apron is not included in the analysis for stress and stability, the downstream slope being continued to intersect the base, as at 11. The total uplift on the base 6–11 is represented by the area of the pressure diagram 6–9–10–11, in which the unit uplift pressure at 6 is uwh and at 11 is uwh_t. The line of action of the total uplift U passes through the center of gravity of the trapezoid 6–9–10–11.

The choice of a value of u between the limits of 0.67 and 0.50 is a matter of judgment and experience for which no rules can be laid down. Mr. Arthur P. Davis [1] has made the following observations on relative perviousness of dam foundations:

"The determination of the perviousness of natural formations is one of the most difficult things in nature. Any examination of such formations, which disturbs them, changes the conditions which it is desired to know. For this reason, it is necessary to allow a large factor of safety in any estimates which involve this factor.

"In general, it may be said that water will more readily follow seams or bedding planes than devious paths through the material of the rock. It follows that it will pass more readily and in larger volume in the direction of stratification than in a direction normal thereto. Similarly, stratified rock will permit percolation more easily and in greater volume than good massive rock, such as granite.

"Granular rock, such as sandstone, is likely to transmit more water through the rock itself that one of denser or finer grain, such as lime-

[1] Transactions, Am. Soc. C. E , vol. 75, p. 208

stone or shale, but no exact rule of this nature can be laid down, because there are many varieties of each kind of rock, with various percolating capacities.   In general, however, the following rules may be taken as a rough guide:

"1. Massive or crystalline rocks, such as granite, gneiss and schists will transmit water less freely than those of sedimentary origin.

"2. Stratified rocks will transmit water much more readily in the direction of stratification than transverse thereto.

"3. In the direction normal to stratification, sandstone will generally transmit water more readily than limestone, and the latter more readily than shale.

"4. Stratification on a plane approximately horizontal is the worst possible condition for introducing upward pressures beneath a dam. Conversely, the most favorable position in this respect for stratified rock is in vertical beds."

Uplift pressure may also be exerted within the dam, either at or between horizontal construction joints or in cracks in the masonry. It is customary to make the same assumptions for all horizontal sections, whether at the base or within the body of the dam.

Three methods of limiting the amount of uplift pressures are used in practice:

1. The use of cut-off walls under the upstream toe, which increase the length of the path of percolation through the foundation and thereby reduce the uplift pressure.

2. The use of drainage channels between the foundation and the dam in consolidated foundations.   They are usually located between the upstream third point and the cut-off and drain toward the downstream toe.

3. The foundation rock is sealed with cement grout under pressure, which closes some of the porous and seamy channels through which the water could flow.

A conservative and safe attitude toward these practices is to use them whenever they are feasible, but to consider them as additional factors of safety in the design, rather than to use them as reasons for reducing the design requirements.

2. *Silt pressure.*—Nearly all streams carry an appreciable amount of silt during floods, and some streams carry large amounts even when the flow is normal.   Where silt is present in a stream, it will eventually find its way to the reservoir and will be deposited in the still water above the dam.   If allowed to accumulate against the upstream face of the dam there will be additional pressure because of the additional weight of the silt in the water.   Sluiceways can be built through the lowest point of a dam, to be opened periodically to sluice out the accumulated deposits.   This arrangement, if effective, will limit the depth which the silt can build up to cause pressure against the dam,

but it depends on manual operation which cannot always be relied upon to function when needed. In cases of this kind, it is safest to design the dam for a silt deposit which might occur if the sluice gates were not used.

A determination of the horizontal component of earth pressure, P_e, exerted on a dam may be made from Rankine's equation:

$$P_e = \frac{w_2 h^2{}_2}{2} \frac{(1-\sin \alpha)}{(1+\sin \alpha)} \quad (1) \text{ (See Appendix B)}$$

This equation uses soil functions which are commonly known or can be estimated within conservative limits to give sufficiently accurate results for this purpose.

The pressure, P_e, acts at a distance of $\frac{2}{3} h_2$ below the surface of the silt deposit. The term w_2 is the unit weight of the silt of other soil material, in the dry state if the deposit is on the downstream side, and in the wet state if it is on the upstream side of the dam. The submerged weight, w_2', of a soil with dry weight w_2 and k percentage of voids is

$$w_2' = w_2 - w(1-k) \tag{2}$$

For example, a dry soil weighs 106 pounds per cubic feet in air and has 32 percent voids.

$$w_2' = 106 - 62.5(1-0.32)$$

$$= 63.5 \text{ pounds}$$

The values of w_2, k, and α vary for different materials, and should be determined for each case for accurate results. The common range of values for w_2' is 60 to 70 pounds per cubic foot. For coarse sand, the angle of repose, α, is about 30°. For finer material, this angle decreases and may become zero for very fine silts and clays that are very plastic under water. When $\alpha = 0$, the material is commonly called liquid mud because in this state it follows the laws of fluid pressure.

Material of this kind is usually very impervious, and it is reasonable to believe that it will prevent percolation at the base, without which there will be little or no uplift pressure. The common practice, where impervious silt deposits are anticipated, is to design for these two conditions independently and use the one which gives the greatest dam section.

Table 1 gives approximate average values for the angle of repose, unit pressure, and total horizontal pressure, P_e, for the common classification of materials. Such a table should be used with extreme caution because of the difficulty in determining the true character of the material and the correct angle of repose.

TABLE 1.—*Horizontal silt pressure on dams*

Soil	Angle of repose (α)	Unit pressure	Total pressure (P_s)
	Degrees	*Pounds per square foot*	*Pounds* [1]
Liquid	0	70 $0h_2$	35 $0h_2^2$
Clay	5	58 $7h_2$	29 $4h_2^2$
Silt	10	49 $2h_2$	24 $6h_2^2$
	15	41 $2h_2$	20 $6h_2^2$
Sand	30	34 $4h_2$	17 $2h_2^2$
	35	28 $3h_2$	14 $2h_2^2$
Gravel	30	23 $3h_2$	11 $7h_2^2$

[1] Per foot of length of dam.

3. *Ice pressure.*—The necessary allowance to be made for ice pressure is another indeterminate factor. No dams are built to withstand the maximum thrust that expansion from freezing action could exert. This maximum thrust is approximately 70 tons per square foot, or 140 tons per linear foot of dam with an ice sheet 2 feet thick. This pressure could occur only if the ice sheet were restrained on all sides and the span were so short that it would not buckle. This case is far different from the ordinary conditions, where the distance from the dam to the opposite shore line is great in proportion to the ice thickness and where the contact of the ice sheet with the shore slopes in such a way that the thrust would be relieved by movement of the ice up the bank. Furthermore, a fluctuating water surface frequently prevents the ice sheet from making contact with the bank.

All these factors indicate why it is not necessary to provide for the maximum pressure. In many localities no provision for ice pressure is needed because freezing temperatures do not prevail long enough to form thick ice sheets. Where the ice sheet will not exceed 2 feet in maximum thickness, its pressure on the dam need not ordinarily be considered. Where this thickness exceeds 2 feet, values of from 10,000 to 24,000 pounds per linear foot of dam should be used if the other contributory factors indicate the need. In many storage reservoirs, the water level will be relatively low during the winter months, and the ice thrust, if present, will be applied at a considerable distance below the top of the dam.

4. *Earthquakes.*—The design of dams for earthquake effects is highly specialized and highly technical. It should never be attempted by an inexperienced designer but must be considered when the structure is to be located in a region where earthquakes are common, particularly if failure would involve loss of life. The following notes are intended to give an idea of the design approach.

The additional loads which earthquakes exert on dams are due (1) to accelerations of the mass of the dam and (2) to the changes of water pressure. For the maximum load condition the pressures caused

by horizontal acceleration are assumed to act normal to the face of
the dam, or at right angles with the axis of a straight dam. The
vertical acceleration loads may be neglected for small dams within the
scope of this manual.

The load on a dam due to the acceleration of its mass imparted by
the oscillating action of the earthquake is represented by the horizontal
inertia of the dam. It is determined from the equation

$$P_q = W \ (\lambda) \qquad\qquad (3)$$

in which P_q is the horizontal inertia force of the dam, W is the weight
of the dam, and λ is the maximum horizontal acceleration of the
foundation divided by g; λ is usually assumed to be 0.1. This force
acts through the center of gravity of the cross section.

There is also an increase in horizontal water pressure against the
dam as a result of the earthquake motion. The magnitude of this
pressure varies as a function of certain characteristics of the earthquake, the two most important ones being the frequency of vibration
and the amplitude. The diagrams of figure 98 show the amount and
distribution of water pressure caused by earth tremors as computed
from approximate formulas developed by Prof. H. M. Westergaard.[2]

Other minor changes in loading which may be introduced by
earthquakes include the effect of vertical acceleration, the vertical
component on inclined surfaces of changes in water pressure, and
changes in the uplift pressure from the same source. These forces
are not usually included in the analysis, as they are negligible.

The most severe assumption which could be made would be that in
which the earthquake loads occurred at the time of maximum flood.
However, for gravity dams, the usual assumption is that the earthquake shock may occur when the reservoir is at normal operating level.

5. *Weight of the Structure.*—The unit weight of masonry and concrete varies with the stone of which it is made. Table 2 gives the
average weight per cubic foot of masonry and concrete made of the
common kinds of stone.

TABLE 2 —*Average weight of masonry and concrete*

[Pounds per cubic foot]

Stone	Rubble masonry	Ashlar masonry	Concrete
Sandstone	130	140	140
Limestone	150	155	145
Granite or trap	155	160	155
Granitic gravel			150

[1] "Water Pressure on Dams During Earthquakes", by H. M Westergaard, M. Am Soc. C. E , in Trans.
of Am. Soc C E , Vol 98, 1933, p 413.

In the analysis of a dam, the weight of the masonry is assumed to act vertically downward through the center of gravity of the cross section.

6. *Reaction of the Foundation.*—Under stable condition the resultant of the horizontal and vertical loads on the dam will be balanced by an equal and opposite force which constitutes the reaction of the foundation. This can be resolved into a vertical load causing compressive stresses and a horizontal force introducing shearing and frictional resistance, as illustrated in the diagrams of figure 81. The points C and D divide the base into three equal parts, CD representing the middle third and BE the total width of the base b. Assuming a straight-line variation, the stress distribution is represented by the diagrams (a), (b), and (c) for three different locations of the intersection of the resultant with the plane of the base.

In (a) the resultant falls downstream from the midpoint but within the middle third. The stress diagram is represented by the trapezoid $B1\ 2E$. For a dam of unit length, the stresses $B1$ and $E2$ are determined by the use of eccentric loading formulas. For case (a) the stress is represented by

$$B1 = \frac{R_v}{b}\left(1 - \frac{6e}{b}\right) \tag{4}$$

$$E2 = \frac{R_v}{b}\left(1 + \frac{6e}{b}\right) \tag{5}$$

Case (b). $E3 = \frac{2R_v}{b}$ $\tag{6}$

Case (c). $E5 = \frac{2R_v}{3n}$ $\tag{7}$

Tension should not exist at the upstream toe of a dam, so that in all cases the resultant must fall within the limits of the middle third of the dam. In equation (7) the compressive stress is determined without consideration of any of the load being carried by tension.

3. **Requirements for Stability.**—A gravity masonry dam should be designed to resist, with ample factor of safety, these three tendencies to destruction: 1, overturning; 2, sliding; and 3, overstressing.

1. *Overturning.*—There is a tendency for a gravity dam to overturn about the downstream toe at the foundation, or about the downstream edge of any horizontal section. Overturning will actually occur if the resultant of the external forces (exclusive of the foundation reaction) passes outside of the base and there is no provision to resist tensile stress at the upstream toe. As previously stated, it is the custom to design the dam so that there will be no tension under the most severe loading condition. When this provision is met, there will be ample factor of safety against overturning.

2. *Sliding.*—The horizontal force, ΣP, is the component of the resultant (see fig. 81) which tends to displace the dam in a horizontal direction. This tendency is resisted by the frictional and shearing resistance of the masonry or the foundation. If f represents the static friction between two sliding surfaces, the resistance to sliding is fW, which, in order to maintain equilibrium, must be equal to or greater than P, the horizontal force.

$$f \Sigma W = \Sigma P \qquad (8)$$

$$f = \frac{\Sigma P}{\Sigma W} \qquad (9)$$

But $\frac{\Sigma P}{\Sigma W}$ is the tangent of the angle of inclination (θ) of the resultant, R, with the vertical. Therefore, tan θ must be less than or equal to the coefficient of static friction between the upper and lower surface of horizontal joints and at the foundation.

For concrete on concrete, masonry on masonry, and concrete on rock, tan θ should not exceed 0.75. The factor of safety is increased by the mechanical resistance of irregularities in the surface of the foundation and by the stepped surfaces of construction joints, but this does not usually permit any reduction in the value of tan θ.

When masonry dams are built on other than rock foundations the allowable value of tan θ is greatly reduced, depending on the values of f and the safety factor assumed.

In table 3 are given the values of f for gravel, sand, and clay foundations, and corresponding values of tan θ to obtain a factor of safety of 2½.

TABLE 3.—*Values of f and tan θ for earth foundations* [1]

Material	f	Safety factor	Tan θ
Concrete or masonry on gravel	0 5	2 5	0 20
Concrete or masonry on sand	.4	2 5	16
Concrete or masonry on clay	.3	2.5	12

[1] From Masonry Dams, William P. Creager.

3. *Overstressing.*—The unit stresses in the masonry and at the foundation should be kept within prescribed maximum values. These stresses include vertical and inclined compression, horizontal shear, and tension in vertical planes near the downstream toe.

A conservative limit for the vertical compressive strength is that it shall not exceed one-twentieth of the ultimate 6 months' compressive stress of the masonry. The method for determining the intensity has been given. Since this is derived only from the vertical component of the load, it is obvious that the horizontal component will introduce stresses on an inclined plane which exceed the vertical stresses. However, for dams up to 30 feet maximum height, a cross section which is safe against sliding and which keeps the resultant within the middle third will have compressive stresses which will not exceed allowable conservative values. Where the foundation is on sound rock, the crushing strength of the rock will frequently exceed that of the masonry, so that the strength of the latter is the controlling element. There will be some locations where the permissible bearing pressure on the foundation will be less than that of the masonry, in which case it will be necessary to determine safe values by field investigation and test.

It is evident that a horizontal shearing stress is introduced by the horizontal component of the resultant. The shearing resistance of the masonry will obviously exceed the frictional resistance of plane surfaces, so a section which is safe against sliding will be well within the safe limits of horizontal shear requirement. The usual limiting value is 500 pounds per square inch for concrete or rock.

Tension in vertical planes near the downstream face of a dam may be introduced if the slope of the downstream face is too flat to permit transmission of the load by shear and compressive resistance. A dam will be safe against this condition if the angle the downstream face makes with the horizontal is 45° or more.

4. Design of Nonoverflow Dams.—The design of nonoverflow dams is started at the top, assuming an elevation which gives safe freeboard above maximum high water in the reservoir. The free board should be sufficient to allow for maximum wave height, as determined by the Stephenson or other acceptable formula.[3] An approximate value of the proper freeboard to allow is $\sqrt{h}$, where h is the height of the dam. The width of the top is somewhat dependent on the height, but may be determined in some instances by other requirements, such as the width needed for travel across the dam or for access to gate operating mechanism. Ordinarily, a ratio of 15 percent of the height will be satisfactory for low dams.

The width of the base and the slope of the downstream face will be determined by the loading assumptions and are subject to analysis for sliding and overturning resistance. The usual method is to assume a section and modify it as required by the analysis until it meets the requirements. Figure 84 shows an analysis of a typical section. Note that it is necessary to batter the upstream face to eliminate tension at the downstream toe when the reservoir is empty. In cases where a nonoverflow dam is used in combination with an overflow dam or spillway section, the slope of the downstream face of one or the other is sometimes increased so that the two coincide.

Figure 86 shows relative dimensions for triangular gravity dam sections for various uplift assumptions from zero to full uplift. This figure is useful for preliminary estimates and to show the effect of uplift upon the cross section of the dam.

---

[3] See ch 7:13.

FIGURE 83 —Coursed masonry, overflow gravity dam No. 1 for flood control, on Fountain Creek, Pike
National Forest.

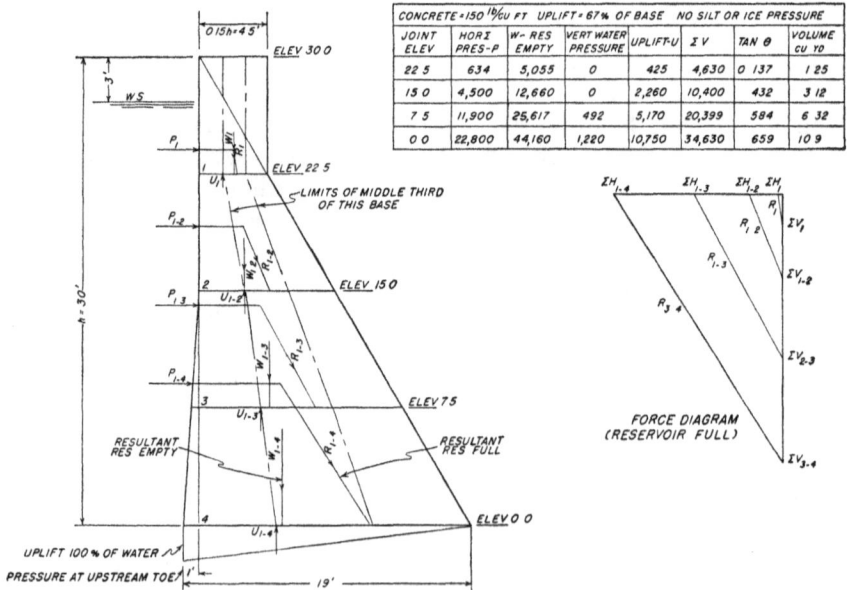

CONCRETE=150 lb/cu FT UPLIFT=67% OF BASE NO SILT OR ICE PRESSURE							
JOINT ELEV	HORI PRES-P	W-RES EMPTY	VERT WATER PRESSURE	UPLIFT-U	Σ V	TAN θ	VOLUME CU YD
22 5	634	5,055	0	425	4,630	0 137	1 25
15 0	4,500	12,660	0	2,260	10,400	432	3 12
7 5	11,900	25,617	492	5,170	20,399	584	6 32
0 0	22,800	44,160	1,220	10,750	34,630	659	10 9

STABILITY ANALYSIS OF NON-OVERFLOW GRAVITY DAM

FIGURE 84.

5. Design of Overflow Dams.—The basic section of an overflow dam will be a trapezoid, with the top surface modified to give increased discharge capacity and to cause the overflow sheet to adhere to the face of the dam at all stages.

FIGURE 85.—Laurel Hill reclamation demonstration project, Pennsylvania, SP-8   Jones Mill Run, masonry gravity dam with naturalistic stone facing.

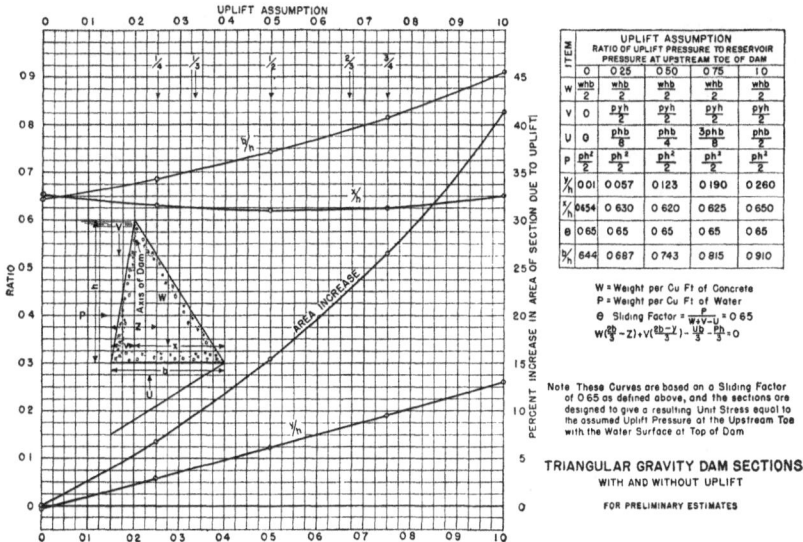

ITEM	UPLIFT ASSUMPTION RATIO OF UPLIFT PRESSURE TO RESERVOIR PRESSURE AT UPSTREAM TOE OF DAM				
	0	0 25	0 50	0 75	1 0
W	$\frac{whb}{2}$	$\frac{whb}{2}$	$\frac{whb}{2}$	$\frac{whb}{2}$	$\frac{whb}{2}$
V	0	$\frac{pvh}{2}$	$\frac{pvh}{2}$	$\frac{pvh}{2}$	$\frac{pvh}{2}$
U	0	$\frac{phb}{8}$	$\frac{phb}{4}$	$\frac{3phb}{8}$	$\frac{phb}{2}$
P	$\frac{ph^2}{2}$	$\frac{ph^2}{2}$	$\frac{ph^2}{2}$	$\frac{ph^2}{2}$	$\frac{ph^2}{2}$
X/h	001	0 057	0 123	0 190	0 260
Y/h	0454	0 630	0 620	0 625	0 650
θ	065	0 65	0 65	0 65	0 65
Z/h	644	0 687	0 743	0 815	0 910

W = Weight per Cu Ft of Concrete
P = Weight per Cu Ft of Water
θ = Sliding Factor = $\frac{P}{W+V-U}$ = 0 65
$W(\frac{ph}{3}-Z)+V(\frac{ph-y}{3})-Up\frac{Ph}{3}=0$

Note These Curves are based on a Sliding Factor of 0 65 as defined above, and the sections are designed to give a resulting Unit Stress equal to the assumed Uplift Pressure at the Upstream Toe with the Water Surface at Top of Dam

TRIANGULAR GRAVITY DAM SECTIONS
WITH AND WITHOUT UPLIFT

FOR PRELIMINARY ESTIMATES

FIGURE 86.

In assuming a trial section, approximate values of the width of the trapezoid at top and bottom can be found from the following equations:[4]

$$t = \sqrt{h} \qquad\qquad (10)$$

$$b = \frac{h - 0.4H}{\sqrt{s}} \qquad\qquad (11)$$

in which t = the effective top width.

　　　　h = the height of the maximum water surface above the base.

　　　　b = the effective base width.

　　　　s = the specific gravity of the masonry.

　　　　H = the depth of overflow.

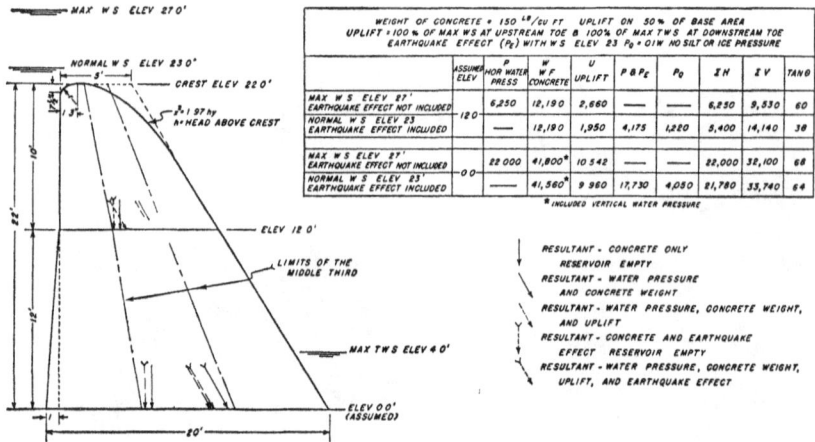

WEIGHT OF CONCRETE = 150 LB/CU FT   UPLIFT ON 50% OF BASE AREA
UPLIFT = 100% OF MAX WS AT UPSTREAM TOE & 100% OF MAX TWS AT DOWNSTREAM TOE
EARTHQUAKE EFFECT (P_e) WITH WS ELEV 23 P_o = OIW NO SILT OR ICE PRESSURE

	ASSUMED ELEV	P HOR WATER PRESS	W W.F CONCRETE	U UPLIFT	$P \& P_e$	P_o	ΣH	ΣV	TAN θ
MAX WS ELEV 27' EARTHQUAKE EFFECT NOT INCLUDED	120	6,250	12,190	2,660	—	—	6,250	9,530	60
NORMAL WS ELEV 23 EARTHQUAKE EFFECT INCLUDED		—	12,190	1,950	4,175	1,220	5,400	14,140	38
MAX WS ELEV 27' EARTHQUAKE EFFECT NOT INCLUDED	00	22,000	41,800[*]	10,542	—	—	22,000	32,100	68
NORMAL WS ELEV 23' EARTHQUAKE EFFECT INCLUDED		—	41,560[*]	9,960	17,730	4,050	21,780	33,740	64

[*] INCLUDED VERTICAL WATER PRESSURE

RESULTANT - CONCRETE ONLY RESERVOIR EMPTY
RESULTANT - WATER PRESSURE AND CONCRETE WEIGHT
RESULTANT - WATER PRESSURE, CONCRETE WEIGHT, AND UPLIFT
RESULTANT - CONCRETE AND EARTHQUAKE EFFECT RESERVOIR EMPTY
RESULTANT - WATER PRESSURE, CONCRETE WEIGHT, UPLIFT, AND EARTHQUAKE EFFECT

MAX WS ELEV 270'
NORMAL WS ELEV 230'
CREST ELEV 220'
$\frac{H}{3}$ = 1.97 hp h = HEAD ABOVE CREST
ELEV 120'
LIMITS OF THE MIDDLE THIRD
MAX TWS ELEV 40'
ELEV 00' (ASSUMED)
20'

STABILITY ANALYSIS OF OVERFLOW GRAVITY DAM
FIGURE 87.

The dam as determined by these equations should be checked and revised as need is indicated by detailed analysis. Figure 87 gives the analysis of design of a dam of this type.

Modification of the shape of the crest is desirable for two reasons: (1) To improve the flow coefficient by rounding the upstream corner and (2) to keep the under side of the overflow sheet in contact with the downstream face of the dam for all conditions of overflow. There are numerous formulas and equations for defining these modifications, which accomplish much the same result. A scheme which will be found acceptable for most low dams is presented in figure 88. The curve on the downstream side is defined by a parabola, which can be approximated by a compound curve of two radii for simplified construction if desired.

The shape of the overflow section near the downstream toe of the dam will be governed by the requirements for dissipating the energy of the overfalling water. The usual method is to provide a concave curved bucket and apron which will induce a hydraulic jump. (See fig. 43.) The theory of the hydraulic jump is discussed in Chapter **5:1**.

[4] "The Design of Dams," by Hanna and Kennedy, McGraw-Hill Co.

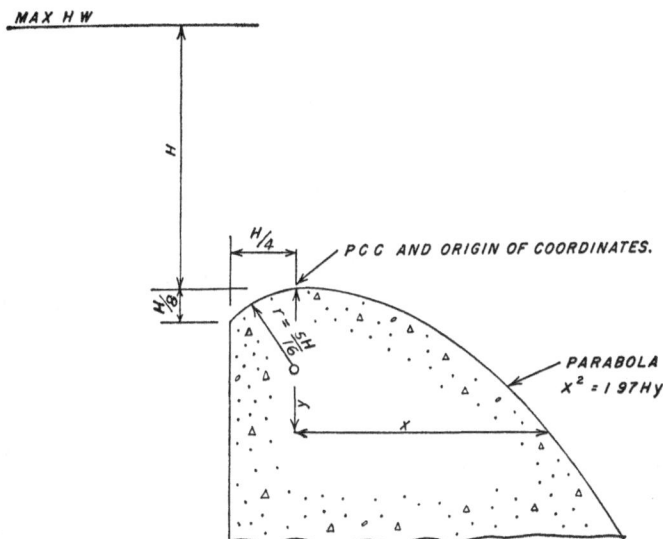

OVERFLOW DAM CREST CURVE
FIGURE 88

6. Design of Dams on Pervious Foundations.—The design of a gravity dam which is to be built on a previous and erodible foundation brings in additional considerations.

Particular attention is directed to the paper and discussions, Security From Under-Seepage, Masonry Dams on Earth Foundations, by E. W. Lane (Trans. Am. Soc. C. E., Vol. 100; p 1235).

The essential difference in loading is in the assumption of uplift pressure as outlined in the previous pages. The foundation on the downstream side must be protected by an apron and cut-off walls to such point that the velocity is reduced to values that are safe against piping.

An essential feature of this type of dam is the cut-off device, which may be of concrete poured in open trenches or one of several kinds of sheet piling. If piling is used an interlocking type is desirable. When wood is to be used, the pattern commonly known as the Wakefield type is recommended. It is made up of laminated units of three planks each, as shown in figure 72. In driving wood piling, a follower block should be used under the pile hammer to reduce splitting of the timber.

The determination of the necessary depth of the sheet piling or other cut-off device and of the width of the apron is dependent on the head and the permeability of the foundation material. The method of solution of this problem is outlined in the chapter on earth dams, and in appendix B. In some cases the use of impervious blankets on the upstream side can be introduced effectively for the reduction of underseepage in this type of dam.

Figure 88 Douthat State Park, Va , of 4 Earth-fill dam, Ogee spillway with stilling pool Sluice
tower at toe of upstream embankment

BIBLIOGRAPHY

A. H. Gibson, *Hydraulics and Its Applications*, John Wiley & Sons.

H. A. King, *Handbook of Hydraulics*, McGraw-Hill Book Co.

Riogel and J. C. Beebe, "The Hydraulic Jump as a Means of Dissipating Energy," *Tech. Reports*, Pt. III, Miami Conservancy District, Dayton, Ohio, 1917.

Julian Hinds, "The Hydraulic Jump and Critical Depth in the Design of Hydraulic Structures," *Eng. News-Record*, Vol. 85.

S. M. Woodward, "Theory of Hydraulic Jump and Backwater Curves," *Tech. Reports, Pt. III*, Miami Conservancy District, Dayton, Ohio, 1918.

Wm. P. Creager, *Masonry Dams*, John Wiley & Sons.

H. K. Barrows, *Water Power Engineering*, McGraw-Hill Book Co

Frank W. Hanna and Robert C. Kennedy, *Designs of Dams*, McGraw-Hill Book Co.

E. M. Burd, "High Dams on Pervious Glacial Drift," *Trans. Am. Soc C. E.*, 99: 792, 1934.

D. C. Henny, "Stability of Straight Concrete Gravity Dams," *Trans. Am. Soc. C. E.*, 99: 1041, 1934.

Fred A. Noetzli, "Gravity and Arch Action in Curved Dams," *Trans. Am. Soc. C. E.*, 84: 1, 1921.

H. deB. Parsons, "Hydrostatic Uplift in Pervious Soils," *Trans. Am. Soc. C. E.*, 93: 1317, 1929.

"Ice Thrust on Upstream Face of Dams," *Trans. Am. Soc. C E.*, 95: 1143, 1931.

E. W. Lane, "Security From Under-Seepage, Masonry Dams on Earth Foundations," *Trans. Am. Soc. C. E*, 100: 1235, 1935.

L. F. Harza, "Uplift and Seepage Under Dams on Sand," *Trans. Am. Soc. C. E.*, 100: 1352, 1935.

CHAPTER 10

SINGLE ARCH DAMS

1. Definition.—A masonry or concrete dam which in plan is curved upstream and which depends upon arch, or arch and cantilever action for its stability is commonly called an arch dam.

2. Design.—Due to the fact that an arch dam is restrained by its contact with the foundation, its action is not the simple action of an arch, but the combined action of arch and cantilever beam, and consequently is of an indeterminate nature. An accurate stress analysis depends upon assumptions as to the deformation of the dam and the distribution of the loads between the arches and cantilevers Design methods for arch dams are highly technical and final designs should be made only by specialists. The following notes are presented to point out salient features of design for the guidance of engineers not familiar with the subject, and are intended to be used only in connection with preliminary studies. A method of determining approximate arch thickness by an application of the cylinder theory, which may be useful for preliminary studies, is presented in section **1**. A method of applying stress analysis to a proposed arch dam is set out in section **2**. The method as there presented is greatly simplified, assumes the stress to be entirely carried in the arch rings, and does not take into account certain of the factors described in section **2** as important in the design of high arch dams. The application of this method to dams of less than 30 feet in height should produce an arch design of safely conservative dimensions.

1. *Cylinder theory.*—For preliminary economic studies and for calculating approximate dimensions preparatory to an accurate stress analysis by a more accredited method, the simple cylinder method will be found useful. This method gives only approximate results, since it ignores arch fixation and deformation and assumes the line of pressure to coincide with the center line of the arch section. It must not be used for final design purposes.

If an arch dam is considered to be a segment of an unrestrained thin cylinder subjected to external pressure, the equation for the stress in the ring is $f=\dfrac{p_e\,r_e}{t}$ in which—

f=stress in pounds per square foot.

p_e=load on the arch in pounds per square foot at the extrados.

r_e=outside or upstream radius of cylinder in feet.

t=thickness of the arch ring in feet.

The load p_e is the summation of the water pressure, wh, plus ice, silt, and earthquake pressures where encountered.

Arch thickness. For minor arch dams with a maximum height of 30 feet, equation (1) may be used to determine the approximate arch thickness, but only with the use of conservative values of f. This stress should be not greater than one-eighth of the ultimate compressive strength of the concrete masonry. Thus, for concrete that is designed for a 28-day breaking strength of 2,500 pounds per square inch, $f = \frac{1}{8}(2,500 \times 144) = 45,000$ pounds per square foot.

APPROXIMATE CONCRETE YARDAGE
SINGLE ARCH DAM

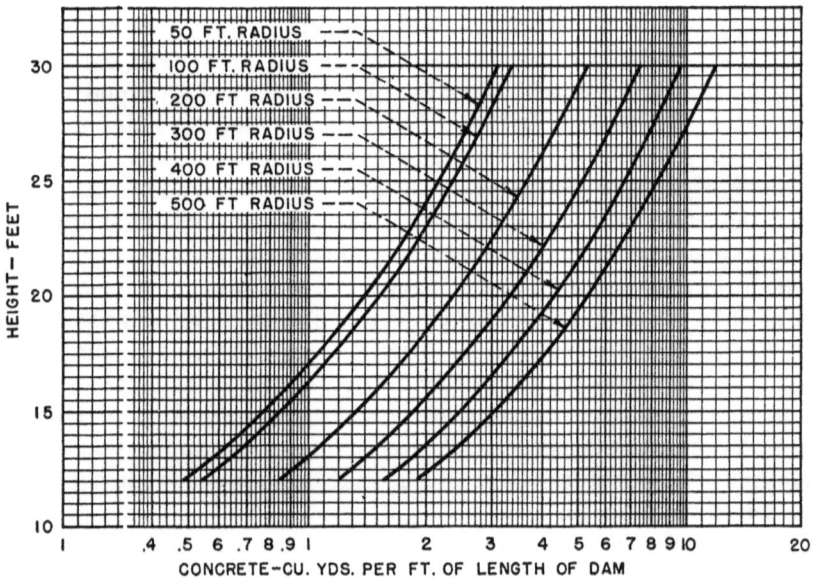

FIGURE 90

As in other types of dam, the most economical arch dam layout is the one which fits the site best, taking into consideration the shape and quality of the foundation and abutments. Equation (1) indicates that the thickness, and therefore the volume per foot of length, is directly proportional to the radius. The shorter the radius, the thinner the dam, but at the same time, the shorter the radius the longer the arc between two fixed points, and the greater the total volume of the arch. To determine the most economical structure, preliminary designs for three or more conditions should be made, using the yardage curves of figure 90 for estimating purposes. These curves are based on the use of a constant radius arch, the radius for the downstream face being equal to the radius of the upstream face

r_e minus the thickness of the concrete t which, as computed from equation (1), may be stated as $t=\dfrac{p_e r_e}{f}$.

It will be noticed that the thickness of the arch ring as determined by this formula will be the same for equal loadings and radii regardless of the length of the arch between abutments. This obviously incorrect relationship results from ignoring the fixation of the arch ring at the abutments and the deformation of the structure. Good design practice indicates that limiting values of the ratio of length of arch

National Park Service
FIGURE 91 —Fillmore Glen State Park, N Y , SP-33   Excavation of abutments for arch dam

ring to thickness, which is equivalent in significance to the column slenderness ratio, should not exceed 75 at the top of the dam, should be reduced to 25 at mid-height, and be maintained at 25 from that point down to the base. Therefore, where this formula is used for preliminary dimensioning of arch thickness, the resulting thickness should be checked with the length of the arch ring, and, if the slenderness ratio then exceeds the values above, the thickness should be modified so as not to be less than one seventy-fifth of the length of the arch ring at the top, or one twenty-fifth of the length in the lower part of the dam height.

2. *Stress Determination by Arch Analysis.*—It is not within the scope of this manual to present design methods which make proper allowance for vertical cantilever action in arch dams, nor to present methods of accurate analysis of dams of varying thickness in hori-

zontal section. The method of design outlined in this section is, however, sufficiently accurate for most dams of less than 30 feet in height; the neglect of cantilever action will act as an additional factor of safety. The solution presented is believed to be sufficiently clear that it may be safely applied by an engineer who has specialized to some extent in arch analysis and who understands fully the limitations of the method. Its use should not be attempted by those without previous experience in this particular field of structural design.

The method set out here not only makes no allowance for the distribution of stress between arch rings and cantilever sections, but ignores other factors which are often important in the design of high dams, such as the effect of shrinkage of the concrete, the effect of plastic flow, of moisture-volume changes, of Poisson's ratio, and of the yield of the abutments. The effect of these factors grows relatively less for dams of low height, and may be ignored for dams of less than 30 feet where conservatively low values of f are used.

Instructions are given here for the determination of moment, thrust, and shear of a circular arch having fixed ends and subjected to normal loads. The assumption of fixed ends at the abutments as contrasted with a hinged end assumption has been found to be the most practicable approach to the average condition.

The analysis presented here is based on the Cain formulas, the derivation of which took into consideration the effect of shear on the deformation of the arch. A series of diagrams has been prepared to simplify the solution of these formulas. The use of the diagrams facilitates the determination of the thrust and moment at the crown and abutments and the radial shear at the abutments. The moments and thrusts are determined by assuming the condition that the arcs forming the upstream and downstream faces are concentric and hence the arch is of uniform thickness. Arches for which this analysis is applicable are limited to those having a ratio of crown thickness to mean radius (t/r) of less than 0.2, and therefore include all arch rings likely to be encountered within the 30-foot height limitation.

The t/r ratio, shown on all plates, is the ratio of the crown thickness (t_o) to the mean radius (r_m) for each specific horizontal ring under analysis.

To obtain the thrust in pounds at crown and abutments, P_o and P_1, and the shear at the abutments, it is necessary to multiply the factors for these forces as given on figures 92 and 93 by the radius of the upstream face times the depth of water times 62.5.

To obtain the moments in foot-pounds at crown and abutments, M_o and M_1, it is necessary to multiply the factors as given on figures 94 and 95 by the radius of upstream face times depth of water times 62.5 times mean radius.

After P_o and M_o or P_1 and M_1 are obtained, the maximum and minimum stresses (S) in pounds per square foot on radial planes at the

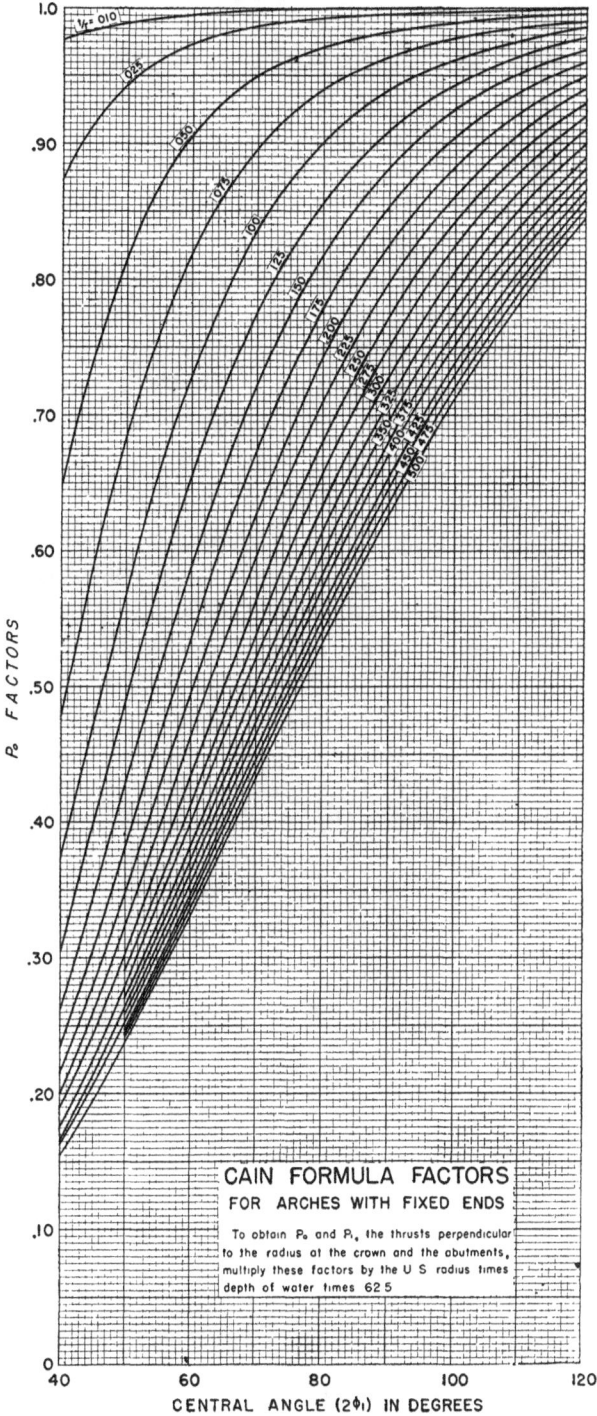

CAIN FORMULA FACTORS
FOR ARCHES WITH FIXED ENDS

To obtain P_o and P_1, the thrusts perpendicular
to the radius at the crown and the abutments,
multiply these factors by the U S radius times
depth of water times 62 5

CENTRAL ANGLE ($2\phi_1$) IN DEGREES

FIGURE 92.

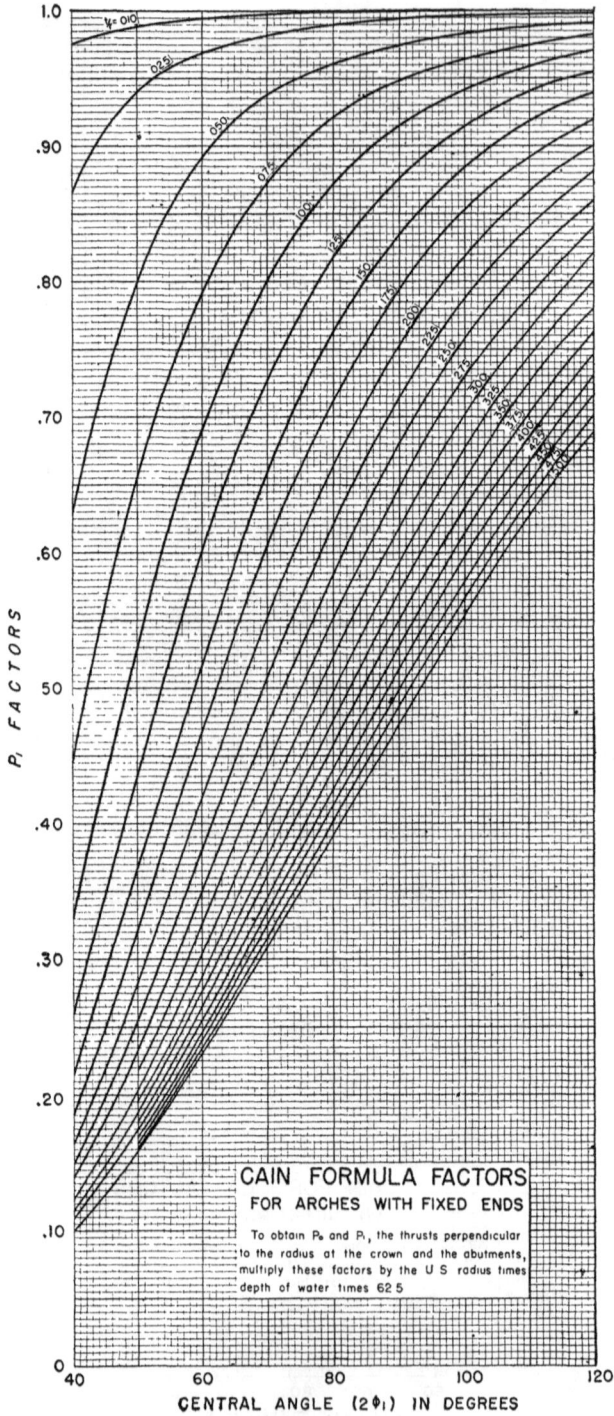

CAIN FORMULA FACTORS
FOR ARCHES WITH FIXED ENDS

To obtain P_o and P_i, the thrusts perpendicular
to the radius at the crown and the abutments,
multiply these factors by the U S radius times
depth of water times 62 5

P, FACTORS

CENTRAL ANGLE (2Φ₁) IN DEGREES

Figure 93.

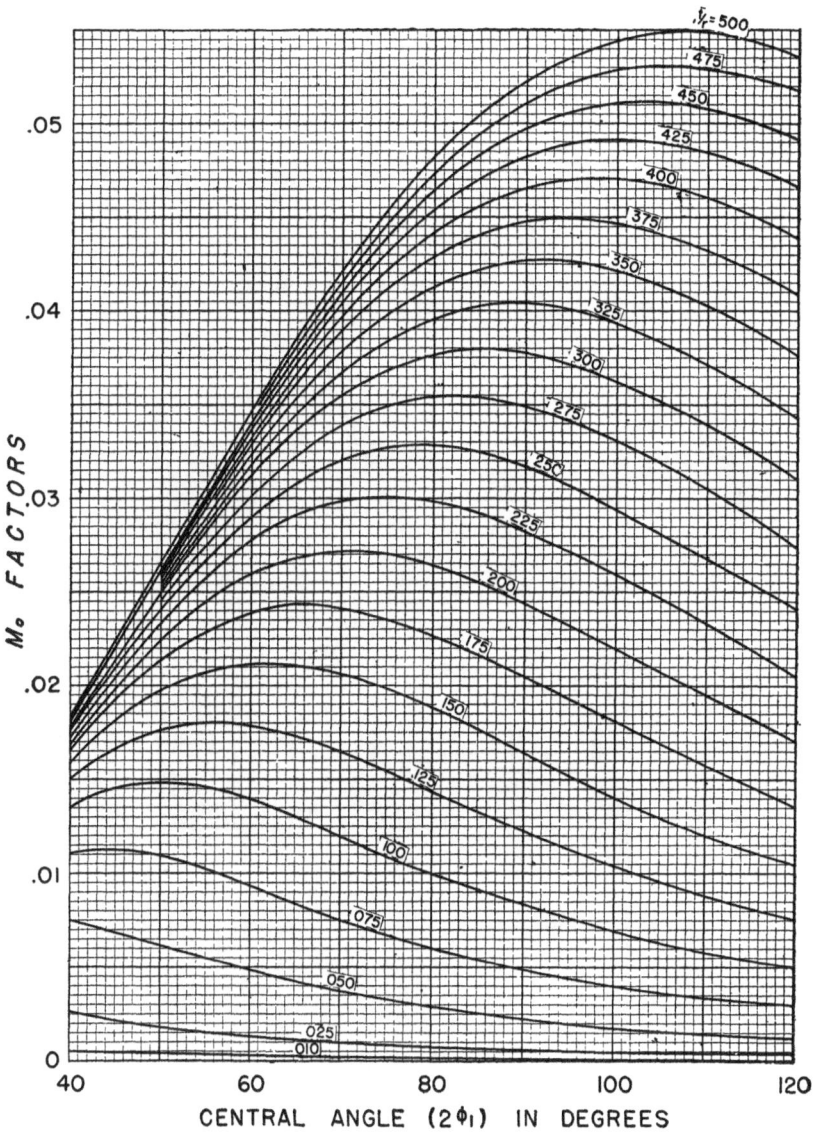

FIGURE 94 —Cain formula factors for arches with fixed ends   To obtain M_o and M_1, the moment at crown and abutments, respectively, in foot-pounds, multiply factors by depth of water times 62 5 times radius of upstream face times mean radius

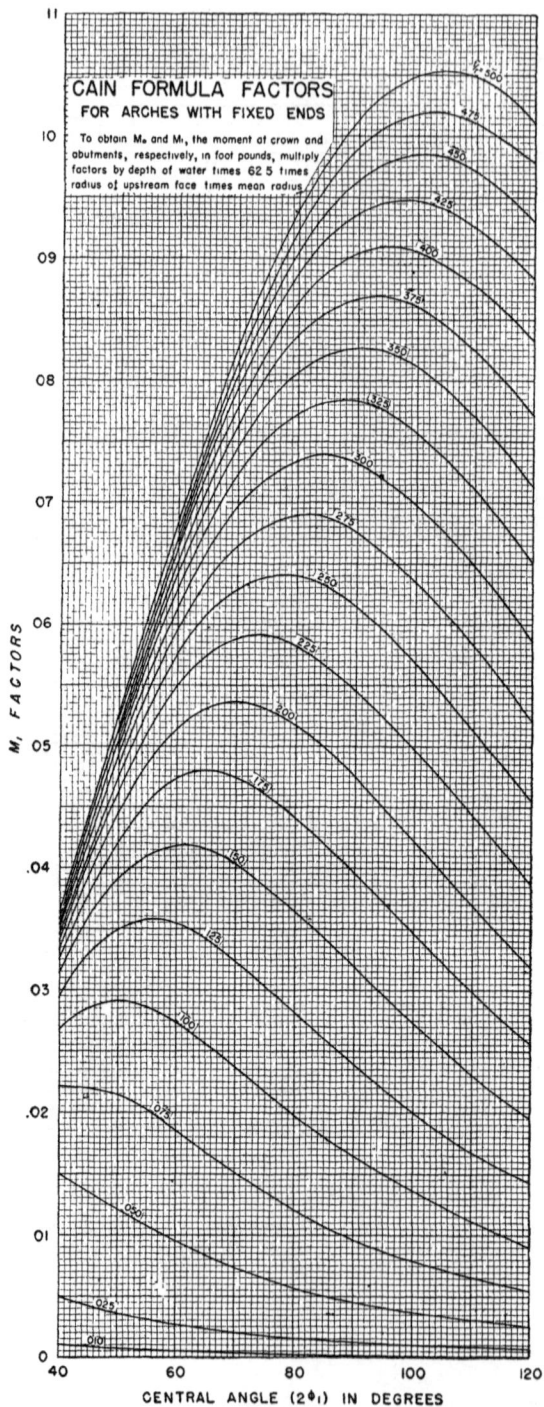

FIGURE 95.

crown and abutments can be obtained by substituting in the formula

$$S = \frac{P}{t_o} \pm \frac{6M}{t_o{}^2},$$

using P_o with M_o and P_1 with M_1.

The amount and direction of the resultant at the abutments can be determined by geometrically resolving P_1 as determined by the foregoing methods, with the radial shear as determined from figure 96.

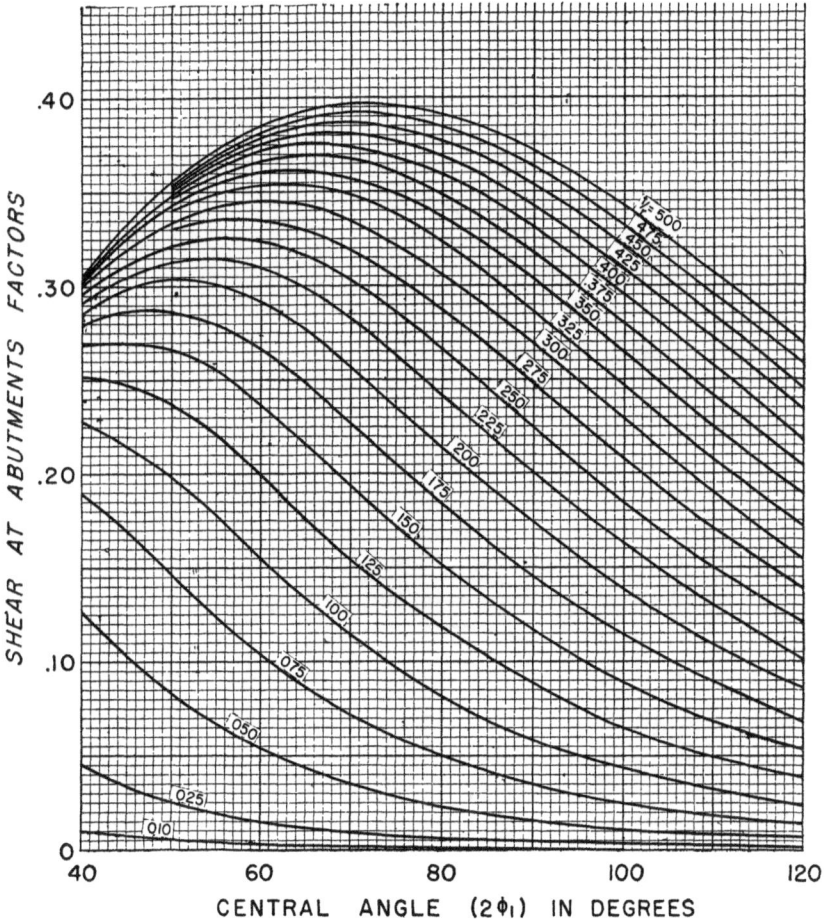

FIGURE 96.—Cain formula factors for arches with fixed ends   To obtain shear at abutments in pounds, multiply shear factor by depth of water times 62 5 times radius of upstream face

The nomenclature of terms used and the Cain formulas on which the curves are based are as follows:

Figure 97 represents a horizontal arch, 1 foot thick perpendicular to the plane of the paper.

t_o=radial thickness of arch at crown, in feet;
r=radius of center line of arch, in feet;
r_e=radius of extrados, in feet;
r_i=radius of intrados, in feet;
p_e=normal radial pressure, in pounds per square foot, on extrados;

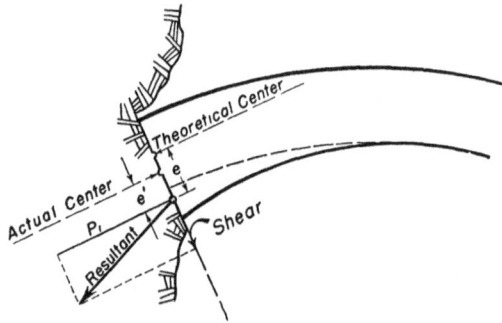

FORCES AT ABUTMENT

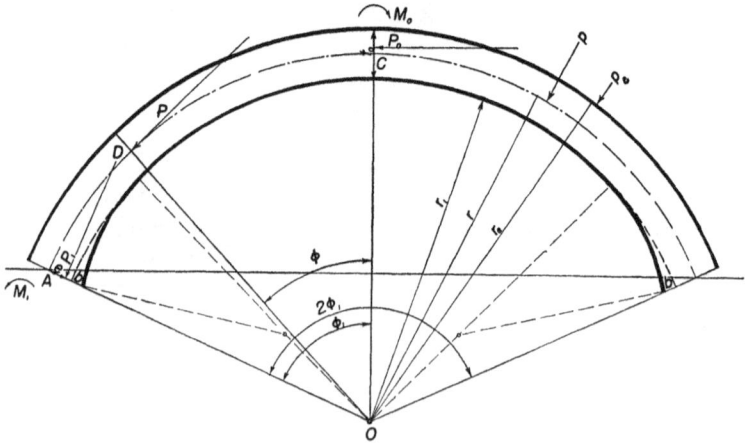

CIRCULAR ARCH FIXED AT THE ENDS AND
SUBJECTED TO A UNIFORM NORMAL RADIAL PRESSURE
FIGURE 97.

p=normal pressure, in pounds per square foot, on center line=

$$\frac{p_e r_e}{r}$$

ϕ=angle with radius of crown for any point, D;
ϕ_1=half central angle, $A\ \ O\ \ B$;
M_o=moment at crown, taken positive clockwise;
P_o=thrust at crown;
M_1 and P_1 are, respectively, the moment, and tangential components of the thrust at abutments.

S_o and S_1 are unit stresses at crown and abutments, respectively.

e=eccentricity of P_1 considering arch as of uniform thickness of t_o;

K=radius of gyration of a radial section.

Cain Formulas for Thin Arches with Fixed Ends with Shear Taken into Consideration

$$pr - P_o = \frac{pr \times 2 \times \dfrac{K^2}{r^2} \times \sin\ \phi_1}{D_s}$$

where $D_s = D + 2.88\dfrac{K^2}{r^2}\phi_1\ (\phi_1 - 1/2\ \sin\ 2\phi_1)$

and $D = \left(1 + \dfrac{K^2}{r^2}\right)\phi_1\ (\phi_1 + 1/2\ \sin\ 2\phi_1) - 2\ \sin^2\ \phi_1$

$$M_o = -(pr - P_o)r\left(1 - \frac{\sin\ \phi_1}{\phi_1}\right)$$

$$P_1 = pr - (pr - P_o)\ \cos\ \phi_1$$

$$M_1 = (pr - P_o)r\left(\frac{\sin\ \phi_1}{\phi_1} - \cos\ \phi\right)$$

radial shear at abutments$=(pr - P_o)\ \sin\ \phi_1$

Problem Illustrating Methods of Using Curves

To illustrate the method of using the curves for thin arches take the following example of a horizontal section of a constant angle arch dam which is 45 feet below maximum water level:

(1) Radius of upstream face (r_e)=80.5 feet. (2) Central angle $(2\phi_1)$=110°. (3) Thickness at crown (t_o)=7.0 feet. (4) Thickness at abutments (t_1)=8.0 feet. (5) Mean radius (r_m)=$80.5 - \dfrac{7.0}{2}$=77.0 feet. (6) t/r=7.0/77.0=0.091.

The depth of water times 62.5 times radius upstream face= $45 \times 62.5 \times 80.5$=226,500 in round figures, and this product times mean radius (r)=226,500$\times$77.0=17,440,500. For a $2\phi_1$ value of 110° and a t/r value of .091 the factors 0.97 and 0.98 are obtained from figures 92 and 93 for P_o and P_1, respectively. Multiplying these factors by 226,500 gives 219,700 pounds and 221,970 pounds for P_o and P_1, respectively.

In the same manner the factors 0.005 and 0.010 are obtained from figure 3 for M_o and M_1, respectively. Multiplying these in turn by 17,440,500, gives 87,200 foot pounds and 174,405 foot pounds for M_o and M_1, respectively.

The radial shear at abutment factor is obtained in the same manner from figure 96 and is .029 which multiplied by 226,500 gives 6,570 pounds for the radial shear at abutments.

The eccentricity (e) of P_1 at the abutments, without taking the flare (b) into consideration, is $e = \dfrac{174,405}{221,970} = 0.794$ feet.

After the moment at the abutment has been determined and the location of the line of thrust computed by dividing the moment by the thrust, it may be desirable in some instances to reduce the maximum compressive stresses at the abutment by an arbitrary flare or increase in thickness in the vicinity of the abutment. Inasmuch as the moment at the abutment will be negative, i. e., such as to give compression in the downstream face of the dam, the thickening will involve the placing of additional concrete on the downstream face. To be effective in reducing maximum compressive stresses, the flare should be a gradual one, extending out along the arch ring for a distance of at least 10 times the dimension of the increased thickness.

The arch analysis up to this point has been carried out as for an arch ring of seven-foot uniform thickness. The moments, thrust, and the eccentricity are those applying to such an arch. The effect of introducing a flare at the abutment, where the length of the flared section is an appreciable proportion of the total length of the arch ring, would produce to some extent an arch ring of varying thickness. For such an arch, the moment at the abutment would probably be somewhat greater and the moment at the crown somewhat less than those resulting from the analysis above of a uniform arch ring. Also the eccentricity of the P_1 with respect to the center of the thickened abutment might be appreciably different from the eccentricity of 0.794 above calculated.

For small arches not justifying a more rigid stress analysis, an approximation of the stresses at the abutment may be made by inserting in the formula—

$$S = \frac{P}{t_0} \pm \frac{6M}{t_0{}^2}$$

the values of P_1 and M_1 resulting from the uniform thickness arch ring analysis, and taking t_0 as the thickness of the flared abutment or 8 feet.

The stress at the abutments $S_1 = \dfrac{221,970}{8.0} \pm \dfrac{6 \times 174,405}{(64)} = 27,700 \pm$ 16,350 $= +44,050$ pounds per square foot or 306 pounds per square inch compression at intrados and 11,350 pounds per square foot or 80 pounds per square inch compression at extrados.

The resultant of P_1 and the radial shear is obtained by resolving the forces 221,970 (P_1) and 6,570 (radial shear) at the point of application of P_1 which has been determined as 0.794 feet from the center. The tangent of the angle the resultant makes with the radius is $\dfrac{P_1}{shear} = \dfrac{221,970}{6,570} = 33.78$ and the angle equals 88°18′.

3. Provision for Earthquakes.—In localities where earthquakes may be expected to occur, provision is made to carry the additional loads caused by them. Where the dam site does not cross a fault, the additional load due to the earthquake vibrations can be computed. Movement of the foundation imparts a similar movement to the dam. This movement will have a horizontal component in a direction normal to the principal chord of the arch dam which will give the condition of maximum pressure. The source of the pressure is the movement of the dam alternately toward and away from the mass of water behind it. The magnitude of this pressure varies as a function of certain characteristics of the earthquake, the two most important ones being the frequency of vibration and the amplitude. The diagrams of figure 98 show the amount and distribution of pressure caused by earth tremors as computed from approximate formulas developed by Prof. H. M. Westergaard.[4]

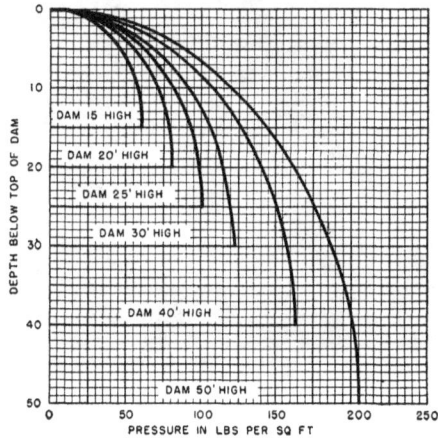

EARTHQUAKE PRESSURES
APPROXIMATE PRESSURES FROM AN EARTHQUAKE
OF AMPLITUDE = 0 IG AND PERIOD = I SEC

FIGURE 98.

These values are for a very severe earthquake, and so probably represent the extreme condition. The application of the loading to stress analysis is the same as for pressures from other sources.[5]

A dam should never be built across a fault line unless approved by acknowledged specialists in engineering and seismology.

4. Abutments.—The principal load on the foundation is the horizontal one from the arch rings to the abutments. The arch thrust is tangent to the arc, and there must be sufficient mass or rock in line

---

[4] "Water Pressure on Dams During Earthquakes," by H M Westergaard, M A M Soc C E , in Trans of Am Soc C. E . vol 98. 1933   [5] See also ch 9 2-4

with it to furnish the necessary shearing resistance.   Figure 101 illustrates unsafe and safe abutment conditions.   In (a) force R represents the direction of the arch thrust.   The angle Θ is the angle of friction of rock on rock (approximately 35°) and the angle α is the angle which the resultant makes with the average plane of the abutment.   The component of R which would produce shear along the plane A–B is resisted by an area of indefinitely large and safe proportions if the rock formation is solid.   The condition in (b) is quite different, for the plane of maximum shearing stress, A–B, is inclined so that it passes

FIGURE 99 —Concrete arch dam with gravity abutments.   San Dimas Experimental Forest, Calif.

through only a small portion of the abutment, representing an unsafe condition.   This leads to the general requirement that the angle α between the rock face of the abutment and the line of the resultant of the arch thrust shall not be less than 35°.

By similar reasoning the inclination of the concrete abutment at contact with rock should be restricted, preferably to a minimum inclination of 45° from the horizontal, and a maximum of 90° (see fig. 102).   The depth of the excavation and the number of steps will depend on the character of the rock and particularly on the inclination of the bedding or stratification planes.

With these limitations to be observed, the choice of the final location and alignment will be a problem of finding the layout that will best fit the site.   This can only be done by trial, computing the excavation and masonry yardages for several combinations and selecting the most economical one.

Bureau of Reclamation

FIGURE 100.—Concrete arch diversion dam, East Park Feed Canal, Orland project, California.

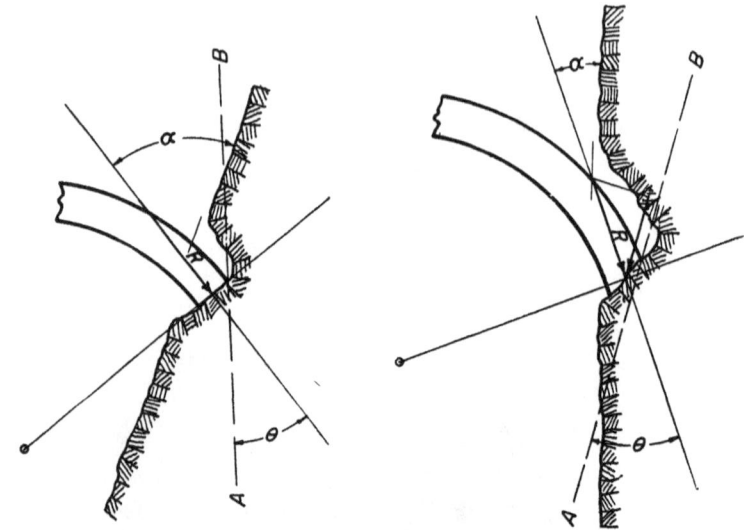

(a) SAFE ARCH ABUTMENT   (b) UNSAFE ARCH ABUTMENT

SINGLE ARCH DAM – SECTIONAL·PLAN

FIGURE 101.

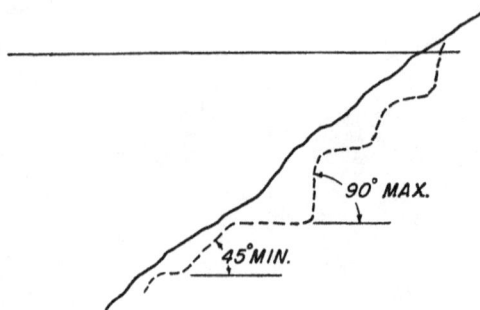

PART ELEV. OF ABUTMENT EXCAVATION

FIGURE 102.

5. Materials.—From a structural standpoint, concrete is the most satisfactory material to use. It should be particularly dense and strong, because the comparatively thin section cannot offer as much resistance to percolation and weathering as heavy gravity sections. Reinforcing steel is not ordinarily used because the principal stresses are compressive. Temperature changes may introduce tension, but in small structures this can be taken care of by the introduction of vertical contraction joints.

If the dam is built of stone masonry, the stone blocks should be cut to reasonably rectangular faces, and laid with their natural beds horizontal. All joints should be completely filled with good sand-cement mortar and made as thin as possible to provide for uniform and continuous arch action.

6. Crest.—For small overflow arch dams the spillway will be over the top of the dam, with the water breaking contact with the concrete at the lip and having free fall to the foundation just below the toe. It is obvious that the erosive power of the falling water will be great, and it should not be permitted to fall on unprotected rock unless solid enough to be safe beyond question. In cases where protection is necessary, a heavy concrete apron or, better, a cushion pool of water formed by a small dam downstream, can be used (see ch. 5:7).

In all cases where free fall occurs, provision should be made to admit air under the nappe to break the vacuum. Pipe air ducts of clay or other materials set in the concrete can be used to conduct air from the abutment to several points on the downstream face of the dam, just below the lip. This is illustrated in figure 103.

For greatest efficiency of overflow, the shape of the crest should conform to the underside of the theoretical nappe. Exact conformation is an unjustified refinement in a thin crest, which for practical use can be patterned after that shown in figure 103.

7. Specifications and Construction Methods.—Suggestions for specifications for single arch dams are given in Appendix G. Information regarding construction methods will be found in the same appendix (see Appendix A, secs. **1** to **8**, inclusive, and also sec. **9–4**).

BIBLIOGRAPHY

C. H. Howell and A. C. Jaquith, "Analysis of Arch Dams by the Trial Load Method," *Trans. Am. Soc. C. E.*, 93:1191, 1929.

Frederick H. Fowler, "A Graphic Method of Determining the Stresses in Circular Arches Under Normal Loads, by the Cain Formulas," *Trans. Am. Soc. C. E.*, 92:1512, 1928.

William Cain, "The Circular Arch Under Normal Loads," *Trans. Am Soc C E.*, 85:233, 1922.

A. V. Karpov, "The Compensated Arch Dam," *Trans Am Soc C E*, 98·1309, 1933.

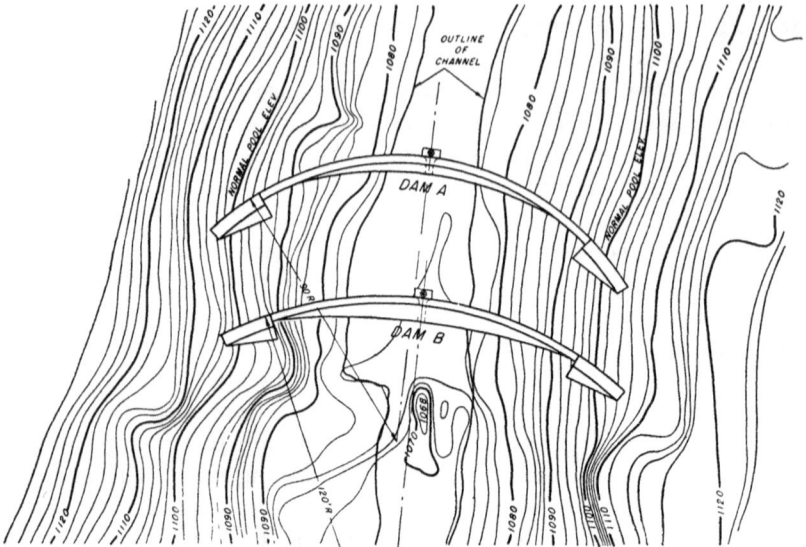

PLAN AT DAM SITES

SCALE - 1 IN 20FT

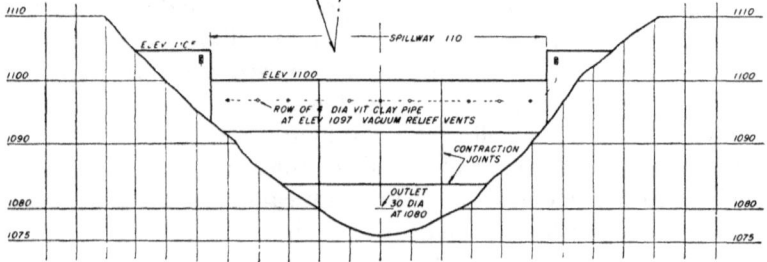

DEVELOPED PROFILE AT UPSTREAM FACE OF DAM A

SCALE -   HOR 1 IN = 20FT
          VER 1 IN = 10FT

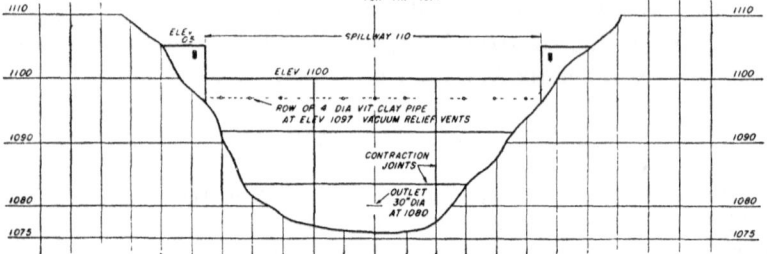

DEVELOPED PROFILE AT UPSTREAM FACE OF DAM B

SCALE -   HOR 1 IN = 20FT
          VER 1 IN = 10FT

FIGURE 103a —Single-arch concrete dam, plan and profiles

The Engineering Foundation, "Report of the Committee on Arch Dam Investigation."

Recommendations of the Engineering Foundation Committee on Arch Dam Investigation. *Civil Engineering*, June 1933.

Stresses in Arch Dams, U. S. Bureau of Reclamation, *Technical Memorandum No. 454.*　　-

F. C. Hanna and R. C. Kennedy, *Design of Dams*, McGraw-Hill Book Co., Inc.

Wm. P. Creager, *Masonry Dams*, John Wiley & Sons.

Edw. Wegmann, *The Design and Construction of Dams*, John Wiley & Sons.

I. E. Houk, *Trial Load Method of Analyzing Arch Dams* in "Dams and Control Works" by U. S. Bureau of Reclamation, 1938. (Obtainable from Supt. of Documents for $1.00.)

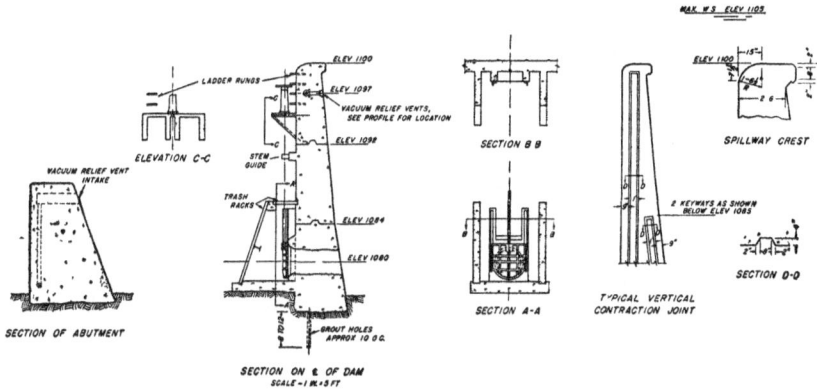

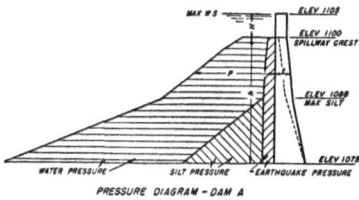

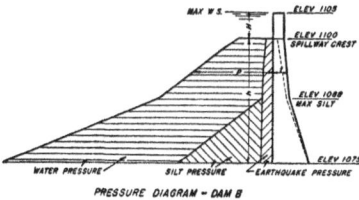

ANALYSIS

MAXIMUM W S ELEV 1105 — SPILLWAY CREST ELEV 1100
SILT = 85 LBS/CU FT FLUID PRESSURE TO ELEV 1088
EARTHQUAKE AMPLITUDE = 0.1g — PERIOD = 1 SEC
f = 45,000 LBS/SQ FT MAX — MINIMUM THICKNESS = 0.1h
MAXIMUM SLENDERNESS RATIO L/t = 75 AT TOP & 25 AT MID-HEIGHT

DAM-A — RADIUS = 90 FT — VOLUME = 305 CU YD

ELEVATION	1100	1095	1090	1088	1085	1080	1075
TOTAL PRESSURE (P) LBS/SQ FT	360	716	1044	1172	1560	2203	2843
LENGTH OF ARC (L) FT	140	120	100	93	79	53	20
THEORETICAL THICKNESS (t) FT	0.8	1.4	2.1	2.3	3.1	4.4	5.7
SLENDERNESS RATIO (L/t)	175	86	48	40	25	12	4
FINAL THICKNESS FT	2.5	*3.0	3.5	3.7	4.2	4.9	5.7
FINAL STRESS LBS/SQ FT	13700	21500	26900	28500	33400	40500	45000
FINAL SLENDERNESS RATIO	56	40	29	25	19	11	4

DAM-B — RADIUS = 120 FT — VOLUME = 315 CU YD.

ELEVATION	1100	1095	1090	1088	1085	1080	1075
TOTAL PRESSURE (P) LBS/SQ FT	360	716	1044	1172	1560	2203	2843
LENGTH OF ARC (L) FT	126	109	94	88	83	65	20
THEORETICAL THICKNESS (t) FT	1.0	2.0	2.8	3.1	4.2	5.9	7.6
SLENDERNESS RATIO (L/t)	126	54	34	28	20	11	3
FINAL THICKNESS FT	2.5	2.9	3.3	3.5	4.4	6.0	7.6
FINAL STRESS LBS/SQ FT	18200	29600	38000	40200	43000	44000	45000
FINAL SLENDERNESS RATIO	50	38	28	25	19	11	3

FIGURE 103b —Single-arch concrete dam, cross section and analysis

CHAPTER 11

BUTTRESS DAMS

1. Definition.—The term "buttress dam" is used here to designate the flat-deck and multiple-arch dams, in which highly stressed reinforced concrete structural members replace the much greater mass of concrete used in solid gravity dams.

2. Flat-Deck Dam.—This type of dam is made up of a flat reinforced concrete slab which transmits the water pressure to a series of parallel buttresses which rest directly on the foundation or upon a concrete slab which rests on the foundation material. The buttresses are usually parallel but in a few instances they have been made radial. The slab is built on a slope of approximately 45°, which makes it possible to utilize the vertical component of the water pressure in analyzing the dam for sliding and overturning resistance. It is the advantage gained by the use of the water load to increase stability, plus the benefit of higher design stresses, which permits the use of the hollow pattern. In the nonoverflow dam, the deck slab covers only the upstream side of the buttresses, the downstream side usually remaining open. It is not uncommon to use the hollow dam for spillways, in which case the openings between the buttresses on the downstream side are closed by slabs, which are rounded in outline to conform to the shape of the falling water, as discussed in Chapter 5.

3. Multiple-Arch Dams.—In the multiple-arch type the slabs of the flat-deck dams are replaced by concrete arches which span from buttress to buttress. The use of the arch principle makes possible a wider spacing of the buttresses, but the general advantages and disadvantages of the two types are similar. The multiple-arch dam, however, is not ordinarily used for overflow crests.

4. Relative Merits.—The greatest advantage in favor of the buttress dam is the economy in concrete as compared with a solid masonry dam. Figure 104 gives the approximate yardage of this type of dam for various heights for both overflow and nonoverflow designs. Because of the smaller amount of concrete to be placed, less time will ordinarily be required for construction of the buttress dam than for a solid concrete or an earth dam of equivalent size. However, the unit cost of the thin reinforced concrete sections will be much higher than that of the solid dam due to the cost of reinforcement, forms, and placing. Which type will hold the advantage depends on the relative

194

BUTTRESS
NON-OVERFLOW DAM
APPROX. CONCRETE YARDAGE

BUTTRESS
OVERFLOW DAM
APPROX. CONCRETE YARDAGE
(INCLUDES BUCKET AND CURVE EXCESS)

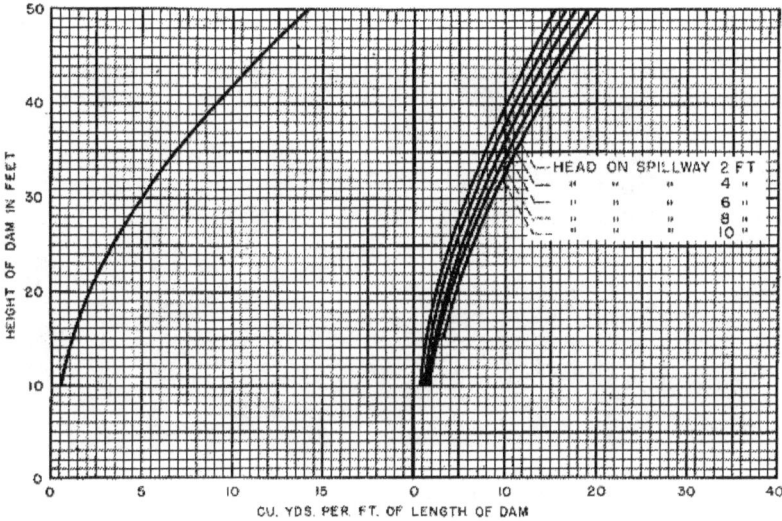

costs of materials and labor in individual cases and the suitability of type for given locations.

The use of buttress dams may be somewhat restricted by the possibility of deterioration of thin reinforced concrete sections in hydraulic structures, either from the corrosive action of acid-bearing water, or from severe frost action.

5. Design.—The design of flat-deck or multiple-arch dams is based on a type of knowledge and judgment that comes only from experience in that field. It should not be attempted by inexperienced men with-

FIGURE 105 —Reinforced concrete slab and buttress dam, Link River, Klamath project, California-Oregon

Figure 100.—Downstream face of Finer River, flat deck concrete dam, Baker project, Oregon

out supervision and advice from an engineer who has the necessary qualifications.

6. Construction.—Suggestions for specifications for buttress dams will be found in Appendix G.   Information in regard to construction methods is also given in that appendix.   See sections G–1 to G–8, inclusive, and section G–9–5.

BIBLIOGRAPHY

W. H. Holmes, Determination of Principal Stresses in Buttresses and Gravity Dams, *Trans. Am. Soc. C. E.*, 98: 971, 1933.
F. C. Hanna and R. C. Kennedy, *Design of Dams*, McGraw-Hill Book Co.
Wm. P. Creager, *Masonry Dams*, John Wiley & Sons.
Edw. Wegmann, *The Design and Construction of Dams*, John Wiley & Sons.

CHAPTER 12

TIMBER DAMS

1. General Features.—Timber dams can be designed to fit almost any foundation condition. Rock is the most desirable foundation, but by increasing the width of the base or of the toe fill, or by the use of sheet piling, the bearing values and the percolation losses can be reduced so that this type of dam can be fitted to gravel, sand, or other pervious foundations.

Water tightness in timber dams is usually attained by the combination of a tight timber plank deck plus a partial backfill of impervious soil against the upstream face. The deck planks are frequently driven into the foundation as sheet piling to form a cut-off against leakage or to increase the length of the path of percolation. In very pervious foundations, two or more rows of piling may be used, as shown in figure 107. The method of determining percolation losses is the same as is used for earth dam foundations under similar conditions. If the water supply is adequate, percolation can be permitted to a limited extent, provided the velocity does not become great enough to cause piping.

The useful life of a timber dam is limited to the life of the timber in it. It is generally known that wood which is kept constantly dry or continuously submerged in water will not decay. It is not possible to keep dam timbers dry, but in many dams it is possible for the timber in the lower part to be submerged or continuously wet, and the life of this timber may be as great as 50 or 75 years. However, timber in the downstream face, and at the top of the dam near or above the water line, will often be alternately wet and dry, which condition is favorable to rapid decay. These parts of the dam, however, being more or less accessible, may be replaced as often as necessary. Besides being exposed to conditions which accelerate decay, the crest of an overflow dam is often subject to the pounding action of floating logs, debris, and ice. The top of the dam should therefore be made of strong, durable wood, framed as solidly as possible and shaped so that it will fend off or pass floating objects readily.

The following paragraphs on decay resistance are quoted from the "Wood Handbook," prepared by the Forest Products Laboratory of the United States Forest Service, Department of Agriculture:

"The natural decay resistance of all common native species of wood lies in the heartwood. When untreated, the sapwood of substantially all species has low resistance to decay and usually has a short life under decay-producing conditions. The decay resistance or durability of heartwood in service is greatly affected by differences in the character of the wood, the attacking fungus, and the conditions of exposure. A widely different length of life may therefore be obtained from pieces of wood cut from the same species or even the same tree and used under apparently similar conditions. Further, in a few species, such as the spruces and the white firs (not Douglas fir), the colors of the heartwood and of the sapwood are so similar that frequently the two cannot be easily distinguished.

"Comparisons of the relative decay resistance of different species must be estimates. They cannot be exact; and they may be extremely misleading if erroneously considered mathematically accurate and universally applicable. Such comparisons may be useful, however, if regarded as approximate averages only, from which individual pieces or lots of a given species may vary considerably, and if they are understood to apply only where the wood is subject to conditions that favor decay.

"The following grouping divides some of the more common native species into five classes listed in accordance with the resistance of heartwood to decay; every grouping of this nature is subject to the preceding limitations. The classification is based on service records, when they are available, and on general experience:

Heartwood durable even when used under conditions that favor decay:
  Cedar, Alaska.
  Cedar, eastern red.
  Cedar, northern white.
  Cedar, Port Oxford.
  Cedar, southern white.
  Cedar, western red.
  Chestnut.
  Cypress, southern.
  Locust, black.
  Osage-orange.
  Redwood.
  Walnut, black.
  Yew, Pacific.
Heartwood of intermediate durability but nearly as durable as some of the species named in the high-durability group:
  Douglas fir (dense).
  Honey locust.
  Oak, white.
  Pine, southern yellow (dense).
Heartwood of intermediate durability:
  Douglas fir (unselected).
  Gum, red.
  Larch, western.
  Pine, southern yellow (unselected).
  Tamarack.

Heartwood between the intermediate and the nondurable group:
    Ash, commercial white.
    Beech.
    Birch, sweet.
    Birch, yellow.
    Hemlock, eastern.
    Hemlock, western.
    Hickory.
    Maple, sugar.
    Oak, red.
    Spruce, black.
    Spruce, Engelmann.
    Spruce, red.
    Spruce, Sitka.
    Spruce, white.
Heartwood low in durability when used under conditions that favor decay:
    Aspen.
    Basswood.
    Cottonwood.
    Fir, commercial white.
    Willow, black.

"There are no adequate service records from which to evaluate the heartwood of the white pines and ponderosa pine in decay resistance. There is a common opinion, as the result of general experience with the use of these two species, that the heartwood of the white pines has more decay resistance and therefore will give longer service under conditions favoring decay than the heartwood of ponderosa pine."

The life of timbers placed in a dam under conditions favorable to decay can be considerably extended by proper treatment with a suitable preservative and the cost per year of service greatly reduced below that of similar timbers without treatment. Coal-tar creosote is the outstanding preservative for the treatment of timbers used in dams. The most effective method of treating wood with preservatives is by a pressure process. For the treatment of dam timbers with creosote the empty-cell process is suggested, with a retention of 8 or 10 pounds of oil per cubic foot. The object of the empty-cell treatment is to obtain a deep penetration with low retention of preservative. Timbers which require framing, cutting, or boring should have this done before treatment so as not to expose untreated wood. If cutting of treated lumber is necessary on the job, the cut surfaces should be thoroughly brushed with creosote. Federal specifications are available covering the treatment of wood used in Federal projects. Should preservatives other than creosote be considered for the treatment of dam timbers, it is suggested that the matter be taken up with the Forest Products Laboratory as to desirable specifications.

The timber used for dam crib work may be either round logs or square-sawed. If round logs are used, all the bark should be removed because many kinds of decay fungi are fostered beneath the bark.

The deck planks should be strong and durable to resist the impact and abrasion of ice and debris.   This is particularly important for the downstream slope of certain types of crib dams where the overflow is carried down the face in steps, as in figure 107.

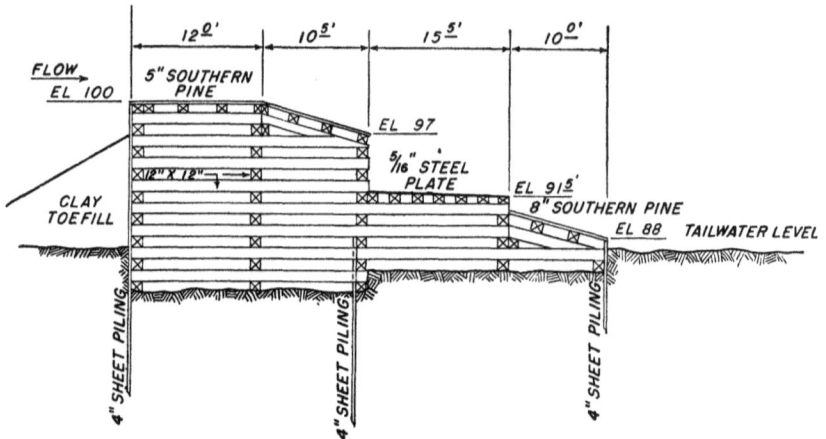

ROCKFILLED CRIB DAM

FIGURE 107.

2. Types.—Timber dams may be divided into three structural groups, as follows:

1. The *rock-filled crib* dam, in which the stability is provided for by the weight of the rock fill.

2. The *frame and deck* dam, in which the water pressure on a sloping deck is instrumental in holding the dam in place.

3. The *crib and deck* dam, which is a combination of the first two often used for very low dams up to 6 to 8 feet in height.

3. Design.—1. *Rock-filled crib.*—This type of dam is a gravity structure and is subject to the same analysis against overturning and sliding as other types of gravity dam.   The unit weight of the rock-filled crib varies under different conditions and should be investigated in each case.   Rock fill composed entirely of field stone in sizes ranging from 6-inch cobbles up to two-man stone can be deposited into a crib to give a unit weight of approximately 110 pounds per cubic foot, but this figure must be corrected for the amount of space which is occupied by the wooden crib.   The percentage of wood will vary with the spacing of the cribwork and the size of the timber, but for the average condition it is safe to assume a net weight of 90 to 95 pounds per cubic foot.   If the rock is to be packed in place by hand, a higher unit weight can be obtained, but it should be determined by test in each case.   Gravel might be added to reduce the voids in the fill and increase the unit weight, but much of it might soon be lost through the openings in the cribwork, so that it is not advisable to depend on it.

There is no uplift pressure to be considered in the analysis of a timber dam, but in locations where the tailwater level is above the base of the dam the submerged weight of the dam must be used.

Submerged unit weight = (unit weight of rock — 62.5) (1 — void ratio)

The buoyant force of the submerged timber must be deducted from the submerged weight of the rock to determine the net weight of the structure. For safe resistance against sliding, the sum of the vertical loads of water pressure and effective weight of the dam should be not less than 2.5 times the horizontal pressure of the water and toe fill under favorable foundation conditions. For important dams on smooth foundations the ratio should be increased to 3.5 or even 4.

Figure 108 shows an analysis of a conventional section of this type of dam.

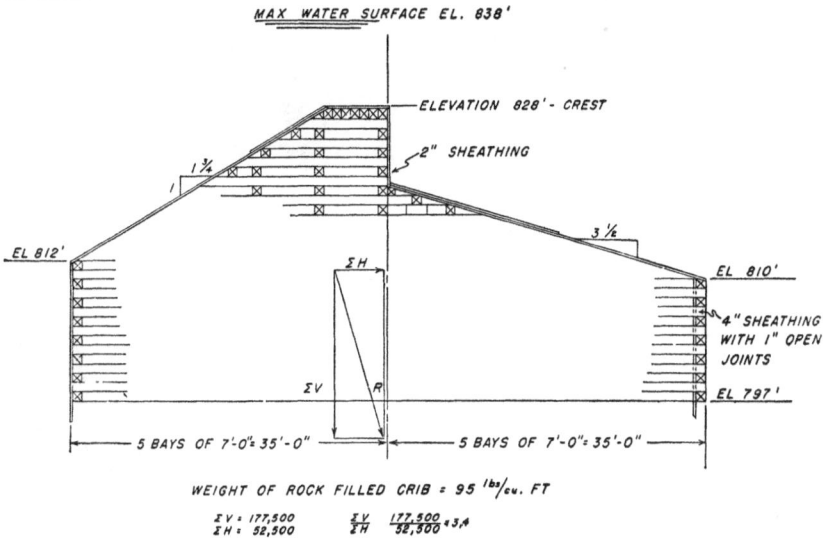

MAX WATER SURFACE EL. 838'

ELEVATION 828' - CREST

2" SHEATHING

$1\frac{3}{4}$

$3\frac{1}{2}$

EL 812'

EL 810'

ΣH

4" SHEATHING WITH 1" OPEN JOINTS

ΣV R

EL 797'

— 5 BAYS OF 7'-0"= 35'-0" —   — 5 BAYS OF 7'-0"= 35'-0" —

WEIGHT OF ROCK FILLED CRIB = 95 $^{lbs}/_{cu}$. FT

$\Sigma V = 177,500$ $\frac{\Sigma V}{\Sigma H}$ $\frac{177,500}{52,500} = 3.4$
$\Sigma H = 52,500$

A TYPICAL ROCKFILLED CRIB DAM

FIGURE 108.

2. *Frame and Deck Dam.*—This design resembles the concrete flat-deck buttress dam in its structural features, the timber deck being supported by timber bents rather than by concrete buttresses. The water pressure on the deck is relied upon to help hold the dam in place. Because the structure has very little weight, resistance against sliding is provided for by some mechanical key or bond to the foundation. On rock the sills are usually pinned down by anchor

National Park Service

FIGURE 109.—Mohawk Trail State Forest Park, Mass., SP-6. Timber frame dam and abutments.

Bureau of Reclamation

FIGURE 110.—Downstream face of rock-fill timber crib dam showing spillway at Lake Keechelus, Yakima project, Washington.

rods set in drill holes.   On soft foundations the sills are buried in trenches and backfilled so that resistance is provided through horizontal shear in the foundation.   The structural design of the deck and frame is a simple application of beam and column loading and will not be detailed here.

Figure 111 shows two typical cross sections of this type of timber dam.

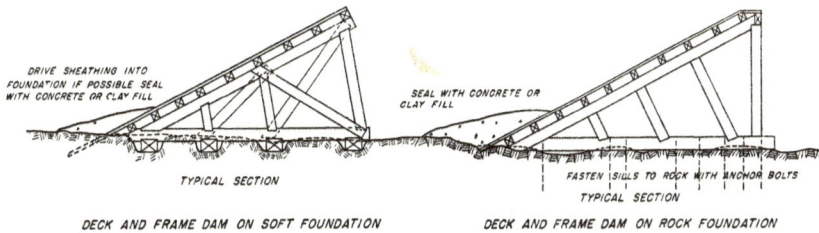

DRIVE SHEATHING INTO
FOUNDATION IF POSSIBLE SEAL
WITH CONCRETE OR CLAY FILL

SEAL WITH CONCRETE OR
CLAY FILL

TYPICAL SECTION

FASTEN SILLS TO ROCK WITH ANCHOR BOLTS
TYPICAL SECTION

DECK AND FRAME DAM ON SOFT FOUNDATION   DECK AND FRAME DAM ON ROCK FOUNDATION

FIGURE 111.

3. *Crib and Deck Dam.*—This is a combination of the crib and frame dams that is often suitable for low dams up to approximately 8 feet in height.   Figure 112 shows the loaded crib substructure with a moderately flat slope of $1\frac{3}{4}$ to 1. The stability of this section is derived partly from the weight of the rock in the crib and partly from the water load on the deck.

The typical designs that are illustrated are for overflow conditions.   In most cases it will be necessary to supplement these designs with non-overflow sections of similar nature to protect the stream banks at the abutments.

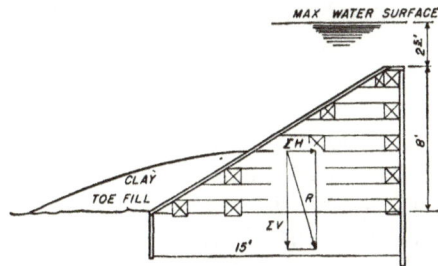

MAX WATER SURFACE

CLAY
TOE FILL

WEIGHT OF ROCK FILLED CRIB = 95 lbs per cu ft

$\Sigma V = 11,790 \quad \frac{\Sigma V}{\Sigma H} = \frac{11,790}{3,420} = 3.4$
$\Sigma H = 3,420$

CRIB AND DECK DAM

FIGURE 112.

4. **Construction.**—Suggestions for specifications for timber dams are given in Appendix G. Information concerning construction methods will be found in the same appendix. See sections G–1 to G–8, inclusive, and section G–9:**6**.

CHAPTER 13

MAINTENANCE AND OPERATION

Arrangements should be made immediately following the completion of a dam for frequent inspection of the structure and all the operating equipment. Adequate measures to accomplish this end are usually taken on the more important structures but are frequently neglected on small dams. General responsibility for such structures may lie with a State, county, municipality, or a special board or commission endowed with administrative powers. Such authority should be advised of the prime importance of making definite arrangements for periodic inspection and report by a responsible person informed of the hazards. In remote locations arrangements may be made with a forester, a minor county official, or a nearby rancher for inspection of a group of small stock water dams at nominal expense.

1. Inspection of Structure.—The first thing to be noted in the inspection of a dam is the amount and character of leakage. Some leakage will occur from nearly every dam. If clear water emerges in small volume from one or more points; it should be watched from time to time to see whether it increases in volume or becomes muddy. On the initial filling of a reservoir the water pressure may penetrate small seams that have been overlooked when the foundation was treated or that have been hidden in the bottom of the reservoir. Such seams are generally blanketed by sediment in a short time. The water pressure may develop a flow through crevices in the reservoir slopes and find outlet in the bottom of the stream where it is covered by the lower slope of the dam. Increases in volume or muddiness of such flows should be promptly reported, the cause determined, and necessary action taken without delay.

In times of long drought or by reason of unequal settlement the surface of an earth dam or the earth blanket of a rock-fill dam may develop deep cracks. Such cracks should be puddled promptly with clay or filled with grout pumped through pipes.

Cracks in a concrete dam, an arched masonry dam, or a buttress or hollow concrete dam may be harmless temperature or shrinkage cracks, but any of them should be regarded with suspicion and observed carefully for possible enlargements.

A wave in the surface indicating a settlement of the foundation may be a matter of great importance. Any unusual circumstance observed by the inspector should be reported promptly and followed

204

immediately with a thorough examination by someone competent to suggest a remedy.

Spillways and outlet structures should also receive periodic inspection. Unlined spillways may develop holes in the bottom or in the banks through which the flow of a small volume of water may cause a great deal of damage, particularly in limestone formations. Outlet structures built through the dam offer a path for seepage that may never be completely shut off and should be watched at the discharge end for evidences of erosion or deposition of fine material. The channels developed along such structures may be closed by application of clay grout under pressure at the upstream end.

2. Inspection of Equipment.—Periodic inspection of spillway gates and tests of operating equipment should be made by an engineer or mechanic familiar with the purposes of the equipment. Inlet and outlet gates and valves should be tested regularly to see that they work freely. Trash racks should be cleared of debris and accumulated sediment. Mechanism should be oiled and all metal surfaces kept painted to protect them from rust.

3. Operation and Maintenance.—1. *Normal Use.* The dams discussed in this manual will normally be used to store water for supplementary irrigation, domestic water supply, recreational purposes, stock ponds, or auxiliary flood control in tributaries of main streams. Their operation will rarely require continuous attention except at seasonable intervals. If warranted there should be an operator's house at or near the control works of dams; telephone service and a supply of small tools, sandbags, and other maintenance and emergency equipment may be necessary or desirable.

Besides control for purposes of distribution, the water level in the reservoir may require a change at regular intervals to prevent propagation of malarial or pest mosquitoes and to abate algae or other aquatic growths. The pond level on storage dams may also have to be drawn down on receipt of storm warnings to provide storage for flood waters.

The stimulation and protection of growth of vegetative cover to retard erosion on the slopes of the reservoir, of the borrow pits used in construction, and on the downstream face of earth dams are important items of maintenance to which careful attention should be given. This cover is an essential item of protection against erosion and sloughing of banks, as well as of beautification of the structure, and may have an important influence on the cost of repairs.

Expert advice on suppression of algae growth in the reservoir should be obtained and followed, and no chemicals should be introduced into the reservoir without competent advice.

2. *Changes in Operating Plan.*—A dam built for purposes of flood

control may be diverted from its intended use by reason of the demand of the community for full storage for irrigation or water supply.

Such demands, if acceded to, may result in a dangerous situation and possibly in the complete loss of the dam by overflow in the event of excessive rainfall.

Raising of the height of a dam is frequently undertaken without due consideration of the relation of the increased pressure to the limitations of the original design. No structural changes should be made without reference to the original plans nor without the advice of an experienced engineer.

4. Diversion Dams.—The diversion dam broadly covers dams built for the purpose of raising the level of the stream and not for purposes of storage or equalization of flow.

The dam may divert the flow into a canal for irrigation of the lowland in the stream valley or to spreading grounds for repletion of ground-water storage.

Diversion dams are in whole or in part almost invariably overflow dams. Control gates are usually supplied so that the required diversion level may be maintained in spite of fluctuations in stream flow, or in order to pass portions of the flow as needed to satisfy downstream water rights.

Diversion dams are generally of concrete or rock-filled timber cribs and are apt to be founded on sandy or gravelly soils. In such cases their stability may be insured by a broad base with frequent cut-off walls. Where possible, the cut-off wall at the upstream toe is extended down to an impervious stratum to reduce percolation. Where such treatment is not practicable core walls may be required.

Such dams must be safeguarded by frequent inspection for evidence of piping, boils below the dam, and any increase in volume of seepage appearing at the downstream toe.

The downstream apron, whether of concrete or of timber, should be protected at the toe by heavy riprap. After floods, the stream bed should be examined and the riprap renewed and repaired if necessary.

5. Retarding Reservoirs.—Retarding reservoirs serve to reduce flood peaks by the temporary storage of that part of the flow which is in excess of the capacity of the spillway or outlet works of the dam. All reservoirs or pools produce such retarding effects.

Structures built for the specific purpose of flood control by retardation may be built with outlets which will automatically control the rate of release within safe limits. Overflow spillways are also provided in order to protect the dam even at the expense of possible flood damage below the structure.

In addition to general inspections, the outlet works of such structures should be inspected frequently and cleared of soil deposits and debris which might affect their proper functioning.

6. Coordination of Multiple Use.—Storage dams may be operated for more than one purpose.   Multiple use may be made of the same storage space or various head ranges in the same reservoir may be utilized for two or more purposes, such as for flood control, power, irrigation, recreation, water supply, navigation.   Such combined operation requires very careful planning and control, as some of the uses are not compatible with other uses.   For example, a power user, in order to be sure of the maximum amount of firm power, may wish to have the storage full when flood hazard is imminent and the need for available reservoir capacity greatest.   Such combined operation is possible only with loss of some measure of benefit to one or all of the participants.   In spite of these difficulties, conditions often make it desirable to permit such multiple use.   Where such allocations are known in advance they may have an influence in the design of the control devices.   Careful management is very important in the operation of multiple-use reservoirs in order to maintain a balanced perspective in the matter of relative values.

APPENDIX A

MODIFIED RATIONAL METHOD OF ESTIMATING FLOOD FLOWS [1]

1. The Modified "Rational" Equation.—The rational method was developed and has found its greatest usefulness in the field of city storm sewer design. In 1932, R. L. Gregory and C. E. Arnold presented in the Proceedings of the American Society of Civil Engineers their paper "Run-off—Rational Run-off Formulas." This work develops a general formula for surface run-off from rainfall, based upon the equation $Q = C \; i \; A$, but takes into account such factors as watershed shape and slope, the pattern of the stream system, and the elements of channel flow. It also demonstrates the applicability of the method to natural watersheds of considerable area.[2]

2. The Run-Off Coefficient, C.—An unfortunate aspect of a study of relationships between rainfall and run-off lies in the meager knowledge of what are commonly referred to as "losses." Experimental data now available are largely limited to run-off measurements from small plots which have no areal significance and cannot be applied directly to larger areas.

The coefficient, C, in the rational equation not only depends upon the principal losses of infiltration and transpiration, but is the means of adjusting for variations in contributions from the given areas. The coefficient, C, is the ratio of the maximum peak flow per acre expressed in cubic feet per second to the average rate of rainfall in inches per hour throughout the period of concentration.

Figure 113 presents suggested values of *limiting* run-off coefficients $C_{\max}$ for that portion of the United States in which rainfall can appropriately be used as the basis of flood estimates. The chart takes into account limitations fixed by—

(1) The geographical location of the watershed;
(2) The seasonal occurrence of excessive rainfall.

Progressing generally from the south and east to the west and north, changing climatic and physiographic conditions have the net effect of reducing the proportionate part of rainfall which finds its way into the streams during the comparatively short period in which flood peaks are created. As the coefficient value reduces rapidly to the west and northwest, it is apparent that geographical variation cannot be disregarded.

Despite the recognized differences in soils, cover, and topography, the greater portion of the central, northern, and eastern United States will, under the conditions prevailing in the late winter and spring months, produce a run-off coefficient reaching and exceeding unity. With equal certainty, it is known that excessive storms do not occur in certain well-defined regions before certain months, reckoning from the beginning of the calendar year. The limiting run-off coefficient chart (fig. 113), eliminates from consideration limiting coefficients in those months in which it would seem wholly improbable that excessive storms could occur. The maximum coefficient developing in any of the remaining months, up to and including unity, has been adopted as the limiting value for the locality.

[1] By Merrill Bernard, Chief, River and Flood Division, U S Weather Bureau
[2] See Glossary of terms in section 6 for these and other symbols used in text of this appendix.

209

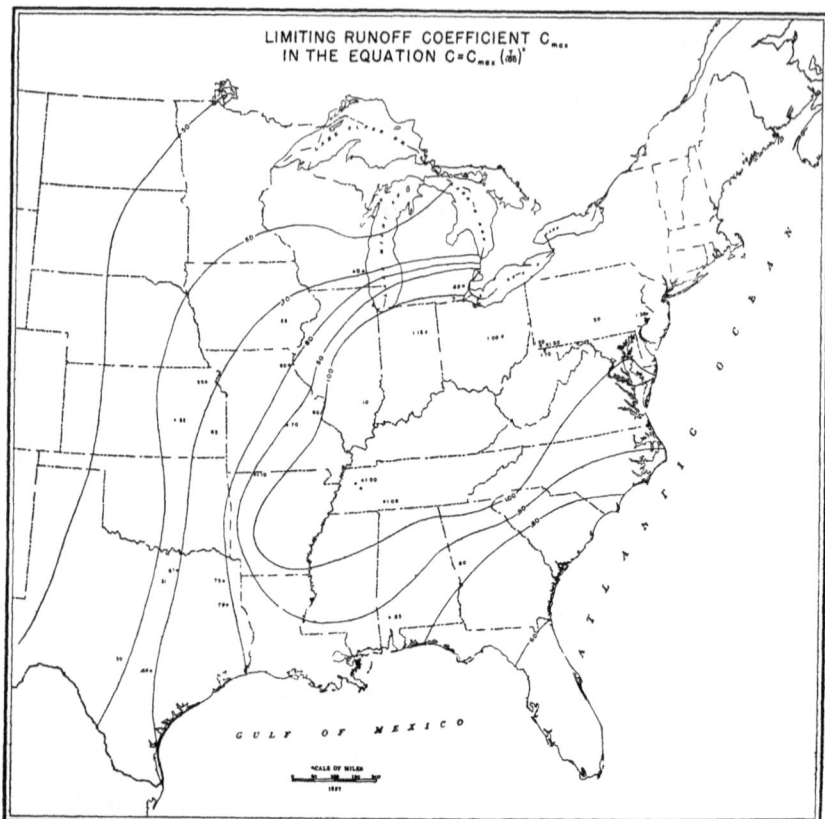

LIMITING RUNOFF COEFFICIENT C_{max}
IN THE EQUATION $C = C_{max} \left(\frac{T}{100}\right)^x$

FIGURE 113.

The relation between a run-off coefficient, expressed in any terms, and storm rainfall is an obscure one. There is reason to believe that, if basic data were available, a duration curve [3] of the run-off coefficient would be found to have a characteristic common to both streamflow and rainfall for all durations up to 30 days. When, for either rainfall or streamflow, frequency is plotted logarithmically against magnitude, the slope of the plot is consistently between 0.15 and 0.23. This slope is the exponent x in the rainfall equation,

$$ i = \frac{KT^x}{t^e} \quad \text{(See item (2) section 3)} $$

For use in the Rational Method it is proposed to reduce the value of the limiting coefficient to that of the selected frequency by a similar equation as follows:

$$ C = C_{max} \left(\frac{T}{100}\right)^x $$

under an assumption that the limiting value of the coefficient has a frequency of once in 100 years. The frequency-discharge curves shown in figure 3 were computed using a coefficient derived from the foregoing equation.

―――――――――

[3] Showing the percentage of time for which specific values are exceeded

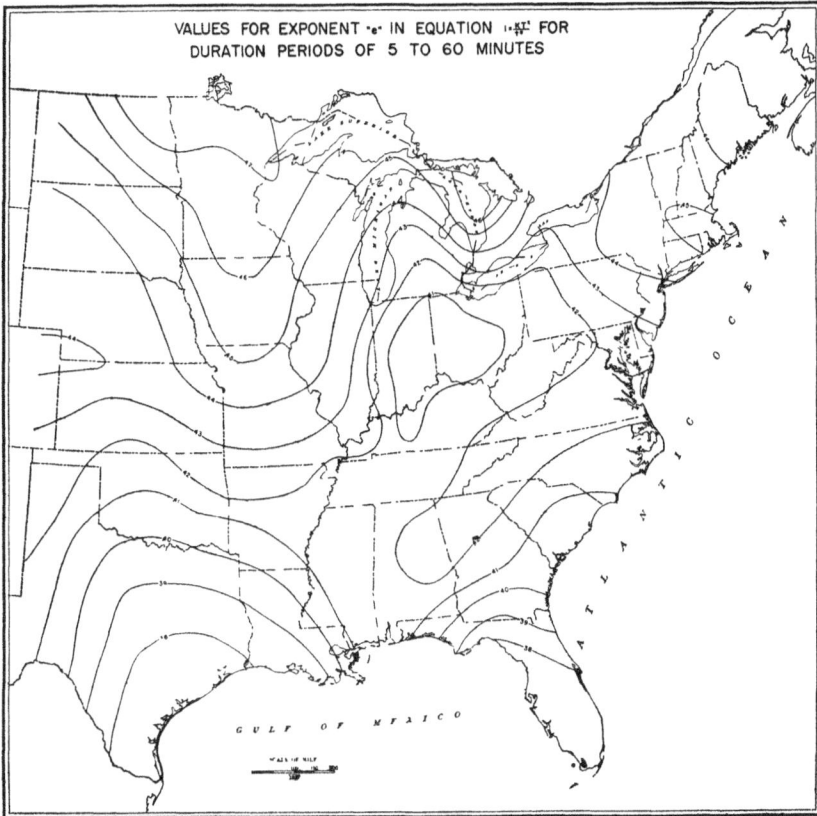

VALUES FOR EXPONENT "e" IN EQUATION ᵢ₌ₙᵗ FOR
DURATION PERIODS OF 5 TO 60 MINUTES

GULF   OF   MEXICO

FIGURE 114.

3. Average Rainfall Intensity, i.—Basic rainfall data for eastern United States have been analyzed as to the relationship of intensity, duration, and frequency in several studies, the outstanding of which are:

(a) Rainfall Intensity-Frequency Data, by D. L. Yarnell. (U. S. Department of Agriculture, Miscellaneous Publication No. 204.)

(b) Storm Rainfall of Eastern United States, published by the Miami Conservancy District.

(c) Rainfall Rate Equations, by Adolph F. Meyer, presented in his Elements of Hydrology.

(d) Formulas for Rainfall Intensities of Long Duration, by Merrill Bernard, Trans. A. S. C. E , Vol. 96, p. 592.

Only the first work listed covers the whole of the United States, the remaining three serving only the country lying east of the 101st meridian. For this portion of the country there is reasonable accord in the results produced by the four studies. It is proposed to utilize the Yarnell charts, considering their agreement with the other studies a check on their dependability for the area commonly served by all.

Average rainfall rate, i, in the rational equation is taken as rainfall depth divided by the duration period, and is stated in inches per hour. It is difficult to interpolate, from the Yarnell charts, rainfall rate throughout any duration. Also, it

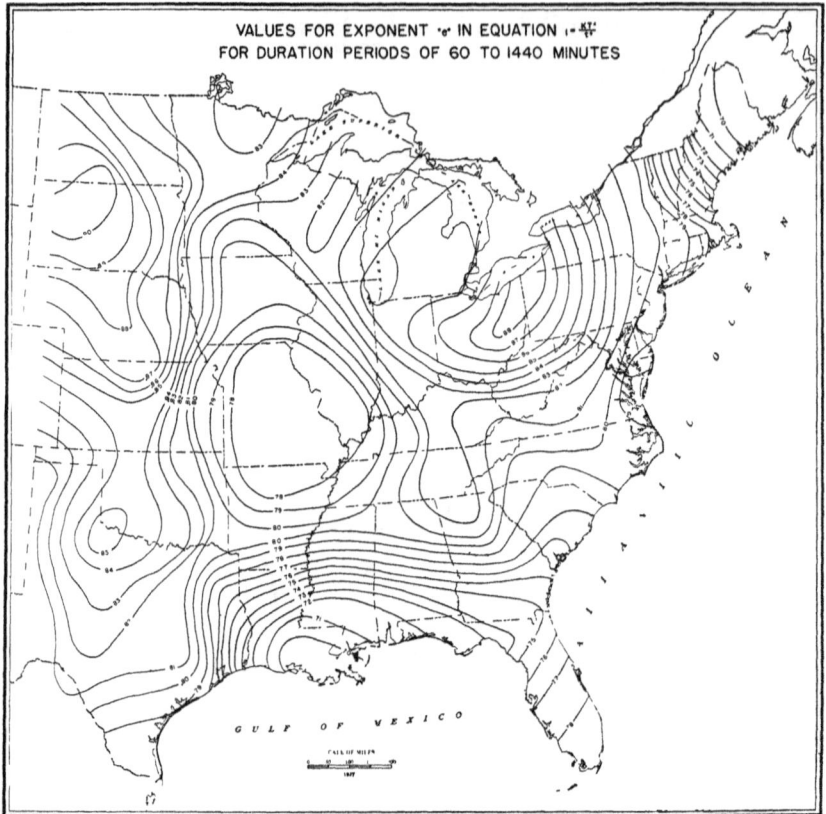

FIGURE 115

is desirable in several of the uses to which these data are put, to express mathematically the relationships of rate, duration, and frequency.   For these reasons
the Yarnell values have been converted to rates and are presented in figures 114 to
118, inclusive.   These charts were developed in the following manner:

1. Rates were plotted logarithmically against frequencies, each duration in the
table being represented by a curve.

2. These curves were adjusted to the plotted points by adopting a slope of best
fit to the curve group, such slope of the curve group becoming the exponent, x, of
the frequency factor, T.

3. The rainfall rate-frequency curves were then extrapolated to an intersection
with the 1-year frequency ordinate, these values, plotted as duration against rate,
becoming the 1-year rainfall curve.

4. The intensity-duration curves will consistently be found to have a pronounced break at about 60 minutes. Therefore it is considered best to treat the
plot as two separate curves, although all durations could be expressed in a single
equation by introducing an additive factor in the denominator.

5. The coefficient, K, and the exponents, e and x, in the general rainfall equation.

$$i = \frac{KT^x}{t^e}$$

were computed and charted for values of t greater than and less than 60 minutes.

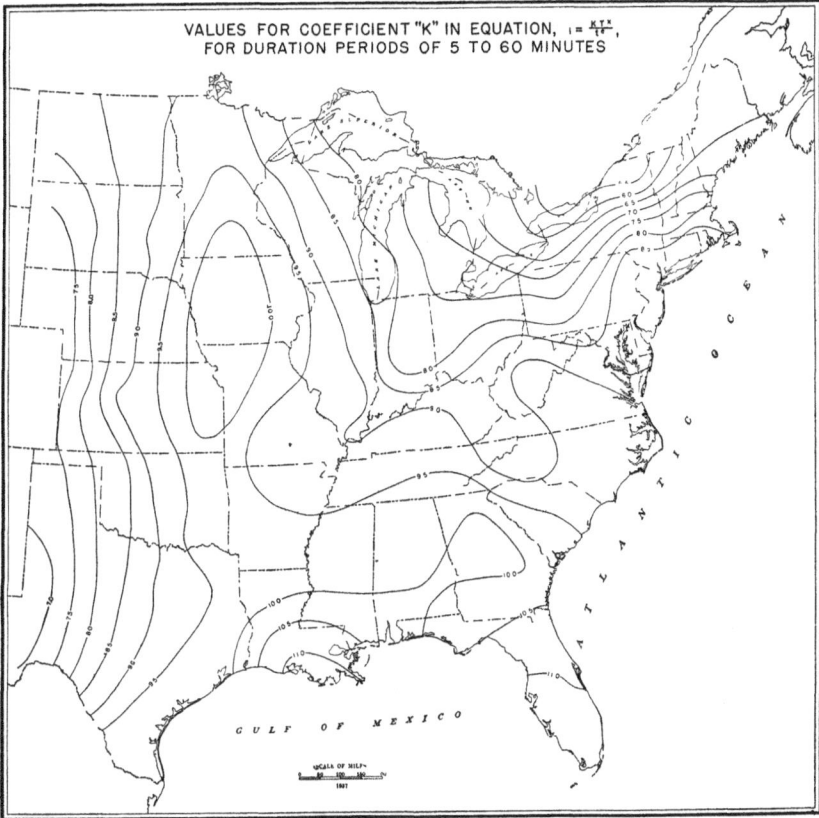

VALUES FOR COEFFICIENT "K" IN EQUATION, $i = \frac{KT^x}{t^n}$,
FOR DURATION PERIODS OF 5 TO 60 MINUTES

FIGURE 116

4. Concentration time, t_c.—The rational method is based upon the theory that, for a given frequency, maximum run-off rate at the channel location being studied, results from a rainfall of duration equal to the time of concentration of the particular watershed. In this, the simplicity of the CiA equation is misleading, for the critical value of the rainfall intensity i, through the medium of concentration time, entails a consideration of such factors as watershed size, shape, and slope; channel length, shape, slope, and condition; as well as variation in rainfall rate, duration, and frequency; all of which can and should be considered in determining its value. The subsequent discussion will illustrate how these various factors are reflected in the Modified Rational Method.

5. The method applied.—The solution of a problem by the method here proposed becomes mechanical after the basic factors have been selected; and the use of figures 119 to 127 inclusive materially reduces the computations in solving the base formula. (Equation 3 below.)

The frequency design factor has been selected to meet the economic and social aspects of the problem as previously discussed.

The rainfall factors are taken from figures 114 to 118 inclusive. The abrupt change in the relation between rate and duration at about 60 minutes makes it necessary to classify the problem by a preliminary approximation of concentration time as outlined in Chapter 2:9 before deciding upon the charts to be used

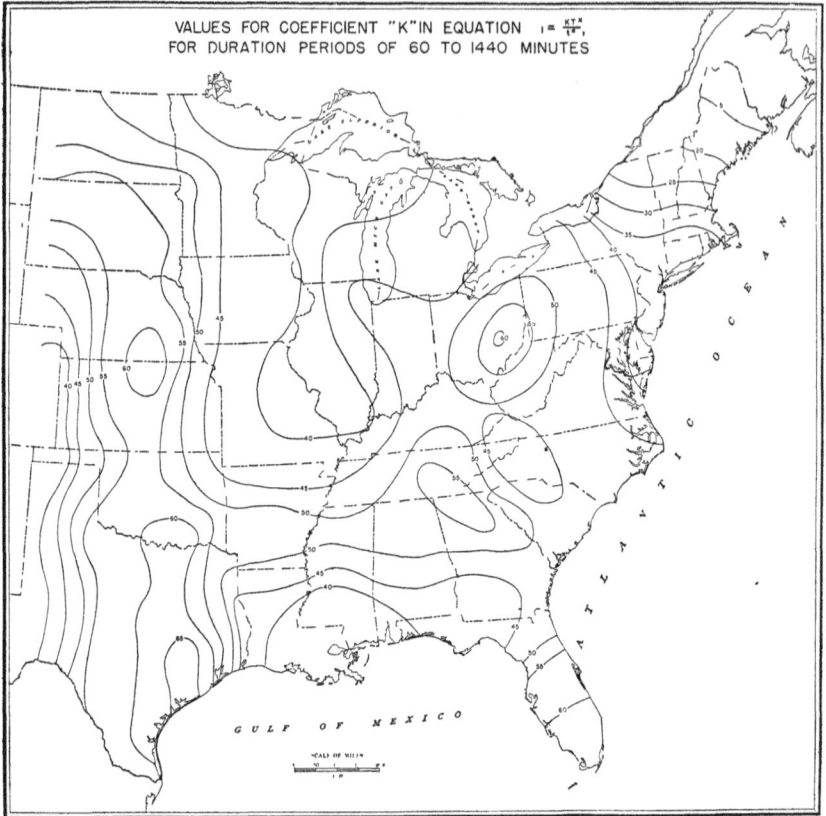

VALUES FOR COEFFICIENT "K" IN EQUATION $i = \frac{KT^x}{t^e}$, FOR DURATION PERIODS OF 60 TO 1440 MINUTES

FIGURE 117.

The topographic factors are available on any reasonably accurate topographic map.

The channel factors are understood to be the average for the whole of the stream system at flood stages. Thus, side slopes, bottom width, average depth, and n are values for the floodway and not for channel of normal flow. A reconnaissance survey should yield sufficient data regarding channel shape and condition to fix the values of these factors.

The profusely illustrated bulletin, "Flow of Water in Drainage Channels," by C. E. Ramser, is an excellent guide to the proper selection of the roughness factor, n. Table 1 and the accompanying photographs are from this bulletin. (fig. 128.)

The slope factors are available on the topographic map of the watershed.

The run-off factor is determined from the C_{max} chart, figure 113, reduced to the adopted frequency of the design factor using the exponent x, taken from figure 118.

It is to be noted that concentration time does not appear in the determination of the Q by the proposed formula, being inherent in the factors K and the exponent e. However, it is necessary to know its value in connection with other phases of the problem. The solution of equation 4 for t_c provides a check on the solution of Q when compared with the value of t_s for equation 1.

6. Glossary of terms.—

A = Area of drainage basin in acres.

L = Length of principal channel, in feet, following general meanders.

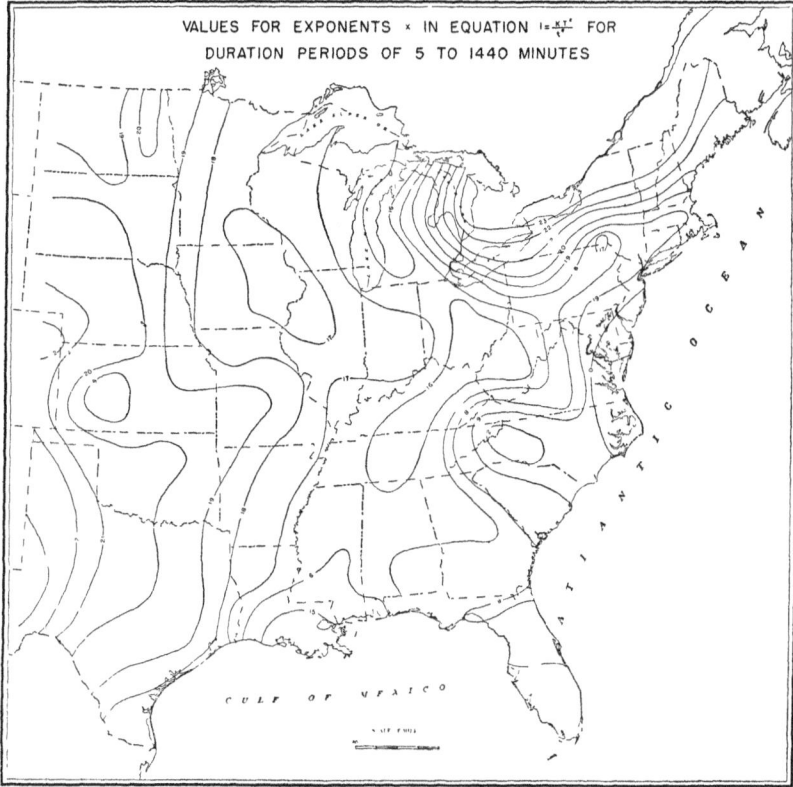

FIGURE 118

$E=$ Difference in elevation between headwaters and outlet, in feet

$D=\dfrac{1{,}000E}{L}.$ Fall of principal channel in feet per thousand

$W=\dfrac{43{,}560A}{L}.$ An index figure; approximately the average width of the basin.

[4] $J=\dfrac{1}{PFS}.$ A watershed factor

[4] $P=$ A shape factor (fig. 119).

[4] $F=$ A channel factor (fig. 120).

[4] $S=$ A slope factor (figs. 123–124).

$i=\dfrac{KT^{x}}{t_{i}^{e}}.$ Average rainfall rate in inches per hour for a duration of t minutes which is reached or exceeded with an average frequency of once in T years.   (Equation 1.)

$K=$ A coefficient, depending for value on locality (figs. 116 and 117).

$t_{i}=$ Duration of rainfall intensity i taken as equal to,

$t_{e}=$ the concentration time of the watershed, in minutes.

[4] An understanding of the meaning of these factors or of the derivation of equations 3 and 4 is not essential to the application of the Rational Method as here outlined, but may be obtained by reference to the Gregory and Arnold paper cited and to the paper "An Approach to Determinate Streamflow" by Bernard Trans. A S. C E , Vol 100—p 347

$T=$ A frequency factor, since i inches per hour for a duration of t minutes will be reached or exceeded on the average only once in T years

$e=$ The exponent of duration, t, depending for value upon locality (figs. 114 and 115).

$x=$ The exponent of frequency, T, depending for value on locality (fig. 118).

$g=\dfrac{1}{4-e}.$

$C_{max}=$ A limiting run-off coefficient depending for value on locality (fig. 113)

$C=C_{max}\left(\dfrac{T}{100}\right)^{x}.$ A run-off coefficient of adopted frequency depending for value upon locality. (Equation 2.) (Fig. 113 and fig. 118.) (For computed values of $\left(\dfrac{T}{100}\right)^{x}$ for values of x see table 10.)

[4]$Q=\dfrac{(CAK)^{4g}T^{4gx}}{J}.$ Maximum flood flow in cubic feet per second (fig. 125 and fig. 127). (Equation 3.)

[4]$t_{c}=\dfrac{J^{1/e}}{(CAK)^{g}T^{xg}}.$ Concentration time, in minutes (figs. 125, 126, and 127). (Equation 4.)

7. An Example.—The headwater area of Mill Creek, lying approximately 15 miles north of Coshocton, Ohio, will be used to demonstrate the application of the rational method as here presented

Design factor:

Spillway to accommodate a peak flow expected to be reached or exceeded on the average of once in 50 years.

Rainfall factors:

$e=.88.$ (From fig. 115.)
$x=.16.$ (From fig. 118.)
$K=59.$ (From fig. 117.)

$i=\dfrac{59T^{0.16}}{t_{i}^{0.88}}$ for $t>60$ minutes.   (Equation 1.)

TABLE 10.—*Values of* $\left(\dfrac{T}{100}\right)^{x}$ *in equation* $C=c_{max}\left(\dfrac{T}{100}\right)^{x}$

x (exponent of frequency)	T (frequency in years)			
	10 years	25 years	50 years	100 years
0 15	0 70795	0 81225	0 90125	1 00000
16	69183	80107	89503	1
17	67608	79004	88884	1
18	66069	.77916	88270	1
19	64565	76843	87661	1
20	63096	75786	.87055	1
21	61660	74742	86454	1
22	60256	.73713	85857	1
23	58885	72699	85264	1

NOTE.—Value of $\left(\dfrac{T}{100}\right)^{x}$ may be found by direct interpolation for intermediate values of x

Topographic factors:

$A=5,620$ acres.
$L=23,000$ feet.

$W=\dfrac{43,560 A}{L}=10,640$ feet.

$\dfrac{L}{W}=2.16$

[4] See footnote p 215

$P = .000575.$ (Interpolated from fig. 119.)

Channel factors:

Side slopes $= 2:1$

$n = .04$

$\dfrac{\text{bottom width}}{\text{average depth}} = 60.$

$F = 17.$ (Interpolated from fig. 120–122)

Slope factors:

$E = 245$ feet.

$D = \dfrac{1,000 E}{L} = 10.65$ feet per thousand.

$S = .0552.$ (Interpolated from figs. 123–124.)

Watershed factor:

$J = \dfrac{1}{PFS} = \dfrac{1}{.000575 \times 17 \times 0.0552} = 1853.$

Run-off factor:

$C_{max} = 1.00$

$C = 1.00 \left(\dfrac{50}{100} \right)^{16} = (1.00)(.895) = .895.$ (Equation 2)

Maximum flood discharge

$Q = \dfrac{(CAK)^{4g} T^{4gx}}{J}.$ $(CAK)^{4g}$ from figure 125; T^{4gx} from figure 127.

$g = \dfrac{1}{4 - e} = 0.3205.$

$4g = 1.282.$

$4gx = 0.205.$

$gx = 0.05128.$

$Q = \dfrac{(0\,895 \times 5,620 \times 59)^{1\,282}}{1,853} 50^{0\,205}.$ (Equation 3.)

$= \dfrac{296,800^{1\,282} \times 50^{0\,205}}{1,853}$

$= \dfrac{(10,368,000)}{1,853} (2.23) = 12,480$ second feet.

Concentration time, t_c.

$t_c = \dfrac{J^{1/e}}{(CAK)^{g} T^{gx}}$ $(CAK)^{g}$ from figure A–10 $J^{1/e}$ from figure 126. T^{gx} from figure 127.   (Equation 4)

$= \dfrac{1,853^{1\,136}}{296,800^{0\,3205} \times 50^{0\,05128}}$

$= \dfrac{5,156}{56.74 \times 1.222}$

$= 74$ minutes.

Rainfall duration, $t_i =$ concentration time.

$Q = CiA$

or, $i = Q/CA$; also, $i = \dfrac{KT^x}{t_i}$

whence

$$\frac{KT^x}{t_i'} = Q/CA$$

or $t_i' = \dfrac{(CAK)T^x}{Q}$. T^x from figure 127.

and $t_i = \left[\dfrac{(CAK)T^x}{Q}\right]^{1/6}$

$\quad = \left[\dfrac{296,800 \times 1.87}{12,480}\right]^{1\,136}$

$\quad = 44.5^{1\,136}$

$\quad = 75$ minutes; note that this checks closely the concentration time, t_c as determined from equation 4.

This example was worked with a log-log slide rule, rather than with the aid of figs. 125, 126, and 127; however, a result not materially different would be obtained by use of those figures.

L = LENGTH OF PRINCIPAL CHANNEL IN THOUSAND FEET

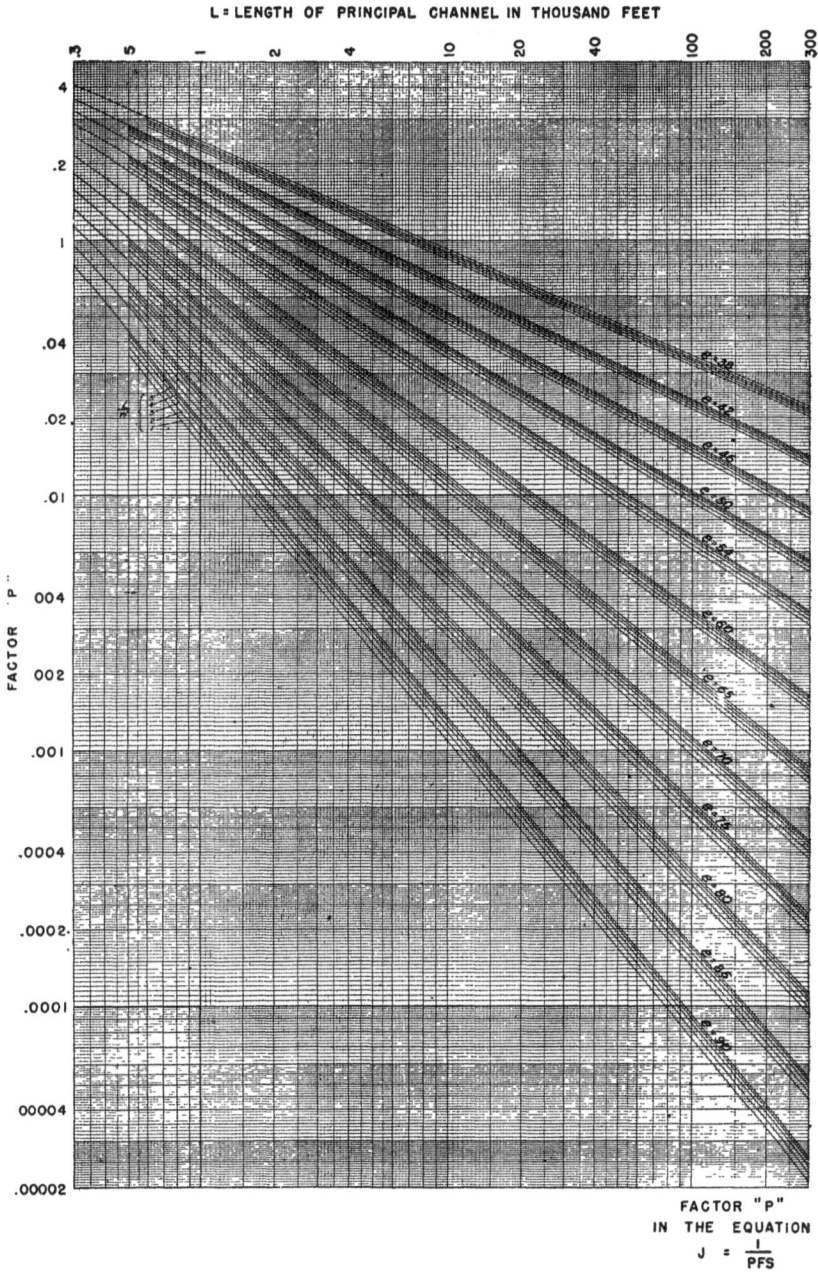

FACTOR "P"
IN THE EQUATION
$$J = \frac{1}{PFS}$$

FIGURE 119.

FACTOR "F"

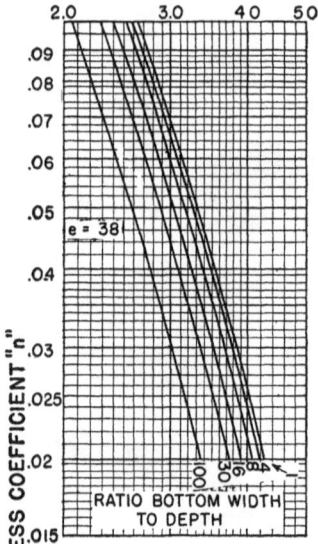
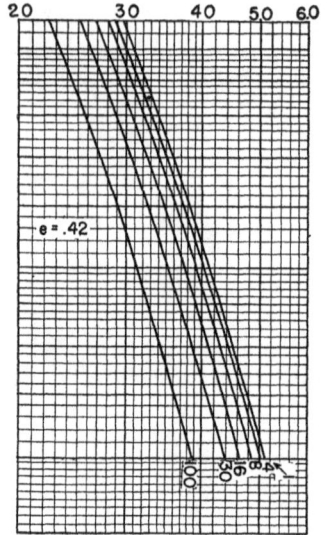
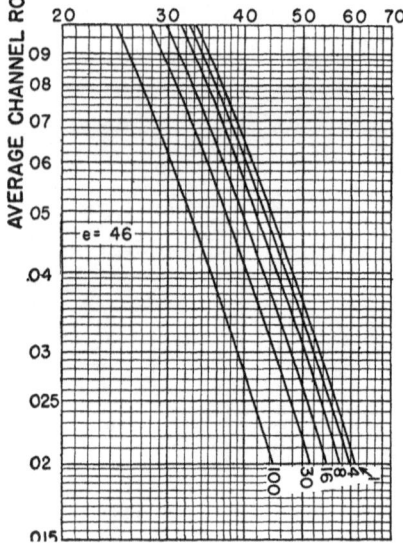
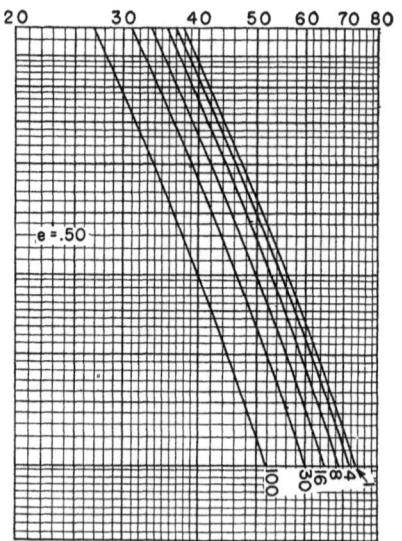

FACTOR "F"

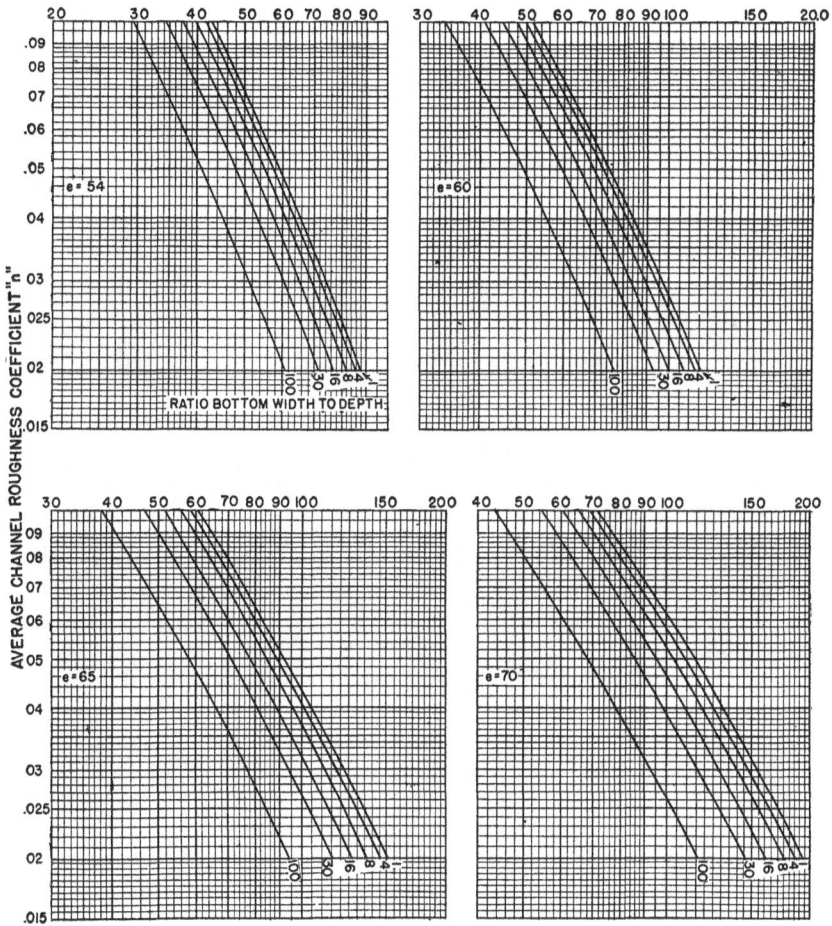

FACTOR F IN THE EQUATION U=PFS

FIGURE 121.

FACTOR "F"

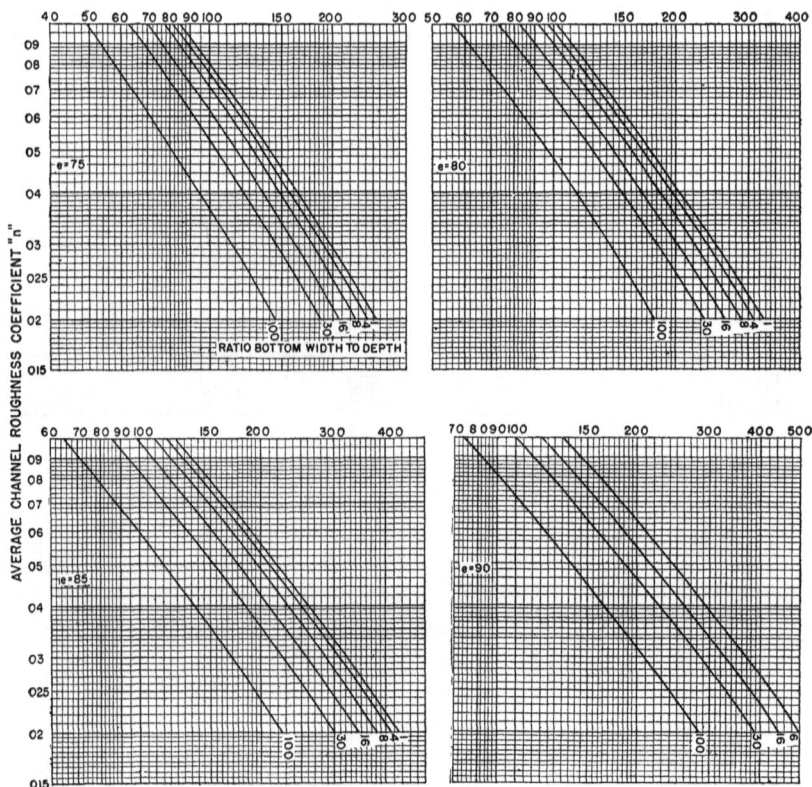

FACTOR F IN THE EQUATION U = PFS

FIGURE 122.

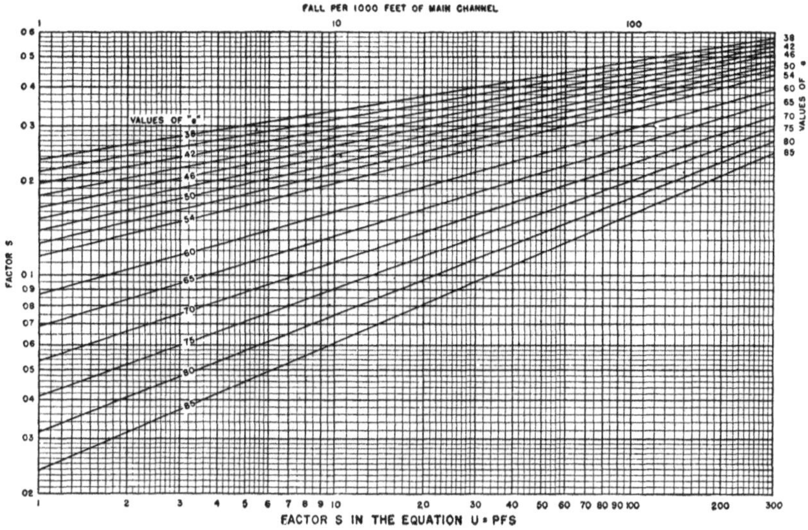

FIGURE 123.

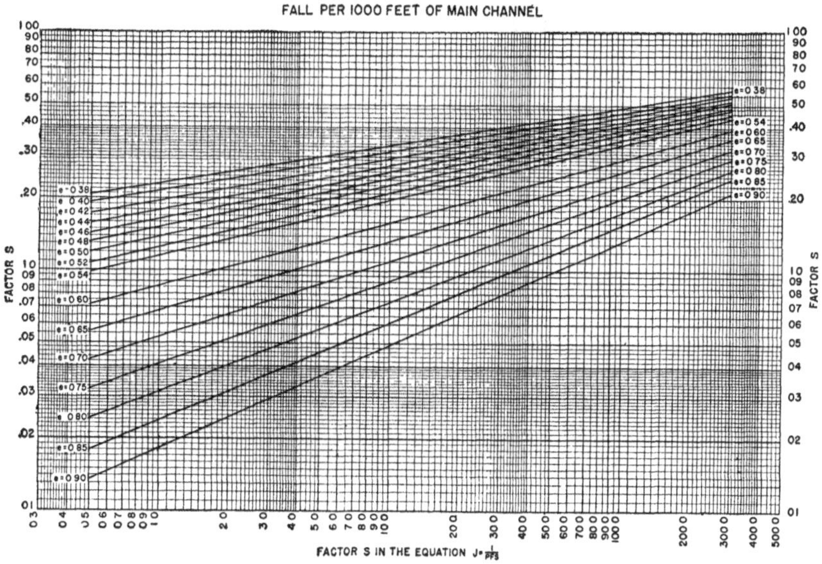

FIGURE 124.

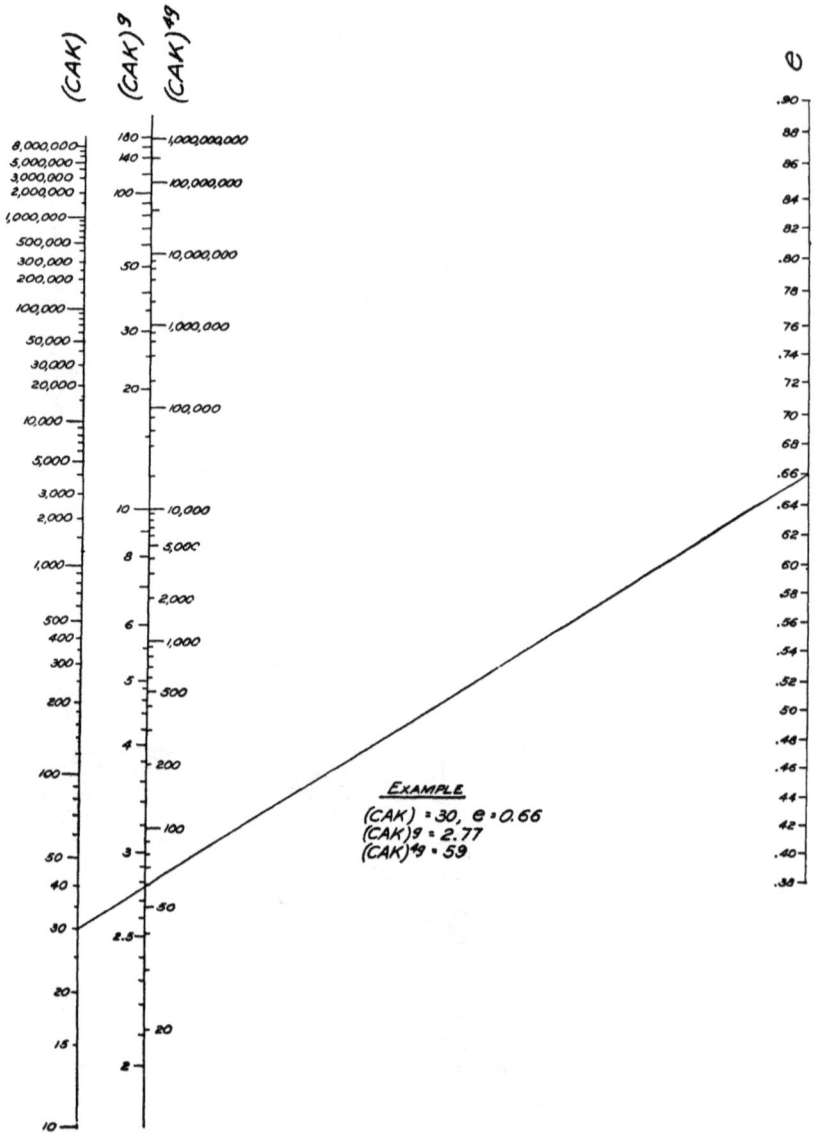

FIGURE 125 —Nomograph for solution of $(CAK)^9$ and $(CAK)^{49}$

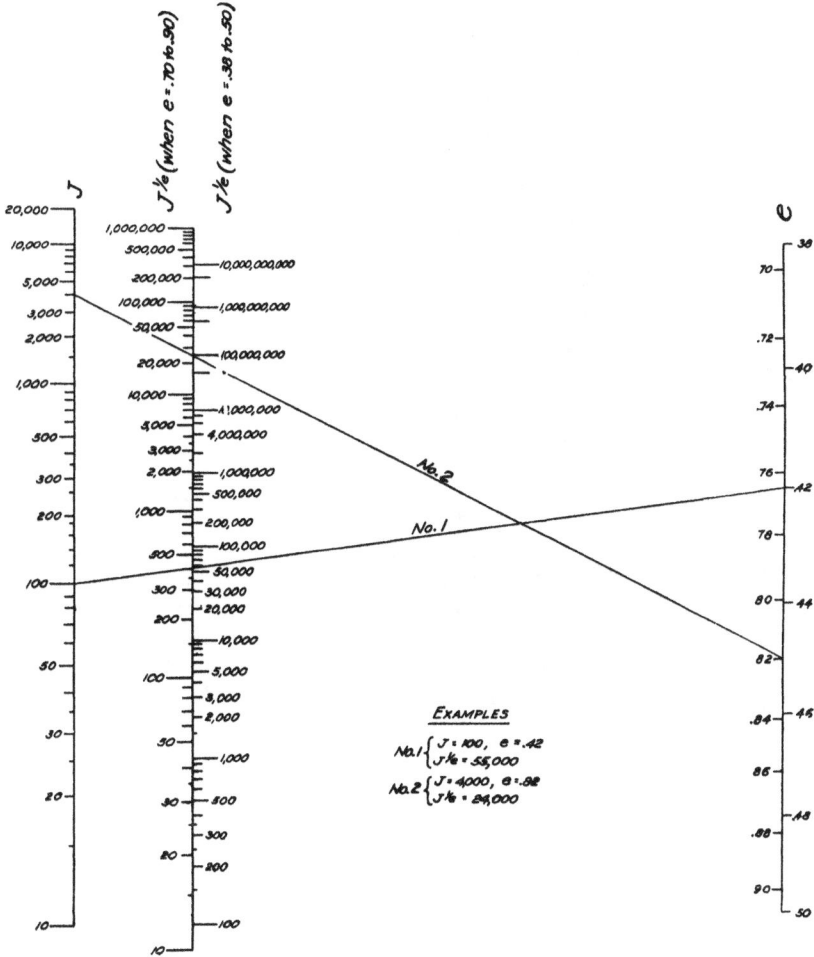

FIGURE 126 —Nomograph for solution of $J^{1/e}$.

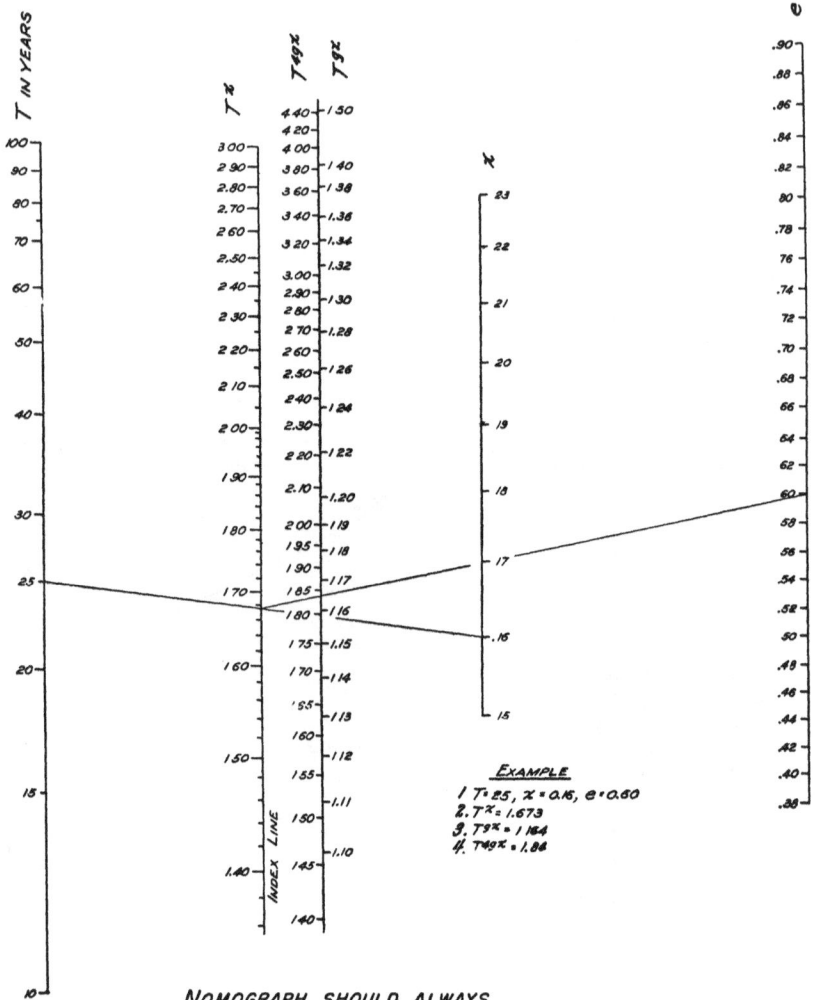

NOMOGRAPH SHOULD ALWAYS
BE USED IN THE ABOVE SEQUENCE

FIGURE 127 —Nomograph for solution of T, T^{4gx}, and T^{gx}

TABLE 1 Channel roughness factors [1]

NOTE.—Numbers refer to photographs comprising figure 128.

PHOTOGRAPH 1

Name.—Embarrass River near Charleston, Ill.
Course.—Straight, 1,000 feet.
Side slopes.—Somewhat irregular.
Bottom.—Fairly even and regular.
Soil.—Lower part light gray silty clay; upper part, light tan silt loam.
Condition.—Bottom comparatively clean and smooth.
Cross section.—Very little variation.

PHOTOGRAPH 2

Name.—Ditch No. 19, near Winchester, Ark.
Course.—Straight, 1,600 feet long.
Side slopes.—Irregular.
Bottom.—Rather irregular.
Soil.—Clay, sandy loam, silty clay.
Condition.—Some silting, grass on slopes, lower part of channel comparatively free from vegetation.
Cross section.—Some variation in shape

PHOTOGRAPH 3

Name.—Kaskaskia River Dredged Channel near Sadorus, Ill.
Course.—Crooked.
Side slopes.—Irregular and uneven.
Bottom.—Fairly even and regular.
Soil.—Lower part hard, waxy, slippery clay; upper part, gray silt loam.
Condition.—Very young growth and stubble on upper part, none on lower part, vegetation cut every two years, channel cleared year preceding flow measurement.
Cross section.—Considerable variation.

PHOTOGRAPH 4

Name.—Old Town Creek Dredged Channel near Tupelo, Miss.
Course.—Straight, 1,224 feet long
Side slopes.—Irregular.
Bottom.—Irregular and uneven.
Soil.—Black, waxy clay at top to yellow clay at bottom.
Condition.—Sides covered with small saplings and brush.
Cross section.—Slight and gradual variations.

PHOTOGRAPH 5

Name.—West Bogue Hasty Dredged Channel near Shaw, Miss.
Course.—Straight, 897 feet long.
Side slopes.—Very irregular.
Bottom —Very irregular.
Soil.—Dark colored waxy clay.
Condition.—Weeds and grass.
Cross section.—Slight variation in shape for variation in size.

[1] From C E. Ramser, "Flow of Water in Drainage Channels," U S Department of Agriculture, Technical Bulletin No 129—1929

PHOTOGRAPH 6

Name.—Ditch No. 18 near McGehee, Ark.
Course.—Straight, 810 feet long.
Side slopes.—Irregular and uneven.
Bottom.—Irregular and uneven.
Soil.—Heavy silty clay.
Condition.—Practically entire section filled with large size growth of trees, principally willows and cottonwoods.
Cross section.—Quite uniform.

PHOTOGRAPH 7

Name.—Kaskaskia Mutual Dredged Channel near Bondville, Ill
Course.—Nearly straight.
Side slopes.—Irregular.
Bottom.—Irregular.
Soil.—Lower part, black clay; upper part dark gray silty clay loam.
Condition.—Badly obstructed by trees, 2 to 12 inches diameter.
Cross section.—Some variation.
Note.—Winter condition.

PHOTOGRAPH 8

Name.—Same as No. 7.
Note—Summer condition.

PHOTOGRAPH 9

Name.—St. Francis River Floodway near Marked Tree, Ark.
Course.—Straight, 8,000 feet.
Side slopes.—None.
Bottom.—Fairly even and regular with occasional flat bottom sloughs.
Soil.—Varies from medium fine sand to fine clay.
Condition.—Practically virgin timber, very little undergrowth except occasional dense patches of bushes and small trees; some logs and dead fallen trees.
Cross section.—Variation in depth only.

PHOTOGRAPH 10

Name.—South Fork Deer River near Jackson, Tenn.
Course.—Very crooked.
Side slopes.—Very irregular.
Bottom.—Very irregular and full of holes.
Soil.—Sandy clay loam.
Condition.—Many roots, trees and bushes, large logs and other drift on bottom, trees continually falling into channel due to bank caving.
Cross section.—Large variations.

BIBLIOGRAPHY

"Floods in the United States," U. S. Geological Survey, 1936, *Water Supply Paper* No. 771.
"Rainfall and Run-off in the United States," U. S. Geological Survey, *Water Supply Paper* No. 772, 1936.
C. S. Jarvis, "Flood Flow Characteristics," *Trans. Am. Soc. C. E.*, vol. 89, 1926.
R. L. Gregory and C. E. Arnold, "Run-off—Rational Run-off Formulas," *Trans. Am Soc. C. E.*, vol. 96, 1932.

1.—n = 0.035.

2.—n = 0.043.

3.—n = 0.040.

FIGURE 128.—Roughness factors.

4.—$n = 0.045$.

5.—$n = 0.050$.

FIGURE 128.—Roughness factors (continued).

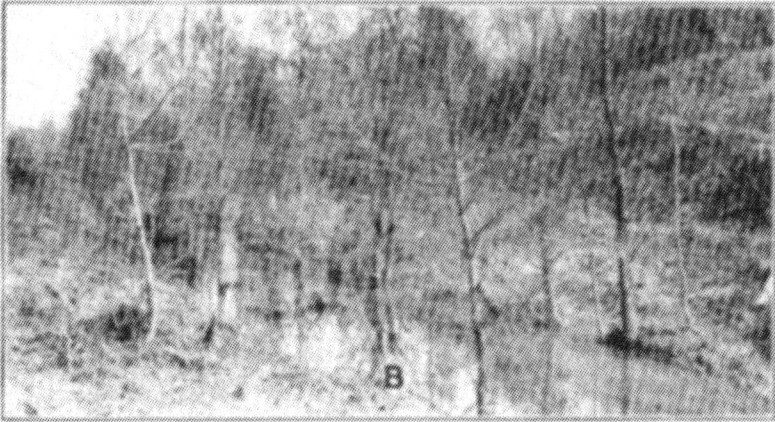

6.—$n = 0.060$.

7.—$n = 0.070$.

8.—$n = 0.110$.

FIGURE 128.—Roughness factors (continued).

9.—$n=0.125$.

10.—$n=0.150$.

FIGURE 128.—Roughness factors (continued).

C. E. Ramser, "Run-off from Small Agricultural Areas," *Journal of Agricultural Research*, Vol. 34, No. 9, 1927.

"Storm Rainfall of Eastern United States" published by the *Miami Conservancy District*.

David L. Yarnell, "Rainfall Intensity-Frequency Data," U. S. Dept. of Agriculture *Misc. Pub.* No. 204, 1935.

Merrill Bernard, "Formulas for Rainfall Intensities of Long Duration," *Trans. Am. Soc. C. E.*, vol. 96, 1932.

William A. Liddell, "Stream Gaging," McGraw-Hill Book Co., Inc., 1927.

Don M. Corbett, "Gaging the Flow of Streams," an unpublished manuscript, U. S. Geological Survey.

"Equipment for Gaging Stations for Measuring River Discharge," prepared by the *U. S. Geological Survey*.

C. E. Ramser, "Flow of Water in Drainage Channels," U. S. Department of Agriculture, *Technical Bulletin*, No. 129, 1929.

"Instructions for Cooperative Observers," Weather Bureau *Circulars B and C*, 8th Edition revised, U. S. Government Printing Office, 1935.

"Measurement of Precipitation," Weather Bureau *Circular E*, Revised, U. S. Government Printing Office, 1922.

Horace W. King, "Handbook of Hydraulics."

Robert E. Horton, "Some Better Kutter's Formula Coefficients," *Eng. News* vol. 75.

APPENDIX—B

SOIL MECHANICS [1]

SECTION A—GENERAL

1. Foreword.—The term "soil mechanics" is now accepted quite generally to designate that science which deals with the character and behavior of soils for engineering purposes. To Dr. Terzaghi belongs the credit for the application of mechanics and his work forms the base upon which the more rational approach to the problem of soil behavior is founded. Moreover, most of the present technique of soil character investigation was devised by him.

It is of the utmost importance that the state of development of the science be thoroughly understood. Development has proceeded at a rapid pace during the past 10 years and positive accomplishments have been recorded. Nevertheless, the science is in its infancy and many of the factors and relations are as yet obscure. It is not now, and probably never will be, a subject which can be tabulated in the fashion of a steel handbook, from which a definite solution may be obtained when certain facts are known. Soil mechanics will always require broad understanding and experience for the most reasonable evaluation of any given set of data. Soil is not a perfectly homogeneous material, although that assumption is frequently required for a solution. Any solution may, therefore, be in error as much as one or two hundred percent. Even with this possible error, however, the solution will be more dependable in the great majority of cases than that obtained by the older strictly empirical rule-of-thumb methods.

It is emphasized that the possibility of error is not a justification for the elimination of precision and refinement of method. There is a tendency to assume that extreme care is useless because the results must often be evaluated rather broadly. This is not the case. Dr. Casagrande has pointed out that an error in the specific gravity of a soil greater than ± 0.02 will result in an error in the hydrometer analysis beyond the tolerable limits. To assume that a soil has an average value of 2.65 is apparently an unwarranted lack of precision.

Little has yet been accomplished in the standardization of soil mechanics technique and soil classification. Committee D-18 of the American Society for Testing Materials, "On Soils for Engineering Purposes," has been functioning for a very short while and it will be some time before that committee will be in a position to recommend tentative standards. It is, however, necessary for the purpose of this treatise to give certain classifications and methods as preferred. It should be understood that the terms and technique recommended herein may be altered more or less in the interest of a common language when the committee presents its recommendations.

2. Symbols.—The following symbols have been chosen with two thoughts in mind, first, that they should conform as closely as possible to generally accepted usage, particularly with respect to dimensional analysis, and, second, that they should conform to the limitations of the ordinary typewriter since in general use they will be typewritten. In order to accomplish the latter all foreign lettere with the exception of theta, phi, and pi have been omitted. Of these three tht first two are easily performed on the typewriter; pi is too deeply fixed to permit either its omission or change despite the inconvenience it introduces. There is,

---

[1] By E. F. Preece, Assistant Chief Engineer, National Park Service. (Not reviewed by editors)

234

of course, much more latitude in printer's type and the subscripts and powers used herein are smaller than the other letters. They can, however, be made quite satisfactorily with but a single size of type. In order to accomplish these objectives a decision had to be made in many cases where conflict existed whether to hold to common usage or to typewriter limitations. Having experienced the difficulty of leaving blank spaces in typewritten reports for Greek letters to be filled in afterward by hand the writer believed that ease of typing was more important than following pedantic precedent. Where possible precedent has been accepted, where inconvenient, it has been departed from

A = area in square feet unless stated otherwise.

   (dimensionally, $A = L^2$).

a = acceleration in ft/sec ^{2} (dimensionally L/T^2).

B = width in feet unless stated otherwise.

  $_b$ = bulk (as bulk specific gravity, G_b).

C = Chezy's coefficient.

  = centigrade.

  = total cohesion.

c = any coefficient.

  = unit cohesion.

c' = unit cohesion required for equilibrium (ϕ circle method)

  $_c$ = critical (as critical height, H_c).

  = corrective (as temperature correction, t_c).

D = diameter in ft. unless stated otherwise.

d = diameter in in. unless stated otherwise.

  $_d$ = days (as time in days = T^d).

E = energy (dimensionally, $E = F\,L$).

  = effective size in millimeters (E_{10} — than which 10% is finer.

                                  E_{20} — than which 20% is finer, etc).

e = void ratio = Q_e/Q_s.

  $_e$ = voids (as volume of voids, Q_e).

$(_e)$ = effective.

F = force in pounds unless stated otherwise.

  = Froudes number = V^2/Lg.

  = Fahrenheit.

f = friction.

G = any specific gravity — G_s = specific gravity of solids.

                          G_w = specific gravity of water.

                          G_b = bulk or apparent specific gravity.

g = gravitational acceleration, in ft/sec^2; (32.174 ft/sec^2 = 980.665 cm/sec^2)

H = total effective head = $z + h_p + h_v$.

  = any total height.

h = any head, in feet, unless stated otherwise.

  $_h$ = hours (as time in hours = T^h).

I = moment of inertia.

i = hydraulic gradient = h/L.

  $_i$ = initial or beginning (as initial head, h_i).

J = absolute viscosity (dimensionally, $J = FT/L^2$) = 1 lb.sec/ft^2.

j = kinematic viscosity in ft^2/sec (dimensionally $j = L^2/T = \dfrac{\text{absolute viscosity}}{\text{density}}$.

K = any constant.

  = porosity.

k = coefficient of permeability in ft^3/day for $i = 1$ and $A = 1^2$, unless stated otherwise.

L = length or distance in ft. unless stated otherwise.

M = water content: M_0 = optimum in percent of weight of solids.

          M'_0 = optimum in percent of weight of bulk.

m = meniscus correction in hydrometer analysis.

$_m$ = model (used in problems of geometrical similarity of model and prototype or natural scale).

N = any number.

   = ratio of model to natural scale.

n = coefficient of roughness.

$_n$ = nature or prototype.

^{n} = any abstract number.

$_0$ = optimum;

   = calibration (as calibration temperature, t_0).

P = total pressure, in lb., unless stated otherwise.

   = wetted perimeter.

   = power in ft-lbs/sec (dimensionally, FL/T).

p = unit pressure, lb./ft.^{2} unless stated otherwise.

$_p$ = pressure (as pressure head, h_p).

Q = total volume in ft^3, unless stated otherwise.

     NOTE.—Q is used as volume for solids as well as fluids, as volume of solids, Q_s).

q = unit volume in ft^3/sec, unless stated otherwise.

$_q$ = volumetric (as volumetric water content, M_q).

R = simplified hydrometer reading corrected for meniscus error = $R' + m$

   = Reynolds Number.

R' = simplified hydrometer reading = $(r - 1)$ (10^3).

r = hydrometer reading to upper rim of meniscus.

   = hydraulic radius.

S = total shear resistance in lb. unless stated otherwise.

$s\%$ = saturation.

s = unit shear resistance, in lb./ft.^{2} unless stated otherwise.

$_s$ = solids, as weight of solid particles, W_s.

T = time.

   = tons.

   = Taylor's Stability Number = $\dfrac{C}{UwH}$.

t = temperature.

$_t$ = total.

U = Factor of safety with respect to both cohesion and friction.

     (U_f, with respect to friction only; U_c, cohesion only).

u = micron.

$_u$ = ultimate.

V = any velocity in ft/sec (dimensionally, $V = L/T$), unless stated otherwise.

v = mean velocity in ft/sec, unless stated otherwise.

W = weight; # = pounds; T = tons.

     NOTE.—W_s, weight of particles is always taken to mean in an oven-dry condition.

w = specific weight in lb/ft^3.

w_t = total unit weight of submerged soil = $w_s + w_w$.

$_w$ = water.

y = any distance, in feet = L.

Z = depth factor (ϕ-circle method).

z = elevation above any datum in feet.

ϕ = effective angle of internal friction.

$\phi' =$ weighted value for $\phi, = \left(\dfrac{G_s-1}{G_s+e}\right)\phi.$

$\phi_c =$ generalized value for $\phi = \dfrac{\phi}{U}.$

$\theta =$ slope angle.

SECTION B—CHARACTER OF SOIL

3. Void Ratio.—For engineering purposes the entire envelope of fragmental materials covering the solid rock of the earth's interior is considered as soil. In this respect the definition is at variance, more or less, with the definitions used by the agronomist, the geologist, and the ceramist. Literature dealing with soil from the viewpoint of these other fields must, for this reason, be interpreted in terms of the engineers' definition when applied to engineering problems.

From the foregoing definition it may be seen that soil is a structure composed of particles of matter, Q_s, and voids, Q_e. This concept must be firmly fixed in mind. The particles form a skeleton or framework enclosing the voids. The designations "pores" and "pore space" are commonly used to identify the soil voids. By practice they may be used synonymously although there is a distinction in their meaning. In general, voids refers to that portion of the soil volume not occupied by the soil particles. In this sense it indicates space only. Pore space has the same meaning and might better be eliminated. Pores, on the other hand, refer to the character of the voids implying intercommunication of individual voids forming a series of tubes. This conception is essential to an understanding of the characteristics of percolation. A dense soil is one in which the particles are arranged, usually by pressure, so as to afford a minimum of voids. Figures 129, (a) and (b), are containers having a cross section of the dimensions indicated and a depth perpendicular to the plane of the page of 12 inches in both cases. Each contains 216 spheres 2 inches in diameter, the total volume of which must be 904.8 cu. in. in each case. The volume of con-

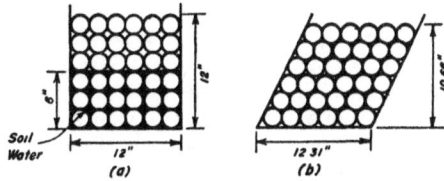

FIGURE 129a-b —Effect of particle arrangement on voids.

tainer (a) is 1,728 cu. in. and of container (b), 1,575 cu. in. These volumes correspond to the definition of soil since they include both the particles, the spheres, and the voids. The difference in total volume may be seen to depend upon arrangement of the particles, the entire structure being less dense in (a).

In the foregoing the volume of particles did not vary but the volume of voids did. In (a) the total volume of voids, $Q_e = 1,728 - 904.8 = 823\ 2$ cu. in. In (b) they are reduced to $1,575 - 904.8 = 670.2$ cu. in. The voids furnish a means, therefore, of identifying a character of the particular soil. The less the total volume of voids the denser the soil.

The ratio of the volume of voids to the volume of particles is called the void ratio, e.

Hence,

$$e = Q_e/Q_s \tag{1}$$

For figure 129, (a)

$$e_a = 823\ 2/904.8$$

$$= 0.91$$

and for (b)

$$e_b = 670.2/904\ 8$$

$$= 0.74$$

It has been stated that a volume of soil is the total of the volume of particles and voids. If Q_b is the volume of soil, this may be written,

$$Q_b = Q_s + Q_e \tag{2}$$

In figure 129a the container was taken as 1 cu. ft. If the volume of spheres had been taken as unity instead of the container, then

$$e = Q_e/1, \text{ from which}$$

$$Q_e = e$$

Substituting these values of Q_e and Q_s in equation (2) we have, when Q_s is unity,

$$Q_b = 1 + e \tag{3}$$

In the case of figure 129a, then, if a unit volume of spheres had been taken the total volume of the container would have been $1 + 0.91 = 1.91$ cu. ft. and for (b) it would have been $1 + 0.74 = 1.74$ cu. ft. This manner of stating the relationship will be found of considerable assistance in certain problems to be encountered.

Care must be exercised to avoid confusing the terms denseness, density, and specific weight. Dense or denseness are relative terms used to describe the degree of compactness of the particles. The smaller the value of the void ratio the greater the denseness of the soil. Density, on the other hand, is mass per unit volume, W/g, and is expressed in slugs per cubic foot in the engineers' system. Dimensionally it is FT^2/L^4. Specific weight is weight per unit volume, usually weight in pounds per cubic foot, W/Q, or, in dimensional units, F/L^3.

4. Specific Weight.—Since the weight of the air occupying the voids in a thoroughly dry sample of soil is sensibly zero, the weight of the soil is also the weight of its particles, that is $W_b = W_s$, the subscript bulk being taken to indicate soil. It is apparent that Q_b will not be equal to Q_s except in the case of a completely consolidated sample, a state impossible to obtain, even approximately, with soils. There will be, then, two values for the specific weight, one for the soil, w_b, and another for the particles, w_s. Consequently, there will also be two values for specific gravity since this is the ratio of the specific weight of the material to that of water at maximum density.

The specific weight of a soil is

$$w_b = \frac{W_b}{Q_b} \tag{4}$$

while that of the particles is

$$w_s = \frac{W_s}{Q_s} \tag{5}$$

5. Specific Gravity.—The term "Specific gravity of the soil" is often loosely used to mean the specific gravity of the particles. Actually the specific gravity of a soil is the bulk specific gravity, G_b, which may be written

$$G_b = w_b/w_w \tag{6}$$

This relation is used but infrequently and then only in problems dealing with submerged soils. The specific gravity of the particles, on the other hand, G_s, is of more practical value and it is this that is often incorrectly stated. In much of the soil mechanics literature "specific gravity of the soil" is taken to indicate the specific gravity of the particles and "bulk" or "apparent specific gravity" that of the soil. If clearly indicated this usage should occasion little confusion. It is, nevertheless, a use of the term "soil" quite opposed to the usual definition and since it introduces no particular advantage its use is not recommended. Looseness of statement will be avoided if the terms "specific gravity of the soil" and "specific gravity of the soil particles" are habitually used.

Assume that the spheres in figure 129 weigh 86.65 pounds   Their volume is 904.8/1,728=0.52 cu. ft.   Their specific weight,

$$w_{s,} = \frac{W_s}{Q_s}$$

$$= 86.65/0.52$$

$$= 165.36 \text{ lb./ft.}^3$$

Since the weight of the voids=0, then the total weight of the soil is also 86.65 pounds.   The soil volume is 1 cu. ft. hence. the specific weight of the soil,

$$w_{b,} = \frac{W_b}{Q_b}$$

$$= 86.65/1$$

$$= 86.65 \text{ lb./ft.}^3$$

Specific gravity is, as has been stated, the ratio of the specific weight of the particular material to that of water, hence the specific gravity of the soil particles may be written

$$G_s = w_s/w_w \tag{7}$$

For figure 129a,

$$G_s = 165.36/62.4$$

$$= 2.65$$

and, from equation (6),

$$G_b = 86.65/62.4$$

$$= 1.39$$

In figure 129b the specific weight of the particles is unchanged and the value of G_s, must, therefore, be identical in both cases.   The soil volume, Q_b, is, however, not the same.   Hence, a different value will be found for w_b and, consequently, also for G_b.

6. Effective Specific Weight $w_{(e)}$, is the weight of the soil particles when submerged, $w_{s(e)}$, or the soil mass is a saturated state $w_{b(e)}$.   In the case of the soil mass the effective weight may be considered in two ways: (1) as the sum of the specific weight of the particles and the weight of a volume of water equal to the volume of voids; or, (2) as the sum of the effective weight of the particles plus the weight of a volume of water equal to the volume of the entire soil mass.   This may be seen from the following:

(1) The total volume of soil $= 1+e$.
     Effective weight of particles $= w_s$.
     From equation (7) $w_s = w_w G_s$.
     The total weight of the mass—

$$W_b = (1)(w_w G_s) + w_w(e)$$

$$= w_w(G_s + e) \tag{8}$$

(2) The total volume, as before, $= 1+e$.
     The effective weight of the submerged soil particles—

$$w_{s(e)} = w_w(G_s - 1) \tag{9}$$

Volume of particles $= 1$.
Volume of water $= 1+e$.
Then.

$$W_b = w_w(G_s - 1)(1) + w_w(1 + e)$$

$$= w_s G_s - w_w + w_w + w_w e$$

$$= w_w(G_s + e),$$

the same as before.

It is more convenient in the following discussion to use the metric relation of specific weight and specific gravity. In this case the unit volume is 1 cubic centimeter (1 cc.), and the unit weight is 1 gram (1 gm). Since 1 cc of water may be taken as weighing 1 gm., the specific weight=the specific gravity=1. In the English units on the other hand specific weight=62.4 (specific gravity). Since the specific gravity is identical no difficulty will be introduced because the relations to be discussed in metric units need only to be multiplied by w_w to convert them to English units. The exponent m will be used to indicate metric units. Where the exponent is not used the units have the same value in both systems.

From (3), $Q_b{}^m = 1 + e$,

by definition $w_w{}^m = G_w = 1$; $w_s{}^m = G_s$; and $w_b{}^m = G_b$

For the general case, then, equation (4) may be written

$$G = \frac{W^m}{Q^m} \qquad (10)$$

and

$$G_b = \frac{W_b{}^m}{Q_b{}^m} \qquad (11)$$

Moreover, in a volume of soil$=1+e$, the weight of particles is $(1)(G_s) + eG_w = G_s + e$. Equation (11) may, therefore, be written

$$w_b{}^m = G_b = \frac{G_s + e}{1 + e} \qquad (12)$$

Hence, the specific weight of a saturated soil in English units is

$$w_{b(e)} = w_w \frac{G_s + e}{1 + e} \qquad (13)$$

The effective specific weight of a submerged solid is $w_s - w_w$.

In metric units

$$w_s{}^m{}_{(e)} = G_s - 1 \qquad (14)$$

since $G_w = 1$, hence, the submerged specific weight of soil mass

$$w_b{}^m{}_{(e)} = G_b - 1 \qquad (15)$$

The submerged weight is, therefore,

$$W_b{}^m{}_{(e)} = Q_b{}^m(G_b - 1) \qquad (16)$$

If the value for G in equation (15) be replaced by its equivalent in (12), equation (15) may be written

$$w_b{}^m{}_{(e)} = \frac{G_s + e}{1 + e} - 1$$

$$= \frac{G_s - 1}{1 + e} \qquad (17)$$

Hence, in English units,

$$w_{b(e)} = w_w \frac{G_s - 1}{1 + e} \qquad (18)$$

7. **Porosity,** K, is the ratio of the volume of voids to the volume of soil expressed as a percent, hence,

$$K = (Q_e/Q_b)(100) \qquad (19)$$

In figure 129a,
$$K = (823,2/1728)(100)$$
$$= 47.6\%$$

From equation (3), $Q_e = e$ and $Q_b = 1 + e$ when $Q_s = 1$. Equation (6) may, therefore, be written

$$K = \frac{e}{1+e}(100) \qquad (20)$$

For the conditions given in figure 129a,
$$K = (0.91/1.91)(100)$$
$$= 47.6\%,$$

8. Water Content, M, unless specifically stated otherwise is the ratio of the weight of water, W_w, held in the voids to the weight of solid particles, W_s, expressed as a percent.   It is written

$$M = (W_w/W_s)(100) \qquad (21)$$

(Note.—In soil mechanics literature, water and moisture are often used interchangeably.   The latter is, however, associated more with the effect than the cause and for this reason the term water is to be preferred.)

In figure 129a water is indicated as filling one-half of the voids.   Its volume, then, is 411.6 cu. in.   The total weight of the water is 14.85 pounds; that of the spheres has been given previously as 86.65 pounds.   Then,

$$M = (14.85/86.65)(100)$$
$$= 17.2\%$$

If, in this case, the entire void space were filled with water, the total weight of water would be 29.70 pounds, the weight of the spheres would be the same as before, 86.65 pounds, from which M_{max} would be 34.3 percent.

Since, in the latter case, the voids are completely water-filled the sample is saturated.   The value of M_{max} will vary for different states of consolidation but will seldom exceed the value determined in the foregoing example.   As will be more apparent later in the discussion of optimum water content it is important to keep in mind that saturation exists at a water content the numerical value of which is relatively low and that a water content approaching or exceeding 100 percent in absolute value is a suspension.

It was stated that M_{max} varies with the state of consolidation.   Hence it is also a function of the void ratio, e.   When $M = M_{max}$ the volume of water is obviously the volume of voids, that is $Q_w = Q_e$ and its weight, W_w, is 62.4 Q_e. The weight of the particles W_s, $= 62.4\, Q_s G_s$.   Equation (8) may, then, be written

$$M_{max} = \left(\frac{62.4\, Q_e}{62.4\, Q_s\, G_s}\right)(100)$$

from which $M_{max} = \left(\dfrac{Q_e}{Q_s}\right)\left(\dfrac{1}{G_s}\right)(100) \qquad (22)$

But, from equation (1), $Q_e/Q_s = e$.

Hence
$$M_{max} = (e/G_s)(100). \qquad (23)$$
For figure 129a, $e = 0.91$ and $G_s = 2.65$, from which
$$M_{max} = (0.91/2.65)(100)$$
$$= 34.3\%.$$

9. Degree of Saturation, $s\%$, is the ratio of the volume of water in the voids to the total volume of voids expressed as a percent. This may be written

$$s\% = (Q_w/Q_e)\ (100.).\qquad(24)$$

For figure 129a, $Q_e = 823.2$ in³ and $Q_w = 411.6$ in³ from which

$$s\% = 411.6/823.2$$
$$= 50.$$

Plummer [2] uses the term "relative humidity" for this quality and gives the following designations for varying degrees of wetness:

	$s\%$		$s\%$
Humid	0–25	Wet	75–100
Damp	25–50	Saturated	100
Moist	50–75		

10. Effective Size, E, is a term used to designate the upper limit of those particles which, it is believed, determine certain qualities with respect to a particular soil. As a result of a series of experiments to determine the character of the flow of water through filter sands, Allen Hazen [3] stated that the 10 percent of the entire sample by weight which contained the finest particles determined the resistance of the sample to the flow of water through it. Lately the Corps of Engineers soil laboratory at Muskingum expressed the opinion [4] that the finest 20 percent by weight determines the permeability. In any case, it is a wholly empirical value and at present inadequately supported by experimental data. Since the limit is not rigid it is advisable to show the limit intended in any particular case by designating the percent by weight used. This is most conveniently done by showing this quality as a subscript, thus: E_{10}, E_{20}, etc.

11. The Atterberg Limits [5] are more or less arbitrary definitions limiting the various states assumed by a soil as it progresses from one limit of the degree of saturation to the other.

When a cohesive soil is completely saturated it acts as a liquid. If the sample is permitted to dry it will shrink and gradually acquire the property of self-supporting form although it will offer little resistance to a change of shape under a load. The transition from the liquid to the plastic state is a gradual one, with no marked division between the two. Atterberg arbitrarily defined the limit between them as that water content at which two separated portions of the sample would flow together when struck a certain number of times. An apparatus now generally used, figure 130, was developed by the United States Bureau of

---

[2] "Notes on Soil Mechanics and Foundations," by Fred L Plummer, Edwards Bros , Inc , Ann Arbor Mich , 1936

[3] Report, Massachusetts State Board of Health, 1892, p 539

[4] "Practical Soil Mechanics at Muskingum," by Theodore T Knappen and R R Philippe, Engineering News-Record, Apr 9, 1936

[5] Die Plastizitat der Tone, by A Atterberg, Internationale Mitteilungen fur Bodenkunde, 1911

Public Roads to minimize the effects of the personal equation. With the improved apparatus the dish is raised and dropped by turning the hand crank. Beginning with a low water content, the sample is dropped until the groove formed by the tool closes over a distance of one-half inch. More water is added and the procedure carried out again. The water content is plotted against the number of blows, water content being plotted arithmetically and the number of blows plotted logarithmically. Additional points are determined experimentally and

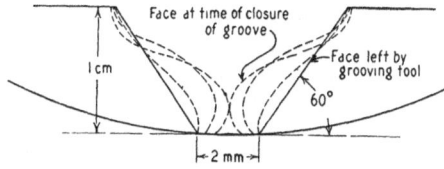

FIGURE 131 —Typical closure, liquid-limit test

a curve plotted. The liquid limit is taken from the curve and is the water content for closure at 25 blows

As the soil continues to lose its water content it passes from a plastic to a semi-solid condition. It will now crack if deformed. The limit between the two states is taken as the water content at which a thread of the sample will crack when rolled to a diameter of one-eighth inch. The test is started from the wet side, rolled to the proper diameter, re-kneaded, and again rolled out, see figure 132. This

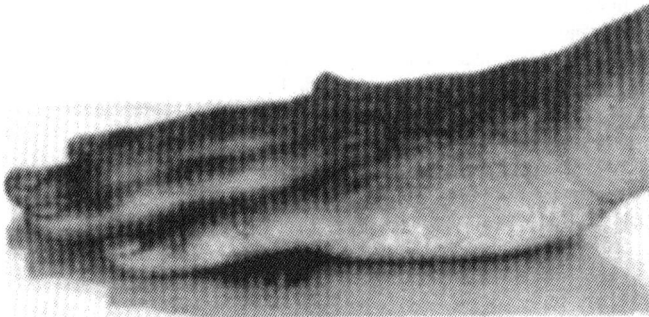

routine is continued until the thread cracks at one-eighth inch diameter, at which point a water-content determination is made. This water content is the plastic limit.

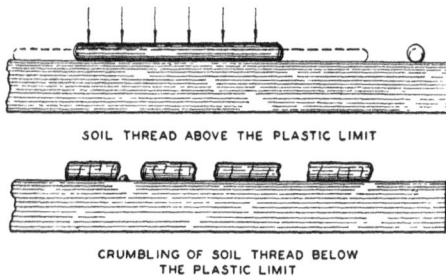

SOIL THREAD ABOVE THE PLASTIC LIMIT

CRUMBLING OF SOIL THREAD BELOW
THE PLASTIC LIMIT

FIGURE 133.—Soil thread, above and below plastic limit

As the sample dries from a wet state it continues to shrink until eventually it changes color, and further shrinkage ceases even though drying continues. The water content at this point is the shrinkage limit. Dr. Terzaghi, member American Society of Civil Engineers, has given the following method of determining the shrinkage limit. Where Q_1 and W_1 are the wet-soil volume and weight, Q_0 and W_0 the dry-soil volume and weight,

$$\text{Shrinkage limit} = \frac{(W_1 - W_0) - (Q_1 - Q_0)}{W_0} \qquad (25)$$

12. The Plasticity Index is the arithmetical difference of the liquid and plastic limits.

13. Colloids.—When soil particles are suspended in a liquid they tend to settle under the force of gravity. At the same time the molecules of the liquid, which are in a state of agitation, impinge on the settling particles. This has an insignificant effect on those particles that are large in comparison with the molecules but it prevents the settlement of the finer particles. Dr. Casagrande [6] has defined the upper limit of particle size effected in this manner as 0.0002 mm (=0.2 micron). Particles of a magnitude less than this are designated "colloids." A colloidal suspension, also called a "sol," is a suspension of colloids or colloidal particles, the terms being synonymous.

In a colloidal suspension each of the colloidal particles has a small electrical charge of the same sign. As the particles dart about in the suspension they are mutually repelled by this similarity of charge and in this state they will remain in suspension indefinitely. However, if an electrolyte is added to the suspension the charges of the particles are neutralized by the process of ionization. Lacking this charge the particles are no longer mutually repellant and collision takes place. Molecular attraction, which is not sensible with respect to the larger particles, is sufficient to cause the colloids to stick together, forming a floc. When the accumulations become large enough so that the molecular agitation of the liquid no longer affects them sufficiently to prevent settlement, they will, like the large particles, settle under the force of gravity. The formation of the floc is called "flocculation," and the product of flocculation is called a "gel." In some instances it is desirable to prevent flocculation as, for example, in the hydrometer determination of particle sizes. It is apparent that any flocculation in this case would result in the determination of sizes of accumulations rather than individual particle sizes. In order to prevent this a deflocculating agent is added to the suspension, and the reduction of the floc to its constituent particles is called deflocculation.

14. Physico-Chemical Phenomena.—Adsorption is the adhesion of the molecules of gases or fluids to the surfaces of solid bodies, resulting in a relatively high concentration of the gas or fluid at the surface of contact. Adsorb is to condense by adsorption. If moisture is not available the solids will adsorb a film of air. A common demonstration of this occurs when drops of water fall on a surface of thoroughly dried dust particles. Lacking moisture the particles have adsorbed a film of air which temporarily repels the globules of water causing them to roll over the dust without wetting it. Solids have a greater affinity for some liquids than they have for others, and in the foregoing example, since soil particles have a greater attraction for water than air, the air films are quickly replaced by water films.

The thickness of water films varies considerably. Studies by the Bureau of Public Roads and George Washington University reported by C. A. Hogentogler, Jr.,[7] indicate that they may vary from 1/1,000,000 inch to 1,100/1,000,000 inch, both approximate, for different types of materials and between almost as wide limits for a particular material under different conditions.

The film has been described [8] as consisting of an inner and an outer layer of

[6] "The Hydrometer Method for Mechanical Analysis of Soils and Other Granular Materials," by Dr Arthur Casagrande, unpublished manuscript

[7] "Essentials of Soil Compaction," by C A Hogentogler, Jr, paper, annual meeting, Highway Research Board, 1936

[8] "Discussion D20," by C. A. Hogentogler, Proceedings, International Conference on Soil Mechanics and Foundation Engineering, Vol. III, Harvard University, 1936

water, figure 134. Due to the mechanics of adsorption the layer nearest the surface of contact with the solid is under tremendous pressure and because of this is in a semisolid state. Dr Terzaghi has stated [9] that this condition exists in a film thickness of less than 2/1,000,000 inch. The outer layer, cohesive water, is less dense and consequently less stable. In all probability the transition from most

dense at the surface of the solid to least dense at the outer limit of the cohesive film is a gradual one. It is, however, convenient to consider the film as two distinct hulls. That portion of the film at the outer limits of the cohesive film is rather indifferently held and it breaks down readily under limited stress. The cohesive film varies in thickness with the amount of water available for film formation, increasing with the water content until saturation is

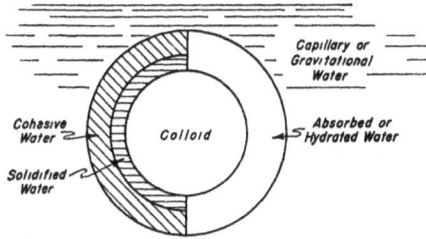

WATER FILM

FIGURE 134 — Water film

reached, at which stage the cohesive fraction of the individual particle films apparently breaks down to become free or gravitational water once more. This phenomenon is quite apparent when water is added to sand which is well below the point of saturation. Below saturation the water is completely taken up as

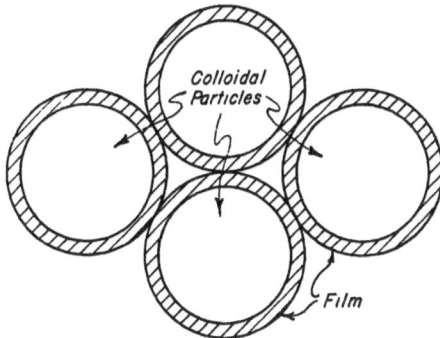

FILM AND AIR VOIDS

FIGURE 135 — Film and air voids

films. The films during this stage are well defined and the enlarged particle functions as a unit, voids existing between particles just as though the enlarged particle were a homogeneous unit, figure 135. As more and more water is added the films become thicker and bulking occurs. This is a familiar experience with sand. When sufficient water is added to cause saturation the cohesive fraction of the films appears to break down, the voids become water-filled in place of air-filled and the sand returns to its original bulk.

It may be shown by simple mathematics that the total of the surface areas of the individual particles composing a unit volume depends upon the grain size of these particles. A single cubical particle 1 cubic foot in volume would have a total surface area of 6 square feet. The total of the areas of the individual particles of a cubic foot of fine clay may approach 4 or 5 acres. Since the volume of film water depends on the area of the particles and not on their diameter it is apparent that a very fine clay soil below the saturation point may have a considerable water content without the voids being water-filled.

It has been stated that the cohesive water is more or less unstable. Disturbance of the soil appears to overstress its cohesion to the inner shell of semisolid water, and it then becomes free water occupying the voids. This is quite a common occurence when dry concrete is overpuddled. If the concrete be dry enough it will appear mealy, with little or no free water apparent. If the concrete is then

[9] "New Facts About Surface Friction," by Charles Terzaghi, Physical Review, Vol 16, 1920.

puddled sufficiently the moisture films surrounding the particles of sand and cement are partially reduced to free water, which quickly works to the surface.

Water films have a major influence on the behavior of a soil. The force involved in the formation of the semisolid portion of the film is evidently of considerable magnitude. If a dry soil is densely compacted the stress in the soil due to the bulking of the particles when water becomes available for film formation must also be of similar magnitude. This is easily demonstrated by compacting dry particles in a cylinder one end of which is solid and the other a piston and pressure gauge. If water then be introduced the fact that stress exists may be observed. On the other hand, a soil having an excess of cohesive water may, if strain occurs, become completely unstable. The latter concept will be discussed more fully under stability of embankments.

The chemical character of soil particles depends largely on the character of the rock from which they were derived and the conditions under which that rock disintegrated. Hogentogler has stated [10] that weathering in humid northern climates results in the removal of iron and aluminum oxides, leaving a preponderance of silica, while in humid southern climates the silica leaches out leaving the iron and aluminum oxides. The ratio of silica to iron oxide plus aluminum is designated the "silica sesquioxide ratio." The value of this ratio is to some degree indicative of the character of a soil. Podsols are clays having a ratio of 2 to 6; laterites are clays having a ratio of 2 or less. The former are high in silica and consist largely of scalelike particles; the latter consist largely of spherical particles. Their compaction and elastic characteristics are not at all similar. While much work on this subject has been accomplished by the agricultural chemists and physicists, the application of these theories to engineering problems is not well enough understood at the present time. It appears very possible, however, that eventually this phase of soil character may account for much of the behavior that today appears erratic and unpredictable. The relation of adsorbed ions on the surface of particles to Brownian movement is understood. It has also been demonstrated that the adsorbed ions have an effect on the plasticity of a soil and the water films. The use of common salt, calcium chloride, and other chemical compounds in soil stabilization is quite generally practiced, and though not generally understood is an example of film behavior controlled by chemical treatment. As Dr. Winterkorn has pointed out [11] a purely physical chemical analysis may always be limited to semiempirical relationships because of the complexity of any soil system, but it will indicate in what direction the physical character of a soil will be altered by chemical treatment and will undoubtedly be a very valuable tool in soil classification.

SECTION C—CLASSIFICATION AND IDENTIFICATION

15. Classification.—There have been many attempts to classify soils, most of which have been predicated upon a single characteristic. These have included classifications based upon the grain size distribution, the voids ratio, the water content, the colloid content, the mineral composition, etc., in an effort to establish a relation between a particular characteristic and the mass character. Soil is a complex material and does not lend itself to so simple a cataloguing.

Dr. Terzaghi has stated [12] that the properties which determine the behavior of soil are:

[10] "Subgrade Soil Constants, Their Significance, and Their Application in Practice," by C. A Hogentogler, A. M. Wintermyer, and E A. Willis, Bureau of Public Roads, U S Department of Agriculture, vol 12, Nos. 4 and 5 1931.

[11] "Physico-Chemical Testing of Soils and Application of Results in Practice," by Dr Hans F. Winterkorn, paper, annual meeting, Highway Research Board, 1936

[12] "The Science of Foundations" by Charles Terzaghi, Transactions, American Society of Civil Engineers, Vol. 93, 1929, p. 296.

(a) The volume change produced by an increase of pressure acting on the soil;

(b) The permeability of the soil; and

(c) The cohesion of shear resistance of the soil at zero load.

As time goes on and more data are available it will probably be found that this list of characteristics, or a modification of it, includes the behavior determinants. Unfortunately, those data are not now available and the theory cannot be developed.

Among the classifications illustrated in figure 136 is that of the Bureau of Public Roads.   This is the only one so far developed that takes soil behavior into consideration   The characteristics indicated by the Atterberg limit tests form the

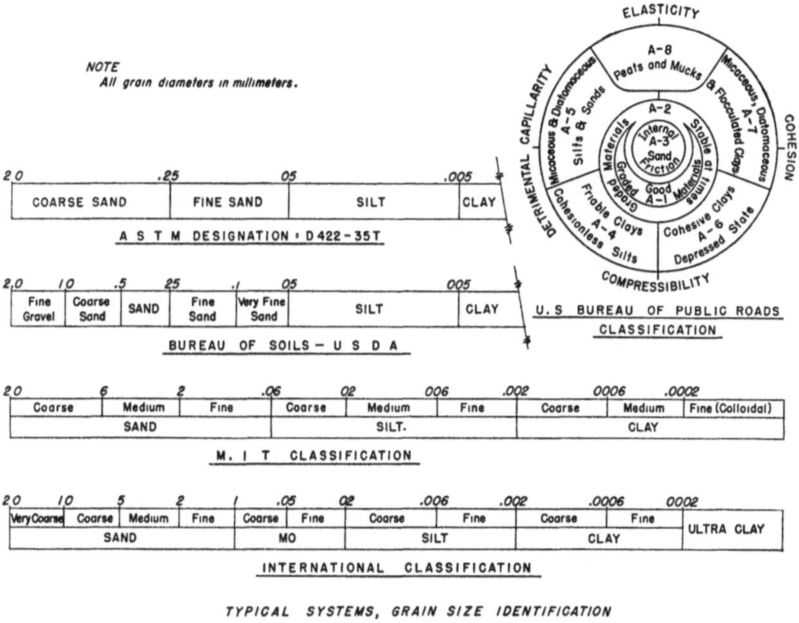

FIGURE 136 —Typical systems, grain size identification

basis for this method of classification.   According to Hogentogler [13] the following are indicated:

Liquid limit: Capillary capacity of soil when thoroughly manipulated.
Plasticity: Cohesion.
Shrinkage limit: Combined effect of cohesion and resistance to consolidation.

Different materials possess the foregoing qualities in varying degrees.   Mixtures of various materials will proportionately reflect the qualities of the ingredients. For example if two materials of equal plastic limit are mixed the resulting mixture will also have that plastic limit regardless of the proportions of the ingredients. Again, if two materials having different plastic limits are mixed the plastic limit of the mixture will depend upon the proportion of each ingredient used.   From this it follows that the proportion of various materials permissable in a soil to

---

[13] "Engineering Properties of Soil," by C A Hogentogler, C E   McGraw Hill, 1937

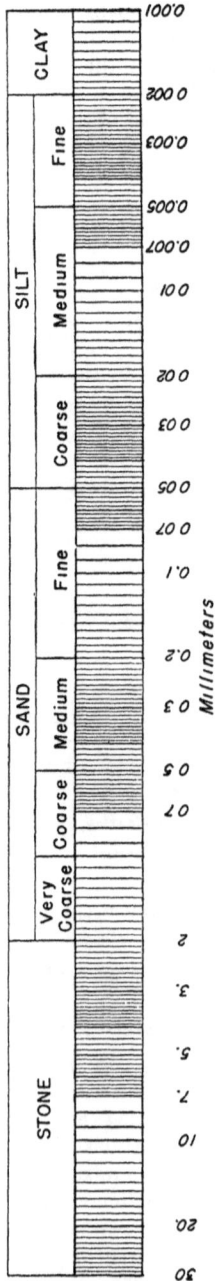

FIGURE 137.—Grain size identification, National Park Service

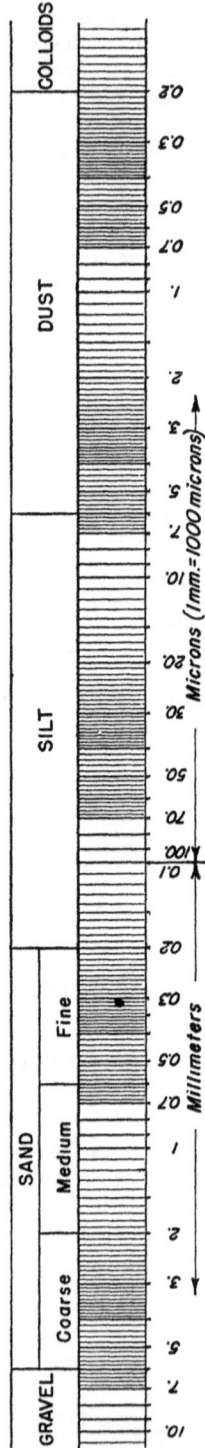

FIGURE 138.—Grain size identification, National Park Service modified.

be used for a certain purpose depends upon the permissable characteristics of the grouped qualities. The group qualities as defined by the Bureau of Public Roads [14] follows:

Group A-1.—High internal friction; high cohesion; no detrimental shrinkage, expansion, capillarity, or elasticity;

Group A-2.—High internal friction and high cohesion only under certain conditions. May have detrimental shrinkage, expansion, capillarity or elasticity;

Group A-3.—High internal friction; no cohesion; no detrimental capillarity or elasticity;

Group A-4.—Internal friction variable; no appreciable cohesion; no elasticity; capillarity important;

Group A-5.—Similar to A-4 and in addition possesses elasticity in appreciable amount;

Group A-6.—Low internal friction; cohesion high under low moisture content; no elasticity; likely to expand and shrink in detrimental amount;

Group A-7.—Similar to A-6 but possesses elasticity also;

Group A-8.—Low internal friction; low cohesion; apt to possess capillarity and elasticity in detrimental amount.

Typical gradings are as follows:

Group A-1.—Material retained on No. 10 sieve, not more than 50 percent. Fraction passing No. 10 sieve composed as follows: Clay, 5 to 10 percent; silt 10 to 20 percent; total sand 70 to 85 percent; and coarse sand (retained on No. 60 sieve) 45 to 60 percent. Effective size, E_{10}, approximately 0.01 mm and uniformity coefficient, $= E_{60}/E_{10}$, greater than 15. Liquid limit not less than 14 nor greater than 25; plasticity index seldom larger than 8; shrinkage limit seldom smaller than 14 or larger than 20.

Group A-2.—Not less than 55 percent of sand in soil mortar (the material passing the No. 10 sieve is designated the "soil mortar"). Liquid limit not greater than 35; a plasticity index of zero with a significant shrinkage limit, or, a plasticity index greater than zero and less than 15 with or without significant shrinkage limit.

This classification system was developed principally from the viewpoint of highway construction. It has not been applied to many cases of dam design so far as is known. It may deserve more study in this connection than it has received although it would appear to be over-complex for this purpose. It may be stated, then, that at the present time there is no satisfactory system of classification applicable to impounding structures of earth.

16. Laboratory Identification.—Several nomenclatures have been developed among which those illustrated in figure 136 are the most widely used. Figure 137, the National Park Service nomenclature, is essentially the same as the International Classification, figure 136, with the exception that the lower limit of sand and the upper limit of silt are both extended to 0.05 mm in order to eliminate the term "Mo."

Figure 138 is a proposed nomenclature which it is believed eliminates most of the inconveniences and weaknesses of those of figures 136 and 137. In the first place, the scalar differences between the abscissas of 6 and 2 is approximately

---

[14] "Subgrade Soil Constants, Their Significance, and Their Application In Practice" by C. A. Hogentogler, A. M. Wintermyer, and E. A. Willis, Public Roads, Bureau of Public Roads, U. S. Department of Agriculture, vol 12, Nos 4 and 5, 1931.

equal to that between 2 and 0.6.   Sub-divisions based upon these values, then, will be separated by an approximately uniform distance.   The M. I. T. nomenclature, figure 136 is subdivided in this manner.

The next consideration is the nomenclature.   The term gravel is reasonably well defined.   True, it is quite often loosely used as, for example, "bank-run gravel" which is understood to include sand as well as gravel, but it is correctly used generally and its use limited to particles greater than one-fourth inch should occasion no serious difficulty.

Sand and silt offer no difficulties.   Both terms are well defined in general use with the possible exception that organic matter is sometimes thought of as an essential element of silt.   As may be noted from all the nomenclatures, silt is defined only by particle size.   It has been customary to designate the size group immediately below silt as clay.   This term has introduced errors, in some cases of considerable consequence.   It must be realized that in any grouping based solely upon particle size no other character of the material is considered.   Clay, on the other hand, is usually indentified as a cohesive, plastic material.   The proposed Boston Building Code [15] defines clay as "A fine-grained, inorganic soil possessing sufficient cohesion when dry to form hard lumps which cannot be readily pulverized by the fingers."   Hogentogler [13] defines clay as a chemically reactive consitutent of soil which is more or less plastic when wet.   The difficulty is apparent when rock flour is considered.   This material is completely lacking in the qualities associated with clay although its grain size distribution falls within the same limits.   For this reason the term "dust" is introduced in Fig. 138.   This term implies particle size only and not behavior.

The term "colloids" is generally taken to mean those particles which remain in suspension indefinitely due to Brownian movement.

The third consideration is the arbitrary limits of the classes.   Eliminating those characteristics which are not related to engineering uses of soil there are but two well fixed boundaries.   These have been given by Dr. Casagrande [6] as the limiting particle diameters whose fall through water is expressed by Stokes Law.   The upper limit is 0 2 mm.   Diameters greater than this cause turbulent disturbance in falling and do not, therefore, follow this relation.   The lower limit as given by this authority is 0.0002 mm (0.2 microns).   Below this Brownian movement takes place.   As may be seen in figure 138, the upper limit of 0.2 mm is approximately defined by the No. 65 Tyler seive.   This seive is given as the limit between coarse and fine sand in the proposed Boston Code [15].   In figures 136 and 137 the lower limit of sand is given as 0.05, 0.06, 0.1, and 0.05.   It can be appreciated that the distinction between a very fine sand and a coarse silt is a rather vague one and in view of the fact that is of no value in a field classification and of extremely doubtful value in the laboratory there seems to be no particular object in holding to the vicinity of the present values if a logical reason exists for doing otherwise.   The limits established by Dr. Casagrande, on the other hand, do determine a characteristic based largely on particle size.   Moreover, those limits, 0.2 and 0.0002 coincide with the 2–6 system of abscissas.   In the proposed nomenclature, figure 138, this range of sizes has been divided into two equal sections, one designated

[6] "The Hydrometer Method for Mechanical Analysis of Soil sand Other Granular Materials," by Dr Arthur Casagrande, unpublished manuscript

[13] "Engineering Properties of Soil " by C. A. Hogentogler, C E   McGraw-Hill, 1937.

[15] "Proposed New Boston Building Code—Chapter on Foundations," reported by Gilbert Small, Chairman, Boston Code Committee, Paper Z–17, Vol. II, Proceedings, International Conference on Soil Mechanics and Foundation Engineering, Harvard University, 1936   Note —Further references to these proceedings will be indicated by "Proceedings, Internat. Conf."

silt and the other dust, merely to permit of finer identification without the use of such modifying terms as coarse, medium, and fine.

It is doubtful that the terms medium silt and fine silt convey distinctive properties to the average engineer. Since, in general, subdivision of the classes is of little value to the laboratory and of no value to the field it has been omitted except in the case of sand for which material such a differentiation has a practical meaning.

The upper limit for sand of one-fourth inch has been chosen for the purpose of figure 138 principally because it more clearly definies gravel as a material composed of coarse aggregate only. Extending the gravel span to 2 mm diameter tends to confuse gravel with a gravel-sand mixture.

Attention is called to the fact that the nomenclature given in figure 138 is used throughout this appendix unless specifically stated otherwise.

It was stated that the grain size distribution may furnish an indication of the character of a soil although not a positve one. Usually it will be more indicative of permeability than stability   Various investigators have attempted to develop a classification for soils based upon a grain size analysis of samples taken from apparently satisfactory dams. Figure 139 illustrates three of these. It will probably be found that a material having a distribution curve which lies entirely within the limiting curves B and B_1, and approximately of the same form is satisfactory although this is not rigidly so. On the other hand, a distribution curve not entirely within these limits or not approximately of the same form is not, *per se*, unsatisfactory. It follows from this that the grain size distribution is of extremely doubtful value as an indication of the behavior to be expected of the soil. In the absence of a better means the curve does serve as a means of identification but little more can be claimed for it.

17. Field identification.—Certain soil types, such as clay, hardpan, and alluvial silts for example have characteristics sufficiently apparent and well enough understood so that use of the terms conveys within the limits of refinement expected of field approximations, some understanding of the soil's behavior. There are many instances in which greater refinement of identification is not justified. The following system of identification is given for use within this limitation.

Shale —Fine grained sedimentary rocks composed of silt and clay which has undergone solidification. They may be divided into two types (1) compaction shales and, (2) cemented shales.[16] The latter type behaves as a consolidated material when saturated and further consideration is not given it herein. Compaction shales are, as the term implies, merely highly compacted clay and silt particles, and, having no cementing matrix, they disaggregate when immersed after having been dried. This disaggregation is due to the formation of the semisolid films which has been discussed in section 14. In this condition the material behaves as a highly plastic clay and should be so considered with respect to competency under load.

Clay.—A fine inorganic material consisting largely of a coarse fraction predominantly dust and a colloidal fraction, the latter being for the most part hydrous aluminum silicate derived as a product of chemical weathering. The colloidal fraction imparts the characteristic plasticity which varies with the water content. Pure clay contains no particles larger than the lower range of dust, and as will be

---

[16] "Geology of Dam Sites in Shale and Earth" by Warren J. Mead, Affiliate, American Society of Civil Engineers, Civil Engineering, vol. 7, No. 5, May 1937

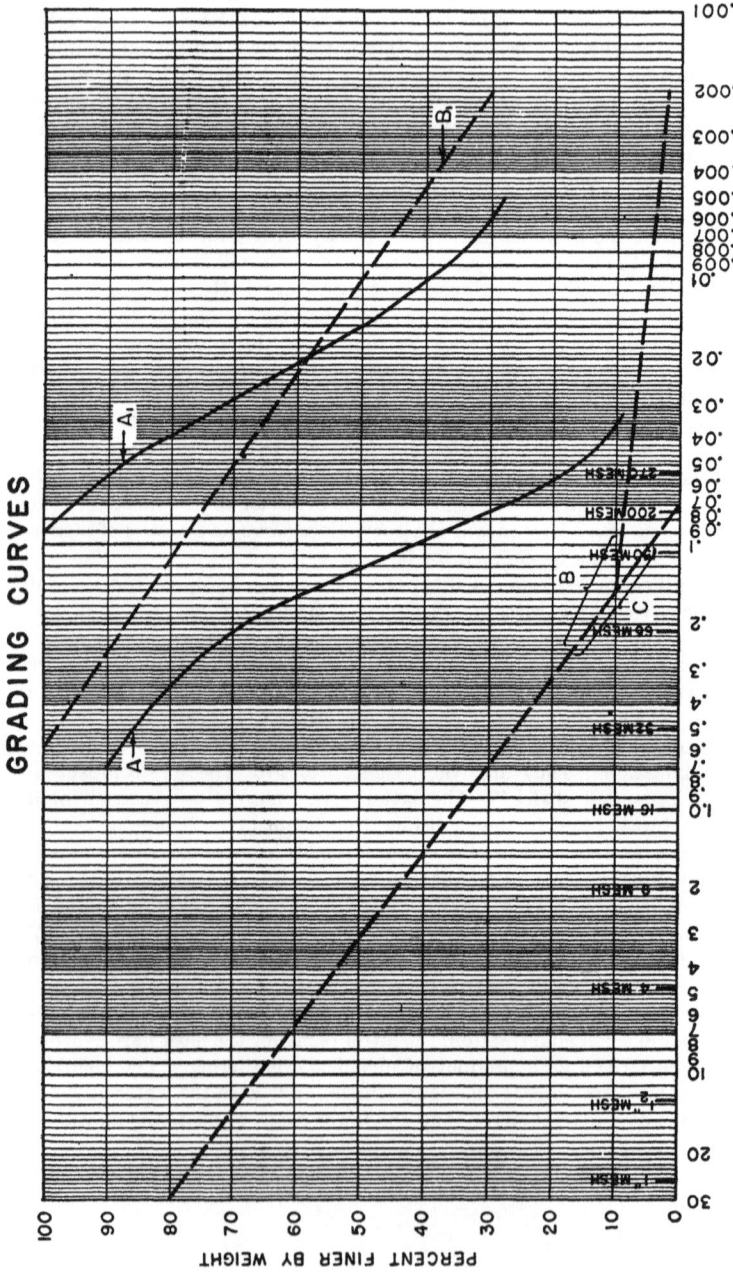

GRADING CURVES

FIGURE 139.—Typical grading curves.

A   Possible limit of satisfactory core material
A_1 Probable limit of safe core material, L. C. Urquhart (Civil Engineering Handbook, McGraw-Hill, 1934)
B–B_1 Proposed limits for ungraded materials suitable for impervious section for rolled fill dams, Charles H Lee (Proc Amer Soc C E, vol 62, No 7, September 1936)
C–B_1.   Approximate limit of materials suitable for impervious sections of rolled fill dams, Joel D Justin (Earth Dam Projects, John Wiley & Sons, 1932)

discussed under capillary water, has a considerable volumetric change when drying from a wet state or when wetted after thorough drying. The volumetric change varies with the coarse fraction, and as the sand fraction is increased the volumetric change is decreased. Clays may be described as "hard" when a fresh sample can be molded but very slightly or not at all in the fingers; "medium" when a fresh sample is molded in the fingers with considerable difficulty; [15][17] and "soft" when molded with little difficulty. Wet clay has the properties of a viscous fluid and when unrestrained will flow from beneath an applied load. Although highly plastic it possesses a certain amount of resistance to deformation. This resistance is much higher in the undisturbed state than it is after disturbance, that is, the load which a clay will support after it has been disturbed by working or over-stressing is very much lower than the same clay in an undisturbed state will support. This phenomenon is attributable to the moisture films. Disturbance reduces part of the cohesive water to free water. Transfer of the load from the soil skeleton to this free water causes a condition of saturation and reduced shear resistance which may result in viscous flow. If the stress causing failure is removed, it is highly probable that the free water eventually returns to the state of cohesive water and the clay will regain an "undisturbed" condition although not necessarily that preceding failure. This undoubtedly requires a very long time, depending upon the material.

Adjustment of the strain within a massive body of overstressed clay may be imperceptible for a long time. When it has reached a sufficient state, however, the reduction of capacity from the higher to the lower value will take place rapidly. This explains why a clay may appear to be competent under an excessive load for a long period and then fail suddenly. This does not apply, of course, to those situations in which the degree of saturation has been increased by other causes outside the body of the material.

Sand.—A granular material defined arbitrarily by grain size lacking in both plasticity and cohesion although it may show considerable stability when wet due to film phenomena. This is demonstrated in the suitability of sands on various tidal beaches for automobile traffic for a period following the recession of the tide and before drying has progressed too far.

Gravel.—A granular material arbitrarily defined as having a grain diameter not less than one-fourth inch nor more than approximately 6 inches, noncohesive and nonplastic. Rarely found completely devoid of other soils.

Sand-gravel mixtures, bank run.—Sand and gravel are very often found together. Both mark stages in the disintegration of the parent rock. Assuming both to be present in a mixture in an appreciable quantity, the stability under loads depends but little on the proportions of the two ingredients. The state of compaction is much more important in this respect. Bank run is a convenient general term for the designation of such mixtures.

Hardpan.—A densely compacted mixture of gravel, sand, and clay or a cemented sand with or without gravel and boulders. Extremely difficult or impossible to remove by picking.

Silt.—The smallest particles that can be identified by visual inspection are silt. Settling in water as practiced in the field rarely discloses any finer fraction because flocculation causes the accumulations of finer particle to settle out with the coarser fraction. This introduces no difficulty since greater refinement is meaningless for

---

[15] "Proposed New Boston Building Code—Chapter on Foundations," reported by Gilbert Small, chairman, Boston Code Committee, Paper Z-17, vol. II, Proceedings, International Conference on Soil Mechanics and Foundation Engineering, Harvard University, 1936. Note —Further references to these proceedings will be indicated by "Proceedings, Internatl. Conf."

[17] "Exploration of Soil Conditions and Sampling Operations" by H. A. Mohr, Soil Mechanics Series No. 4, Graduate School of Engineering, Harvard University, June 1937.

field identification. If silt contains an appreciable amount of disintegrated vegetable matter it is usually called "organic" silt. If completely lacking in organic matter it is called "inorganic" silt. Both organic and inorganic silt are extremely dangerous materials under a load. They are usually poorly compacted and often are the cause of destructive settlement of structures. They possess no appreciable cohesion and in the presence of water go into saturation readily with complete loss of stability. Loads should never be supported on silt without the advice of an experienced foundation consultant. If such advice is not obtainable the material should invariably be accepted as unsatisfactory.

The unsuitability of the term clay for particle size designation has been discussed in section 16, Laboratory Identification. Mistaking inorganic silt for clay has resulted in numberless experiences that were unfortunate, to say the least, and, in many cases, disastrous. It must be borne in mind that silt is not plastic, clay is; silt has practically no cohesion when dry, clay has considerable; silt may be highly permeable, clay is relatively impermeable. If a sample of soil is thoroughly dried it will be extremely difficult if not impossible to crush it in the hands if it is clay. If it is silt or silt with a low clay content it will crush quite easily under the same pressure.

Other soil mixtures —There are other soil mixtures, some of which are abundant within a limited locality and others that are quite generally encountered. For the most part they can be identified, approximately, within the limitations of one or the other of the foregoing groups. Marl, for example, is a clay containing a high calcium carbonate content. Little is to be gained by subdividing the foregoing types into several subtypes when the approximate values in bearing and shear resistance can be stated for the general types only.

Types identified by manner of deposition.—Such terms as fill, alluvium, talus, etc., designate the manner in which the material has been deposited rather than the kind of material. On the other hand, such terms indirectly describe the material. Fill is any man-made deposit. It may, therefore, range from a heterogeneous mass of household wastes and other rubbish to a carefully constructed embankment. Unless the history of the fill is known it should be classed as doubtful and adequately explored.

Alluvium is material deposited out of water. Where highly erosive streams enter quieter water the material deposited may range through sand to fine gravel. In general, however, the particles are mostly silt and dust, and the cautions applicable to silt apply. Whether or not the silt is organic depends largely on the flood characteristics and plain.

Talus is the accumulation of debris at the foot of cliffs and is composed of the products of weathering of the material of which the cliff is composed. Such deposits may have considerable stability depending upon the parent material and the degree of weathering.

Glacial fill, like fill, may be extremely complex in its composition and should be adequately explored.

Nomenclature of mixtures.—If a single type predominates and the fraction of other types is insignificant, the type name is used as, for example, gravel, sand, clay, etc.

If two or more types are mixed approximately equally, the both type names are used, predominant one placed first, as

60% sand; 40% gravel   =Sand-gravel.
60% gravel; 40% sand   =Gravel-sand.
40% sand; 30% gravel; 30% silt=Sand-gravel-silt.

If one type is less than approximately 20 percent but still in sufficient quantity to affect the mixture, the adjective form of the type name should be used, as

85% clay; 15% sand   = Sandy clay.
50% sand; 45% gravel; 5% clay = Clayey sand-gravel.

NOTE.—Bank run is a sand-gravel mixture which may have an appreciable clay or silt content. Usually a reasonably well-graded mixture which has considerable stability when compacted.

SECTION D—CAPILLARY AND GRAVITATIONAL WATER

18. Soil Structure.—As has been stated, soil is composed of both the particles and the voids. The behavior of the water in the soil will depend largely on the character of the voids. In figure 129a uninterrupted channels exist both horizontally and vertically. In figure 129b on the other hand, the channels are more or less interrupted. If the soil were composed of flat, instead of spherical, particles the channels would be most marked in the direction of the long axis of the particles, regardless of the inclination of the stratification. In the problem of the behavior of water in soil it is essential to think of the voids as continuous so that they form tubes through the soil. For this reason the voids are usually called pores in this connection.

19. Classes of Soil Water.—Water in soil may take one of three forms, (a) film water; (b) capillary water; or (c) gravitational water. The film water acts as an integral part of the enlarged soil particles and for this reason has already been discussed under section B, Character of Soil. Capillary water is subjected to force greater than that of gravity and for this reason is not defined by the principles of fluid flow, whereas gravitational water, on the other hand, differs from other examples of fluid flow only in the character of its conduct. Except in problems dealing with static forces with respect to the stability of embankments capillary water is usually of no particular concern in dam design. Gravitational water introduces problems with respect to stability involving both static and dynamic forces and, in addition, those of subsurface storage and loss.

20. Capillary Water.—Because of the character of the affinity of the molecules of solids and liquids, certain solids attract one liquid more than another. For example, quartz attracts water much more than it does kerosene; for copper the opposite is true. The angle of contact, θ, between the liquid and solid will depend upon the degree of attraction. For water on metal or glass $\theta = 0°$; for mercury in contact with the same solids $\theta = 143°$. That is, regardless of the relative position of the planes, the surface of the water and of the surface of the solid, at the point of contact a meniscus will form so that the surface of the water at the point of contact is tangent to the surface of the solid and the angle of contact, θ, consequently will be zero. Due to the cohesion of the water particles, there is a dragging of the interior particles as those in direct contact assume a condition of $\theta = 0°$ with the result that the meniscus is a curved surface. This is illustrated in figure 140. If the diameter of the tube, $2r$, is less than 0.1 inch the meniscus is spherical [18] with a radius

$$b = r/\cos\theta \tag{26}$$

Since $\theta = 0°$ for the contact of water and glass, then $b = r$.

Although it is an incorrect conception, it is convenient to visualize the surface of the liquid as a membrane capable of resisting tensile stress. From this follows the conception of surface tension which, for the purpose of this section will be designated T. From elementary mechanics it may be shown that the pressure against the concave side of the curve exceeds that against the convex side by some

[18] "Hydraulics," by R. L. Daugherty, A. B., M. E., McGraw-Hill Book Co., Inc., New York, 1937

amount which may be designated p_c for convenience and further, that this excess results in a tangential stress, T. This relation may be shown to be

$$p_c = 2T/b \qquad (27)$$

and for water is approximately 75 dynes per cm, varying inversely with the temperature.

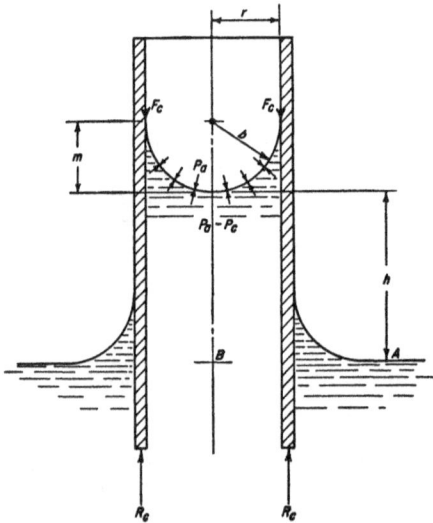

DETAIL OF CAPILLARY RISE

FIGURE 140 —Detail of capillary rise.

In figure 140 the pressure against the concave surface of the meniscus is atmospheric pressure, p_a. As has been stated, this exceeds the pressure against the convex surface by the amount p_c. Since the first is p_a, then the latter must be $p_a - p_c$. It may be shown that all points in a connected body of fluid at rest are under the same intensity of pressure. Since the pressure at A, figure 140, is atmospheric, then from the foregoing it must be atmospheric at point B also. It is obvious that the bottom of the meniscus cannot be at point B since the intensity of pressure cannot be both p_a and $p_a - p_c$ when p_c is greater than zero. The fluid in this case will rise in the tube to some height, h, so that the pressure is

$$p_c = h \ w \qquad (28)$$

and the intensity of pressure at B will then become $p_a - p_c + hw = p_a$. From equations (27) and (28) the relation may be written

$$hw = 2T/b \qquad (29)$$

from which

$$h = 2T/wb \qquad (20)$$

The height of capillary rise, then, is a function of the pore radius and vaiies inversely with respect to it.

Consider the conditions in the vicinity of the meniscus, figure 140. Capillary rise having taken place the forces which caused it are in a state of equilibrium. If the inside area of the tube is A, then the pressure on the surface of the liquid is pA. Also, the upward pressure on the same surface is $(p_a - p_c)A = p_a A - p_c A$. When p_c is greater than zero it is apparent that the former exceeds the latter by the quantity $p_c A$. It has been stated that the excess pressure, p_c, is caused by the mutual attraction of the particles of the water and the tube. Since this excess may be resolved as a tangential force at the point of contact it is apparent that an equal and opposite stress must exist with respect to the tube wall, F_c. The tube is in equilibrium, hence, there is a reaction, R_c. It has been shown that the total excess pressure is $p_c A$, hence from equation (28) it may be written

$$p_c A = hwA \qquad (31)$$

from which the compressive force in the tube may be seen to be equal to the pressure caused by a column of liquid having an area equal to the pore area and a height equal to the capillary rise.

Inasmuch as the force due to surface tension is tangential at the point of contact of the liquid and the tube wall its resultant is in the direction of the axis of the tube. Capillary force is not, therefore, the cause of bulking. This is due to film phenomena. Indirectly capillarity may contribute to such a state by supplying moisture for film formation. Even though capillary and film phenomena are generated from the same source, surface tension, they are individual phenomena and that behavior of the soil ascribable to each one is separate and distinct.

Two conditions are necessary for the formation of a meniscus, (a) a free water surface in contact with (b) a solid surface extending from within the body of the water to some distance above the water surface. These conditions cannot exist when the solid surface is completely submerged, hence, when a capillary soil is submerged the capillary forces which may have existed previously are dissipated. The significance of this will be more obvious in the following discussion.

The pores of a soil composed entirely of very fine particles will, for example, be much smaller of diameter than those of coarse sand. As a matter of fact, the capillarity of coarse sand is relatively insignificant. The capillarity of clay is high. If a lump of thoroughly saturated clay is permitted to dry, the water loss progresses inward. In the wet state the clay resists consolidation because both the soil particles and the water completely occupying the voids are incompressible. As the void water is progressively reduced, however, realignment of the particles becomes possible and under the compressive force of capillarity the soil shrinks. As continued drying and shrinking take place the internal resistance to consolidation increases and eventually further shrinkage is prevented even though drying continues. The water content at this state is called the shrinkage limit. Shrinkage causes cracks when the capillary force exceeds the cohesion of the soil particles. These cracks are simple tension failures. Again, as the soil shrinks under the influence of capillary force, the particles are brought tightly together, resulting in increased molecular cohesion of particles and a consequent increased stability. If the same soil is wetted again, the reformation of the films may result in the almost complete loss of this stabilizing phenomenon so that the soil is less stable than it was before drying, although of approximately the same water content. The cause of this is not clear but it is probably associated with the time influence on water films. A somewhat similar phenomenon is observable over a short period of time in bentonite. If a saturated sample of bentonite is permitted to stand undisturbed, it will pass from a fluid to a gelatinous state. In the first condition it has no rigidity; in the second, it actually displays cohesion measurably. If the mass next be agitated, it will once more become fluid, regaining stability once more when permitted to remain undisturbed. It is probable that film solidification is a continuing phenomenon caused by physico-chemical and electro-chemical or thixotropic processes. The thickness of the film is probably constant from its formation if undisturbed solidification, however, is progressive. As first formed the cohesion of the more or less unstable outer shells of the films is considerably less than that of the denser form which occurs with time. There is, therefore, less cohesion of particles in a soil whose films are young than there is in a similar soil of the same water content but in which the films have become dense through aging.

21. Fluid Flow.—Because the behavior of gravitational water is expressed by the principles of fluid flow, the following concepts are emphasized because of their application to the flow of water through soil. They are necessarily abridged and it is recommended that reference be made to standard hydraulic texts for a more adequate treatment.

For the purpose of this appendix, water may be considered an incompressible fluid having a specific weight of 62.4 pounds, which does not vary sensibly within the range of pressures and temperature to be encountered in the design of low dams.

Whenever motion of the fluid takes place there is a resistance within the body of fluid which tends to oppose the movement of the individual particles with respect to one and another. This resistance is due to a property called viscosity, J. As the temperature of a fluid increases, the force of cohesion decreases and, therefore, the viscosity also decreases. In the engineers' system, the unit of viscosity is 1 lb /sec. per square foot. In the absolute system, the unit is 1 dyne second per square centimeter ($=1$ gram per cm-sec.$=1$ poise). In the English units of measurement the latter is 1 pound per foot/sec. ($=1$ poundal sec. per sq ft.)

$$1 \text{ poise} = 100 \text{ centipoise.}$$
$$0.0672 \text{ poundal sec. per sq. ft.}$$
$$0.00209 \text{ lb. sec. per sq ft.}$$
$$1 \text{ lb. sec. per sq. ft.} = 478.8 \text{ poises.}$$

If w be taken to designate the specific weight of water (in other sections of this appendix the specific weight of water$=w_w$), then the pressure at any point within the fluid mass is given by the relation

$$p = wh \qquad (32)$$

in which $p=$ unit pressure, and $h=$ the vertical distance from the free surface (or its equivalent) to the point under consideration. Furthermore, the intensity of pressure acting at different points $a, b \ldots n$ within a continuous body of fluid at rest varies directly as the depths, $h_a,\ _b \ldots\ _n$ of those points. Hence, in fluids at rest all points in the same horizontal plane are subjected to the same intensity of pressure, that is, when $h_a = h_b \ldots = h_n$, $p_a = p_b \ldots = p_n$. This is not true for fluids in motion.

When the intensity of pressure is 1 pound per square inch then $z = 2.308$ feet; if $h = 1$ foot then $p = 0.433$ lb./sq. in. Hence, the depth to a given point may be written as

$$h = 2.308p \qquad (33)$$

and the pressure for a given depth as

$$p = 0.433h \qquad (34)$$

in which $h=$ ft. and $p=$ lb./sq. in.

There are two types of fluid flow, laminar and turbulent. If laminar flow takes place in a straight parallel-sided conduit every particle in a cross section of the flow will move in a straight line parallel to the sides and to each other. Turbulent flow is completely lacking in orderliness. Although the net movement is in a given direction, individual particles may vary in direction and velocity and there is no definite relation between the paths of succeeding particles.

If a fluid flow be started at a very low velocity the motion will be laminar. With a gradual increase of the velocity, flow will remain laminar until at some velocity, depending upon the viscosity of the fluid and the conduit dimensions, the motion will change abruptly to the turbulent state. If, now, the velocity is reduced, turbulence will continue until a velocity is reached, much below that at which turbulence started, at which the flow once more becomes laminar. The former point of transition is called the "upper critical velocity," the latter, the "lower critical velocity " At velocities below the lower critical velocity, the flow will always be laminar; at velocities above the upper critical velocity, it will always be turbulent; at velocities between the upper and lower limits, the flow may be either laminar or turbulent.

In flowing through soil, the motion may be either laminar or turbulent depending upon the conditions which exist. Through very porous gravel, the flow might very well be turbulent. In dams, however, such a condition would exist only in the vicinity of drains, or at least should not exist at any other location, and the body of the structure which actually determines the character of flow is invariably sufficiently dense to permit flows well below the lower critical velocity only. This being the case, only laminar flow can occur and no further attention is given to turbulent flow in this respect. In three-dimensional flow the velocity components lie in two mutually perpendicular planes; in two-dimensional flow the components lie in a single plane. Usually it is assumed in soil applications that the flow is two-dimensional. This is a practical assumption.

The rate of discharge, "rate of flow," or simply "discharge," q is the quantity of fluid flowing past any section per unit time. In this appendix it is expressed as cubic feet per second, unless stated otherwise. The mean or average velocity, v, is a function of the discharge and conduit area and may be written

$$v = q/A \tag{35}$$

It may be shown that both q and v are linear functions of the hydraulic gradient, i, for laminar flow but not for turbulent flow.

In "steady" flow the conditions at all points in the stream remain constant with respect to time; that is, the rate of flow at any cross section of the stream is constant. If the flow entering a section of conduit is constant and no water is added or taken away then the rate of discharge from the section must be constant also and equal to the flow into the section. This concept is expressed by the "Continuity Equation"

$$q = A_1v_1 = A_2v_2 = \quad . = Av = \text{constant} \tag{36}$$

From this equation it may be seen that the mean velocity, v, varies inversely as the cross-sectional area, A.

A flow line is the trace of the path taken by a particle of the flowing water. Since we are concerned only with laminar flow, the flow lines will be parallel when the sides of the conduit are parallel and uniform when the latter are not or an obstruction is encountered within the path of the stream. Figure 141 illustrates laminar flow through a flume having a constant width and a variable depth. The curves $a-a'$, $b-b'$, etc., are the traces of the paths taken by the particles entering the flume at point a, b, etc. respectively. Since the flow is laminar, every particle entering the flume will take the same path as that taken by the particle preceeding it at the same point of entry. Only five flow lines are shown. There could be, however, as many flow lines as there are particles in a vertical section. No advantage is gained by plotting a great number of paths and much of the clearness of the illustration would be lost. Assume that the fluid flowing through the flume is frictionless so that v, the mean velocity, is constant throughout the full depth of the fluid with respect to any vertical section. From the continuity equation, (36), the rate of flow at sections $B-B'$, $C-C'$ and $D-d'$ is constant and equals q, hence,

$$q = A_Bv_B = A_Cv_C = A_Dv_D \tag{37}$$

In the illustration $Z_B = Z_c = 2Z_D$.
Since v varies inversely with A, $v_B = v_c = \frac{1}{2}v_D$.

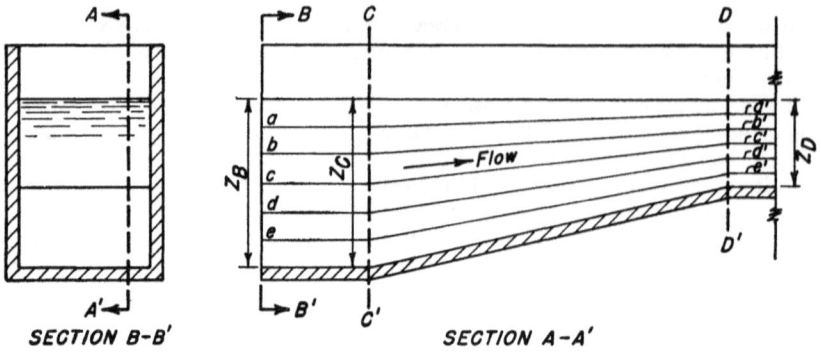

LAMINAR FLOW

FIGURE 141 —Laminar flow.

Consider any two flow lines, a–a' and b–b' for example. From section B–B' to C–C' the two flow lines are parallel and since all particles are flowing in a parallel direction no particles can be added to or taken away from that volume bounded by two curves. The volume is constant, then, throughout this length. It may be shown that the volume bounded by these flow lines is also constant between section C–C' and D–D'. From this it follows that the continuity equation applies to flow between flow lines just as it does between the conduit boundaries, the velocity varying inversely with the area. In two-dimensional flow, the depth normal to the plane containing the flow lines is uniform, therefore, the velocity varies inversely as the distance between the adjacent flow lines.

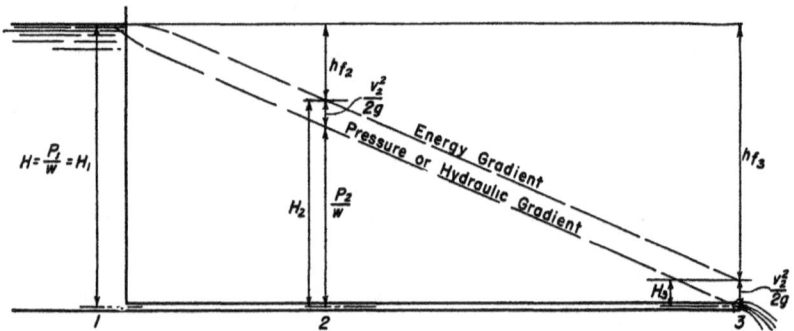

FIGURE 142 —Flow through a uniform conduit

Figure 142 illustrates flow through a conduit from a reservoir which is sufficiently large so that the discharge causes no appreciable lowering of the head, H. It is further assumed that the diameter of the conduit is very small by comparison with its length, L; that it is uniform in cross-section throughout the entire length and that datum is the center line of the pipe.

At point 1 there is no flow hence there is neither velocity nor will any head loss have occurred. For this set of conditions the energy equation may be written

$$H = \frac{p_1}{w} \cdot \ \cdot \ \cdot \tag{38}$$

At point 2 a portion of the initial pressure will have been used to maintain flow and still more will have been used to overcome the friction due to the flow. Both the pressure head, $\frac{p}{w}$, and the flow pressure, $\frac{v^2}{2g}$, are forms of energy.

Because they may be converted one to the other that part of the pressure head converted to flow pressure is not a loss. That part of the pressure head used in overcoming friction, on the other hand, is a loss to the system. At point 2, then, the available energy is

$$\frac{p_2}{w} + \frac{v_2^2}{2g} = H_2. \tag{39}$$

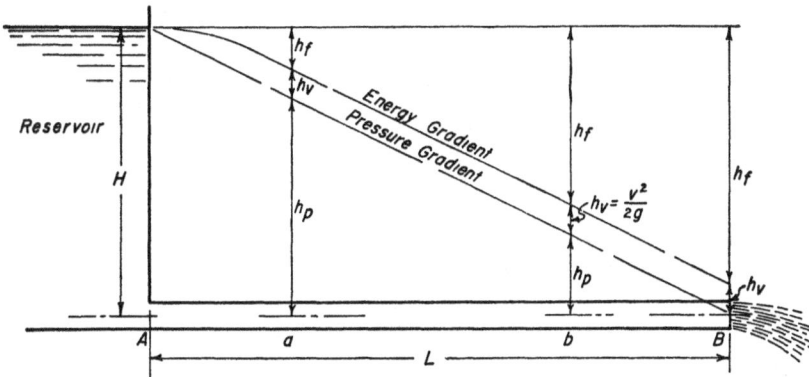

ENERGY GRADIENT, UNIFORM AREA

FIGURE 143 —Energy gradient, uniform area.

At point 3 the flow discharges into the open therefore the pressure head must be zero. Atmospheric pressure exists of course both at points 1 and 3, but since it is identical at both points in figure 142 it is not considered ordinarily.

At the point of discharge there is velocity, hence there is a velocity head. As has been stated, this head may be reconverted to pressure head. It is for this reason that the velocity at emergence of percolating water is considered as a pressure with respect to its ability to remove soil particles at the point of emergence. This is discussed under "Flow pressure," section D (24).

The continuity equation (36) for the conditions illustrated in figure 142 may be written:

$$q = A_2 v_2 = A_3 v_3 \ . \ \ . \ \ . \tag{40}$$

Since it was assumed that $A_2 = A_3$ then $v_2 = v_3$. From this it follows that

$$\frac{v_2^2}{2g} = \frac{v_3^2}{2g}. \tag{41}$$

Furthermore, it was assumed that the conduit is uniform in character throughout therefore the friction loss per unit length is constant and the hydraulic gradient, i, is a straight line.

Assume now that figure 142 represents a section through a compact, homogeneous soil mass in which percolation is taking place. The reservoir now becomes the subsurface flow and the conduit becomes a series of interconnected voids in the soil. Of course there would be as many conduits one above another as there are what we might term void tubes. It has been explained in connection with flow paths that the flow between two adjacent flow lines may be considered as though it were bounded by impermeable walls. In this case the sides of the void tube take the place of the flow lines and the single tube may be considered as isolated from the rest of the flow.

If the soil is homogeneous and rather densely compacted the velocity will be very low, so low in fact that the value of $v^2/2g$ becomes relatively insignificant and for practical purposes the energy gradient will coincide with the hydraulic gradient. In such a case

$$H = \frac{p}{w} + h_f. \tag{42}$$

The hydraulic gradient, then, will define the limits of the pressure available at any point to maintain the flow toward the outlet and the pressure which has been dissipated in maintaining the flow from the inlet to the point in question.

When, as in figure 144, there is a variation in the area of the conduit there must be a variation in the flow also. For the same volume of water to flow through a

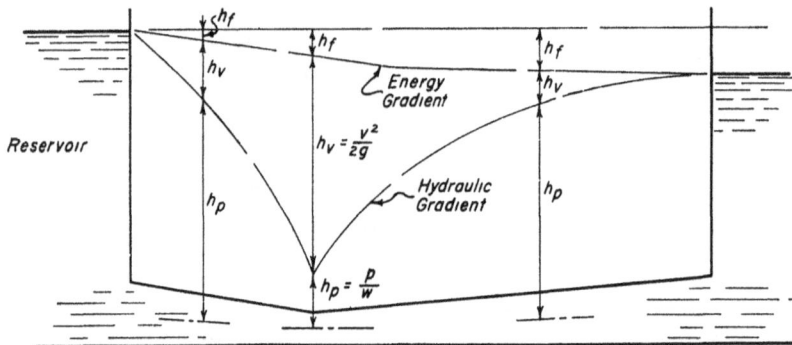

Note :
Entrance and other
minor losses ignored.

ENERGY GRADIENT, NON UNIFORM AREA

FIGURE 144 —Energy gradient, nonuniform area.

smaller conduit the velocity must be increased; through a larger conduit it must be decreased. Velocity head and pressure head are interchangeable. When the velocity head is increased the pressure head is decreased proportionately; when the velocity head is decreased the pressure head is increased proportionately.

The conditions illustrated in figure 144 will fit several cases. The section of the conduit having the least diameter might represent a particularly dense section of embankment adjoined on both sides by less dense material. The same point might represent also the bottom of a file diaphragm, or the conduit might be a passage for water beneath a concrete foundation and underlying, impermeable rock.   In either case it is seen that the hydraulic gradient is not an indicator of the total available energy on either side of any point.

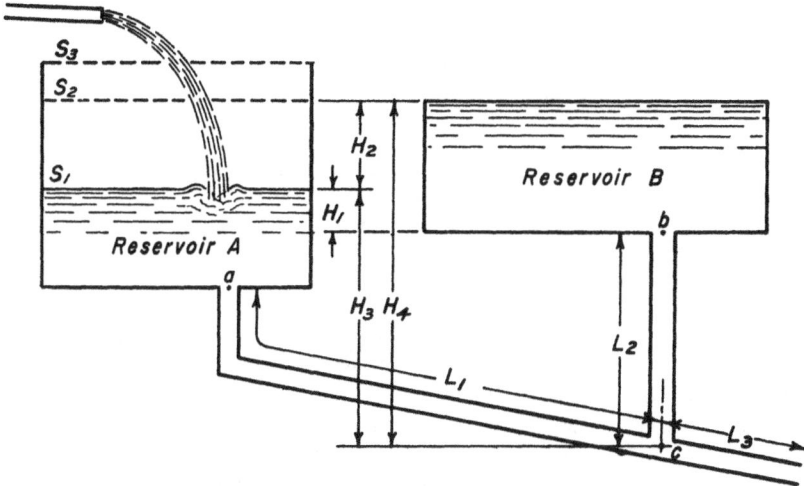

BRANCHING CONDUIT

FIGURE 145 —Branching conduit.

Figure 145 illustrates a set of conditions which may be encountered where springs exist on the floor of a proposed reservoir and is of particular interest where the groundwater elevation is determined by a subsurface spillway.   Assume that Reservoir A represents the source of a subsurface flow and that Reservoir B is empty.   Flow will occur under the influence of H_3.   At c the pressure head will be $H_3 - h_{f(a-c)} - h_{v(a-c)} = h_{p(c)}$ Under the influence of this head flow will occur in the direction of L_2 and L_3 inversely as the unit resistance to flow in each conduit. Assume that conduit L_3 is closed off.   Flow will then occur through L_1 and L_2. The net head causing flow will be H_1 and the velocity at point b will be $H_1 - h_{f(a-c-b)}$. Suppose that reservoir B is now formed by the construction of a dam.   If L_3 is still closed flow will cease when the water surface reaches elevation S_1 since the pressure head at c will be equal in both directions.   When the water surface in reservoir B reaches elevation S_2 flow will occur in the direction of reservoir A. If there is an outlet in reservoir A between the levels S_1 and S_2 sufficient to pass the flow an equalizing head in reservoir A will not be attained and flow will continue in reverse of that which took place before water was impounded in reservoir B.   If, on the other hand, there is no outlet in reservoir A above elevation S_1 and supply to reservoir A continues a head will be built up sufficient to cause flow in the direction of a–c–b once more.   This additional head will equal $h_{f(a-c-b)} + h_v$.

22. Ground-Water Flow.—Ground-water flow may be either laminar or turbulent depending for the most part upon the resistance to flow offered by the material through which the flow is taking place. The resistance of soils composed largely of grains smaller than gravel is sufficient to prevent turbulent flow if enlargement of the natural channels is prevented. Hence, for practical purposes, only laminar flow need ordinarily be considered. This being so, the velocity of ground-water flow is proportional to the slope of the hydraulic gradient. The upper surface of water flowing through soil is a free water surface. The hydraulic gradient, therefore, coincides with this surface.

The continuity and energy equations apply to flow through soil just as they apply to any other conduit. Because the character of the pores through which flow takes place is unknown and extremely variable the energy equation cannot be applied directly.

The fact that the velocity of ground-water flow is proportional to the hydraulic gradient was verified by Darcy in 1856. The relation of hydraulic gradient, velocity and soil characteristic, usually referred to as Darcy's Law, may be written

$$V_o = ki \tag{43}$$

in which, V_o = seepage velocity
              k = coefficient of permeability
              i = hydraulic gradient = h/L

Seepage velocity, V_o, is the rate of flow with respect to the area of the soil column. Unit or average velocity is the rate of flow with respect to the area of voids. For example, assume that 1 cubic foot per second is flowing through a column of soil having a cross-sectional area of 1 square foot. Assume further that the void ratio e, of the soil is 0.91 (fig. 129a). In this case, the seepage velocity, V_o = 1 foot per second. The unit or average velocity,

$$V = V_o \frac{(1+e)}{e} = 2.1 \text{ ft. per sec.} \tag{44}$$

The total volume flowing through soil is given by the relation

$$q = v_o A \tag{45}$$

Darcy's equation may be written—

$$q = kAi \tag{46}$$

Various investigators have endeavored to define the coefficient of permeability, k, in terms of soil characteristics. Allen Hazen,[3] derived the following formula—

$$v = c(E_{10})^2 \frac{h}{L}(0.70 + 0.03t) \tag{47}$$

in which c = constant and t = temperature in °C. If we let $v = v_o$ and $\frac{h}{L} = i$, this equation may be written, $v_o = [c(E_{10})^2(0.70 + 0.03t]i$ which by inspection is seen to be Darcy's equation (43). This being so, it follows that—

$$k = c(E_{10})^2(0.70 + 0.03t) \tag{48}$$

Thus, k was stated as a function of the effective size and the temperature. It must be remembered that Hazen's experiments had for their purpose the determination of the rate of flow through filter sands and for fine, clean, and graded sand the formula still holds.

In 1902 Charles Slichter stated permeability to be a function of the porosity

---

[3] Report, Massachusetts State Board of Health, 1892, p 539

of the soil as well as its grain size and the temperature of the percolate.[19] His relation is written

$$v = 11.3 \frac{h}{L} \frac{(E_{10})^2}{k'} [1 + 0.0187(t - 32°\text{F.})] \qquad (49)$$

from which, $k = (11\ 3\ (E_{10})^2 k')\ [1 + 0.0187(t - 32°\ \text{F.})]$ (50)

and in which k' = a constant depending upon porosity, (a measure of compaction) and t = temperature °F.

Tables 1 and 2 are after those given by Slichter.[20]

TABLE 1

[Coefficient of permeability, k, in ft /min   (For $t = 60°$ F)]

E_{10}	Porosity					
	30 percent	32 percent	34 percent	36 percent	38 percent	40 percent
0 01	0 00003	0 00004	0 00005	0 000060	0 000072	0 000085
0 02	00013	00016	00020	000239	000286	000339
0 03	00030	00036	00045	000538	000645	000763
0 04	00053	00065	00079	000958	001145	001355
0 05	00082	00101	00124	001495	001790	002120
0 06	00118	00146	00178	002150	002580	003050
0 07	00161	00198	00243	002930	003510	004155
0 08	00211	00259	00218	003825	004585	005425
0 09	00266	00328	00402	004845	005800	006860
0 10	00328	00405	00496	005980	007170	008480
0 12	00473	00583	00713	008620	01032	01220
0 14	00643	00794	00972	01172	01404	01662
0 15	00739	00912	.01115	01345	01611	01910
0 16	00841	01036	01268	01531	01835	02170
0 18	01064	01311	01605	01940	02320	02745
0 20	01315	0162	01983	02390	02865	03390
0 25	020	0253	03100	03740	04480	05300
0 30	0296	0364	04460	05380	06450	07630
0 35	0403	0496	0608	07330	08790	1039
0 40	0527	0648	07940	09575	1145	1355
0 45	0665	0820	1005	1211	1450	1718
0 50	0822	.1012	1240	1495	1780	2120
0 55	0994	1225	1500	1810	2165	2565
0 60	1182	1458	1784	2150	2580	3050
0 65	1390	1710	2095	2530	3030	3580
0 70	1610	.1983	2430	2930	3510	4155
0 75	1850	2278	2785	3365	4030	4770
0 80	2105	2590	.3175	3825	4585	5425
0 85	2375	2925	3580	4325	5175	6125
0 90	2660	3280	4018	4845	5800	6860
0 95	2965	3650	4470	5400	6460	7650
1 00	.3282	4050	4960	5880	7170	8480
2 00	1 315	1 620	1 983	2 390	2 865	3 390
3 00	2 960	3 640	4 460	5 380	6 450	7 630
4 00	5 270	6 480	7 940	9 575	11 45	13 55
5 00	8 220	10 12	12 40	14 95	17 90	21 20

[19] "Motion of Underground Waters" by Charles Slichter, Water Supply and Irrigation Paper, No 67, U S Geological Survey, U S Department of Interior.
[20] "The Rate of Movement of Underground Waters" by Charles S Slichter, Water Supply and Irrigation Paper No. 140, U. S. Geological Survey, U S Department of Interior

TABLE—Continued

[Coefficient of permeability, k, in ft/min   (For $t=60°$ F)]

E₁₀	Porosity				
	42 percent	44 percent	46 percent	48 percent	50 percent
0 01	0 000101	0 000123	0 000149	0 000181	0 000215
0 02	000405	000492	000597	.000724	000861
0 03	000911	00111	00134	.00163	00194
0.04	00162	.00197	00239	.00290	00344
0 05	00253	.00307	00373	00452	00538
0 06	00364	00442	00537	.00652	00775
0 07	00496	00602	.00732	00887	.0105
0 08	00647	00787	.00956	.0116	.0138
0 09	00820	.00995	0121	.0147	.0174
0 10	0101	.0123	0149	.0181	.0215
0 12	0146	0177	0215	0261	.0310
0.14	0198	0241	0293	.0355	.0422
0 15	0228	.0277	.0336	.0407	.0484
0 16	0259	0315	0382	0463	.0551
0 18	0328	.0398	0484	.0586	.0697
0 20	0405	.0492	0597	.0724	.0861
0 25	0632	0768	0933	.113	.134
0 30	0911	.111	134	.163	.194
0 35	124	.151	183	222	.264
0 40	162	.197	239	.290	.344
0 45	205	249	302	366	436
0 50	253	.307	373	452	.538
0 55	306	.372	452	547	.651
0 60	.364	.442	537	652	.775
0 65	428	519	631	.765	909
0 70	496	.602	732	887	1 05
0 75	569	691	840	1 02	1 21
0 80	648	787	956	1 16	1 38
0 85	731	888	1 08	1 31	1 55
0 90	820	995	1 21	1 47	1 74
0 95	913	1 11	1 35	1 63	1 94
1 00	1 01	1 23	1 49	1 81	2 15
2 00	4 05	4 92	5 97	7 24	8 61
3 00	9 11	11 1	13 4	16 3	19 4
4 00	16 2	19 7	23 9	29 0	34 4
5 00	25 3	30 7	37 3	45 2	53 8

TABLE 2 —*Temperature correction*

t	t_c	t	t_c	t	t_c
°F		°F		°F	
32	0 64	55	0 93	80	1 30
35	67	60	1 00	85	1 39
40	73	65	1 08	90	1 47
45	80	70	1 15	95	1 55
50	86	75	1 23	100	1 64

The following example will illustrate the use of the tables
Given:

    Elevation ground water surface at station $1+00=95\ 00$
    Elevation ground water surface at station $5+00=93\ 00$
    Temperature of water, $t=50°$ F.
    Cross sectional area of flow $A=1$ square foot.

Effective size of soil, $E_{10} = 0\ 20$ mm
Porosity, $K = 40$ percent.
$h = 95.0 - 93.0 = 2.0$ feet.
$L = 500 - 100 = 400$ feet.
k, 60° F. $= 0.0339$ (table 1).
t_c 50° F. $= 0.86$ (table 2).
k (50° F.) $= (0.0339)\ (0.86) = 0\ 029$.

From equation (46) $q = kA\dfrac{h}{L} = (0.029)\ (1.0)\ (2/400) = 0.000145$ cubic feet per minute.

From equation (35), the seepage velocity $v_o = \dfrac{q}{A} = 0.000145/1 = 0.000145$ foot per minute.

This velocity is with respect to the entire cross section of 1 square foot and it is this velocity that is proportional to the hydraulic gradient. The actual flow however, is taking place through the pores and the actual velocity may be determined from the relation $v = \dfrac{v_o}{K}$. Hence $v = \dfrac{0.000145}{0.40} = 0.00036$ foot per minute.

In table 1 it is to be noted that the variation of k with respect to porosity is very small as compared to its variation with respect to effective size. If the dimensions of a sample are known with reasonable accuracy a coefficient can be approximated from a badly disturbed sample by determination of the effective size and the use of tables 1 and 2. The value obtained in this way may be quite satisfactory for preliminary purposes.

If, on the other hand, the most accurate determination possible is required from a badly disturbed sample it will be more satisfactory to determine the porosity of the sample of known volume by means of the specific weight of the soil particles. The sample should then be compacted to various degrees of porosity, as will be explained under Compaction of Soil, section H, and a permeability test made to determine a k for each porosity These may be plotted against the porosity and a curve drawn from which the approximate k for the undisturbed porosity may be interpolated. This method is by no means as accurate as that in which an undisturbed sample is used, but there are occasions when it becomes the most accurate practicable.

While the refinements added by Hazen, Slichter, and Kresnik [21] give added workability to the method of computing permeability, it is

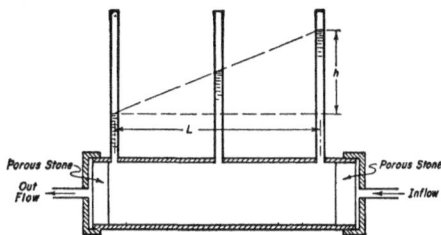

FIGURE 146.—Flow through soil (after Schoklitsch).

much more satisfactory to determine k by actual measurement, and if a reasonably undisturbed sample of the material can be obtained the measurement method is by far simpler.

Figure 146 [21] is a simple form of permeameter The volume of discharge, Q, and the head loss, h, are measured The length of path, L, and area of sample, A, are known. From these

$$k = Q/A\imath \tag{51}$$

[21] "Hydraulic Structures," by Ing Dr Techn A Schoklitsch, translation by Samuel Shulits, Freeman fellow, American Society of Mechanical Engineers, vols I and II, 1937

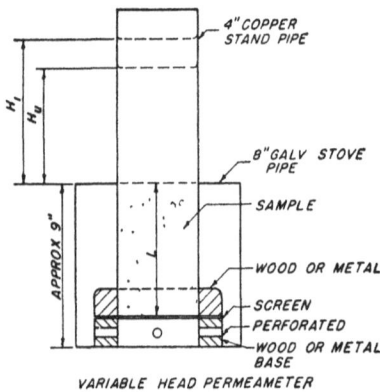

FIGURE 147 —Details of variable head permeameter

The variable head permeameter is probably the most practicable type now available. Figures 147 and 148 illustrate a form suitable for coarse soils such as sands. In this apparatus the column of liquid and the soil sample are of the same area. Where the tailwater is maintained at a constant level (shown in fig. 147 as level with the top of the soil sample, although this relation is not a fixed one), $H_i=$ an initial head, and H_u is the final head after an elapsed time of t, then, by Dr. Gilboy,[22]

$$k = \frac{L}{t} \log_e \frac{H_i}{H_u} \qquad (52)$$

Soils having a low rate of percolation give but a slight and often immeasurable difference between H_i and H_u for time t that may extend beyond a day. In order to magnify the difference in heads the area of the standpipe above the sample may be decreased and the ratio of areas introduced, from which

$$k = \frac{aL}{AT} \log_e \frac{H_i}{H_u} \qquad (53)$$

In the latter form the relation permits the use of many very convenient types of apparatus. Figure 149 illustrates its application to a small undisturbed sample, figure 150 a compaction cylinder, and figure 151, the Casagrande consolidometer. Figure 152 is the Proctor consolidation apparatus as used by the Bureau of Reclamation. While the variable head can be used with this apparatus, that Service makes use of a constant head device with various sized reservoirs

---

[22] "Soil Mechanics Research" by Dr Glenmon Gilboy, Transactions, American Society of Civil Engineers, vol. 98, 1933.

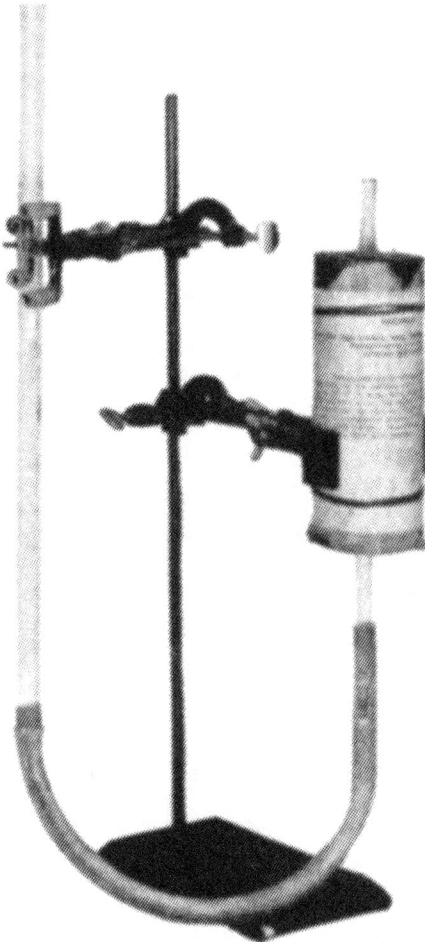

FIGURE 149.—Variable head permeameter using undisturbed sample.

The value of k is not constant for a given soil, although it is customary to so consider it. The flow of a fluid through a capillary tube depends, among other things, on the viscosity and the specific weight of the fluid. The specific weight of water varies from 62.42 at 32° F. to 62.0 at 100° F., both under atmospheric pressure. Under a pressure of 1,000 pounds per square inch it varies from approximately 62.64 at 32° F. to 62.19 at 100° F. The variation in specific weight is, therefore, slight and may be disregarded. Viscosity, on the other hand, varies from 1.9 centipoises, to 0.7 centipoise both approximate, between 32° F. and 100° F. respectively. In view of the fact that a higher temperature usually obtains during the test than is to be found in situ, the omission of correction for variation of absolute viscosity is generally on the safe side. It should be considered, however, and the correction applied whenever the test temperature is lower than that of the ground water.

Where k = coefficient determined by test at temperature t.

k_c = coefficient for ground water flow, at temperature t_c.

J = absolute viscosity of water at $t°$.

J_c = absolute viscosity of water at $t°_c$,

then $k_c = \dfrac{Jk}{J_c}$ 　　　(54)

FIGURE 150 —Proctor compaction cylinder used as variable head permeameter.

The coefficient of permeability, k, may be stated in terms of both metric and English units. Usually the former is given as centimeters per second. This unit does not appear to have any particular advantage over the English units and does have the disadvantage of being different from the units of measurement commonly used with respect to fluid flow by American engineers. Certainly, to the latter 1 foot per minute is more understandable than 0.5 centimeter per second, although they are practically equivalent. If, in equation (43), the hydraulic gradient = 1, then $v_o = k$. For conditions of unity, therefore, k = velocity, and in this connection it should be stated as feet per second to conform to general practice. With respect to water loss it is probable that feet per day is perhaps the most convenient. Whatever the volume and time used, it is inconvenient to have to use a decimal fraction of a great many places. The following tabulation indicates this.

> 1 foot per day = 0 000694 foot per minute.
> = 0.000012 foot per second.
> = 0.000352 centimeter per second.

variable head permeameter

Because of the inconvenience of writing the long decimals it is common to raise the coefficient by some power of 10 and express it as the product of the increased coefficient value and 10 to a negative power of the same degree of magnitude as that by which the coefficient was multiplied. For example, $0.000036 = (0.000036) (10^4) (10^{-4}) = 0.36 \times 10^{-4}$.

In two-dimensional flow the velocities components lie in a single plane. If the path followed by a particle in figure 141 were traced for a certain time, T, the line which resulted would be the velocity vector with respect to T. The velocity vector would be in coincidence with the flow line. From B–B' to C–C' the flow lines are horizontal, hence the velocity vectors have no vertical component. From CC' to DD' the flow lines slope upward, hence the velocity vectors have a horizontal and a vertical component. If, however, the flow line were viewed in plan it would be straight from BB' to DD' because the flow is laminar and the width of the conduit is constant. If now the flume became narrower between CC' and DD' as well as shallower, the flow line would have vertical and horizontal components between CC' and DD' as before and, in addition, would have components both parallel and normal to the longitudinal axis of the conduit. In the latter case the components would lie in two planes, one vertical and the other horizontal. It would, therefore, be three-dimensional flow. Ground-water flow

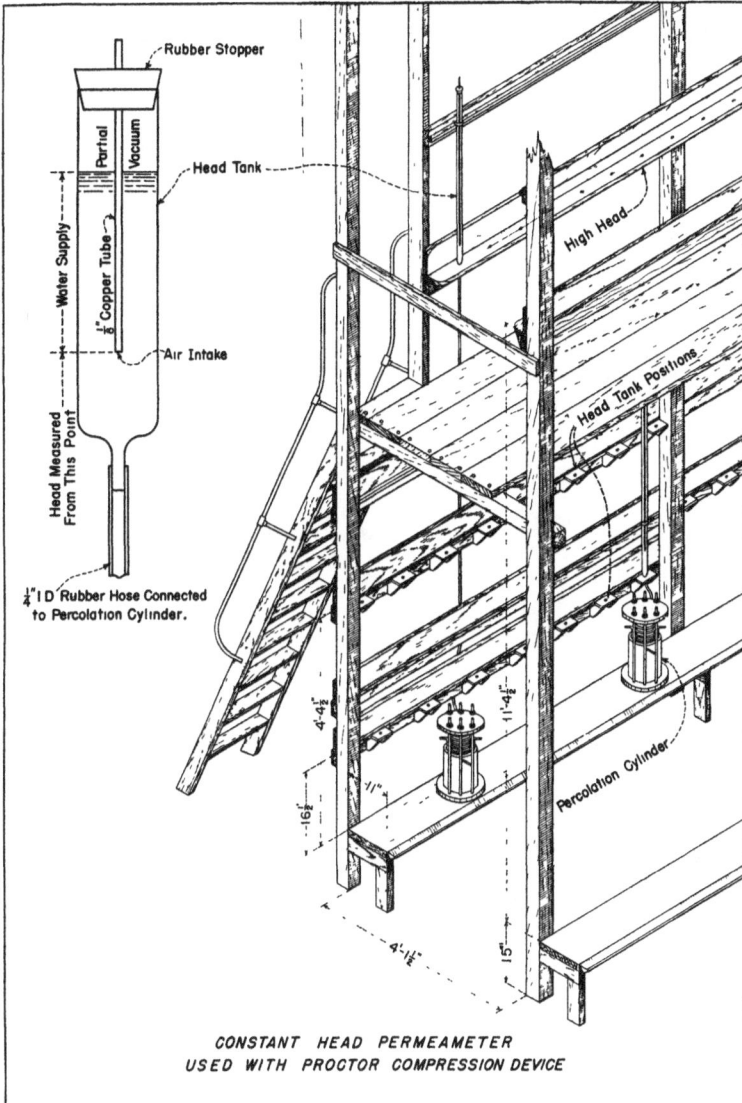

Rubber Stopper

Partial Vacuum

Head Tank

$\frac{1}{8}$" Copper Tube

Water Supply

Air Intake

Head Measured From This Point

$\frac{1}{4}$" I D Rubber Hose Connected to Percolation Cylinder.

High Head

Head Tank Positions

Percolation Cylinder

4'-4$\frac{1}{2}$"

1'-4$\frac{1}{2}$"

16$\frac{1}{2}$"

$\frac{1}{2}$"

4'-1$\frac{1}{8}$"

15"

CONSTANT HEAD PERMEAMETER
USED WITH PROCTOR COMPRESSION DEVICE

FIGURE 152 —Constant head permeameter used with Proctor compression device.

through an infinite, homogeneous stratum would closely approach true two-dimensional flow. There are many situations in which foundation strata are sufficiently uniform in stratification so that two-dimensional flow takes place in a practical sense, although consideration of the complexity of the commonly encountered geological formations will indicate that two-dimensional flow in nature is not usual.   Nevertheless, it is necessary to assume that it does take place, just as it is necessary to consider the soil as homogeneous; otherwise, it would be

ımpossible to obtain even an approximate solution.   (An application of the principles of three-dımensional flow has been developed in Russia, [23] but it has not been
used in the United States to any great extent.)

When two or more strata having different values for the coefficient of percolation are superimposed one upon another a weighted average value is determined
to apply over the entire depth.   The following example illustrates the method
of computation:

Given (surface of soıl assumed elevation 100 ın order to avoid negative elevations):

Elevation	Depth of stratum	k
1. From elevatıon 50 to elevation 65	15 feet	0. 008
2. From elevation 65 to elevation 82	17 feet	. 0004
3. From elevation 82 to elevation 91	9 feet	. 15
4. From elevation 91 to elevation 98	7 feet	. 05
5. From elevation 98 to elevation 100	2 feet	. 01

	Depth of stratum	k
1.	15×0.008	=0.120
2.	17× . 0004	= . 0068
3.	9× . 15	=1. 35
4.	7× . 05	= . 35
5.	2× . 01	= . 02
Totals	50	1. 8468

$$k_c = \frac{1.8468}{50}$$

$$=0.0369$$

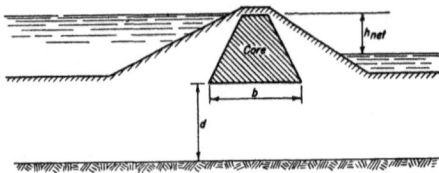

FIGURE 153 —Dımensions for Terzaghı's subsurface flow approxımation.

Such a value is at best a rough
approximation and should be used
without further correction only
where the strata are extensive horizontally and stratification reasonably uniform.   Such a value, where
strata are lenticular, one soil merg-
ing horizontally into another quite
different one in a short distance with
respect to the length of area under consideration, would be much more apparent
than real.   In the latter case only broad experience can dictate what modificatıon
should be made in the method, and even then the result can be considered little
more than a wishful guess.

Dr. Terzaghi has developed two approximate formulas for the flow of water
whıch were reported by Knappen [24]

---

[23] ' Electrıc Analogy Applied to Three-Dımensıonal Study of Percolation Under Dams Buılt on Pervious
Heterogeneous Foundatıons," by B F Reltov, Unıon of Soviet Socıalıst Republıcs, paper, Second Congress
On Large Dams, Washıngton, D C., 1936

[24] "Calculatıon of the Stabılıty of Earth Dams," by Theodore T Knappen, C. E , paper on Questıon VII,
Second Congress on Large Dams, Washıngton, D C , 1936

In figure 153,

b=base width of upper relatively impervious layer;

d=depth of pervious stratum;

k=coefficient of permeability with respect to pervious stratum; and

$h=h_{net}$.

Case I—where b is greater than $2d$

$$Q=\frac{hk}{0.88+(b/d)} \qquad (55)$$

Case II—where b is less than $2d$

$$Q=\frac{hk\sqrt[3]{(2d/b)-1}}{2} \qquad (56)$$

In equations (55) and (56) Q will depend for its value on the system of units used for h, k, b and d. It equals the rate of flow through a cross-sectional area of d depth and 1 foot wide, normal to the direction of flow.

Fig. 154 is a somewhat similar condition with the exception that the width of the base of the sheet piling in the direction of flow is relatively insignificant. Schoklitsch [21] has given the following solution for this case:

Where q=cubic feet per second.

k=feet per second.

$h=h_{net}$.

Φ=a constant, table 3.

a=cross-sectional area under diaphragm.

A=cross-sectional area of pervious stratum.

$q=100$ $k\Phi$ $(h/2)$. (57)

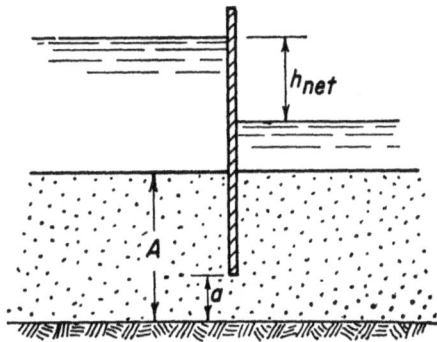

FIGURE 154 —Dimensions for Schoklitsch' subsurface flow determination

TABLE 3 —*Values for Φ in equation (57)*

Ratio of areas:

a/A=	0.1	0.2	0 3	0.4	0.5	0.6	0.7	0.8	0 9
Φ=	.49	.62	.74	.86	1.00	1.16	1.35	1 62	2 06

Both of these approximations are based upon such broad assumptions that they constitute little more than a guess in the majority of cases. On the other hand, interpreted in the light of extensive experience they may permit estimation otherwise impossible.

23. Flow Nets.—A flow net is a graphical representation of the direction of flow and the pressure head available in a subsurface flow. It consists of flow lines and equipotential lines (contours of equal pressure head) superimposed upon a cross section of the soil through which flow is taking place. The two families of curves may be derived mathematically by means of the continuity equation written as a differential equation in which the velocity components are expressed in terms of the discharge velocity.[25] Mathematical

[21] "Hydraulic Structures," by Ing Dr. Techn. A Schoklitsch, translation by Samuel Shulits, Freeman Fellow, American Society of Mechanical Engineers, vols I and II, 1937.

[25] "Seepage Through Dams," by Arthur Casagrande, Journal, New England Water Works Association, Vol. LI, No 2, June 1937.

solution indicates that the curves intersect at right angles forming rectangles the sides of which have a definite relation to each other. Ph. Forchheimer [26] developed a graphical method of solving the equation which is practicable for almost all cases of two-dimensional flow.

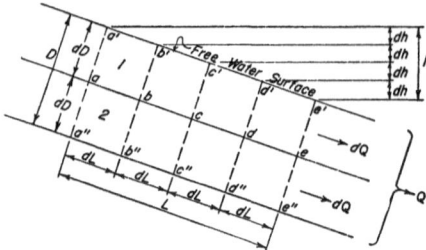

FIGURE 155 —Idealized flow and equipotential lines

There are as many flow lines as there are particles in a cross section of the flow. It is convenient, however, to choose only a limited number so spaced that the rate of flow, Q, between each pair of flow lines is equal. Figure 155 represents two adjacent flow channels each bounded by a flow line. The upper line $a'-b'$ .. e' coincides with the free-water surface. Since the hydraulic gradient coincides with the surface, the line $a'-b'$ ... c' is the hydraulic gradient from which the head loss or potential drop from a' to e' is h. This loss may be subdivided into an infinite number of increments, dh. It is convenient, however, to subdivide h into equal increments of such magnitude that the interval between equipotential lines $a'-a-a''$, $b'-b-b''$, etc., is the same as the interval between flow lines, in which case the flow lines and equipotential lines form a series of squares. If the distance between equipotential lines is dL, the hydraulic gradient is

$$i = \frac{dh}{dL} \qquad (58)$$

If the total distance between the upper and lower boundaries of flow is D then the distance between flow lines is dD. In figure 156 the width normal to the plane of the page is unity. The area through which flow is taking place then is $(1)(dD) = dD$. If the foregoing values are substituted in Darcy's equation (46) the latter becomes

$$dQ = k\frac{dh}{dL}dD \qquad (59)$$

By construction the flow between pairs of flow lines is constant, i. e., $dQ =$ constant. Moreover, since the soil is assumed to be homogeneous k is constant and the increment of head loss, dh, is also constant

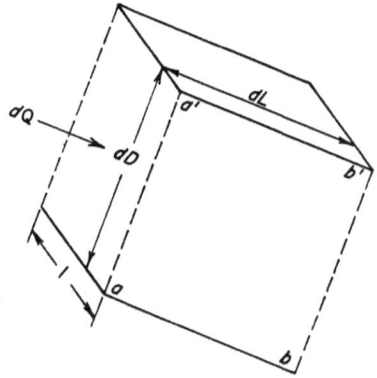

FIGURE 156 —Perspective of a typical flow net unit

Equation (59) may, therefore, be written

$$dQ = C_1 = C_2\frac{dD}{dL} \qquad (60)$$

[26] "Hydraulik," by Ph Forchheimer, Vienna, 1930

where C_1 and C_2 are constants. From this it may be seen that the ratio of the sides of the rectangles of a flow net must everywhere bear a constant ratio to each other if the other conditions with respect to equal rate of flow between flow lines and equal pressure drop between equipotential lines are to be satisfied. Hence, if the values of dQ and dh are chosen so that the rectangles are squares, then the rectangles of the entire net will be rectangles within approximate limits. Actually, many of the rectangles are not squares in the plotting. (See fig. 157.) If, however, the distorted figures are subdivided into smaller areas it will be found that the smaller the subdivisions are made the more nearly will they be true squares. The distortion is introduced, therefore, by using but a few flow and equipotential lines. For practical purposes it is sufficient that the average distances between the flow lines and the equipotential lines be approximately even.

Since the pressure head loss from one equipotential line to another is equal, if the number of increments is n_1, then the drop from one potential line to the succeeding one is $\dfrac{h}{n_1}$, that is,

$$dh = \frac{h}{n_1} \tag{61}$$

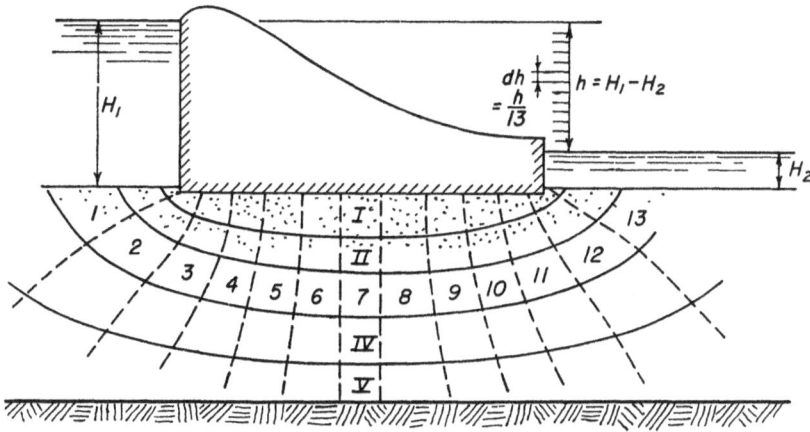

FIGURE 157 —Simplified flow net.

Figure 157 is a flow net for the flow in under a weir with a downstream apron In this case $n_1 = 13$, but it must be understood that the number of rectangles can vary at will for any particular problem. If the flow lines had been chosen closer together so that dQ would have been a smaller increment of Q the distance between equipotential lines would also have been less. If the value for dh by equation (61) is substituted in equation (58) we have—

$$i = \frac{h/n_1}{d} \tag{62}$$

from which

$$i = \frac{h}{n_1 \, dL} \tag{63}$$

The rate of discharge, then, bounded by two flow lines, is

$$dQ = k \frac{h}{n_1 dL} dD \tag{64}$$

But $dD = dL$ by construction (fig. 156),
Hence

$$dQ = k\frac{h}{n_1} \tag{65}$$

By definition the rate of flow between flow lines is equal. Hence, if the number of flow channels bounded by pairs of flow lines is n_2, the total flow is

$$Q = n_2 dQ = kh\frac{n_2}{n_1} \tag{66}$$

This equation may be used to determine the total volume of flow wherever sufficient information is available for the construction of a flow net.

Flow nets may be constructed without the aid of models by determining the upper boundary of flow by mathematical and graphical methods. Where the soil through which flow is considered is composed of elements of varying permeability such a boundary determination requires very broad experience to arrive at a reasonable approximation. Models based on the electric analogy are used to some extent and because it may serve to make the relations involved clearer the following description is given.

Ohm's law which expresses the fundamental relation for the flow of an electric current is expressed by the equation:

$$I = \frac{E}{R} \tag{67}$$

in which I = current in amperes, a measure of quantity of flow,
  E = pressure in volts, and
  R = resistance in ohms.
Equation (67) is in terms of resistance. The Darcy equation (44), on the other hand, is in terms of conductance. Since in Ohm's law, conductance, K', is the reciprocal of resistance, equation (67) may be written:

$$I = K'E \tag{68}$$

The conductance K' varies directly as specific conductivity, k', and the area, a, and inversely as the length L, then $K' = k'\frac{a}{L}$, from which formula (68) may be written:

$$I = k'\frac{E}{L}a \tag{69}$$

If it is noted that I is quantity of flow; k' the coefficient of flow; E the pressure head; L the length of flow; and a the flow area, it may be seen that equation (69) is identical with equation (44). This analogy is the basis for the electrical method of flow analysis which was first proposed by Prof. N. N. Pavlovsky, in Russia in 1920 [27]. Since the principles of flow are similar, with the same conditions as regards pressure and path of flow the flow itself will be similar.

Figure 158 [28] is one type of electric-analogy apparatus. In this two copper-plates represent the upstream and downstream ground surfaces. The head of water is replaced by the electric pressure or voltage impressed on the conductor.

[27] "Theory of Groundwater Flow under Hydrotechnical Structures and its Principal Application," by Prof N. N Pavlovsky, Leningrad, 1922

[28] "The Flow Net and The Electric Analogy," by E W Lane, B F Campbell, and W H Price, Civil Engineering, American Society of Civil Engineers, vol 4, No 10

The soil is replaced by a salt solution. Since the flow of electricity is similar to that of water, the flow in the model will be similar in character to the flow in the prototype. It remains to measure the flow. A high resistance wire is introduced in a parallel circuit. It is assumed that the entire voltage drop across this parallel circuit occurs in the high resistance section. This may, then, be laid off in dis-

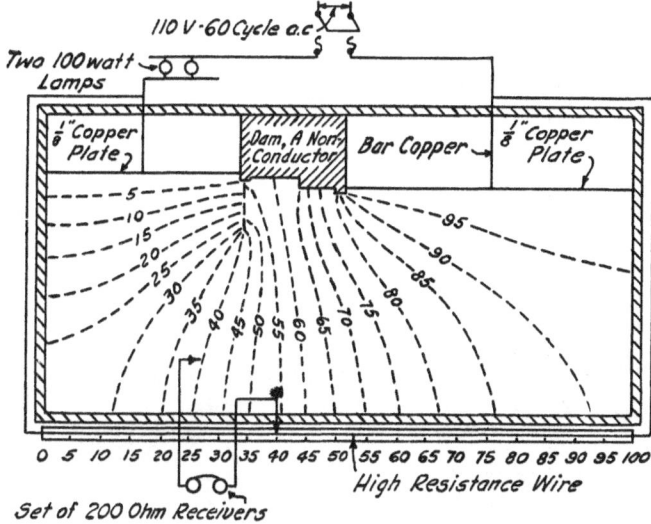

FIGURE 158 —Flow net by electric analogy method

tances representing certain increments of the drop, in this case 5 percent. In the illustration a bridge is shown which includes a pair of receivers. If one end of the bridge is set at any point on the high resistance and the other end inserted in the electrolyte there will be a flow through the bridge unless the potential is the same. In this case one end of the bridge is set on 40 percent on the high resistance the other end used as a probe in the solution. The probe is moved until a point is located where no current flows through the receivers. This point, then, has the same potential as the point on the high resistance. Several points are found and the contour plotted. The bridge is then moved to another potential and the process

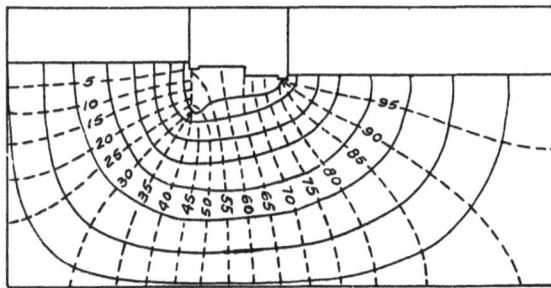

FIGURE 159 —Typical application of the electric analogy.

repeated. When completed as in figure 158, the series of contours show the pressure causing flow at all points in the conductor. By reversing the insulators and terminals the flow paths also may be located in the same manner. It is, however, relatively simple to sketch them in as has been done in figure 159. It follows from the analogy that the flow of water through a homogeneous foundation under the same pressure and linear dimensions will have similar characteristics.

While it must be borne in mind that model study is at best only indicative, very valuable information can often be gained by this means which, properly interpreted, forms a basis for an estimate which otherwise would be little more than a guess—and often a poor one. Figure 160 illustrates a type of flume used quite commonly. It is of shallow depth between the glass faces to encourage simple two-dimensional flow insofar as possible. Models are constructed to a reduced scale with respect to both geometrical proportions and permeability. The geometrical scale ratio is a linear relation and varies with individual models according to the size of the prototype and of the model flume. It is usually inconvenient to use the same permeability of the different elements of the model as will occur in the prototype because the rate of flow in the latter is usually very low, at least for the controlling elements, and this would necessitate considerable time for

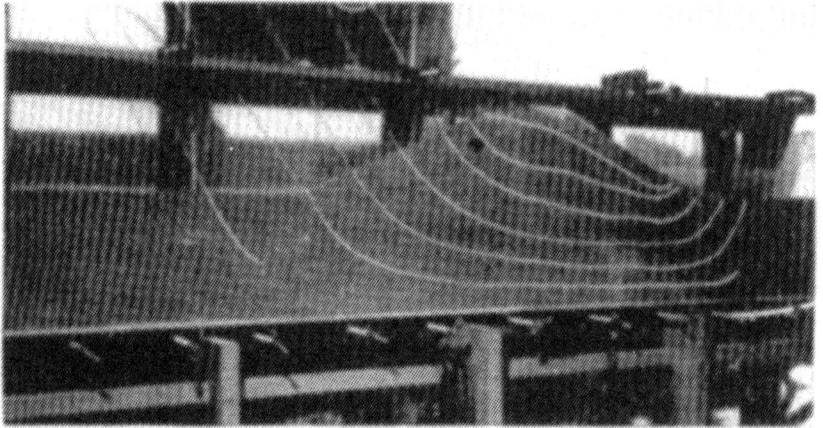

the model test to be performed. For this reason the permeability of the various elements of the model is increased, often 50 to 100 times that of the corresponding element of the prototype. The ratio of the permeability of each element of the model to the corresponding element of the prototype is constant for the particular model so that the ratio of permeabilities of the elements of the model with respect to each other is similar to the ratio of permeabilities of the elements of the prototype with respect to each other. Dye is introduced and since the flow is laminar it is carried along the flow line of the particles entering the soil at the point of injection. The dye tubes are quite apparent in figure 160. The flow lines traced in this manner serve as a guide for the location of those lines chosen as described in the preceding description of the construction of a flow-net. It is not practicable to locate the points of dye injection so that the flow lines traced by the dye will define boundaries of an equal rate of flow.

24. Flow Pressure.—The difference in head between two points represents, at the downstream end, the loss of head due to the resistance to flow offered by the particles of soil just as the inner surface of a pipe offers resistance to flow through it.

Figure 161 represents a soil particle many times enlarged. It is evident that there is a difference in pressure on the upstream and downstream sides of the particle, the intensity of which depends upon the slope of the hydraulic gradient, i

Imagine that the soil particle is lodged in a tube having the same diameter. The particle will act as a piston and cause a reduction of pressure from h to zero, that is, the pressure exerted on the particle is the difference in head on the upstream and downstream faces of the particle. Now imagine the enlarged particle replaced by a mass of small particles. The pressure on each particle will remain proportionately the same. The total pressure on the mass will be the difference in h at its upstream and downstream ends. Harza [29] has stated that the force "exerted by seeping water against a volume of granular material will be the difference in hydraulic head at the

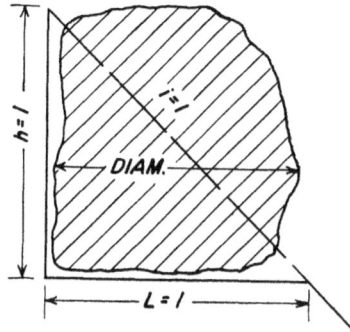

FIGURE 161 —Transmission of pressure to particle

approaching and receding faces of the volume applied to the entire cross section, as on a solid, instead of a granular, piston, although the pressure is actually applied gradually through the material instead of against an impervious face."

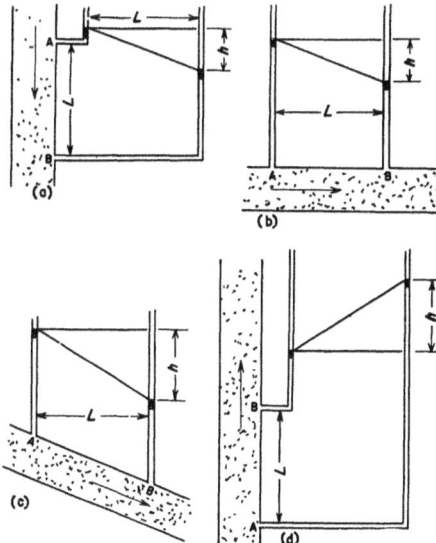

FIGURE 162 —Head loss in directional flow.

Figure 162 illustrates flow in four directions. The pressure exerted on the particles is equal to the loss of pressure head, h, and it is applied in the direction of flow. Although this is exerted on the individual particles throughout the length L the result is the same as though the particles were resting on a piston against which a force equal to the pressure head h were acting. The unit pressure head, h_p, is $w_w h_f$. The unit submerged or effective specific weight of the soil $w_b(_e)$, from equation

(18) is $w_w \dfrac{G_s - 1}{1 + e}$. In Figure 162d the dynamic force of the flow is upward; the gravitational force of the submerged soil is downward. It may be seen that if the two forces were equal the soil particles would be in a state of indifferent equilibrium. The particles would be at the point of flotation and any slight addition to the dynamic force would cause

[29] "Uplift and Seepage under Dams on Sand," by L F Harza, Transactions, American Society of Civil Engineers, vol 100, 1935.

them to move upward.   Hence, the condition

$$h_p = w_w \frac{G_s - 1}{1+e} \qquad (70)$$

is a critical one.

When $h_f = 1$, under which condition the hydraulic gradient, i, would also equal 1 since L was assumed as unity in equation (70), the pressure head is $h_p = w_w = 62\ 4$ per square foot.   Again, the average value of G_s is 2.65 approximately from which the average value for $G_s - 1 = 1.65$.   The average value of the void ratio, e, is approximately 0.6, hence the term $1+e$ has an approximate value of 1.6 from which the fraction $G_s - 1$ approximates unity and it follows that the effective specific weight is w_w.   It may be seen from this that as the hydraulic gradient approaches unity it approaches a critical value which will depend upon the particular soil.   It may be stated then that a hydraulic gradient of unity is a dangerous one with any soil.

Figure 163 is similar to figure 157.   The velocity vectors (a), (b), (c), (d), (e) indicate the direction of the dynamic force.   The conditions are similar to those illustrated in figure 162.   If a critical area exists it will always be located at the point of emergence for two reasons: (1) because the particles must have a free path of escape and (2), because the pressure vector usually has its sharpest upward slope in this region.   In connection with reason (1) it is pointed out that unless the particles are free to move the ratio of pressure to effective specific weight has no significance. In figure 156 the length of each elementary rectangle is dL and since the rectangles are squares $dL—dD$.   In figure 157 the pressure across each square is dh.   The hydraulic gradient, then, is

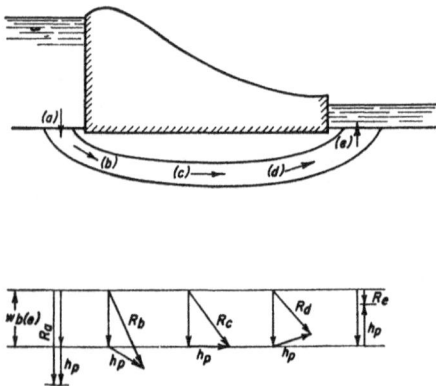

FIGURE 163 —Force diagrams, directional flow.

$$i = \frac{dh}{dL} \qquad (71)$$

which may also be written

$$i = \frac{dh}{dD} \qquad (72)$$

The greatest value of i for a particular set of conditions will be, therefore, at that point where the interval between two flow lines is least. This point should be investigated.

25. Flow of Water Through Embankments.—Figures 164–175 inclusive are a series of flow patterns obtained in the manner illustrated in figure 160. No attempt was made to locate the flow lines with respect to each other with a constant relation. This may best be accomplished by sketching the related flow

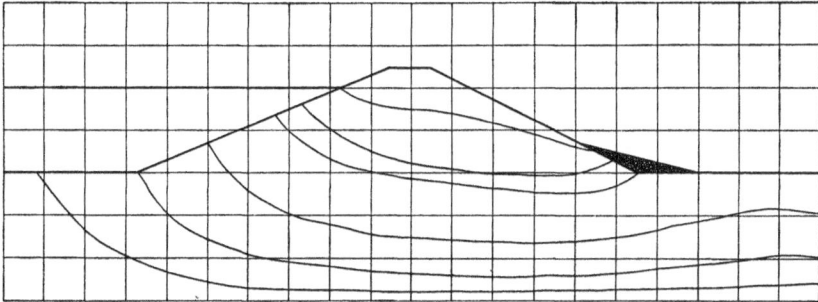

FIGURE 164 —Flow pattern, homogeneous foundation and embankment, no tail-water

Rectangular grid 10 cm x 10 cm actual
Material—foundation—sand,

$$k = 0\ 00189 \text{ cu ins /min}$$
$$= 18\ 9 \times 10^{-4} \text{ cu ins /min}$$

Dam —Same as foundation
Toe blanket—$\frac{1}{16}$-inch pebbles
Actual flow through model—46 4 cu ins /min.

This case illustrates t he flow pattern for a dam and foundation of the same material   Location of flow paths have been selected to show the direction of flow only and not to designate areas of equal flow   Flow lines (see sec. 4–e) may be interpolated.

The rate of flow necessitated a toe blanket to prevent sloughing   When removed the sloughing progressed up the downstream face approximately 1 inch i n depth and the flow paths flattened out to become almost horizontal

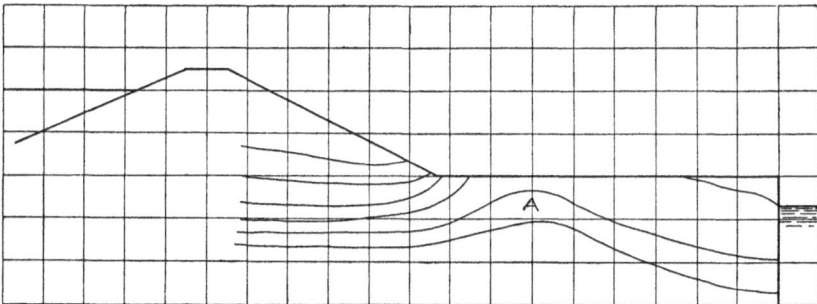

FIGURE 165.—Flow pattern, transition from pressure to gravity flow.

This illustrates the downstream pattern of Figure 152 extended to the end of the model flume   At the downstream end of the flume a screen 10 cm from the end forms a well in which the water is maintained at a constant head at the elevation indicated

It may be noted in several of the patterns that the lower paths tend to rise to the surface just downstream from the toe and then to turn downward again to the well   The dye trace remains uniform in width throughout its length except at the downstream hump at which point it spreads vertically indicating a combined longitudinal flow toward the well and a vertical flow toward the surface   As the path starts downward again it regains its sharp definition and continues to the well with a constant width equal to that upstream from the hump.

lines using the model flow pattern as a guide.   The only purpose of this series of patterns is to indicate in an approximate manner the relative effect of various structural elements having different values for k which are employed in earth dam construction.

Dr. Leo Casagrande [30] has shown that the form of flow-line will be discontinuous at the boundaries of sections of different permeabilities the character of which will depend upon the ratio of the permeabilities and also upon whether the flow is toward a higher or lower value of k.   This is apparent when figures 164 and 175 are compared.   The entire section in figure 164 as well as the upstream and downstream embankments of figure 175 have the same permeability, k_1.   The

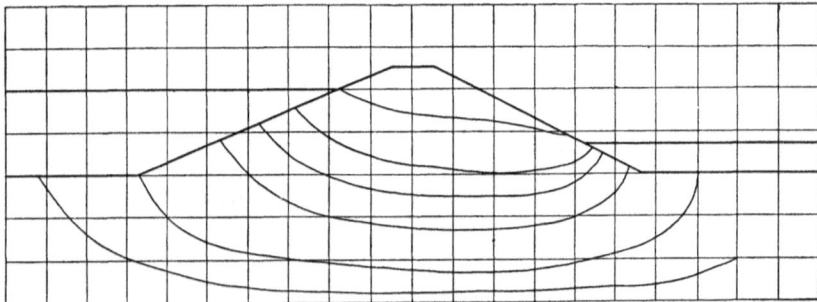

FIGURE 166.—Flow pattern, homogeneous foundation and embankment, with tail water

Grid scale—Same as figure 164
Material—same as figure 164
Tail water—approximately 7 5
Effective head—12 5 cm.
Actual flow through model—36 0 cu. ins /min

It is to be noted that tail water affected the flow in two ways, (1) it created a pressure downstream so that the easiest path for such flow paths as that from the heel is to the surface just downstream from the toe rather than in the general horizontal direction taken by the path from the same origin in figure 145, and (2), it reduced the effective head causing the flow   The flow to the vicinity of the toe was increased and the toe blanket used in figure 145 was necessary to prevent failure   The reduction in flow due to the reduction of effective head more than compensated for the concentration of flow at the toe.

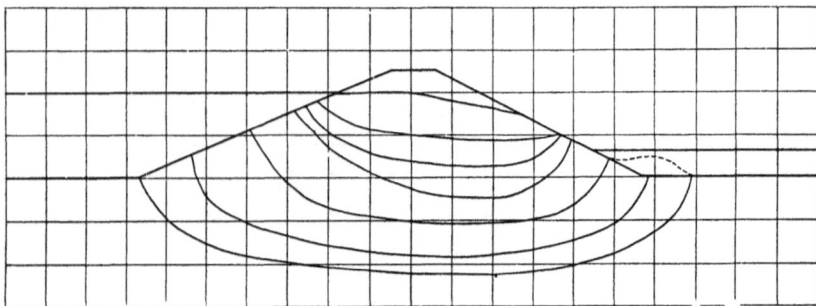

FIGURE 167 —Flow pattern, homogeneous foundation and embankment, beginning of failure.

All conditions as for figure 165.

This illustrates the first stages of failure when the toe blanket was removed   Instead of the sloughing progressing up the downstream face in a relatively thin layer the failure occurred at the toe as a typical flotation.   It may be noted that an immediate readjustment of the flow paths occurred with increased concentration of flow at the toe   Sloughing such as occurred in figure 145 would undoubtedly have taken place at a more rapid rate had the failure been permitted to progress beyond the initial stages

[30] "Naeherungsmethoden zur Bestimmung von Art und Menge der Sickerung durch geschuettete Daemme" by Leo Casagrande, Thesis, Technische Hockschule, Vienna, July 1932, translated by V K Wagner and W. R Perret, U. S Waterways Experiment Station, Vicksburg, Miss.

core of figure 175 has a lower value, k_2. Even in the region of the upstream embankment of figure 175 the pattern is not similar to the corresponding region in figure 164 because the flow in the former is not determined from k_1.

From the continuity equation the rate of flow into a section must equal the rate of discharge from it. It may be seen that the discharge from the upstream embankment into the core is determined by the lower permeability of the latter, hence the rate of flow into the former is also determined by the core permeability. Since the velocity through the embankment of figure 175 is reduced, the hydraulic gradient is also reduced and the section is less effective in reducing the pressure head than is the case in figure 164. In the downstream embankment, on the other hand, the pressure reducing ability of the soil is completely utilized. It

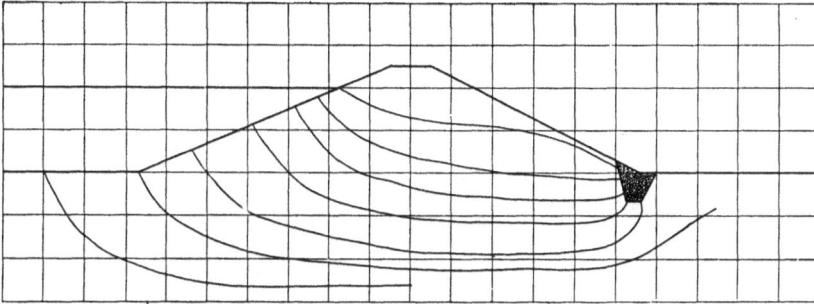

FIGURE 168 —Flow pattern, effect of toe drain

Scale—same as figure 164
Material—foundation—same as figure 164
Dam—same
Toe blanket—omitted
Toe drain trench—⅟₁₆ inch pebbles
Tail water—none
Actual flow through model—50 0 cu ins /min.

It is evident from this pattern that the introduction of any path of less resistance to flow than the surrounding area tends to concentrate the flow on the easiest path The structure gave no indication of being either more or less stable than that illustrated in figure 145 although the flow was increased by the trench. In view of this it would appear reasonable to believe that the toe blanket is as satisfactory in stabilizing soil at incipient flotation as the trench and is less difficult to construct as well as more effective in reducing percolation.

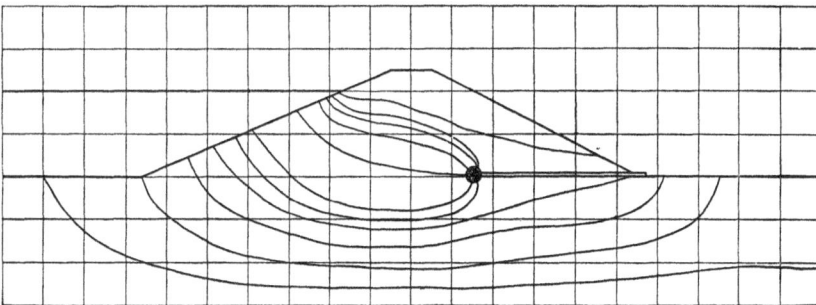

FIGURE 169 —Flow pattern, effect of interior drain, downstream third point

Grid scale—same as figure 164
Material—same as figure 164
Drain with reverse filter from downstream third point to toe
Actual flow in model—55 8 cu ins /min

The drain serves to concentrate the flow on itself and, because of the reduction of the resistance, to increase the rate of flow Since the volume of flow in the area downstream from the drain opening is reduced, there is a marked heightening of the lower flow paths would appear, therefore, that the elements having higher permeability are most effective on the discharge side of elements of lower permeability. Hence, the most effective section is that in which the dense element is farthest upstream. This is also indicated in the relative rate of discharge of figures 171–173 inclusive.

In figure 165 it is to be noted that the two lower lines conform in general to the common pattern to a point, A, downstream from the two where they change direction upstream from A the flow is dictated by the pressure due to the head h Since this pressure gradient is above the flow line flow will occur in the direction of the least resistance whether it be upward or downward. The point A indicates the condition of $dh=0$. Downstream the flow is due solely to the force of gravity and for that reason must always be downward.

Flow upstream from A is similar to flow through a pipe under pressure. If the pipe be shortened, other things being equal, the pressure gradient is steepened

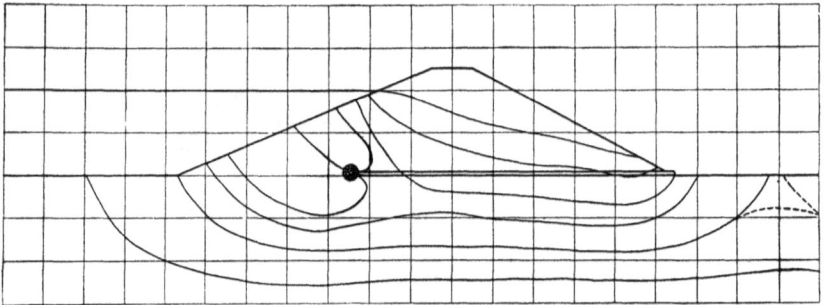

FIGURE 170 — Flow pattern, effect of interior drain, upstream third point.

Grid scale—Same as figure 164
Material—Same as figure 164
Drain—Same as figure 169, extended to upstream third point
Actual flow through model—78 4 cu in /min
This pattern accents the effect of drains In this particular test the flow was limited by the capacity of the drain pipe If this had been larger the concentration would have been more marked It is apparent that drains reduce the amount of flow in the dam downstream from the entrance, but at the same time they reduce the effectiveness of the structure in resisting flow.

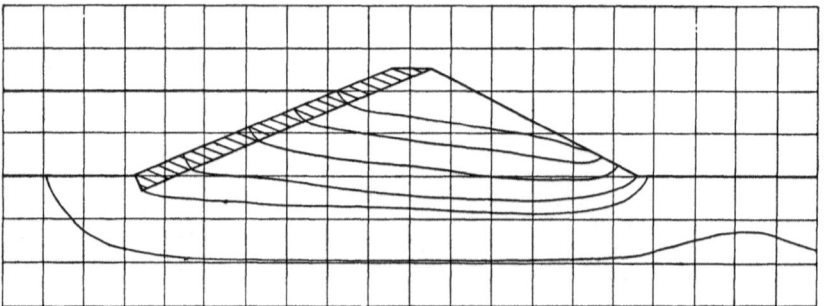

FIGURE 171 —Flow pattern, effect of dense blanket on embankment.

Grid scale—Same as figure 164
Material—Foundation, $k=0 00189$ cu in /min
Embankment, $k=0 00189$ cu in /min
Blanket, $k=0 00011$ cu in /min
Actual flow through model—31 2 cu in /min
In this test the blanket had an effective thickness of 3 5 cm It is particularly noticeable in this pattern that as the flow through the structure is reduced at the upstream face there is a definite flattening out of the flow paths and an increase of concentration at the toe The ground-water level downstream is controlled by a constant head device and was maintained the same for all tests with the exception of figures 165 and 166

and the velocity and rate of discharge are increased. Flow downstream from A is similar to flow in an open flume set on that slope which will just maintain flow. Decreasing the length of the flume has no effect on the pressure gradient and hence no effect on velocity and rate of discharge. The point A is, then, the point of tangency of the pressure gradient and the flow line. By means of adequate dense elements it is possible to move point A to well within the body of the structure. A drain having its entrance downstream from the transition would not increase the rate of flow through the structure; drains opening into the area upstream from A, figures 169 and 170, do increase the rate of flow. However, the drain serves to steepen the slope of flow lines above it by affording a path of less resistance. This tends to draw the flow net into the structure thereby bettering the condition at the surface of the embankment. The immediate vicinity of the drain inlet then becomes the point of emergence and flotation

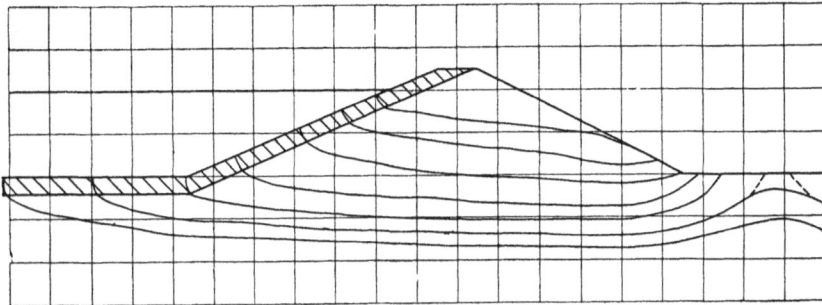

FIGURE 172 —Flow pattern, effect of dense blanket on embankment and reservoir floor.

Grid scale—Same as figure 164
Material—Same as figure 171
Blanket—Extended upstream from heel
Actual flow through model—25 8 cu in /min

The flattening out of the flow paths is particularly noticeable in this pattern This may best be seen by comparing the path from the heel in the various figures The concentration at the toe due to a reduction of flow pressure is borne out in this pattern also This concentration is offset, of course, by the reduction of flow pressure Such a concentration would have been disastrous in the test for figure 164 It would prove disastrous in actual practice also if, through erosion or other means, the blanket should be broken through.

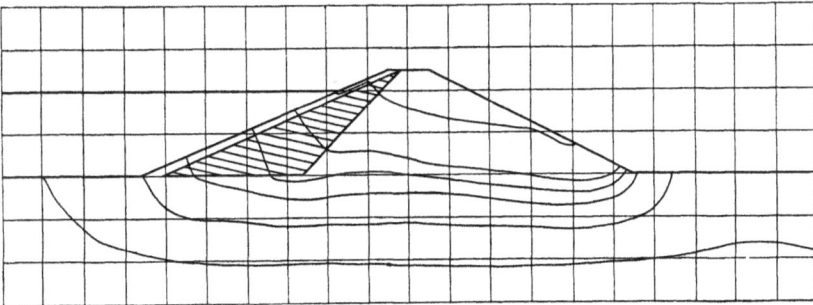

FIGURE 173 —Flow pattern, effect of dense upstream third section.

Grid scale—Same as figure 164
Material, Foundation, same as figure 164, embankment, same as figure 164, core, same as blanket, figure 171
Core—Upstream third of structure
Actual flow through model—31 7 cu in /min

The concentration of flow at the toe is accented in this pattern More important than this, though, the core is much less effective in reducing percolation than the blanket in figure 172 and will be more difficult to construct.

here is as serious as flotation at the surface. Properly protected down-stream drains usually increase the stability of the downstream embankment. On the other hand they also increase the water loss through seepage and in bodies of water having a deficient supply for proper maintenance this is an important consideration. If the material is available it will generally be more satisfactory to increase the dimensions of the dense elements and retain the drains.

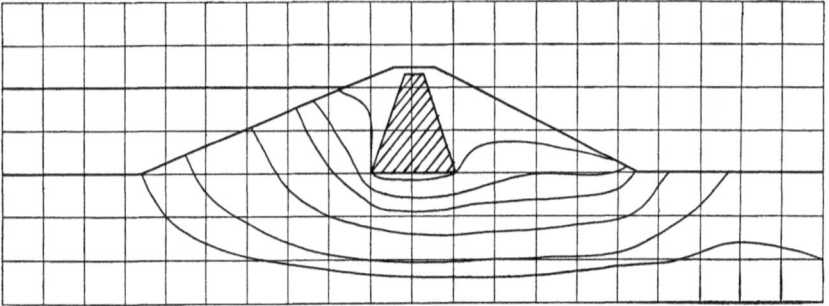

FIGURE 174 —Flow pattern, effect of dense interior section.

Grid scale—Same as figure 164
Material—Same as figure 175
Core—Cross section as indicated
Actual flow through model—34 6 cu in /min

It may be noted that the flow will follow the path of least resistance In figure 173 it penetrated the core material, in figure 174, having a path to do so, it avoids the core. The concentration at the toe is considerably reduced with a slight increase in flow.

The uppermost flow line in figure 165 as well as the next below it emerge on the slope of the embankment. The resultant of the gravitational and pressure components in such a case is similar to case (c) in figure 162. It is to be noted that the resultant in this case is approximately parallel to the slope of the embankment. Moreover, the velocity of the percolate increases upon emergence and its transporting power is thereby increased. Usually, where the embankment is roughly homogeneous, the critical location will be at some point on the embankment above the toe.

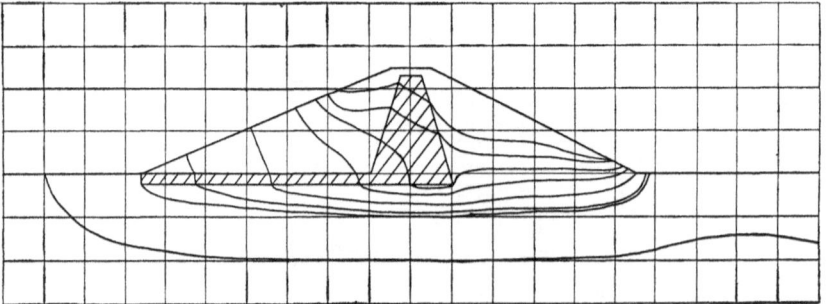

FIGURE 175 —Flow pattern, effect of extending dense interior section to heel.

Grid scale—Same as figure 164
Material—Same as figure 173
Core—Interior blanket from core to heel of dam
Actual flow through model—29 1 cu in /min.

The result of a dense stratum in the upstream section is apparent in this pattern. Lacking the easier path around the core of figure 174, the flow penetrated the core. Very little of the flow is through the foundation It is of particular interest that the flow through this model is but slightly less than that through figure 171 and actually greater than that through figure 172 It would appear that the extended surface blanket is the most effective cut-off

Consider next the two flow lines in figure 165 which emerge just downstream from the toe. This situation is similar to case (e) in figure 162. Assume that the pressure gradient is unity at this point. The particles will be in indifferent equilibrium completely lacking in any degree of stability. This is quicksand. Quicksard is, therefore, a condition of soil rather than a type. As water emerges it will carry with it the fines, particularly those whose velocity of settlement is expressed by Stokes law and may therefore be transported by velocities which will fail to move the larger particles. Assume that the slope away from the point of emergence is sufficient to maintain the emergence velocity. In this case the transported particles would be completely removed.

Beginning at the surface and progressively working deeper there will be a tendency for the flow to remove the particles of a grain size equal to and less than 0.01 mm. Exactly what will happen under these conditions in any particular case is beyond the limits of even an idealized solution. The structure of the soil might be such that there would be no rearrangement of the larger particles; or it might be such that partial collapse of the mass would serve to reduce the enlarged pores. In either case, however, the removal of fines would result in a revised grain size distribution approach curve II, figure 175 and consequently a revised and higher value of k. This action has been referred to as "colloidal erosion" by Krynine [31] and is the reverse of a common cause of increased head loss in filters. Recent tests [32] have shown that fine particles carried in suspension may penetrate into filter sand a distance, termed the "critical depth," which varied from 2 inches to more than 37 inches under a head of 8 feet.

The increase in the value of the coefficient of permeability of an already critical area would lead to a concentration of flow similar to that caused by a drain. If, as in a typical spring, a pool is formed to dissipate the pressure and permit the particles to drop back without being carried away a boil results which may continue indefinitely without further structural failure. (The dissipating pool of a boil may be entirely below the surface of the ground. The flow tube from which the fines have been eroded forms the pool, the uneroded larger particles being in suspension as has been discussed under "quicksand.") The essential consideration is whether particles continue to be removed. If the pressure is not too great and the soil contains particles graded to include the smaller stone sizes the tendency is to form a natural reverse filter.

SECTION E—SHEAR IN SOILS

26. Character of Shear Resistance.—Although the resistance of a soil to shear is a characteristic long recognized, probably less is known about its practical determination and application than any other recognizable factor. Various types of apparatus have been devised, some of them rather intricate, most of them capable of giving results which while quite satisfactory qualitatively are questionable quantitatively. Nevertheless, with careful and understanding application the principles as now understood do provide data which are available from no other source and which for all that they may not be mathematically exact are infinitely nearer that desideratum than is possible through rule-of-thumb methods.

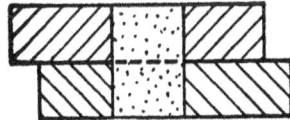

FIGURE 176 —Simplified shear test normal load zero

---

[31] "Soils Mechanics Research," discussion by D P Krynine, Transactions of the American Society of Civil Engineers, vol 98, 1933.

[32] "Filter Sands for Water Purification Plants," Progress Report of the Committee of the sanitary engineering division on Filtering Materials for water and sewage works, Proceedings, American Society of Civil Engineers, vol 62, No 10, December 1936

Figure 176 shows a section through two metal plates having an opening through each one in which a sample of tough clay is inserted. The thickness of the plates is small so that the weight of the sample is insignificant. If the plates are moved horizontally in opposite directions considerable force will be required to shear the clay sample. Imagine now that the openings are realligned and the clay replaced by dry sand. Very little force will now be required to move the plates. That is,

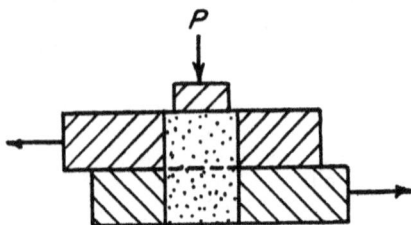

FIGURE 177 —Simplified shear test, normal load = P.

the dry sand will not offer the same amount of resistance to rupture. Figure 177 is similar to figure 176 with the exception that a load, P, is applied to the sand by means of a piston in such a manner that the load bears on only the sand. If horizontal movement of the plates is now attempted it will be found that the sand offers a resistance to rupture by shearing which increases as the load, P, is increased. With the clay there is also an increased resistance to shear when a normal load is applied.

It is evident from the foregoing that there must be some pressure within the clay mass which is almost totally absent in the sand when no normal load is applied. Dr. Terzaghi [33] has designated this "intrinsic" pressure and stated that there are two principal types: (1) that due to molecular cohesion and (2) that due to the surface tension of the moisture film. The first he has called "true" and the second "apparent" cohesion.

The degree of internal pressure due to true cohesion depends upon the area of contact of soil particles one with another. The finer the particles the greater the proportionate contact area. This is greatest in the case of the clays which have flat, scale-like particles and such clays have considerable "toughness," being able to carry a substantial load on an unsupported cylinder without rupture. All soils have this property, even sands, but as the particles increased in diameter their area of contact decreases and the pressure becomes insignificant.

Pressure due to the surface tension of moisture films has been discussed in section 20, Capillary Water. If sand is wetted and molded to form a cylinder it will actually support a slight weight in addition to its own without rupture. If, however, the surface tension is released either by immersion or drying the sand will slump even without any normal load in addition to its own weight.

Figure 178 illustrates the cross section of two plates, the abutting areas of which are shaped as shown. If horizontal forces F and F₂ are applied in the directions indicated, there will be a vertical displacement in conjunction with the horizontal. Imagine the

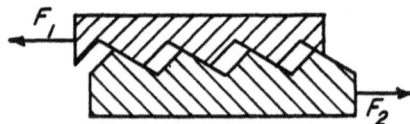

FIGURE 178 —Idealized internal shear resistance.

teeth of each plate replaced by spheres. In this case, also, there will be both vertical and horizontal movement. If a normal load be applied to the upper plate, the horizontal forces necessary to produce movement must be correspondingly increased. When soil is sheared, this same combination of vertical and horizontal displacement occurs. With such an apparatus as illustrated in figure 177 it is actually observable with coarse grained sands. Resistance to shear due to this dual displacement has long been recognized and is designated "internal friction."

[33] "Erdbaumechanik" by Karl Terzaghi, Vienna, 1925

27. Unit Shearing Strength.—The angle of friction has been defined as that angle at which sliding of one mass over another is impending. This angle, ϕ, has been determined experimentally for various materials by Rankine and others. Coulomb's law states that when one body slides over another the frictional resistance equals the normal weight of the sliding mass times a coefficient, called the coefficient of friction, f, which is equal to tan ϕ. (The reader is referred to any of the textbooks of elementary physics for a more detailed development of this relation.) As has already been explained in section 26 the resistance to shear is due to both intrinsic pressure and frictional pressure, hence:

$$s = c + P \tan \phi \tag{73}$$

in which

$s =$ unit resistance to shear
$c =$ the coefficient of cohesion, and
$P =$ the normal load.
$\phi =$ angle of internal friction

If, as in the case of sand, $c = o$ the formula reduces to:

$$s = P \tan \phi \tag{74}$$

It is evident that if either P or tan ϕ in formula (74) is equal to zero then $s = o$ and the material, having no resistance to shear, that is, no resistance to change of shape, will act as a liquid. This condition has already been referred to in the description of quicksand (sec. 25). In this case P is the weight of the overlying material at any point in the mass. Because of the pressure of the upward flowing water, the load is zero and the mass of sand behaves as a liquid.

28. Shear Resistance of Soil.—If the value of the internal friction is determined for two samples of the same soil, one densely compacted and the other loose, the former will have the higher value. This is due to the fact that as the particles ride over each other, the more loosely compacted the soil the more will the vertical displacement be confined to the individual particles within the mass rather than to the mass as a unit. Hence, the state of compaction affects the internal friction. Again, the internal friction would be greater if the surface of the particles was rough than would be the case if the particles were perfectly smooth. Therefore the character of the particles affects the internal friction. Soil moisture acts as a lubricant and the internal friction will be affected, too, by the degree of lubrication. It follows, then, that tabular values of tan ϕ classified as "earth on earth" and "earth on earth, wet clay" can be at best only a rough approximation of the actual value in any particular case.

C. A. Hogentogler has pointed out [13] that materials composed largely of iron and alumina will become much more dense when compacted than materials composed largely of silica. Moreover, the density of a soil may be increased or decreased by the addition or leaching out, respectively, of certain electrolytes by either natural or artificial means. The increased stability of soils treated with common salt or calcium chloride, other things being equal, is quite generally known. Stability is, of course, simply another designation of shear resistance. Because of these and other complicating factors it is customary to determine the value of c and tan ϕ experimentally for each individual type of soil where such determination is economically justified and the necessary apparatus

[13] "Engineering Properties of Soil," by C A Hogentogler, C. E.   McGraw-Hill, 1937.

is available. Figure 179 illustrates the shear apparatus used by the Engineering Laboratory of the National Park Service. In this design the normal load, P, is applied by means of the yoke and weights. The horizontal load is applied by means of a water load in the suspended tank. The water is siphoned from the supply on the scale. The tank itself is counterweighted. Figure 180 illustrates the box in more detail. The upper section is rigidly fixed to the bed plate. The lower section slides. The force F, in figure 178 is, therefore, supplied by the rigid connection; force F_2 is supplied by the water load. Dial gauges indicate the strain due to the vertical load and the horizontal load. The test sample is 2.5 inches in diameter. This size was chosen so that natural samples obtained by means of a 3.5-inch core barrel might be tested.

To determine the shear curve for a particular soil a sample under zero normal load is loaded horizontally to failure. This gives the intercept of the curve and the zero normal load ordinate. In equation 73 the term $P \tan \phi$ must equal zero and the intercept is the measure of the cohesion. Since dry sand has zero cohesion its intercept would be zero. Additional samples are each sheared at a different normal load to obtain sufficient points on the curve to permit plotting it.

The use of the shear curve will be discussed in connection with stability of embankments. It is sufficient here to say that its determination is not yet well understood even though considerable use is made of the results obtained by present methods. The shear curve for highly cohesive soils is particularly difficult to determine.

SECTION F—CONSOLIDATION OF SOILS

29. General Principles.—In section A it was stated that for engineering purposes the term soil is taken to include both the particles and the voids and, that the voids ratio of a given soil is the ratio of the volume of voids to the volume of particles. Within the limits of loads common to dams, soil particles and water are incompressible, that is, they are not subject to change of volume. Hence a reduction in the volume of a given mass of soil, the volume of particles remaining constant, can occur only as a result of a reduction of the volume of voids. Where the voids are filled with water the reduction of volume must be accompanied by a proportional reduction in the volume of pore water and the rate at which the excess water can escape determines the rate of consolidation. These concepts are fundamental.

It was stated above that under the loads to be encountered there will be no change in volume of individual soil particles. This should not be confused with

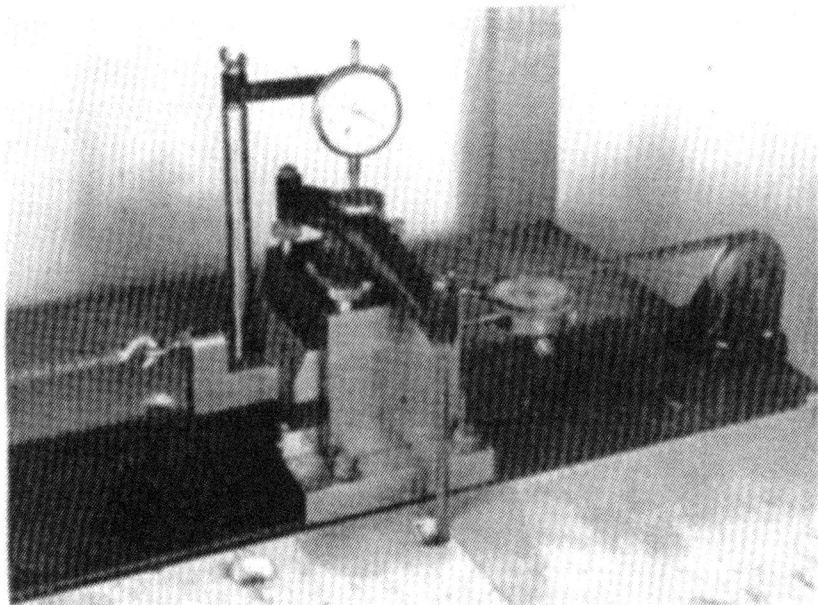

deformation of shape. Soil particles may be deformed appreciably although this is limited principally to the flat, elongated particles, the spherical particles deforming but slightly. It is natural that grains deformed under a load will, if their elastic limit is not exceeded, tend to regain their original condition. Consolidation which is due to such deformation is, for this reason, essentially elastic. Compaction which is due solely to a reduction of the voids is nonelastic.

Figure 182 illustrates a section of a watertight cylinder having a petcock and a tight, heavy piston. Assume that dry soil has been placed in the cylinder loosely to the line A–A and that under the weight of the piston that volume of soil has been compressed so that its upper surface is the line B–B. In the dry state this would happen quite rapidly since the air in the voids offers little resistance. Now assume that the soil has been replaced to A–A so that it has the same voids ratio as it originally had, that those voids have been completely filled with water, the piston replaced, and the petcock at the bottom closed. It must be remembered that since a volume of soil is the sum of the volume of particles and the volume of voids and that in such a case as this, wherein the voids are water-filled, it is the volume of particles plus the volume of water. Since both the particles and the water are incompressible the piston is supported at the line A–A so long as the water cannot escape. If now, the petcock is opened slightly so that the

water flows a trickle, as the water leaves the voids the piston will descend since
the soil alone cannot support it in that position.   The rate at which the piston
descends to line B–B will, therefore, depend upon the rate at which the water
escapes from the voids.   If the petcock is opened wider the piston will descend
more rapidly.   When the piston reaches the line B–B it will descend no farther
whether the voids are water-filled or not since, as has been demonstrated, the
soil alone will support it when it reaches this state of consolidation.

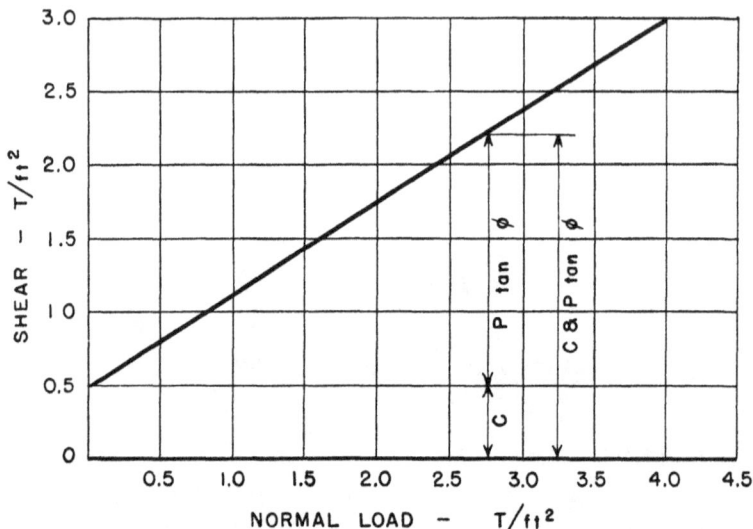

FIGURE 181 —Typical shear curve.

In the foregoing the same conditions have existed as would have been the
case if the soil had first been compressed dry under the piston load to B–B and
then enough water had been added to fill the pores and also the cylinder from
B–B to A–A.   Under these circumstances the piston would once more have been
supported at A–A until enough water had escaped to permit the piston to rest
on the soil at B–B.

The foregoing must be accurately visualized if the following basic concepts of
the nature of consolidation are to be understood:

(a) The rate of consolidation of a soil under a fixed load will be determined
by the rate at which the pore water can escape, that is, the coefficient of percolation of the soil.   For very dense soils, such as clays, percolation is very slight
and consolidation may extend over a period of many years.   Dr. Terzaghi has
stated [31] that in some clays this period may approach 200 years.

(b) When the rate of consolidation is retarded by the inability of the pore
water to escape but slowly during that period which elapses between the application of the load and the completed consolidation of the particles the entire load
is carried by the water and not at all by the particles.

(c) During that period when the normal load is carried by the water and the
load on the soil is zerc, the soil is in a state of practical saturation and has no

---

[31] "Soils Mechanics Research," discussion by D. P. Krynine, Transactions of the American Society of
Civil Engineers, vol 98, 1933

supporting value. This may be difficult to visualize at first but the principles involved are all elementary and have been previously discussed herein. It may be simpler to understand if the cylinder in figure 182 is assumed to be filled loosely with soil to B–B and water to A–A. By elementary hydraulics the water pressure is distributed equally in all directions hence the pressure due to the load on the water is equal on all surfaces of each particle. Since the soil is saturated cohesion, $C=0$ hence, equation (74) applies. The only normal load on the soil particles then is that due to their own buoyed weight. In the discussion of equation (74) it was explained that when either P or ϕ is zero, then $s=0$ and the soil acts as a liquid. (This condition will be referred to under stability of embankments.)

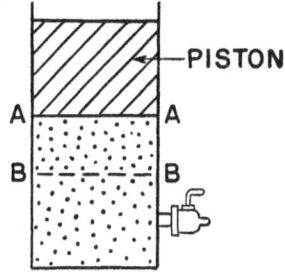

FIGURE 182 —Device for illustrating relation of permeability to consolidation

The consolidation apparatus as generally used was developed by Dr. Terzaghi. Its essential elements are shown in figure 183.   When a load is applied the initial stage of consolidation will be relatively rapid and if the percent of consolidation is plotted against the time, as in figure 184, the curve for the initial stage will be vertical. The rate will gradually lessen and eventually a point is reached at which no further consolidation will take place. If, however, an additional increment of load is applied an additional consolidation will take place the plotted curve for which will be in general similar to the first curve. In actual practice the sample is loaded with progressive increments of the total load that is to be applied to the prototype permitting the sample to acquire stability under each loading before the next is applied.

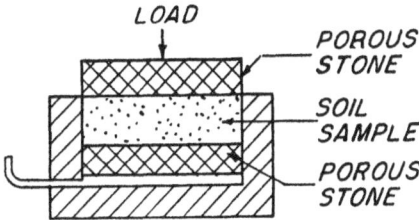

FIGURE 183 —Simplified elements of consolidation apparatus

Figure 151 is an adaptation of the Terzaghi device developed by Dr. Casagrande.   In section it is similar to figure 183.   The vertical load in this case is applied by means of a press.

It has been stated previously that consolidation is due to a change in the void ratio.   This change is computed for the final dimension in thickness of the sample under each increment of load and, following the completion of compaction under the full load.   The load is then reduced by increments, and the void ratio computed for each stage of the load reduction in the same manner.   The values of the void ratios plotted against the respective loads will result in a curve similar to that illustrated in figure 185.   From A to B the curve designates a decreasing void ratio due to the increasing load; from B to C it designates an increasing void ratio due to the decreasing load.   The latter is due principally to the elastic property of the soil.   With fine grained soils, particularly those containing many flat particles, the expansion curve will bend sharply upward signifying considerable rebound; with coarse, spherical grained soils the rebound will be slight and the section of the curve B–C will tend to be horizontal.

Assume that the time-compression curve, figure 184, and the load-compression curve, figure 185, relate to a sample from a stratum 20 feet in thickness. Since the compression of a soil is measured by the reduction of the void ratio it is from the load-compression that the degree of settlement in the prototype may be determined.

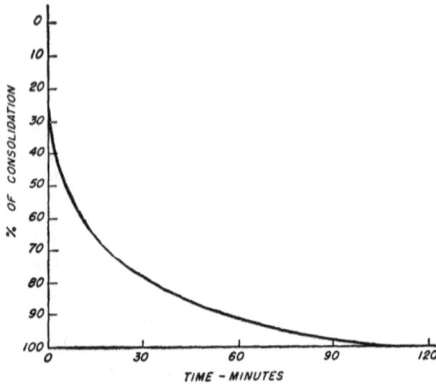

FIGURE 184 —Typical consolidation—Time curve .

Let $C=$ compression in % of D_1,
$e_1=$ voids ratio, before consolidation,
$e_2=$ voids ratio after consolidation,
$d=$ thickness of sample,
$D_1=$ thickness of stratum being tested before consolidation,
$D_2=$ thickness of stratum being tested after consolidation.

Then:

$$C=\left(\frac{e_1-e_2}{1+e_1}\right)(100) \quad (75)$$

and

$$D_1-D =\frac{e_1-e_2}{1+e_1}D_1 \quad (76)$$

from which:

$$D_2=\frac{1+e_2}{1+e_1}D_1 \quad (77)$$

From figure 182, $e_1=1.8$, $e_2=1.225$, hence

$$D_2=\left(\frac{1+1.225}{1+1.80}\right)(20)$$

$$=15.9 \text{ feet.}$$

It should be noted that the rate of percolation has no bearing on the degree of consolidation, provided, of course, that the water can drain away. It does, however, determine the length of time that will be required for the consolidation to take place.

Let $t=$ time required for consolidation of sample from $1+e_1$ to $1+e_2$, and,

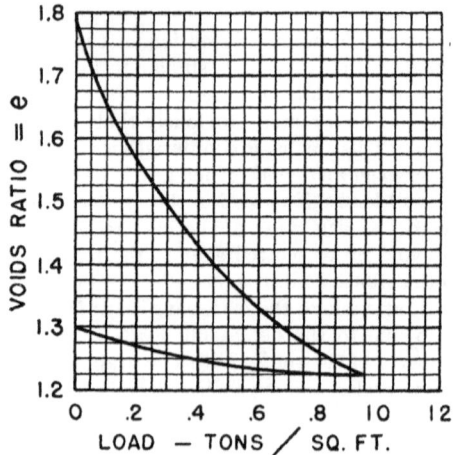

FIGURE 185 —Typical load—Void ratio curve.

$T=$ time required for the corresponding consolidation of the natural stratum.

In the consolidation apparatus the two porous stones furnish a permeable stratum both above and below the sample so that the greatest distance to a permeable boundary will be one-half the sample thickness or $\dfrac{d}{2}$. Hence, for either half the average distance of flow will be $\dfrac{d}{4}$. With the sample saturated in the beginning of the test the water to be displaced must equal the total compression as given by formula (38). From the foregoing it may be shown that the rate of consolidation varies as the square of the distance between drainage boundaries, therefore:

$$\frac{T}{t}=\frac{D_1^2}{d^2} \tag{78}$$

Equation (77) presumes that the stratum being tested has a permeable boundary both above and below similar to the sample. Often, however, the stratum may have one impermeable boundary. This case would occur if the stratum under consideration rested on impervious bed-rock. In such a case the average flow distance would be $\dfrac{D_1}{2}$ from which:

$$\frac{T}{t}=\frac{4D_1^2}{d^2}. \tag{79}$$

In the compression apparatus the load may be considered as applied instantaneously. Dr. Terzaghi has stated that if the load is applied uniformly, as is the usual case in actual construction, the time of consolidation, T, is one-half that for an instantaneous load. For such a case equation (79) may be written:

$$T=\frac{2tD_1^2}{d^2} \tag{80}$$

Let $d=1.5$ inches.
$\quad D_1=20$ feet $=240$ inches.
$\quad\quad t=120$ minutes from figure 184.
From equation (80)—

$$T=(2)\frac{(120)(240)^2}{(1.5)^2}$$
$$=6,144,000 \text{ minutes.}$$
$$=4,267 \text{ days}$$

Figure 186 is a consolidation permeameter cylinder developed by R. R. Proctor. In this apparatus the vertical load is applied by means of a hydraulic jack and calibrated springs. When the required load has been impressed on the spring the top plate is bolted down against the spring and the jack removed. The principles involved and their application in figures 151 and 183 are identical. The latter was developed principally for recompacted material and the size of the required sample limits its value for testing undisturbed samples very materially. It is to be noted also that the load varies as consolidation takes place due to the lengthening of the spring. It is probable that this introduces no particular difficulty if the consolidation which takes place in a sample is slight.

SECTION G—FOUNDATIONS.

30. Distribution of Stresses.—Unless a foundation includes a stratum which is or may become highly plastic, it will be very seldom that foundation competency will be a concern in the design of the embankments of an earth dam less than 30 feet high or a masonry dam less than 20 feet high. The scope of this book does not justify, therefore, the complex and extensive discussion that adequate treatment of the subject would require. Where the structure exceeds the foregoing height limits, or where a stratum of questionable competency occurs in connection with a structure within the limits, it is recommended that competent consulting advice be retained

31. Foundation Investigation and Permissible Loads.—The Foundation Chapter [15] of the new Boston Building Code is an empirical foundation code revised in the light of modern soil mechanics. While many compromises with the old conceptions were made the code does represent a distinct advance. For the design of foundations for outlet and spillway structures it will be found both practicable of application and reliable of result. The following is quoted from that code.

Section 2904—Classification and Allowable Loads of Foundation Bearing Materials

(a) * * *

(b) The maximum pressure on soils under foundations shall not exceed the allowable bearing values set forth in the following table except when determined in accordance with the provisions of sections 2915 and 2916, and in any case subject to the modifications of subsequent paragraphs of this section.

Class	Material	Allowable bearing value (tons per sq. ft.)
1	Massive bedrock without laminations, such as granite, diorite, and other granitic rocks; and also gneiss, trap rock, felsite, and thoroughly cemented conglomerates, such as the Roxbury puddingstone, all in sound condition (sound condition allows some cracks)	100
2	Laminated rocks such as slate and schist, in sound condition (some cracks allowed)	35
3	Shale in sound condition (some cracks allowed)	10
4	Residual deposits of shattered or broken bedrock of any kind except shale	10
5	Hardpan	10
6	Gravel, sand-gravel mixtures, compact	5
7	Gravel, sand-gravel mixtures, loose; sand, coarse compact	4
8	Sand, coarse, loose; sand, fine, compact	3
9	Sand, fine, loose	1
10	Hard clay	6
11	Medium clay	4
12	Soft clay	1
13	Rock flour, shattered shale, or any deposit of unusual character not provided for herein—Value to be fixed by the commissioner.	

(c) The tabulated bearing values for rocks of classes 1 to 3 inclusive shall apply where the loaded area is less than 2 feet below the lowest adjacent surface of sound rock. Where the loaded area is more than 2 feet below such surface these values may be increased 20 percent for each foot of additional depth but shall not exceed twice the tabulated values.

(d) The allowable bearing values of materials of classes 4 to 9 inclusive may exceed the tabulated values by two and one-half percent for each foot of depth of the loaded area below the lowest ground surface immediately adjacent, but shall not exceed twice the tabulated values. For areas of foundations smaller than three feet in least lateral dimension, the allowable bearing values shall be one-third of the allowable bearing values multiplied by the least lateral dimension in feet.

(e) The tabulated bearing values for classes 10 to 12 inclusive apply only to pressures directly under individual footings, walls, and piers. When structures are founded on or are underlain by deposits of these classes, the total load over the area of any one bay or other major portion of the structure, minus the weight of excavated material, divided by the area, shall not exceed one-half the tabulated bearing values.

(f) Where the bearing materials directly under a foundation overlie a stratum having smaller allowable bearing values, these smaller values shall not be exceeded

[15] "Proposed New Boston Building Code—Chapter on Foundations," reported by Gilbert Small, Chairman, Boston Code Committee, Paper Z–17, vol. II, Proceedings, International Conference on Soil Mechanics and Foundation Engineering, Harvard University, 1936. Note—Further references to these proceedings will be indicated by "Proceedings, Internat. Conf."

at the level of such stratum.　Computation of the vertical pressure in the bearing materials at any depth below a foundation shall be made on the assumption that the load is spread uniformly at an angle of 60° with the horizontal; but the area considered as supporting the load shall not extend beyond the intersection of 60° planes of adjacent foundations.

(g) Where portions of the foundation of an entire structure rest directly upon or are underlain by medium or soft clay or rock flour, and other portions rest upon different materials, or where the layers of such softer materials vary greatly in thickness, the magnitude and distribution of the probable settlement shall be investigated as specified in section 2916, paragraph (f) and, if necessary, the allowable loads shall be reduced or special provisions be made in the design of the structure to prevent dangerous differential settlements.

(h) Whenever, in an excavation, an inward or upward flow of water develops in an otherwise satisfactory bearing material, special methods satisfactory to the commissioner shall be immediately adopted to stop or control the flow to prevent disturbance of the bearing material.　If such flow of water seriously impairs the structure of the bearing material, the allowable bearing value shall be reduced to that of the material in loose condition.

Section 2905. Foundation loads.—(a) The loads to be used in computing the maximum pressure upon bearing materials, under foundations shall be the live and dead loads of the structure, as specified in chapter 23, including the weight of the foundations, but excluding loads from overlying soil.

(b) Eccentricity of loading in foundations shall be fully investigated and the maximum pressure shall not exceed the allowable bearing values.

SECTION H—COMPACTION OF SOIL

32. Relation of Moisture Content and Maximum Density.—In 1933 R. R. Proctor [34] demonstrated that there is a fixed relation between the water content of a soil and the denseness to which it may be compacted and, further, that for every soil there is one water content termed the "optimum water content" at which maximum denseness is reached by a specific amount of compacting pressure. This relation of water content to maximum compaction furnishes the most satisfactory method of construction control available.　For this reason it is of very practical value.

33. Compaction Test Method.—The apparatus consists of a compaction cylinder and a hammer.　These are shown in figure 187 together with a sampler and its handle which will be discussed later.　The cylinder is 4 inches in depth, has a volume of one-thirtieth cubic foot, and is composed of a base, a lower and an upper section.　Proctor's hammer has a 2-inch diameter striking face and weighs 5.5 pounds.　The hammer in figure 187 has the same face and weight as Proctor's but is enclosed in a pipe to guide the fall of the hammer and also to gauge the height of fall which Proctor fixed at 12 inches.　Figure 188 shows the compaction-test section of the National Park Service Engineering Laboratory.　Two compaction cylinders are shown, one of which is used for permeability tests of the compacted sample (see fig. 150).　A small compressor supplies air for the spray by which water is added to the sample.　In the foreground is a box filled with sand the top of which is covered with cloth.　All compactions are performed on this, because the inertia and rigidity of the supporting base have been proven to have an appreciable influence on the effectiveness of the compacting blows. The sand box is an easily provided base which will give similar conditions whether in the laboratory or in the field.

The soil to be tested is dried and carefully pulverized.　The soil is then placed in the cylinder to a depth of 2 inches and compacted under 25 blows of the hammer.

---

[34] "Fundamental Principles of Soil Compaction" by R. R Proctor, Engineering News-Record, Aug. 31, Sept. 7, 21, 28, 1933.

A second and third layer each 2 inches deep are added and each compacted. The upper section of the cylinder is then removed and the compacted soil trimmed off even with the top of the lower section of the cylinder. This is next weighed to obtain the net specific bulk weight of the soil and the penetration resistance determined (sec. 36) following which a sample is removed for water content determination; the soil is again pulverized; more water is added and the sample once more compacted. The weight of the dry soil particles per cubic foot of soil for each compaction is then computed and plotted against the respective water content. A curve similar to that of figure 189 results.

Most laboratories have found that the tests are more consistent if started from the dry end and worked toward the higher water content. It should also be pointed out that the specifications for the hammer weight, hammer fall, and number of blows are purely arbitrary. It does not give the maximum compaction possible

FIGURE 186 — Proctor compression apparatus

because if the number of blows or height of fall is increased the denseness will increase also. Moreover, different supports for the cylinder as, for example, a table or a solid concrete floor, will give different degrees of compaction, other things being equal. With respect to the former, it is probable, so far as is known now, that the Proctor compaction specification results in the maximum denseness that can be duplicated by practicable construction methods. With respect to the second the sand box in figure 188 is used by the National Park Service laboratory so that the support can be exactly duplicated in the field. This variable is eliminated in this way and compactions made in the laboratory and in the field are comparable.

34. Optimum Water Content.—This is usually called optimum "moisture" content. For reasons stated in section 8 "Water Content", the term moisture is

not used herein. The weight per cubic foot of the compacted sample is obtained
for each water content tested. This is converted to compacted dry weight per
cubic foot and the two unit weights are plotted as ordinates against the respective water contents as illustrated in figure 188. Inspection of these curves
clearly shows that at a water content of 16.5 percent this particular soil can be
most densely compacted, that is, the greatest number of particles compacted
into a unit volume which is reflected in the higher unit weight. The individual
value for optimum water content-maximum unit weight relation will vary
with almost every soil. All, however, have a curve similar to those in figure 189
with an upward slope on the dry side and a downward slope on the wet side of
optimum, M_o.

35. Relation of plastic limit to optimum water content.—In figure 189 the
optimum water content is 16.5 percent and the plastic limit is 15.9 percent, a

difference of 0.6 percent which is insignificant for practical purposes. In this
case the plastic limit test would provide a convenient water content control.
Unfortunately this relation does not hold for all soils. In some cases it is as
much as several percent under and in other cases as much over optimum with
respect to compaction. It is not therefore, a general criterion of optimum water.

In those cases where it is indicative within practical limits the following caution
is to be emphasized. Inexperienced operators seldom can roll out the thread as
dry as those having considerable experience. As a result the test may indicate
a soil as satisfactory with respect to water content when actually it is much
too wet.

36. Penetration Resistance.—In his procedure R. R. Proctor makes use of a
"Proctor Needle" to determine the relative compaction of the soil. This appara-

tus depends so much on the ability and experience of the operator that different operators often get quite different results. It may be seen that the laboratory determination by an experienced operator may be widely different from the determination by the field operator, particularly if the latter is less experienced. In order to remove at least some of the variables the apparatus shown in figure 190 was developed by C. A. Hogentogler, Jr. In this device the direction of penetration is definitely vertical and the rate of penetration is easily and conveniently controlled. Its principal advantage, however, lies in the fact that the connection from the penetration needle to the operating lever is positive and there cannot be any lag between the rate of penetration of the needle and the operation of the lever as may be the case where the penetrating force is transmitted to the needle through an intermediate spring.

Compaction control by means of the "Proctor Needle" is used by many Federal and State bureaus. On the other hand many equally experienced organizations place no confidence in it. Its use is, therefore, a matter for organizational deter-

mination. The details of the needle are illustrated in figure 191. As may be
noted on the assembled view, pressure is exerted on a calibrated spring enclosed
in the barrel. The load is transferred through the spring to the needle. Sufficient
pressure is exerted to force the needle into the soil at a predetermined rate.
Various sized needles are used, the choice of which is dependent on the penetration
resistance of the particular soil. The large needles are for soils of low resistance;
the small needles are for soils of high resistance. Properly experienced operators
testing screened soils can obtain quite consistent results using this test but it does
have the disadvantage of requiring an operator of a very great deal of experience,
and even with such an operator the results obtained from a soil containing gravel
particles as large or larger than the needle area may be very misleading.

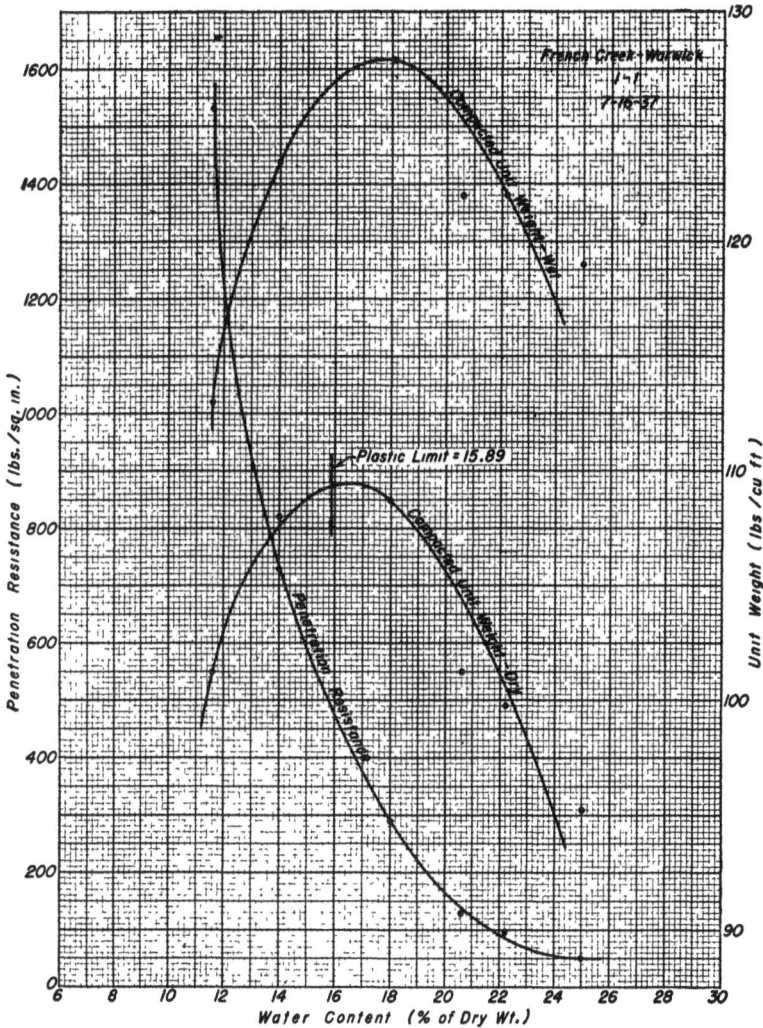

TYPICAL OPTIMUM WATER – COMPACTED WEIGHT CURVES

FIGURE 189.—Typical optimum water—Compacted weight curves.

37. Construction Control.—It is assumed that the soil to be used has been tested in an adequately equipped laboratory and that the field staff has been furnished a graph similar to figure 189 for each individual soil to be used. To use this information the equipment needed consists either of the "Proctor Needle" apparatus or that illustrated in figure 187. For reasons which will be apparent, only the latter is discussed here.

In order properly to construct the embankment, the field staff must know that the water content of the soil is within the tolerable limits and that the proper degree of compaction has been accomplished. These will be discussed individually.

The water content of the fill material may be determined with satisfactory accuracy by means of the water-content-unit weight curve. Provided the cylinder is resting on the same kind of a support and reasonable care is used, the unit weight obtained for a given sample of soil will not vary materially with different operators. Hence, if a soil compacted in the field results in a certain unit weight it must be at that water content at which the laboratory obtained the

Hogentogler).

same unit weight. Hence, for the unit weight obtained in the field the corresponding water content may be taken directly from the unit weight-wet curve. Since the field staff must work with the wet compacted weight only, this curve need be furnished. The dry-unit weight is of interest to the laboratory chiefly because it is this weight that is generally referred to in literature dealing with soil mechanics.

If the sample compaction indicates a water content either too high or too low, this must be corrected before compaction of the fill is undertaken. If the soil is on the dry side, water must be added; if on the wet side, the soil must be dried. In the first case, the soil should be well harrowed after the water is added in order to distribute it throughout the soil as evenly as possible. In the second case, the soil should be spread and harrowed to encourage drying to the proper degree of water content. This may be very difficult or impracticable in some highly cohesive soils. It may be found necessary in this case to spread a thin layer of the wet soil, covering this with dry soil and rolling the two together with a sheeps-foot roller. While most field men accept the idea of adding water to too dry soil, they often ask what should be done if the soil is too wet to dry quickly. There is but one answer to this—wait until it does dry.

A soil which is wetter than it should be is unsatisfactory and should no more be used than unsatisfactory lumber, cement, or any other construction material.

Realization of this should lead to more careful drainage provisions in the borrow pit and in the manner in which the surface of the fill itself is pitched to drain rapidly.

Assume that the soil is at the proper water content, has been spread, and compaction started. The field supervisor now must know when the proper com-

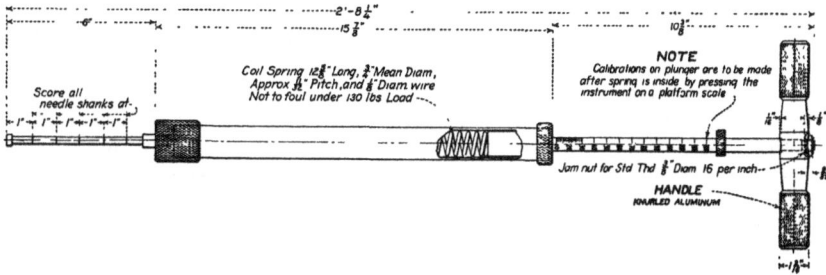

FIGURE 191 —Details of Proctor "needle".

paction has been accomplished both so that compacting will not be carried on longer than necessary and so that sufficient compaction is accomplished. Consider the nature of compaction. It does not alter the character of the soil particles nor does it affect the water content sensibly. The only thing that happens is that the soil particles are adjusted in position under a dynamic load so that more particles can be crowded into a unit volume. This results in an increased stability due to increased shear resistance and also increased unit weight due to the weight of the added particles. The measurement of the increased shear resistance, whether it be by shear machine or some penetration device, is at best uncertain because it depends so much on the condition of the measuring device and on the skill and care of the operator. The measurement of the increased weight, on the other hand, is simple, and individual skill does not enter. It is equally positive because it may be taken for granted that the shear resistance will be uniform for a given soil if the compactness, as indicated by the unit weight, is uniform. Hence, for a given field sample weight, it may be assumed that the other qualities determined by the more accurate methods available to the adequately equipped laboratory also hold. Because of its simplicity and almost complete elimination of the personal element the use of unit wet weight is the most positive method of compaction control.

The sampler in figure 187 consists of a sampling cylinder, a handle, and a spring balance. The field test will consist simply of forcing the sampler into the compacted soil, trimming the excess soil flush with the top and bottom of the sampler, and weighing it. Provided the compaction is uniform, it may be seen that any number of operators of varying degrees of skill should all arrive at the same result. If the proper unit weight has been obtained, compaction is complete regardless of the number of passages of the compacting equipment. If the proper unit weight is not obtained, compaction is not complete. Hence, it is apparent that a specification which states that the compacting equipment shall make six or some other arbitrary number of passages over the soil is questionable. In some cases less might be sufficient; in other cases more might be required. The purpose of the compacting equipment is to compact the soil to a predetermined unit weight. It is impracticable to foretell with exactness the number of passages of the equipment that will be required. For the purpose of estimating, however, the specifications may state a minimum and maximum number of round trips of the compacting equipment and this will very often be four and eight, respectively. The exact number, however, should be based on obtaining the proper compaction. It should be pointed out that should the unit weight remain too low after, say, 10 or 12 passes of the roller, a careful check should be made. The fill material may be different from that for which the unit-weight water-content curve was constructed or the water-content determination may be in error or have changed since the determination was made.

Where the services of an adequately equipped laboratory are available they should be taken advantage of in preference to temporary field laboratories—the equipment for which is often makeshift. With such services available the apparatus illustrated in figure 187 is both competent and sufficient for the construction control of any embankment project which is not large enough to warrant an adequately equipped laboratory manned by thoroughly experienced operators. Only the very largest projects, far above the limits of this manual, fall in the latter category.

38. Significance of Water Content—Denseness Relation.—In a paper presented in 1936,[7] C. A. Hogentogler, Jr., pointed out that if the volume of the moisture content is stated, as a percent of the combined volume of the particles and moisture, M_o', instead of just the dry soil alone, the Proctor curve becomes a series of straight lines, the intersections of which established the limits for certain characteristics of the particular soil being tested. Figure 192 is a typical family of

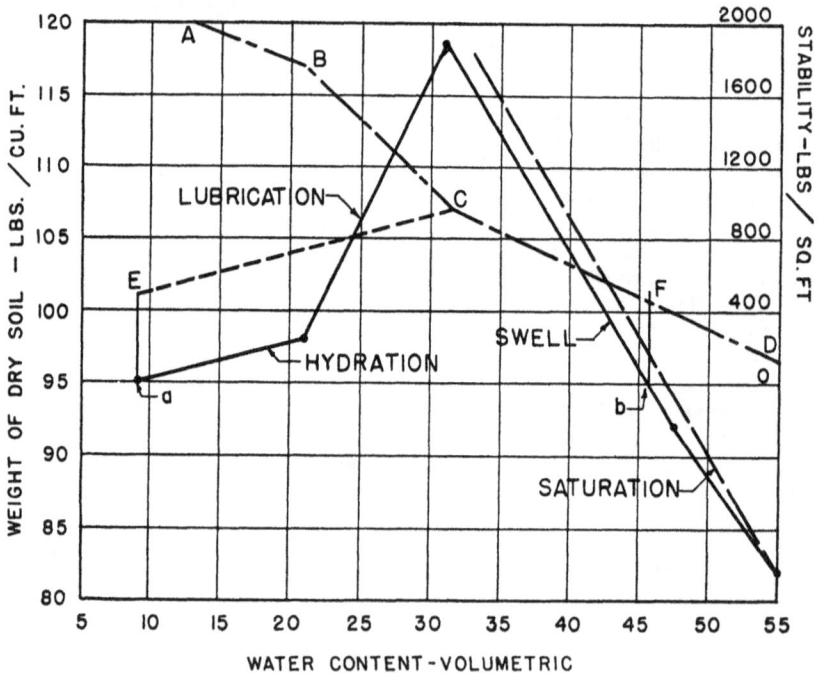

FIGURE 192.—Relation of water content to compacted weight and "Stability"

7 "Essentials of Soil Compaction," by C A Hogentogler, Jr , paper, annual meeting, Highway Research Board. 1936.

curves. It may be noted that the addition of water up to approximately 22 percent increases the compacted weight but little. This is due to the fact that most of the mositure is either absorbed by the dry particles or forms surface film. The addition of moisture from 22 percent to approximately 32 percent, however, is accompanied by a substantial increase in the compacted weight; the added moisture of this stage increases the film thickness and acts as a lubricant which permits the particles to rearrange themselves more compactly for the same compacting load. As more water is added over and above the optimum, which for this particular soil is 32 percent, the additional water replaces part of the soil, and, being less in weight than the displaced soil, unit for unit, the compacted volume naturally becomes less. Since the amount of soil at 32 percent now occupies a greater volume, there has been a swelling. This continues in this case to approximately 45 percent, at which point the soil approaches saturation, arriving at that stage at 55 percent. For other soils the limiting points of the various stages will not be at the same values but the curves will be, in general, similar with definite breaks at the limits.

In Section B (12) it was explained that below saturation the water does not completely fill the voids. The dashed line to the right of the Swell curve in figure 192 is a measure of the volume air in the voids. It was also explained in the section referred to above, that film tension becomes zero under saturation and the air is removed. This is indicated by the junction of the swell and void-air curves at the saturation point. While it is not extended in figure 192, the void-air curve continues upward in the same general direction for water contents below optimum since, naturally, it becomes 100 percent of the void volume at zero water content.

Let the term "stability–lbs./sq. in." designate the force required to penetrate the compacted soil sample by means of a small piston at a uniform rate (called penetration resistance in preceeding discussion). Curve A–B–C–F–D gives the stability for the various water contents. It may be seen that the stability is highest for the minimum water content and decreases in value as the water content increases. It would appear from this that the most stable soil under a load would be that compacted dry. Theoretically, this is correct; practically, it is incorrect. It should be noted that for every point on the Hydration and Lubrication Curves there is a corresponding point on the Swell curve having the same dry weight per cubic foot. Since the dry weight is identical it follows that the void ratio is the same. Hence, the only difference between point "a" for example on the Hydration curve and point "b" on the Swell curve is that at "a" the voids are entirely air filled, at "b" mostly water filled. It is of the greatest importance to realize that the voids at "a" will become water filled at the first opportunity whereupon the soil will be identical in every respect with that at "b", including the stability value. Curve E–C, then, represents the stability values of the Hydration and Lubrication curves whenever the water becomes available. Practically then, except possibly on a desert, the higher values of stability for water contents below optimum cannot obtain for long in the structure.

As yet, the relation of stability value to shear value is qualitative only, a quantitative relation not having been determined. Hence, though they both are a measure of the resistance of the soil to rupture, the stability values cannot be applied rationally as was the shear value. Qualitatively, though, it does furnish a measure of the stability of a submerged, compacted soil.

Consider first the soil compacted at optimum water content in figure 192. Due to its higher degree of compaction, the pores finer, and compression within the mass due to capillary pressure is, therefore, greater than is the case for other points on the curves. When the soil is not submerged, therefore, surface tension has its maximum value for the particular soil. If the soil is permitted to dry, its resistance to consolidation, due to its density, quickly equals the capillary tension and little if any shrinkage can take place. If the soil is next submerged so that capillary tension is released, the swelling will be correspondingly negligible with the exception of one case. If the soil is elastic and in the process of compaction it is over-stressed, capillary tension may exert enough pressure to hold the soil in its over-stressed condition. If this pressure is released by submergence, the soil may expand. This swelling will, of course, be due to the elastic properties of the particles and is quite distinct from swelling due to moisture. Assuming that the soil has not been severely over-stressed during compaction it will, because of its density, offer considerable resistance to rearrangement of its particles. It is evident that the maximum areal contact of particles will occur with maximum density and the molecular forces will have their maximum value for the particular soil. The only force available, then, to cause expansion or swelling of the soil will be that due to increase of the water film thickness surrounding each particle. The optimum water content, in all probability, satisfies the demands of the solidified water, so that any swelling will occur as an increase of the cohesive water. If the loss of stability of some clays when stressed is due, as it probably is, to a breaking down of the cohesive film, it would follow that a stress such as that due to a load would prevent the formation of the cohesive film. Very little is known about the film phenomena, particularly for a submerged soil, but it appears reasonable to believe that a load equivalent to that used in obtaining maximum compaction should be sufficient to prevent the rearrangement of particles, and if this is so, swelling and loss of stability will not occur under this load.

It is pointed out that the immersion of a small unconfined sample is probably not indicative of that soil's ability to retain its stability when submerged, since its surface area is out of proportion to the mass in a practical sense. It must be remembered that the outer layer of particles are held to the mass rather weakly, particularly when submerged, and they can offer little resistance to the formation of the cohesive film which, in turn, further weakens the bond, so that the particles tend to slough off. This progressive failure is to be seen in most immersion tests of well-compacted soils.

If a soil is compacted at a water content below optimum it would, upon being submerged, take the character of a soil compacted at a water content above optimum. It is, therefore, necessary to consider then only that portion of the curve between maximum density and saturation for soils compacted at other than optimum water. Take, for example, the soil at approximately 47 percent water content, as shown in figure 192. Being less dense than that at optimum moisture, the pores will be coarser, from which capillary pressure will be less. Cohesion likewise will be less. Therefore, the resistance to the thickening of cohesive films will be less also. If cohesive films do increase in thickness, swelling must follow, from which the weight of a unit volume of soil will decrease. Such a decrease is in the direction of zero stability. It is in no sense intended to imply that any soil compacted at other than optimum water is a potential failure. This is not the case. A soil compacted to a stability of 1,000 pounds per square inch might, for example, drop to 800 pounds per square inch and still be satisfactory. Until methods are developed to use this information quantitatively, it is not possible to say. There is, however, little question with regard to the dangers inherent in saturation and any condition which approaches that state is potentially dangerous.

SECTION I—STABILITY OF EMBANKMENTS

39. Fundamental Principles.—Shear in soils has been discussed in section E. It was explained that the resistance to sliding of one body of soil over another, the shear resistance, is due to the cohesion of the particles and the frictional

resistance of the individual particles.   From equation (73), $s=c+P \tan \phi$, it is
seen that the shear resistance is a function of the normal load also.

In figure 193 the lines A–C and A'–C' represent shear planes.   (It will be explained later that the shear surface is not a plane.   However, for the purpose of illustrating the basic concept, it is convenient so to consider it.   Assume that the soil is homogeneous throughout the mass, in which case c and tan ϕ of equation (73) will be identical for all points in the mass.   Consider first the prism A–B–C. The force tending to cause this prism to slide along the plane A–C is the component of the weight of the prism parallel to A–C and in the direction of A.   The force tending to prevent sliding is the total shear resistance along the plane A–C and is in the direction of C.   The total shear resistance will depend on the component of the weight of the prism normal to A–C which enters as P in equation (73). Next, consider the prism A'–B'–C'.   Since the prism is larger, the components parallel and normal to A'–C' will be greater than the corresponding components with respect to A–C.   Because the shear resistance happens to be greater than the force tending to cause shear in one case does not mean that this is also true for the other.   If all possible shear planes were investigated, and there could be as many as could be drawn in, it would be found that at some one plane the value of the force tending to cause sliding is greatest.   If, along this plane, the total shear resistance exceeds the total force tending to cause shear, embankment is probably safe.   If the shear resistance does not exceed the force tending to cause sliding, then the embankment probably is unsafe.

It is pointed out at this time that the laboratory determination of the unit shear value of a soil is at best uncertain.   This uncertainty is introduced by two factors; first, the uncertainty of obtaining truly undisturbed samples even under the most experienced supervision and, second, the present state of laboratory technique.   Improvements are being made constantly and it is not too much to hope that a practicable method of testing may be developed or one of the present methods substantiated in the not too far distant future.   (A collaborative experiment is now being carried on by most of the major laboratories in the United States under the direction of Committee D–18, "Soils For Engineering Purposes," of the American Society for Testing Materials, which should remove at least some of the uncertainties within the coming year.)   For this reason it must always be kept in mind that the present methods of determining embankment stability do not furnish a rigid solution. They may be accepted as indicative only and their interpretation

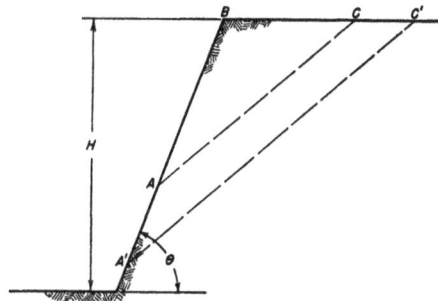

FIGURE 193 —Simplified shear planes in an earth embankment.

and application are matters requiring considerable experience.   Properly applied, however, they are a very valuable tool even now.

40. Embankments of Cohesive and Non-cohesive Soils.—Although it is not strictly true, it is generally assumed that the slope of repose and the angle of internal friction of non-cohesive soils are equal.   As a matter of fact, the slope of repose has the larger value of the two when the material is loose, and the angle of internal friction has the largest value when the material is densely compacted.

The variation is, however, slight and it is sufficiently accurate for practical purposes to consider them equal. Contrary to this, there is no relation between the slope of repose and the angle of internal friction for cohesive soils.

If an embankment of non-cohesive soil is constructed so that its slope, θ, figure 193, does not exceed the slope of repose the height of the embankment, H, is without limit. The embankment can be indefinitely high and stability will in no way be related to the vertical height. Cohesive soils, on the other hand, do have a definite height, called the critical height, for each slope, shear resistance assumed constant. The design of embankments of noncohesive materials involves few difficulties in the determination of the slope of repose. The following is, therefore, devoted only to cohesive materials.

41. Systems for Determining Stability.—The various systems for determining the stability of embankments are based upon certain assumptions with respect to the character of an embankment failure. Among these, the failure surface is assumed to be one of three forms; first, a plane; second, a logarithmic spiral; and third, a circular arc. The first is no longer credited. The second requires fewer simplifying assumptions than the others but involves extensive computations. The third is now quite generally accepted as being the most practicable system and is the only one that will be discussed herein.

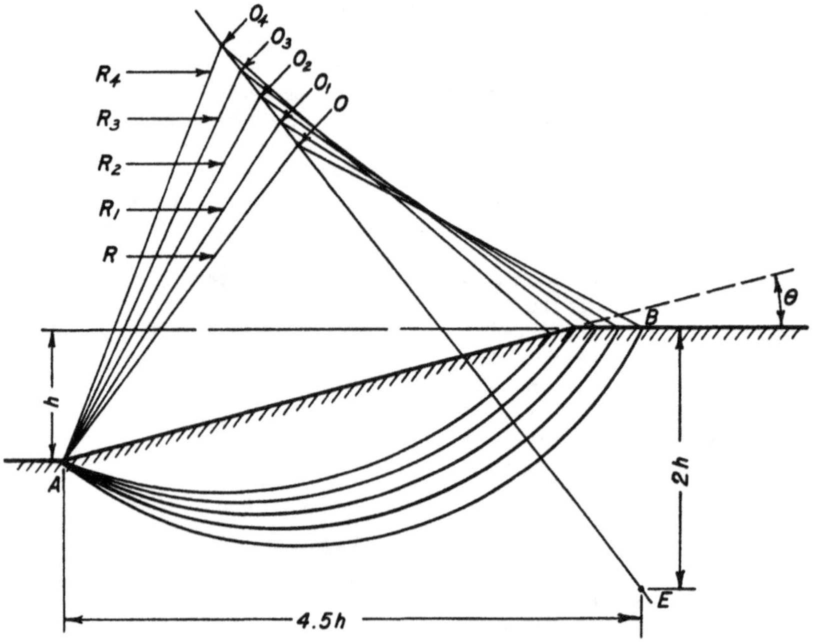

FIGURE 194 —Shear surfaces and centers of rotation, circular-arc method.

The fact that rupture surfaces closely approximate circular arcs was suggested by K. E. Petterson in 1916. This was substantiated by a great many field measurements made under the direction of the Swedish Geotechnical Commission, which was charged with the task of determining the cause of a large number of serious failures of the high embankments of the Swedish railways and a method of designing stable embankments. Various methods have been developed for the solution of the circular arc method, among which are those of the Swedish Geo-technical Commission,[35] Dr. Ing. Jaky,[36] W. Fellenius,[37] and Donald W. Taylor.[38] Only the latter two are discussed herein because they are simpler to use and at least equal to the others in practicable accuracy.

42. The Slice Method.—This procedure for the solution of the circular arc method was proposed by the Swedish Geotechnical Commission and developed to include the graphical solution by W. Fellenius.[37] [39]　In the discussion of figure 193 it was shown that the weight of soil and area of sliding surface varied for planes A–G and A'–C' and that several of an infinite number of such planes must be analyzed to locate the critical one if such a one exists.　This same procedure is followed in the slice method of Fellenius.　Figure 194 illustrates five of an infinite number of circular arcs that might be considered.　The force tending to cause rotation is the tangential com-ponent of the weight of the mass bounded in each case by the arc and the surface.　The force resisting rota-tion is the product of the unit shear resistance by the total length of the arc A–B, figure 196, or L, figure 195.

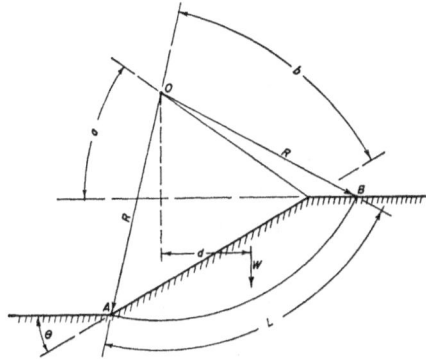

Fellenius has shown that the center of rotation moves upward and out-ward as the value of ϕ increases from zero.　For practical purposes the movement of the centers of rotation is along a line which passes through the center O determined from table 4 and the point E, figure 194, which

FIGURE 195 —Location of center of rotation when $\phi=0$ (after Fellenius).

is $4.5h$ horizontally within the embankment from the toe and $2h$ below the upper surface.　The weight of the mass and the length of sliding surface varies with each radius.　To obtain a solution by this method it is necessary to solve several of the arcs.　Since the procedure for each individual arc is the same, the procedure for one only will be described.

TABLE 4

Slope	θ	a	b
1 0 58	60°	29°	40°
1 1	45°	28°	37°
1 1 5	33°47'	26°	35°
1 2	26°34'	25°	35°
1 3	18°26'	25°	35°
1 5	11°19'	25°	37°

---

[35] Statens Jarnvagars Goetekniska Kommission, 1914–22, Slutbetankande, May 31, 1922

[36] "Stability of Earth Slopes," by Dr Ing. Joseph Jaky, Paper G-9, Vol II, Proceedings, International Conference on Soil Mechanics and Foundation Engineering, Harvard University, 1936

[37] "Calculation of the Stability of Earth Dams," by Wolmar Fellenius, Question VII, Second Congress on Large Dams, Washington, D C , 1936

[38] "Stability of Earth Slopes," by Donald W Taylor, Vol XXIV, No 3, Journal, Boston Society of Civil Engineers, July 1937

[39] "Earth Statical Calculations With Friction and Cohesion and Upon Supposition of Circular Cylin-drical Sliding Surfaces," by W Fellenius, Stockholm. 1926

In order to simplify the graphical solution Fellenius assumes first that $\phi=0$ and that the only force resisting shear is cohesion. For this condition the center of rotation, O, figure 195, is located by means of the values for a and b, table 4, which were given by Fellenius. The mass bounded by the surface of the embankment and the arc A–B is next divided into a number of slices as in figure 196.

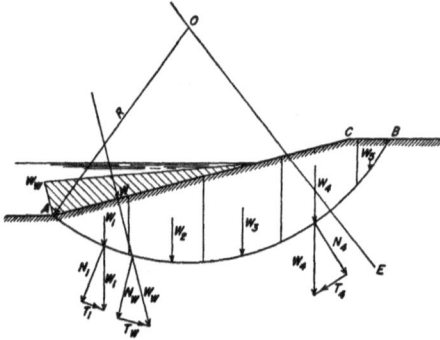

FIGURE 196.—Force diagrams, method of slices.

As is common to all the methods so far proposed it is necessary to introduce a simplifying assumption in order to make the problem statically determinant. Fellenius assumed that the forces on the opposite sides of each slice are equal and opposite. The forces W_1, W_2, etc., are the product of the weight of the slice for a depth normal to the plane of the page of 1 foot by its arm, d, figure 195. This is resolved into components normal and tangential to the arc under consideration at the point of intersection with the arc. The force tending to cause rotation of each individual slice is the tangential component, T, of the force W. The force resisting rotation is the unit shear resistance of the soil times the length of arc intercepted by the sides of the slice. The total force tending to cause rotation is the algebraic total of the tangential components of the individual slices. The total force resisting rotation is the unit shear resistance, $c + P \tan \phi$, by the total arc length, L. The relation may be written

$$T_{total} = L(c + P \tan \phi) \tag{81}$$

It is customary to state the relation

$$\frac{T_{total} - (P \tan \phi)}{L} = c' \tag{82}$$

determining thereby the unit value of shear resistance required for stability. The various values, c'_1, c'_2, c'_3, etc., determined for the respective centers are then plotted as in figure 197 from which the maximum required value for c' and the location of the center of rotation of the most dangerous arc are readily determined. Comparison of the required value with the actual value of c for the particular state of the soil will disclose the state of stability, the ratio of the actual value to the required value being the factor of safety. Atten-

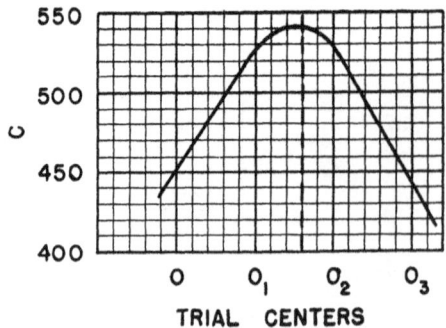

FIGURE 197.—Graphical determination of center of most dangerous arc

tion is called to the fact that the factor of safety in this case is with respect only to cohesion.

This method requires considerable work even when, through experience, the locations of the trial centers of rotation are chosen to give the required points

for plotting the curve, figure 197, with the fewest possible trials.   For this reason
the following method is to be preferred.

43. ϕ Circle Method.—This method, which was proposed some years ago by
Prof. Glennon Gilboy and Arthur Casagrande, has been developed by Donald W.
Taylor, research associate in soil mechanics, Massachusetts Institute of Tech-
nology.   In an excellent paper [38] he has given the theory upon which the method
is based as well as a comparison with the other methods.   The following is ex-
tracted from the paper to give the practical application of the method.   To those
who are sufficiently well versed in mechanics it is recommended that the original
paper be referred to.   To those who are interested only in the application the
following should suffice.

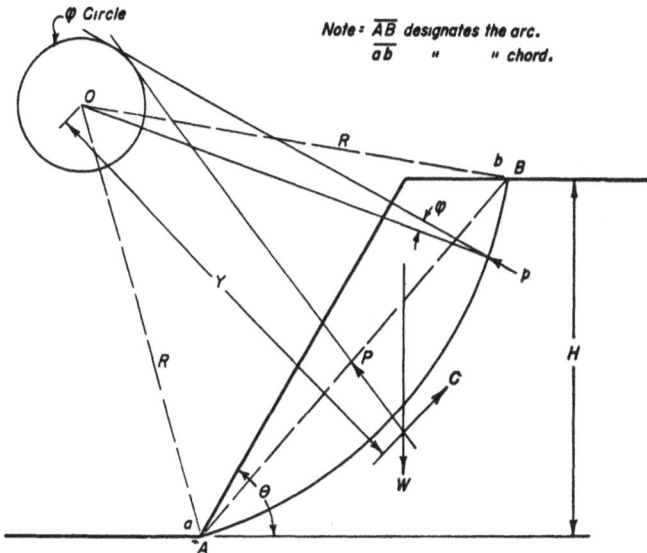

FIGURE 198 —Force diagram, ϕ=circle method

The force P, figure 198, is the force which is transmitted from grain to grain
across the arc $A-B$ and is the resultant of the elementary forces similar to p.
These elementary forces act at an angle, ϕ, from the normal to the arc at the
point of application.   The ϕ circle is a circle drawn with its center at O and tangent
to the directional line of elementary force p.   Since all the elementary forces have
the same obliquity they all will be tangent to the ϕ circle and hence, their resultant
P will be tangent likewise.

C is the resultant cohesion.   Where c' is the unit cohesion required for equilib-

[38] "Stability of Earth Slopes," by Donald W. Taylor, Mem., Journal, Boston Society of Civil Engineers,
Vol. XXIV, No. 3, July 1937.

rium, and $\overline{ab}$ is the chord as in figure 198,

$$C = c'\overline{ab} \tag{83}$$

and its moment arm is

$$Y = \frac{R\overline{AB}}{\overline{ab}} \tag{84}$$

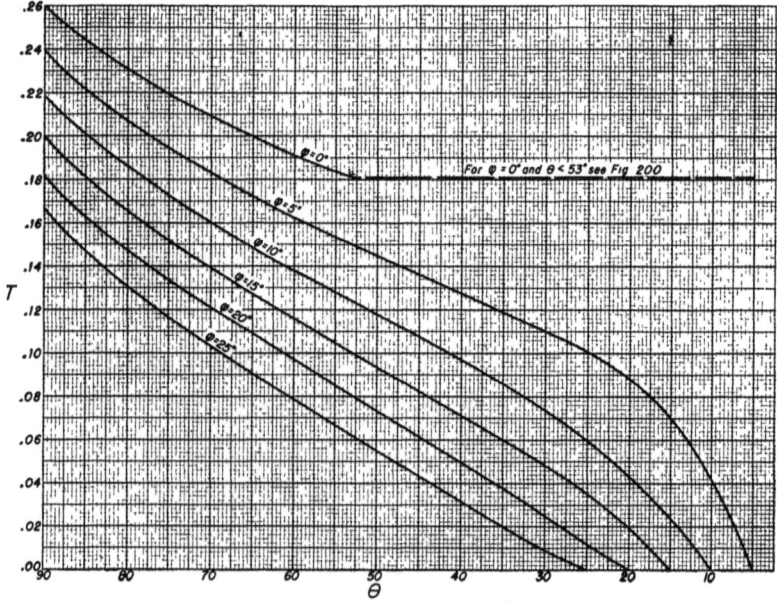

FIGURE 199.—Chart, Taylor number, T (after Taylor).

W, figure 198, is the vector representing the weight of the soil and passes vertically through the center of gravity of the mass.

If the soil is in equilibrium the three forces are concurrent.

It was found that the requirement of equilibrium with respect to the most dangerous circle in any case for a given slope, and angle of internal friction is expressed by the relation $\dfrac{2c'}{w_b H}$, where $w_b =$ dry bulk weight of the soil and $H =$ vertical height of the embankment, figure 198. Taylor stated this relation in the following form:

$$\text{Stability number} = \frac{c}{U_c w_b H} \tag{85}$$

in which, $U_c =$ factor of safety with respect to cohesion and $c =$ actual unit cohesion for the particular soil. As a result of considerable computation of groups of arcs for values of θ between 0° and 90° and of ϕ between 0° and 25° Taylor presented the family of curves shown in figures 199 and 200 by means of which the use of the ϕ circle method is greatly simplified.

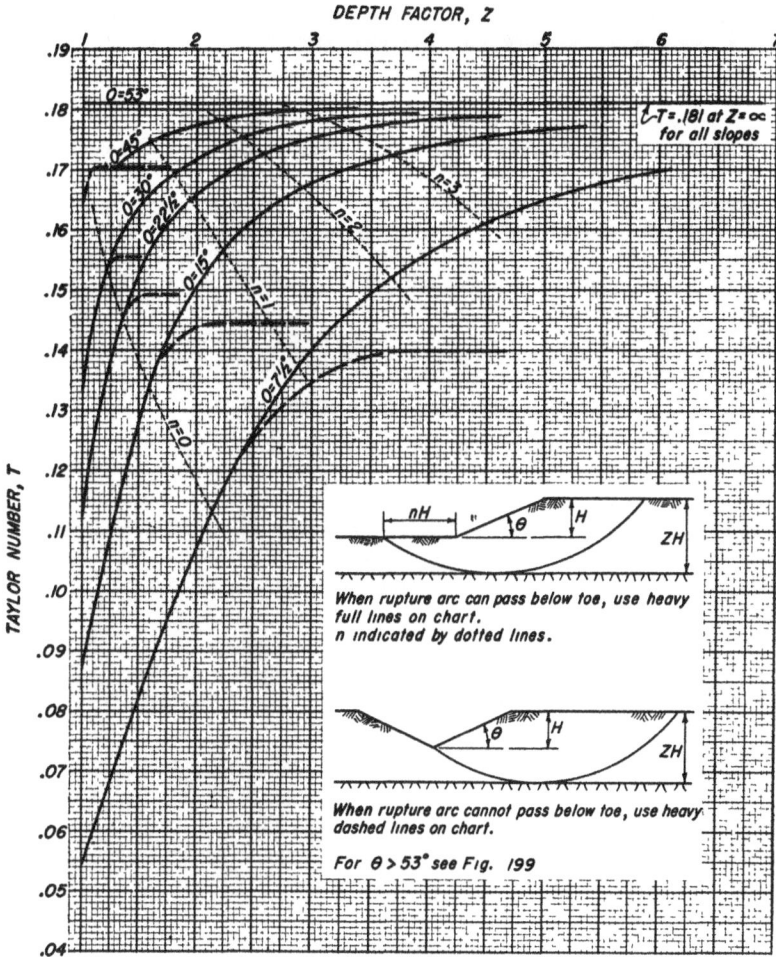

FIGURE 200 —Chart, effect of depth limitation on Taylor number, T (after Taylor)

In equation (85) a factor of safety with respect to cohesion only was used. When a true factor of safety—that is, with respect to both cohesion and friction—is used it has been suggested by Taylor that the following relation is sufficiently accurate:

$$\phi_c = \frac{\phi}{U} \qquad (86)$$

Equation (83) may then be written

$$T = \frac{C}{U w_b H} \qquad (87)$$

in which $T =$ Taylor number, the stability number with respect to the true factor of safety and $U =$ true factor of safety.

The use of this method is illustrated in connection with the following:

44. Embankment Stresses and Stability Determination.—In the following discussion the cases are limited to those in which the soil is homogeneous throughout. Generalized solutions of embankments and foundations in which different strata vary widely in the value for ϕ_c are of course impossible. There are many simple cases in which practical homogeneity may be assumed. In others, competent consulting advice is indispensable. Even in the first case, however, it is recommended that consulting advice be obtained if reasonably possible.

Case I. Simple Embankment, no Saturation, no Suddenly Applied Loads.— This case has been discussed in connection with the principles of the ϕ circle method and the stresses are indicated in figure 198. The following illustrative examples indicate the procedure to follow in this simple and idealized case.

(a) The following values have been determined by laboratory methods for the soil compacted to maximum denseness at optimum water content:

$$c = 350 \text{ pounds per square foot,}$$

$$w_b = 120 \text{ pounds per cubic foot,}$$

$$\phi = 15°.$$

The embankment has a vertical height of 25 feet $= H$, and the true factor of safety, U, is specified as 1.5.

What is the maximum permissible value for θ and the slope?

From equation (86), $\phi_c = \dfrac{\phi}{U} = \dfrac{15}{1.5} = 10°$.

From equation (87), $T = \dfrac{350}{(1.5)(120)(25)} = 0.078$.

From figure 199 for $T = 0.078$ and $\phi_c = 10°$, $\theta = 32°$. and the slope is 1:1.7.

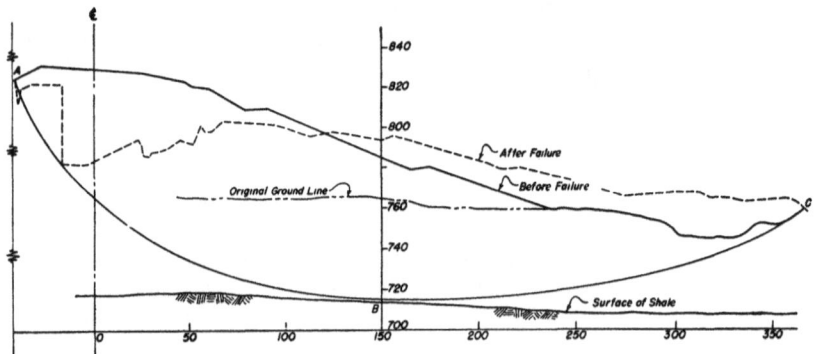

FIGURE 201 —Typical earth embankment failure

Figure 201 illustrates the conditions of an actual dam failure. The structure was not quite completed at the time it failed and the elevation of the water surface upstream at the time was but a few feet above original ground surface. The failure was, therefore, caused alone by the inability of the soil to support the load. The arc $\overline{ABC}$ defines the surface along which failure took place, the entire mass above the arc rotating as a unit about some center which is seen to be above the surface of the embankment. Investigation following the failure showed the material of the dam and the foundation to have an angle of internal friction, ϕ, which varied from 5°30' to 0°0' and an approximate average value for cohesion of 0.16 T per square foot = 320 pounds per square foot. Prior to the failure the structure was completed to elevation 825. From this $H=60$ feet and the depth from the crest to the underlying shale, 111 feet. The average downstream slope was 16°30'. The weight of the material is assumed to be 120 pounds per cubic foot.

It will be shown later that the conditions in this case do not lend themselves to solution by the generalized procedure discussed above. There are, however, certain approximations that may be disclosed which are illustrative of the use of the procedure. It must be borne in mind that the investigation was made after the failure, and the values for c and ϕ may, therefore, be quite different from the actual values prior to the failure. From the foregoing the following values are given:

$H=60$ feet.

$\theta=16°30'$.

$c=0\ 16\ T$ per square foot = 320 pounds per square foot.

$\phi=5°30'$ to 0°30'.

$w_b=120$ pounds per cubic foot.

$ZH=111$ feet.

From figure 199 for $\theta=16°30'$ and $\phi=5°30'$, $T=0.09$.

By equation (87), $0.09=\dfrac{320}{U(120)(60)}$, from which $U=0.43$. Or, to determine the unit cohesion necessary for equilibrium, when $U=1.5$, $0.09=\dfrac{c'}{(1.5)(120)(60)}$, from which $c'=972$ pound per square foot.

The values for c and ϕ which have been used above are known to be lower than must have been the actual case. For the further illustration of this type of failure a more logical set of values is assumed. The general character of the failure, that is, where the rupture arc

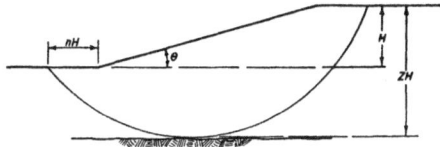

FIGURE 202 — Rupture arc passing below toe

passes below the toe of the embankment, is, however, similar. Assume that the following values hold for figure 202.

$\theta=\ 15°$.

$c=675$ pound per square foot.

$U=1.5$.

$H=25$ feet.

$ZH=50$ feet.

$\phi=0$.

$w_b=\ 120$ pound per cubic foot.

Since θ is less than 53° and $\phi=0$, figure 200 is used. From this, for $Z=\dfrac{ZH}{H}=$ 2 and $\theta=15°$, $T=0.15$ and $n=0.65$.

From equation (87), $0.15 = \dfrac{c'}{(1.5)(120)(25)}$ and $c' = 675$ pounds per square foot $= c$. Had c' been greater than c failure would be imminent, in which case the downstream wave would probably occur in the vicinity of a distance $nH = (0.65)(25) = 16$ feet downstream from the toe. If the slice method analysis is considered in the latter case it will be seen that any load in the area extending a distance nH downstream will have a component tangent to the rupture arc and opposed to the direction of failure movement. Such a condition destroys the validity of the foregoing procedure except for the single case where n is restricted to zero. This case is covered by the dashed curves in figure 200.

Case II. Completely Submerged Embankment.—The effective specific weight of a submerged soil has been given in equation (18), section B (4) as

$$w_{b(e)} = w_w \frac{G_s - 1}{1 + e}.$$

In the case of a submerged embankment the effective specific weight of the soil must be substituted for the unit weight, w_b. Equation (87) for this case is written

$$T_{II} = \frac{c}{U w_{b(e)} H}. \tag{88}$$

For example, assume that the following values apply to a submerged embankment:

$$H = 20 \text{ feet}$$
$$G_s = 2.75 \text{ feet.}$$
$$e = 0.70.$$
$$\phi = 20°.$$
$$\theta = 34°.$$
$$U = 1.5.$$
$$c = 200 \text{ pounds per square foot.}$$

By equation (18)

$$w_{b(e)} = 62.4 \frac{2.75 - 1}{1 + 0.70}$$

$$= 62.4 \frac{1.75}{1.70}$$

$$= 64.3.$$

By equation (86),

$$\varphi_c = \frac{20}{1.5}$$

$$= 13°.$$

For $\phi_c = 13°$ and $\theta = 34°$, $T = 0.068$, from figure 199.

From equation (88)

$$c = T_{II} U w_{b(e)} H \tag{89}$$

$$= (0.068)(1.5)(64.3)(20)$$
$$= 130 \text{ pounds per square foot.}$$

Since c is greater than c' the embankment is stable. Furthermore, it is apparent that a construction saving may be accomplished by increasing the angle of slope. With respect to c,

$$T = \frac{200}{(1.5)(64.3)(20)}$$

$$= 0.104.$$

For $T = 0.104$ and $\phi_c = 13°$, $\theta = 51°$, from figure 199.

Case III. Saturated Embankment, not Submerged, no Percolation.—The dynamic effect of percolating water has been discussed in section D (7), Flow Pressure. Since case III covers static conditions only, it does not apply where percolation occurs in an appreciable degree. However, Taylor has shown that such a case lies between cases II and IV. For approximate preliminary study it will be on the side of safety to treat it as the latter.

When an embankment is saturated by capillary water there is no appreciable movement. This satisfies the requirements of case III. In such a case the total weight of the water and the soil is substituted for the weight of soil, w_b, in equation (87). From equation (13) this is

$$w_{total} = w_w \frac{G_s + e}{1 + e} \qquad (90)$$

and equation (87) becomes

$$T_{III} = \frac{c}{U w_{total} H} \qquad (91)$$

Case IV, Sudden drawdown.— Assume that the embankment illustrated in figure 203 is completely submerged and in equilibrium. The forces to be considered are those which existed prior to submergence, C, W, and P, figure 198, and those which have been added by submergence. The former, considered separately, are in equilibrium from which it follows that the latter also must be in equilibrium when considered separately as a unit. It is assumed that submergence has existed long enough so that the embankment is completely saturated. There are, then, three additional forces, to the added weight of the

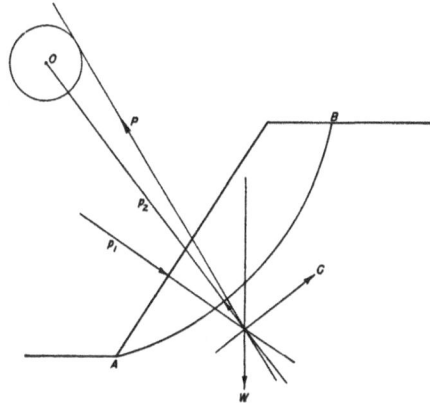

FIGURE 203 —Force diagrams, submerged embankment.

water in the voids of the soil, W_w; the total downward water pressure across the surface of the embankment, p_1; and the total upward water pressure across the rupture arc, p_2. Since the force p_2 must be normal to the arc it passes through the center O and, hence, has no moment arm about this center. The moments of p_1 and W_w about O balance each other.

In section F (1), "Consolidation of Soils," it was pointed out that if a sudden load is placed on a saturated soil that load will be temporarily carried by the water occupying the voids and that the rate at which the load will be transferred to the soil skeleton will depend upon the rate at which the excess water can percolate away and the soil skeleton be consolidated to a condition of equilibrium under the imposed load. In such a case and prior to the transfer of the load from the water to the soil this additional load cannot increase the internal friction of the soil because the load is not on the soil.   Given time for the strains in the soil mass to develop then the additional load is added to P in $s = c + P \tan \phi$ and frictional resistance is increased with the increase of shearing stresses.   The term "sudden drawdown", then, is a relative one and is taken to mean quickly with respect to the rate at which the granular stresses are developed.   If an embankment were composed of very porous sand it may be seen that the drawdown would have to be extremely rapid in which case the failure would probably be progressive and due largely if not entirely to the dynamic effects of the high percolation velocity in the direction of the lowering impounded water surface.   On the other hand, if the mass enclosed by the surface of the embankment and the rupture arc is extremely dense and hence has a very low rate of percolation the drawdown may be relatively slow and still be "sudden" with respect to the transfer of the load from the void water to the granular skeleton.   This concept of the term "sudden drawdown" must be borne in mind in the discussion which follows.

If, when it is submerged, the mass is in equilibrium the forces W_w and p_1 balance each other.   Therefore, if the force p_1 is removed equilibrium is unbalanced by the amount of the force W_w.   If the removal of p_1 is "sudden" then this must be carried by cohesion alone since additional frictional resistance can be developed until the essential readjustment of stresses occurs.   The force W will become $W_s + W_w = W_{total}$, and if equilibrium is maintained C will necessarily be equal to $C_s + C_w = C_{total}$.   The value of ϕ with respect to the soil will be greater than zero if it were so originally.   The value of ϕ with respect to W_w will be zero in any case. Since there may be two values for ϕ an individual analysis is indicated for each set of forces.   Taylor has stated, however, that for practicable purposes it is sufficiently accurate to use W_{total} and a weighted value for ϕ.   If $\phi' = $ weighted value of φ for sudden drawdown conditions, the relation may be written

$$\phi' = \frac{G_s - 1}{G_s + e} \phi \tag{92}$$

and, where the generalized value for φ is used, equation (86), this becomes,

$$\phi' = \frac{G_s - 1}{G_s + e} \phi_c \tag{93}$$

The sudden drawdown case is further complicated by the fact that the unbalanced force p_2 will cause an outward flow.   Moreover, the percolation downward of the void water in the saturated mass above the lowering surface of the impounded water will add to the unbalance.

The instances in which the conditions of sudden drawdown can occur are not overly-frequent and even in many of the instances under which they may occur they can be avoided by proper operation.   The foregoing discussion of this case has been given more to emphasize the seriousness of setting up the conditions than to furnish a procedure for the analysis of such problems.   The latter almost invariably requires consulating advice.

APPENDIX C

SUBSURFACE EXPLORATION

The following and very full description of present practice of underground exploration and soil sampling at a depth below the surface is taken with permission from chapters II and III of the publication by the Graduate School of Engineering of Harvard University, No. 208, entitled "Exploration of Soil Conditions and Sampling Operations," by H. A. Mohr.

Methods of underground exploration for foundation purposes can be classified into the following three groups:

1. Soundings or probings, in which an instrument is forced into the ground to determine the resistance of the soil to penetration at various depths.

2. Borings, in which a relatively small open hole is made in the ground for the purpose of bringing the material to the surface in a greater or lesser degree of disturbance.

3. Test pits and test caissons, which permit the visual inspection of the foundation material in its original position in the ground.

Some of the methods discussed in this chapter are included only because they have been extensively used for underground investigations in foundation engineering, although they are not suitable for this purpose.

Test Rods or Sounding Rods.—This method of soil investigation consists of driving a steel rod or a pipe, usually three-fourths inch or 1 inch in diameter to a definite resistance or refusal by means of a sledgehammer or a drop weight. The rod or pipe consists of sections 4 or 5 feet in length. The bottom section is pointed. They are connected together and made continuous by the use of recessed couplings and are so threaded that the ends will butt together at or near the center of the couplings to relieve the threads of stresses caused by the driving operation. A driving cap containing a hard wood plug to protect the top thread is used at the top.

As the rods are driven into the ground the number of blows required for 1 foot or some other suitable unit of penetration, are counted. The results are translated into an arbitrary basis of comparison established by the one driving the test rods. When the required resistance or refusal is reached, the job is done. The string of rods is pulled by the use of a chain and purchase, to be used for the next test.

Please observe the process. A piece of steel is driven into the ground and then removed. Absolutely no information is obtained regarding the type or types of soil passed through, except from an examination of the very insignificant annular ring of material that might be reclaimed from around the top of the couplings. And that material might come from any position between the surface of the ground and the lowest elevation to which the particular coupling had been driven. There is no way to determine the nature of the soil in any stratum or the thickness of each stratum. Nothing is determined about ground water elevation nor is any information gained relating to ground water conditions that may become hazardous during construction, should they exist. Refusal does not necessarily mean that ledge has been reached. It might be sold rock or it might be a boulder. A definite resistance will mean one thing if a 20-foot rod is being driven, but it will mean something entirely different if the rod being driven is 50 feet long.

In addition, no two men using the same driving tools will deliver the same energy per blow. All of these items are seldom if ever considered by the one driving test rods.

The method is of use to an excavator to determine if digging the material will be easy or hard. Dredgemen use it quite extensively for this purpose.

Occasionally it will give an indication of the length of pile that will be used, if soft material overlies compact granular materials. Each pile contractor, however, will have to do his own guessing since what goes for conical piles will not go for cylindrical piles. A great many other factors enter into the interpretation of the results of driving test rods for this purpose. The results should never be used in this connection without having the results of proper borings for a guide. When the length of rod exceeds 35 or 40 feet, the results are misleading and inaccurate in any event.

With these few exceptions time and money spent to drive rods or pipes or bars, all of which are known as test rods or sounding rods, is time and money wasted.

Borings.—For making borings the soil is removed as the work progresses by means of tools extended to the bottom of the drill hole. Every such method employs what is termed "casing," excepting particularly favorable conditions. One complete description of the use of casing will serve for all boring methods.

Casing, driving weights, and derrick.— Casing is pipe. Manufacturers produce pipe made especially for the purpose. It is of various thickness of wall and of inside and outside diameters to telescope with the least possible loss of size and at the same time to permit the operation of standard size drilling tools. Couplings of various types are made for use with the casing. These special materials are used for core drilling, deep well drilling, and similar classes of work.

Casing used for earth boring and shallow core and well work need be of standard make only. Before borings are started, nothing is known of the amount of pounding and punishment the casing will be required to withstand, so extra heavy black pipe of steel or wrought iron should be used. Some contractors use steel pipe and others use wrought iron pipe, their choice being determined by experience. Use whichever gives the least breakage. It is a question of economy. The use of galvanized pipe costs extra money and has no advantage. Two or three times in and out of the ground and the galvanized plating is destroyed. Because of the working head room furnished by the tripods or four-legged derricks used on boring work, the casing is cut to make up in 5-foot lengths. The type of threading is another question of choice. Some contractors prefer threading that will have the pipe ends come together in the couplings on the basis that this method relieves the threads of the driving stresses. Other contractors use a threading that will keep the pipe ends one-fourth inch or so apart, arguing that this prevents beading of the ends by the driving operations, which causes a reduction of inside diameter. Again experience should determine the choice. Whichever is used, the threading should be done on a power machine so that all threads will be exactly the same and the string of pipe will make up a straight line.

There are many types of standard couplings on the market. A malleable steel recessed coupling gives the best results. It is sturdy, will withstand the punishment of driving and pulling the casing and the recessed ends provide lateral stiffness, all of which are most desirable features in any type of coupling used. For details of various types of couplings and pipe used for casing, the student is referred to manufacturers' catalogues. When it is intended that the casing be seated on rock to provide for subsequent core boring or well drilling work, or where heavy gravel or boulders are anticipated, the bottom end of the casing should be fitted with a steel drive shoe. This attachment strengthens the pipe end and prevents its being damaged. Otherwise its use is unnecessary.

The casing is driven with a drive weight operated as a drop hammer. Two types are in common use. The one type has a rod securely fastened to the bottom of the weight to act as a guide to force a fair blow on the downward stroke. The rod is somewhat longer than the length of stroke used so the bottom end will not leave the pipe at the top of the stroke. The underside of the weight is fitted with a hard wood plug to avoid a metal to metal blow. This arrangement is workable on casing up to 3 inches in diameter. Another type rides a pipe smaller in diameter than the casing, which acts as a guide to force the falling weight to strike a fair blow on the anvil block. The guide pipe and anvil block are firmly connected and the bottom is threaded to screw on the size casing being used. Drive weights for hand operation weigh from 140 pounds to 150 pounds, and for power operation seldom in excess of 800 pounds because of possible damage that might be done to the casing.

To remove the casing from the hole after the boring is finished, a slip weight is used when the casing refuses to move when pulled with blocks and fall. The working stroke is up, instead of down as with the drive weight. To pull the casing the heaviest lift possible with available equipment should be applied. If it refuses to move, the drive weight is operated and the maximum lift is maintained continuously. Occasionally the casing is extremely difficult to pull and various schemes are resorted to before being successful.

In the use of wash boring methods the top of the casing is stopped at about the height of a half barrel or tub above the ground surface or operating platform when the foreman is ready to drill or take samples. A T is attached at this point. The side opening is fitted with a short piece of horizontal pipe to run the returning wash water into the tub for re-use. The top opening is fitted with a short piece of vertical pipe to prevent overflow and waste of wash water which avoids a sloppy condition where the men must work.

The size of casing to be used will depend upon the information wanted and the depth of hole to be made. If it is intended to penetrate rock, the casing will have to be of a diameter that will admit the core or well drill tools to be used for the purpose. If the hole is to be of considerable depth or if heavy gravel or boulders must be passed through, it is wise to start with casing one or two sizes larger than that required at the finished bottom. This will allow for telescoping one or two inner strings of casing in the event trouble is encountered with the starting size. For ordinary earth borings the usual size of casing used is 2½ inches, although just as good results are obtained with the use of 2-inch casing.

The purpose of the casing is to provide an opening through which the actual boring operation is carried on. The casing is driven only so far as is necessary to keep the walls of the bore hole from caving. Sometimes, with but 15 feet of casing driven, it is possible to make a hole 70 or more feet in depth. This condition will usually occur if the soil is organic, silt, peat, or clay. Occasionally it might occur in sand and clay, or sand, gravel and clay or in compact sand when the ground water conditions are favorable. When the soil is clean loose sand the casing must be driven almost as fast as the hole is made. Another use of the casing is to seal off trouble that may develop from unfavorable ground water conditions in porous materials. The latter may be caused by the absence of or an excess of ground water. A thin sand layer in an otherwise stable material may sometimes require the driving of casing.

To start a boring a 5-foot piece of casing is driven. It is cleaned out by whatever method is being used to make the boring. Then another piece of casing is added to the top of the piece in the ground and both are driven so the top of the second piece is at the working grade. Then the casing is cleaned out again. This procedure is repeated as often as is necessary to complete the boring, providing the hole will not stay open without the use of the casing.

All connections between sections of casing must be made up as tight as possible, using large wrenches.   Negligence in tightening connections results in the loosening of these connections during driving and frequent breakage of the casing at the joints with consequent loss of expensive pipe.

After two or three pieces of casing have been driven and cleaned out, it is proper to advance the hole a few feet or many feet below the bottom of the casing depending upon whether or not the walls of the hole stay in place.   The first indication for the successful accomplishment of this procedure will be the nature of soil encountered.   An unsuccessful attempt will be indicated by inability to make progress or the binding of the string of tools.   If progress cannot be made, then the casing must be driven to seal off the material causing the trouble.   If the tools bind it might mean a caving hole or it might be that the tools are being advanced more rapidly than clearance is made.   To determine just what is happening, operate the tools at one elevation for some time.   If the tools run free then too rapid progress was being made.   Proceed without driving casing but at a slower rate of advancing the boring tools.   If the tools still bind then casing must be driven.

The utilization of the casing as outlined applies to earth boring work only.   If the boring is to extend into ledge, which means through every material above ledge, the casing must be driven the full depth of the hole and sealed on the ledge.   Where boulders are encountered, they must be eliminated to pass the casing.   This is accomplished by one of two methods

1. The boulder is drilled through with the largest diameter tool that will operate inside the casing in the hole and a string of the next smaller size casing is used from that point.   This procedure will be successful provided the number of boulders encountered does not exceed the number of changes in size allowed by the starting size of casing.

2. The boulder or boulders may be drilled and blasted out of the way with dynamite to pass the casing already in the hole.   Before blasting the casing must be pulled back 8 or 10 feet to avoid being damaged by the force of the explosion.

For ease of handling the casing and weights, a tripod or four-legged derrick is used.   It may be made of any available material suitable for the purpose.   Those which are to be transported from one job to another are of 1½-inch or 2-inch pipe, pinned at suitable lengths for dismantling, handling, and shipping.   It should be made to provide 12 feet to 14 feet of headroom under the block.   The legs are fastened together at the top by means of a through bolt.   A sheave for rope is lashed to the top or is carried by the bolt.

Classification of borings.—In the past considerable confusion has arisen due to the fact that borings were sometimes classified according to the method of advancing the hole, and other times according to the method of obtaining samples. Since in foundation engineering the recovery of samples of the underlying soil strata is the principal purpose of borings, it appears logical that classification of borings should contain not only a statement of the method used in making the boring, but also a statement of the method by which the samples were obtained. It is to be regretted that such specific designation of borings is not ordinarily employed in contracts, reports, and wherever references to borings for foundation purposes are made.

Wash borings.—The term "wash boring" is used more or less promiscuously.   To the majority of engineers and architects any boring made through a casing, where water is used to float out the cuttings, is a wash boring.   However, among boring contractors the term "wash boring" generally designates that type of boring in which the soil samples are recovered from the wash water.

The essential parts of a hand-operated wash boring outfit are as follows:

1. The casing, weights, and derrick, as outlined under Casing.

2. Wash pipe in 5-foot sections and 1 inch in diameter is used almost universally. The type of coupling used and the threading is usually the same as for the casing. It should be extra heavy black pipe. The bottom section is a hollow chopping bit, where necessary to make progress through hard materials. The top is fitted with a water swivel and a cross pipe or bar for rotating the wash pipe.

3. Chopping bit. There are various types of bits in use. Their purpose is to cut the materials loose by churning the wash pipe up and down and twisting it at the bottom of the stroke. Water ports are provided in the sides of the bit so that the wash water may be forced out at the bottom of the hole and carry the loosened particles up the restricted opening between the wash pipe and the casing. Use the type of bit giving the best results.

4. The water swivel is attached to the top of the wash pipe and provides flexibility. One connection is to the wash pipe, the other to the pressure hose of the water supply. This attachment allows the wash pipe to be churned up and down and rotated for the chopping operation with no leakage of wash water. There are various types of water swivels with patented features, but the purpose is the same in all.

5. Pumps of all kinds are used to circulate the wash water. City water pressure will often be sufficient to do the work. For wash borings the pump that will wash the material out of the hole in the shortest time seems to be the one most used. Both hand pumps and power pumps are used. A pump that will furnish from 20 to 60 gallons of water a minute under a pressure of 50 pounds per square inch will wash out most any material. A good grade of suction and discharge hose should be used as it will last longer and will save time lost, making repairs on a poor grade.

6. A tub or half-barrel is used for water storage and to reclaim the materials washed from the drill hole. Its use also avoids a messy condition at the top of the hole where the men must work.

7. Small tools such as pipe wrenches, hammers, tackle, etc. Occasionally pipe cutters and stocks and dies go with a boring kit. Handthreaded pipe seldom if ever produces a straight string of pipe. All pipe used should be machine-threaded so that all threads will be the same and the pipe will make up straight.

8. A supply of sample bottles with screw metal tops fitted with gaskets to prevent evaporation. The neck section of the bottle should be the full size of the body of the bottle.

9. A notebook, printed especially for recording the boring data. The wanted information should be shown with a blank space in which the foreman is to write his findings for each boring.

Description of wash boring method, hand operation.—With the block fastened at the top, the tripod is set up and centered over the boring location. A rope, usually ⅞ inch in diameter, is reeved through the block and one end made fast to the drive weight. A section of casing is set and driven. The tee and short pipes are attached. The tub is placed under the horizontal pipe and filled with water. The pump is connected, with the suction placed in the tub and the discharge connected to the water swivel. The wash pipe, with or without the chopping bit, is connected to the water swivel. The hand line used on the drive weight is fastened to the water swivel and is used to churn the wash pipe up and down.

The pump is started, taking water from the tub and forcing it through the wash pipe. The wash pipe is churned up and down and rotated at the bottom of the stroke cutting the soil loose. The purpose is to force circulating water through the wash pipe and return it in the annular space between the wash pipe and the casing, to the tub. In its upward flow it carries the loosened soil in suspension. Because of the comparatively large volume of the tub, the water in the tub is reasonably quiet and the larger particles settle out. The pump suction should be kept near the top of the water in the tub.

The boring is sunk to the desired depth by a continuation of the washing process, additional casing being driven when needed, as outlined under Casing. The foreman handling the string of wash pipe determines, by the feel of the pipe as he rotates it and by observing the color and consistency of the returning wash water as it flows out of the horizontal pipe into the tub, when a change of material occurs. If the operator's feel is sensitive to a change, there will be a time lag before the change is noticeable in the returning water. But by some means or any means a sample must be obtained. *And it is obtained by scraping the hand or a receptacle along the bottom of the tub when the foreman decides a sufficient amount has been deposited for a sample.* The samples so obtained and the corresponding changes in strata of the underlying soil, as determined by the foreman's sense of feel, are the result of a *simple wash boring.*

Information obtained by the simple wash boring method is the most misleading and unreliable obtained by any method. The reasons for this are:

1. It is known to almost all engineers, architects and contractors, a great many of whom know nothing of the details of the operation or just how the samples are taken.

2. The sample is a completely disturbed and washed sample containing the coarse grains only. The fine grains are in suspension or have been lost by leakage.

3. The sample obtained is inevitably a mixture of the coarse grains from every stratum passed through. The only way to avoid such a mixture is to change the wash water and thoroughly clean the bottom of the tub at every change of material. And this is never done. Such a requirement could not be enforced because the men engaged in such work would not understand the reason and would find means of noncompliance.

4. The results are impossible of intelligent interpretation by any human and will always be so.

5. The uninformed have confidence in the results, interpret them as if there could be no question of accuracy, and come to grief many, many times more often than they succeed. Such experiences are demoralizing and cause a general feeling of uncertainty in all types of borings.

6. They give a false sense of security which is exploded only when actual construction is under way. Therefore, the results are negative rather than positive, as they should be.

From the above it is obvious that this type of boring should never be used.

Dry-sample borings.—Such borings are made in the same way as ordinary wash borings, as described above, except that the samples are obtained by driving a pipe or spoon into the soil at the bottom of the hole. To identify this method of soil investigation as a dry-sample boring is misleading. The samples of soil obtained below water level are not necessarily dry samples in the usual meaning of the word dry, since the voids in the soil are usually filled with water. Some designation more nearly identifying the results should be used. However, that is the term used in the trade, so it will be followed here.

The changes in procedure from the simple wash boring are outlined in the following. In the first place the foreman must be the most experienced of boring foreman, particularly in his sense of feel and judgment. The boring is started in the same manner as a wash boring is started. When washing out the material the cutting bit is invariably used on the lower end of the wash pipe. For accuracy a hand pump is used as different volumes of water are needed when drilling in different soils. The foreman handles the top of the wash pipe, turning it as the pipe is churned up and down. During this operation he is able to feel any change in material encountered. The instant that a change is felt the wash pipe is raised off the bottom of the hole and pumping is continued until the water in the casing has cleared up. Then the string of wash pipe is removed from the casing and a sample spoon or a piece of open 1 inch diameter pipe is substituted for the chopping bit. The wash pipe is extended to the bottom of the hole and driven for a sample. The wash pipe with the sample in the lower end is removed from the casing. The sample is placed in a bottle which is properly labelled with boring number, the depth from which the sample was taken, and an identification of the soil. Then the chopping bit is replaced on the bottom of the wash pipe and chopping and washing continued until another change in material is found, when washing is stopped and another sample is taken. This method is followed for each change of material from the surface of the ground to the bottom of the hole. From thick strata additional samples may often be necessary.

Please note that the entire operation used in making this type of boring from start to finish is the same as that used for making a wash boring, except that the sample is taken ahead of the washing, that is, before the material has been disturbed by the chopping bit and put into suspension in the wash water. This method of obtaining the soil sample is important. The difference between this method of sampling and the previous methods described should be thoroughly understood.

Samples taken by this method, insofar as the natural state of the material is concerned, are disturbed. However, the material will not be separated by having been placed in suspension in the wash water. All the grains from the finest to the largest that will enter the sampling device will be present in the sample, in natural proportion. The sample will give little or no indication of the compactness or stiffness of the material as it is in the ground. But these important characteristics may be determined in a rough way as the sample is being taken by noting the resistance to penetration of the sampling device in each stratum. The resistance may be on the basis of static load or of foot pounds of energy developed by a drop weight, applied at the top of the wash pipe. Such data should always be obtained on the wash pipe and never on tne casing as the question of side friction does not exist on the wash pipe except for the depth driven for sampling, while friction on the outside of the casing and for its full depth of embedment below the surface of the ground cannot be eliminated.

No difficulty is encountered with this method of sampling except in very clean sand below water level. The water in the wash pipe will often force out a sample of such material. This condition may be overcome by forcing a plug in the lower section of pipe. Dry samples can and should be obtained and one should never resort to wash sampling.

The rate of progress by this method is slower than by the simple wash boring method, being slower by the time required to remove and replace the wash pipe each time a sample is taken. For wash borings the wash pipe need never be removed from inside the casing from the time of starting the washing process to its completion, provided proper arrangement is made for driving additional casing should it be found necessary. The total time of delay for dry sampling will depend upon the depth of hole and the variation in soils encountered.

The cost per foot of dry-sample wash borings is nominal. It is the most positive of all inexpensive exploratory boring methods employed at present for soil investigations. It should be used at all times in preference to other methods. Its use is limited to earth boring, as boulders and rock cannot be penetrated by the tools used. The results will be of inestimable value for guidance, should a more positive or more complete investigation be deemed necessary. The results of such borings should be available before test pits or test caissons or undisturbed sampling is ever attempted.

It should be understood that no known method of making borings now in use will give a complete, accurate picture of the ground conditions nor will they indicate in more than a general way the conditions that will develop when an excavation is made.   Experience gained by actual foundation construction and judgment developed from such experience is the basis of interpretation of all borings.   Ordinarily that answer is sufficient.   Occasionally the work is of such great importance that more accurate data are essential.   Then the answer is obtained from test pits, test caissons, or undisturbed sampling.   Planning any of these operations with any hope of success is futile in the absence of the results of dry sample boring for guidance.

Undisturbed sampling.—Undisturbed samples are merely large "dry samples" frequently taken in a 6-inch diameter casing.   The wash boring process is usually used to clean out the casing and sink the hole to the point where the sample is to be taken.   Samples are generally 4¾-inches in diameter, which is of a size to permit trimming to fit modern laboratory consolidation cylinders.

This method of sampling has been under development for the past seven years. It is not perfected at this time.   No spoon on the market or in private use will produce a completely undisturbed sample.   In addition, with present equipment there is no certainty of reclaiming a sample every time the spoon is brought to the surface.   However, improvements are being made and eventually the desired samples will be produced.

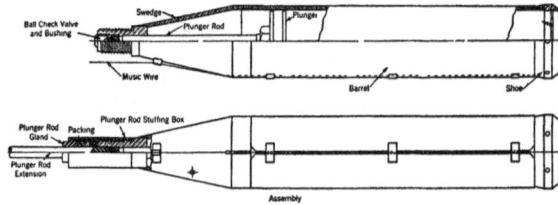

Courtesy of American Instrument Company, Inc

FIGURE 204.—S & H M I T soil sampler for obtaining large-diameter specimens of clay and other plastic materials (not suitable for sand, gravel, or other granular materials).

The latest developments in the spoon and attachments are as follows:

1. A cutting edge about 1 inch in length, with a slightly smaller inside diameter than the body of the spoon, to provide a positive clearance for the sample as it enters.

2. Joints at both the top and bottom of the barrel of the spoon, so that it may be opened to reclaim the sample from either end.

3. A diaphragm at the top of the cylindrical section of the spoon to prevent the sample from entering the reducer section, with consequent distortion.

4. A knife-sharp cutting edge on the spoon.

5. Metal liners fitting within the core barrel and in which the samples are sealed and sent to the laboratory.

6. A piano-wire arrangement to cut off the sample at or near the bottom of the spoon, just before the spoon is raised.

7. A vacuum created in the top of the spoon above the sample and maintained until the spoon is out of the ground.

8. If necessary air pressure supplied below the bottom of the sample to avoid stretching of the sample and loss as the spoon is pulled up.

9. Positive check valves in the head or above the head of the spoon.

10. Means of jacking the spoon for the sample instead of driving with a drop weight.

Recent experiences indicate that these samples should be taken for the full depth of compressible strata, to provide the laboratory technician complete data for a proper study. It has also been found that better results are usually obtained if the hole is kept full of water.

Before starting an undisturbed sample boring, a dry sample boring should be made within a distance of 6 to 10 feet of the location selected for the large boring. The results of the small boring will provide valuable information in planning for the success of the large boring.

Assume that the casing is driven and the hole full of water, is open, admitting the spoon to the bottom, with no side friction. In other words, everything is in readiness to take a sample.

The core barrel is fitted with the liner tubes and the cutting edge and reducer sections are screwed on tightly so that the air opening and cutting wire grooves match up. The cutting wires are fitted into the groove provided for them in the bottom of the spoon and threaded through the slot to the top of the spoon and securely fastened to the

FIGURE 205 —Details of sand pump.

pull rope. At places where the wires do not voluntarily stay in position, they are pasted in with beeswax. The outside of the spoon is covered with a coating of heavy grease. The air pressure and vacuum hoses are made fast to their proper connections in the top of the spoon. The two lines of hose and the pull rope are loosely bound to the drive pipe at the bottom check valve location just above the spoon, with friction tape. Then the spoon is lowered to the bottom of the hole by adding sections of pipe and taping the hoses and rope to the pipe at 10 feet intervals. Extreme care should be exercised to prevent fouling the hoses and pull rope as the spoon is lowered into the hole.

When the spoon is on the bottom it is jacked its inside length into the soil. The jacking may be done by any convenient means. It is advisable to put air pressure into the vacuum line several times during the jacking operation to blow the water out of the head of the spoon and the wash pipe. All water should again be blown out of the wash pipe when the jacking is finished. Clearing the wash pipe and the top of the spoon of water not only permits of an effective vacuum but cleans the check valve seats so that the valves will function.

Block and tackle is rigged to pull the pipe and spoon. While this is being done a vacuum is created in the top of the spoon through the vacuum hose. A small power-driven air compressor is desirable for this purpose. The vacuum is created by connecting the hose to the suction of the compressor. Just enough tension is taken on the tackle which has been rigged to the pipe to create tension in the soil at the bottom of the spoon. At this point the cutting-wire rope is pulled. The cutting wires are placed so that, when the section has been properly

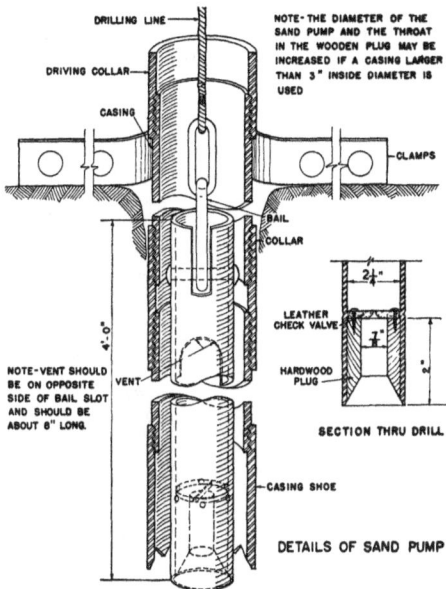

cut, the wires and rope may be pulled freely to the surface. An intact loop in the wires indicates a well cut sample. The spoon is started up and compressed air is put into the pressure line hose. The pressure should be regulated to avoid suction on the bottom of the sample. The hoses are hauled out as the spoon is raised. The drive pipe is disconnected section by section.

When the spoon is at the surface it is laid in a horizontal position, the vacuum discontinued, and the cutting edge and reducer section are removed. The liner

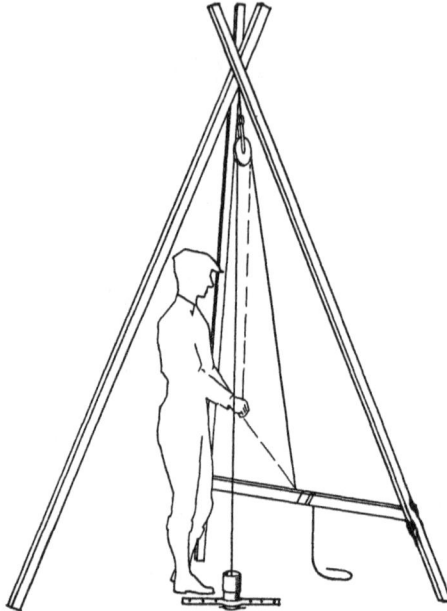

FIGURE 206

tubes are slipped out until the center joint is uncovered, where the sample is cut in two with a wire saw. Then the samples are cut flush with the tubes. The ends are covered with metal discs made to fit the diameter of the tube. These discs are bound to the tube ends with friction tape, which furnishes a seal. The depth at which the sample was taken, the top end of the sample and the boring number should be marked on each tube.

It must be evident that these samples are valuable. They should be acknowledged as such and not be destroyed or damaged by careless handling. They should be cared for in a manner such that they are delivered to the laboratory in as good condition as when taken.

Auger borings.—To make auger borings, and be prepared for every emergency, the job should be equipped with a complete wash boring outfit and casing, as outlined under those headings. In addition, the following equipment will be needed.

1. An auger or set of augers. Various types are used, not only to reclaim different classes of soils that may be encountered, but to satisfy the idiosyncrasies of the particular operator doing the work. The top of the auger will be threaded or shaped to receive the pipe or rods used for turning it, which must be extended as the bore hole is deepened. The auger must be of a diameter that will enter the casing being used. The pitch of the blades or worms is important. Usually

a steep pitch is used to reclaim plastic materials. However, soft plastic material cannot be removed with an auger having too great a pitch as the material will draw out into a straight ribbon shape as the auger is raised.

2. Pipes or rods, in sections of equal length, with couplings or fittings, by means of which the sections are connected together, are used to turn the auger into the soil. As the hole is deepened, sections are added as needed to the section in the ground. There is a cross handle or some such arrangement at the top for turning the auger.

3. Buckets and bailers of several types are used to remove granular materials encountered below water level.

Method of Making Auger Borings.—The casing is started in the same manner as that for making wash borings. The material inside the casing is removed by screwing the auger into it, then pulling out the auger full of material. This material is used for the samples. The augering and sampling may continue to any depth below the bottom of the casing so long as the walls of the hole stay in place and material is reclaimed below the elevation at which the previous sample was taken. If the walls of the hole cave or if no advance in depth can be made, then the casing must be driven to overcome the trouble. When the auger is operated below water level in a granular material no sample can be taken because the material will wash off the auger as it is raised. To overcome this difficulty one resorts to bucketing. If progress cannot be made by the use of buckets or bailers then the wash boring method must be used. It is a case of cut and try, and the set of samples might include samples taken by three different methods of sampling: wash samples, bucket samples, and auger samples. If the method used for sampling is exclusively dry sampling with a spoon or pipe and the augers used only to advance the hole, it should be called a dry sample boring.

It is difficult to understand why this method of making deep borings is ever used. It is positive only in plastic materials, in granular materials containing enough clay to make them cohesive and in granular materials above ground water level. In other words, its use is limited to specific conditions. If it is known beforehand just what the subsurface conditions are and that they are such that the auger method will be successful, then its use is justifiable. But if the subsurface conditions were known to the extent of justifying the use of this method, borings would not be necessary.

Samples taken with an auger are better and more reliable than wash samples. They are badly disturbed, but the grains have not been separated by being put in suspension. Samples taken with the buckets and bailers are in the same class as wash samples and are of just as little value.

The rate of progress is slow and uncertain. Consequently, the cost is high and the results are of little value. It is a hit and miss procedure and should not be considered except as an emergency to give a rough idea of the ground condition. Then proper borings should be made for positive data before proceeding with either the design or construction.

Simple hand augers are successfully used for shallow subsurface investigations, particularly for highway work. When using such augers without a casing in dry sand caving in of the hole may be prevented by pouring water into it. It is not possible to make such borings below the ground water table in cohesionless soils.

Well Drilling or Churn Drilling and Percussion Drilling.—These methods of boring were developed for finding water, oil, and gas and for exploring coal fields, etc. The nature of the materials passed through to reach the objective is of no interest, except as it effects the rate of progress. As a result the purpose in one day's work is to produce footage. Occasionally this method is used for founda-

tion explorations and a description of the results obtained will be given to show
when its use is permissable for such purposes. For a complete understanding of
the machines, tools, appliances and the operation and use thereof, the student
is referred to any one of several excellent books on the subject. No attempt will
be made to outline completely this method of drilling.

Portable power machines are used for shallow well drilling and for deep well
work stationary machines are used. The operators of these rigs are artisans.
They are a composite of most all other skilled mechanics and are ingenious, re-
sourceful, and patient. However, these men have no interest in soils in common
with the engineer's interests.

The effect on the materials drilled through and the nature of the resulting
samples will be outlined merely to demonstrate that the samples are of no value
to the foundation engineer or contractor.

The string of tools used consists of a chopping bit, stem, and rope socket
(on certain work jars are included), assembled in proper sequence and operated
at the bottom of the hole by means of a hawser-laid fiber cable, or a steel cable,
from some form of spudding arrangement on the machine. Water used in the
process is poured into the hole, but is not circulated as is the case in other methods.
The tools churning up and down form a slurry of the displaced soils and the water.
At intervals the slurry is removed from the hole with a bailing bucket. And
that bailed material is the sample. Probably the best way to demonstrate the
inadequacy of this method is to quote from "Drill Work Methods and Costs,"
by R. R. Sanderson (1911), describing the method of procedure when drilling in
both plastic and granular soils.

"As a rule clay will offer no serious obstacles. The main difficulty experienced
is that it mixes with the water slowly, and thus if an attempt is made to force
the drill into it the bit will become plastered with a clay ball which will prevent
the tools from turning; also the bottom of the hole will become filled with clay
balls which will plug up the valve in the bailer.

"In drilling clay, 25 to 30 feet of water should be kept in the hole, as an abun-
dance of water will not only assist in mixing the clay but will also furnish water
enough to fill the bailer, so that any clay balls which collect on the valve will
be forced out of the bailer by the water when the bailer is dumped.

"There are certain kinds of clays which, owing to their greasy nature, will
mix so extremely slow that to make any appreciable progress it is necessary to
handle them in a special manner.

"The most common method employed is to dump sand or fine gravel into the
hole. The sand under the action of the tools is forced into the clay, thus cutting
its greasy texture and making it mix more readily. Usually one gallon of sand
in a 5-inch hole will be found sufficient to mix with 1 foot of clay drilled.

"Always dump in the sand before putting the tools in the hole, otherwise the
tools are liable to be wedged.

"Another means of drilling through clay is by the use of an excavator or clay
socket. The socket is screwed into the bottom of the stem in place of the chopping
bit and is forced into the clay until it is full by raising and dropping the tools,
when it is withdrawn, the clay remaining in the socket until the surface is reached,
when the socket is cleaned."

Samples obtained by intermixing extraneous materials are of no value to the
foundation engineer. Samples taken by the method described in the last paragraph are merely dry samples but in this case are known as percussion samples.
With the heavy tools employed it is possible to obtain such samples in most
materials except ledge rock and boulders.

"As a rule sand and gravel present the most difficult drilling of any materials
encountered above bed rock.

"Owing to the fact that it is impossible to drill but a short distance ahead of the casing, it is necessary, unless the casing is driven ahead of the hole, to drill but a few inches and then drive the casing.

"In sand and gravel, holes can be made most rapidly by driving the casing ahead and then drilling out the plug which had been forced up on the inside of the casing.

"Owing to the fact that sand does not mix with water, in the same sense that clay and water mix, quite a saving in time in drilling out a sand plug can be made if some clay is dumped into the hole before the tools are lowered.   The clay serves the purpose of making a thick solution in the bottom of the hole, and thus holds the sand in suspension.

"Clay can also be made use of when drilling through dry sand and gravel. The tools will force the clay into the spaces between the grains of sand and gravel, making the hole watertight.   Not only will the clay seal the hole and make it hold water, but it will form a wall which will prevent the sides of the hole from caving, and will thus make it possible to drill ahead of the casing."

In each case the only purpose is to make progress.   Sand is dumped in to drill clay and clay is dumped in to drill sand.   The results are of no value to the foundation engineer.

Rotary Drilling.—This method of drilling is used at present for oil well work. The principle should be understood, since preparedness is the best defense.   It is too much to expect that eventually the idea will not be advanced for soil investigations for foundation purposes.

Casing is used at the start only, to avoid scouring a hole at the surface and to direct the circulating fluid.   The fluid is artifically made by use of colloidal material.   It is very heavy and recently has been built up to over 15 pounds per gallon by the addition of iron filings.   This is a specific gravity of about 1.8, which will float most soil grains.

The machines are heavy and powerful.   They deliver a rotary working force in a horizontal plane.   The cutting tolls are fishtail bits for earth and complicated horizontal and vertical reamers for rock.   Special hollow drill rods are used.

The string of tools is rotated and cuts the materials loose.   They are carried out in the circulating fluid and allowed to settle out in settling basins.   Thus the sample is of the same nature as that obtained from a wash boring except it is mixed in a pea soup fluid instead of water.   It is of no interest or value to a foundation engineer or contractor.   The heavy fluid prevents any observation whatsoever relating to ground water conditions and porous strata.

Core Borings.—This method of boring is used to obtain samples of rock, for which earth drilling tools will not produce satisfactory samples.   The result is a machined cylinder of the material cored.   A very good grade of slate is the softest rock from which a reasonable percentage of small diameter cores can be reclaimed. In some instances, because of loose cleavage planes or the angle bedding, very little or no core is obtainable even in good slate.

There is considerable misuse of the term "core boring."   By some it is used to mean dry sampling or percussion sampling.   Sush a usage is incorrect.   The term "core boring" should be used only when referring to results obtained with rotary drilling in ledge or boulders.

As was done with well drilling, no attempt will be made to acquaint the reader with the details of this operation.   To understand roughtly the process, to know where it is to be used and to have an idea of the results which one may expect, is all that interests the engineer.   Actually doing the work is the boring foreman's job.   Trade catalogs describing the machines and tools used are on the market and anyone seriously interested in the details may inform himself by a study of these publications.

Casing is used and the boring is sunk to refusal with earth boring tools.   The casing must be of an inside diameter that will admit the core barrel to cut the size core wanted. It must be seated on the rock and sealed tight. Otherwise control of the wash water might be lost and abrasive soil might enter and damage the coring tool.

Three types of cutting tools are used to make core borings. First, Black diamond (carbon) and Bortz; second, Shot; third, Toothed cutters.

Diamond drills.—The black diamond, known as carbon, is obtained in South America and is used in industry, almost exclusively, because of its peculiar structure. It has no cleavage planes and therefore will seldom fracture. Bortz is a by product or reject, because of off color or flaws, of the jewelry trade. These stones have cleavage planes and are therefore liable to fracture. They are used in coring tools because of their cheapness. Which class of stone to use will be determined by final cost.

Diamond tools will cut a core if this is at all possible in a given material. If the stones are properly set, the core will be as smooth and perfect as if cut in a lathe. The cutting is done by wearing action, that is, the rock wears faster than the diamonds. The tool is turned at a reasonably high rate of speed, it is fed to advance at a positive rate and water—preferably clear water—is supplied in a volume to keep the bottom of the hole and bit free of all cut material.

The smallest core made is ⅞-inch diameter. A casing of 2-inch diameter nominal size is the smallest that will admit the core barrel. This is an extremely small core and should never be attempted in other than solid ledge of good quality, such as marble. The smallest size of core, commercially, is 1⅛-inch diameter. Cores larger than 4-inch diameter are seldom made with diamond tools.

While diamond tools are the fastest of all coring tools, they will at times prove the most expensive. The original cost of the stones is high compared to the cost of other types of cutting tools. In addition, this cost might be increased considerably by the cost of lost stones. A chattering bit, or one improperly fed, or seams in the rock, or a badly seated casing which allows sand to follow the tool will cause loss of stones. One badly set stone that is displaced and allowed to grind around at the bottom of the hole might destroy the whole tool. Theft also enters as an item of cost.

The length of core cut is limited by the length of core barrel used. The core barrel may be of any length but only occasionally will it exceed 10 feet. They are seldom less than 5 feet in length. When the core is cut it is reclaimed by an attachment riding in the bottom end of the barrel and known as a "core lifter." The success of the job will be based upon the percentage of core reclaimed. If the rock is penetrated 20 feet and 20 feet of core is obtained, it is a 100 percent job. Any such job is one in a lifetime. A 70 percent job is good. Occasionally a 10 percent or 5 percent or even a less percentage job will come along. Whenever the results do not furnish the desired information, the engineer must make a decision to improve the results. His decision will be limited to use of the most improved tools, a better boring machine, the most positive feed mechanism, proper operation, or a larger core. If all of these points are of the best, at the time, there is little more that can be done. The nature of the rock is the trouble and it is impossible to obtain cores.

Shot drills.—Shot coring is accomplished with the same tools and equipment as diamond coring, to the point where the casing is sealed off. The same machine used for diamond tools may be used for shot tools except the feed must be adjustable and not positive.

The core barrel is of soft steel and will have one or more throats, depending upon its diameter, cut in the bottom rim. Chilled steel shot of a suitable size, depending in some degree upon the nature of the rock being cored, are fed in with the wash water. The shot settles to the bottom of the hole and imbeds itself in the soft steel core barrel by jamming and otherwise. Some of the shot gets under the tool, some on the outside and some stay inside. The cutting is a combination of wear, abrasion and actual cutting. Too much shot will form a ball bearing under the barrel and spoil the process and too little shot will slow or stop the process. The trick is to supply the proper amount of shot to produce the fastest progress.

Cores less than 2 inches in diameter should not be attempted by this method. Stresses created in the core by the cutting process are severe at times and a reasonably good grade of rock is necessary to allow of coring that size. However, very large cores may be taken by this method. The range at present seems to be from 2 inches in diameter to 52 inches in diameter. There should be no maximum size so long as the machine has the power to turn the core barrel.

Shot cut cores necessarily have a rough surface; they are not smooth like a diamond cut core. The length of core is governed the same as the diamond core. The work is judged in the same manner as the diamond job. The core is reclaimed by packing the barrel with what is termed grout. Grout in this case is an evenly graded, hard, large-grained sand or small-grained gravel.

For a job which goes wrong, the same corrective measures as for the diamond method are taken. But the operator has more of a controlling influence and his inexperience or contrariness may well make a bad job of an otherwise good job. Recognizing such a situation and correcting it is a part of the engineer's job.

Steel-toothed cutters.—Toothed cutters are used in the same manner as diamond tools. The hole is prepared in exactly the same way and the same machine may be used. The feed must be adjustable as for shot tools.

A steel core barrel is fitted on the bottom rim with removable cutting teeth. The teeth are tipped on the cutting edges by welding on a special high abrasive resisting metal. The tool does its cutting the same as a milling machine tool in a machine shop.

At present these tools have not proven successful except in specially selected places. For general application they are not to be considered. The trouble is that rapid wearing of the special metal causes loss of clearance and the tool merely rides around as if it were a smooth pipe. The loss of cutting clearance is not only on the bottom portion of the teeth but on the inside and outside edges as well. The result is that the hole decreases in diameter and the core increases in diameter as the teeth wear. So it is impossible to enter into a hole made by worn teeth with a core barrel fitted with new teeth.

There is a possibility that this method may be successful, when improved, in soft rock, but for granite and rocks of such hardness it is doubtful that it will ever be made to work.

General remarks on rock borings.—A consideration of the results of well drilling and core drilling will identify their fields of application. Well drills are used to drill for water, gas, and oil. They may be used to extend earth borings to determine if a boulder or ledge stopped the earth boring tools. If a determination of the character or structure of the rock is important, very little can be learned from the well drill results. The chopping bit cuts the rock into small pieces, very few of which will not pass a one-fourth inch diameter screen opening, and an examination of the largest pieces will disclose little of value. If the presence of seams is important, well drills should not be used. The cutting action of the tool will, with but a few feet of cuttings in the bottom of the hole, seal off sizeable seams by driving the cuttings into the seam. For dam site explorations no percussion type of drill should be used nor should they be used to drill grout holes.

Core drills will furnish the most complete data from rock drilling. A sizeable and more or less continuous section of the rock is cut and brought to the surface for examination. That is positive data. The hole left in the rock is clean cut. The seams will not be sealed off by the action of the drill. Small seams will

cause the tool to chatter or operate erratically.   Or the string of tools will drop
for the depth of more sizeable seams.   Such occurrences will indicate a lack of
homogeneity and seepage tests can be made to determine the extent of the
condition.

Therefore percussion drills may be used to explore rock conditions for such
structures as bridge piers and other such heavy foundations, but should not be
used to explore sites for water barriers.   Core drills may be used for all rock
explorations but should always be used to explore sites for water barriers.

Test Pits and Test Caissons.—It quite often happens that after a site has been
explored by dry sample or wash borings, additional information is required to
solve properly the foundation problem.   The structure is unusually heavy, or
of particular importance, or the distribution of loading is unusual, or vibrations
might have to be considered, or for one of many other reasons more complete data
are needed.   If rock is within reach, core borings will give its top elevation and
character.   In soft clay undisturbed samples, outlined previously, will provide
samples for laboratory tests.   The question of cost to go to rock and the difficulties caused by bad soil and ground water conditions may need to be investigated
by means of a test pit or a test caisson.

In these cases the money spent on the previous investigations is not wasted.
In all probability it will be earned back as a saving in the cost of the pit.   The
nature of the soil and the depth to rock are known.   An idea of the water conditions
is available.   Arrangements are made to meet these conditions before the pit is
started.   Some changes may have to be made as work progresses but they will be
insignificant compared to the changes usually made when the work is started in
the absence of such preliminary data.

The results of a test pit or a test caisson give a complete and final answer.
Either is an open excavation and has a cross-sectional area large enough to permit
a man to work, which provides access for engineering inspection.   As a result the
engineer and the contractor have full knowledge of exact ground conditions and
the method to follow for successful installation.

For test pits, wood or steel sheeting is used.   If no granular stratum below
ground water level is present, horizontal sheeting may be used.   Otherwise the
sheeting must be vertical and driven as necessary to avoid loss of ground as the
pit is dug.   For test caissons wood lagging or telescoping steel cylinders are used.
Usually pits are square and caissons are circular, in cross section.

The methods and equipment used to make the excavation will be governed to
a great extent by the location of the work, available tools, and the experience of
local contractors.   There is no preference in the way it is done so long as it is done
economically.

Jet Probings, Driven Pipes, and Other Miscellaneous Methods.—Jet probings,
or jet soundings, are made by driving a pipe through which water is pumped under
pressure.   They are a satisfactory and economical method for determining the
extent of soft materials, particularly for dredging work.   If jet probings are made
with the usual ¾-inch pipe, using a hand pump, they will be stopped by any
boulder or other obstruction which is important information in planning dredging
operations.

Occasionally someone unfamiliar with the several methods used to make borings
will decide to drive a pipe into the ground, pull out the pipe with the material it
contains and then clean out the pipe, making a record of the materials removed.
The fallacy of this method should be evident to everyone, but ever so often one
will learn of its having been used again.   Two actual instances of its use will
show why it produces negative results.

A 4-inch pipe was driven into the ground 80 feet.   When the driving was
finished and the pipe was ready to be pulled a tape was lowered inside the pipe

and it was found that what had been the surface of the ground was then 35 feet below its original elevation. In other words, roughly 44 percent of the depth of material driven through was absent from the inside of the pipe, and that 44 percent unaccounted for consisted mainly of the soft materials which are of the greatest interest to all concerned in foundation work. The hard materials will form a plug in the bottom end of the pipe and the soft materials are simply pushed aside as the pipe is driven, once such a plug is formed. The original plug of hard material will not be displaced unless another stratum still harder than that forming the first plug is encountered. And it must be harder to such an extent as to overcome the friction holding the first plug. Not only is the soft material absent from the record, but nothing is known of the ground water condition, which is all-important.

In another instance a 3-inch pipe was driven to a depth of 22 feet and indicated the first good bearing material at that depth. No record was made of the loss in height of material inside the pipe. Dry sample borings indicated good bearing material at a depth of 14 feet. Good bottom was at the 14-foot depth as was proved when the foundation was installed.

Other methods similar in nature to a driven pipe consist of driving structural steel shapes, one piece at a time, until a closure is made, then removing the whole with the enclosed soil. The error might not be so great as with the use of a pipe but an error exists. Besides, being unable to determine ground water conditions, compactness of the various strata, etc., inherent in the use of the pipe, are disadvantages of this method also.

These structural shapes must be used in long pieces as they cannot be joined together as are short pieces of pipe. To handle and drive them individually and pull the assembled form with the enclosed sample requires the use of a power machine having considerable headroom.

All such methods are slow and expensive, and the total results are incomplete.

There are in use a number of soil testing methods utilizing a cone or disk on the bottom end of the apparatus. Usually the arrangement consists of an exterior pipe or casing and an interior rod or wash pipe to which the test plate is attached. The apparatus is driven or pushed into the soil and at predetermined intervals of depth the resistance of the soil to penetration of the test plate is determined by means of compression scales or load beams. From these data the properties of the soil are computed.

These methods might give comparative results of value in specific instances where the performance of previous construction in the immediate vicinity, and of a similar nature to that contemplated, is available. To use the results of such tests as the basis of a foundation design, in virgin territory, would indicate either a lack of judgment or recklessness.

There is just as much wrong with these schemes as there is with the test rod method. Tests at predetermined intervals of depth mean nothing unless the stratification of the ground coincides with the intervals selected. Without the record of previously made borings, it obviously is impossible for anyone to select depths at which to make any kind of tests. But if borings have been made, such tests are unnecessary.

A good dry sample boring will provide many times the information obtainable by any of these methods. Certainly with all of the precise equipment and opportunity of operation afforded by laboratory tests and the difficulties encountered in making such tests, it is impossible that tests made on small plates in a hole can be controlled with a degree of accuracy to be usable. A slight error in the field might be magnified to serious proportions in the final answer as the field results are translated, with assumptions and complicated mathematics, into capacity of the soil to resist applied load.

Geophysical Methods.—In connection with geological and mining investigations, geophysical methods have been developed for the purpose of determining variations in the physical characteristics in underlying strata or the contours of rock underlying sedimentary deposits. These methods have also been successfully applied to the preliminary investigation of dam sites. While these methods offer an economical and rapid procedure for the exploration of large areas, they do not give sufficient information for final decisions. The detailed information required for design purposes must then be obtained from extensive and reliable borings. In general these methods cannot be considered reliable enough for underground exploration for foundation purposes.

The most important of the geophysical methods are the electric resistivity method and the seismographic methods employing either the single impulse of an explosion or continuous vibrations of varying frequency.

Conclusions.—* * * 2. If the site is purchased before the engineer is employed, borings should be made immediately. If the property permits of various locations or layouts of the proposed improvement, it should be explored thoroughly by borings and the structures located or arranged when permissible to effect the greatest economy in the cost of the foundations. * * *

3. Whenever, because of lack of time and isolation of location, it is impracticable to have dependable borings made, it is proper to proceed with any type of boring to provide preliminary information on the ground conditions. Before proceeding with the foundation design, however, proper borings should be made.

4. Ordinary earth borings will suffice for 99 percent plus of all building operations. If the structure is to be exceptional in weight or has other features requiring special consideration of the foundation, the earth borings should be continued into rock with core or well drill borings. If the foundation is to go to rock and the boring results indicate unusual or difficult ground conditions, a test pit or test caisson might be advisable to decide the method to use to install the foundation economically.

5. If plastic soil is encountered and it is intended to found above or on it, undisturbed samples should be taken, laboratory tests made and a settlement analysis developed. If the results indicate dangerous differential settlements, the foundation can be designed to avoid that result. * * *

6. For foundation explorations:

(a) Test rods or sounding rods, driven pipes and similar devices, test plates and rotary drilling should never be used.

(b) Wash and auger borings may be used for preliminary investigations only.

(c) Well drills may be used to determine boulder or ledge condition except for water barrier foundations.

(d) Diamond and shot drills may be used to determine boulder or ledge condition for any type of structure, but must be used for water barrier foundations.

(e) Dry sample borings should be used for all earth boring work. It gives the most positive information of any method used today except, undisturbed sampling, which is expensive and used in specific instances as outlined.

7. One or more borings on every site should be carried to hard bottom. If the hard bottom is to support the foundation, all borings should go into that material. Otherwise the balance of the borings need go deep enough to determine the thickness of the upper crust only.

8. Where advice is needed, obtain the most competent available.

9. Select the boring contractor to do the work on the basis of his record for accuracy. * * *

* * * 12. It should be remembered that no soil exploration can be 100 percent perfect. The best record will be that of the actual construction. The intent of all exploratory work should be to eliminate every possible unknown

TABLE 1.—*Methods of underground exploration and sampling*

Common name of method	Materials in which used	Method of advancing the hole	Method of sampling	Approximate cost per foot.	Value for foundation purposes
Wash borings	All soils except hardpan. Cannot penetrate boulders.	Washing inside a driven casing.	Samples recovered from the wash water.	$0.50 to $1.25 per foot.	Almost valueless and dangerous because results are deceptive.
Dry sample boring	All soils except hardpan. Cannot penetrate boulders or large obstructions.	do	Open end pipe or spoon driven into soil at bottom of hole.	$0.65 to $1.75 per foot.	Most reliable for inexpensive methods. Data on compaction of soil obtained by measuring penetration resistance of spoon
Undisturbed sampling	Samples obtained only from cohesive soils at the present time.	Usually washing inside a 4-inch or 6-inch casing. Augers may be used.	Special sampling spoon designed to recover large samples.	$5.00 to $8.00 per foot.	Used primarily to obtain samples of compressible soils for laboratory study.
Auger boring	Cohesive soils and cohesionless soils above ground water elevation.	Augers rotated until filled with soil and then removed to surface.	Samples recovered from material brought up on augers.	$0.75 to $2.00 per foot.	Satisfactory for highway exploration at shallow depths.
Well drilling	All soils, boulders, and rock.	Churn drilling with power machine.	Bailed sample of churned material or samples from "clay socket".	$5.00 to $15.00 per foot.	"Clay socket" samples are dry samples. Bailed samples are valueless.
Rotary drilling	do	Rotating bits operating in a heavy circulating liquid.	Samples recovered from circulating liquid.	See text	Samples are of no value.
Core borings	Large boulders and sound rock.	Rotating coring tools: Diamond, shot, or steel-tooth cutters.	Cores cut and recovered by tools.	$3.25 to $10.00 per foot.	Best method of determining character and condition of rock.
Test pits, test caissons	All soils. In pervious soils below ground water level pneumatic caisson or lowering of ground water is necessary.	Hand digging in sheeted or lagged pit. Power excavation occasionally used.	Samples taken by hand from original position in ground.	$10.00 and up	Materials can be inspected in natural condition and place. Only method of obtaining undisturbed samples of cohesionless soils.

1. Test rods, sounding rods, jet probings, geophysical methods, etc., are not included in this table, because no samples are obtained

2. The approximate costs per foot of the various methods vary between wide limits because of the large number of factors which govern the cost. A few of these factors are: Character of soil penetrated, depth of hole, number of obstructions encountered, total footage of boring or drilling at given site, accessibility of site, etc.

3. More detailed information on any of these methods will be found in the text under the appropriate section heading.

quantity. While it is to be given consideration, the cost is never money wasted. It is returned as dividends in satisfaction and good work on any job. In practically every instance it is returned many times in savings made by the selection of the proper type of foundation and the actual cost of its installation.

The following table summarizes the various procedure, costs, and value of the methods of exploration of soil conditions for foundation purposes which have been discussed in this paper.

APPENDIX D

SURFACE FEATURES OF WATERSHEDS [1]

Surface features of the watershed, particularly the type of soil, slope, degree of erosion, and present vegetal cover, should be observed and evaluated in as much detail as can be permitted by the necessary limitations in time or funds Two purposes of this investigation are:

1. To ascertain present conditions and their probable effects on volume and rate of run-off, its distribution throughout the year, and the silt load. The latter item is particularly important because in many instances the rate of silting determines or limits the useful life of the reservoirs on the drainage system.

2. To suggest possible modifications, especially in cover or in land-use practice, whereby silting might be reduced or flow might be regulated.

Existing Maps.—*Soil Survey Maps.*—Soil survey maps on a scale of 1 inch to the mile, or reconnaissance maps ½ inch to the mile, are available for more than 50 percent of the United States. A few States have been completely mapped, and others have had only a small area covered in this manner. More detailed information is available on restricted areas or regions, where various Federal, State, or local agencies have had occasion for intensive study and mapping on larger scales. Examples are the Tennessee Valley, the Rio Grande Valley, the Republican River Basin, the "Dust Bowl" of the Great Plains, and numerous other localities where either water or soil conservation, or flood control present major problems. Soil maps show the area and distribution of the various soil types, but ordinarily leave the data on slopes and erosion for brief generalized presentation in the text.

Erosion Survey Maps.—The highly generalized soil-erosion maps which were published by the Soil Conservation Service for most of the States in 1935 are useful mainly to indicate general severity of erosion in a large watershed.

Conservation Survey Maps.—Detailed conservation survey maps on a scale of 4 inches to the mile are available for a number of restricted areas, showing present land use, soil type, slope, and the character and degree of erosion. Such maps, when available for a drainage basin, furnish specific data on surface conditions.

Influence of Surface Features on Run-off and Erosion.—The influence of vegetation, soil type, and slope on soil and water conservation has been investigated by the Department of Agriculture for a number of years; likewise, the influence of different land-use practices and cultural methods, including depth of cultivation, direction of furrows with respect to contours, and crop rotations. The results of certain investigations made by the Division of Research of the Soil Conservation Service are shown in the following tabulations. The plots used in these experiments were generally 6 feet by 72 6 feet, representing 0.01 acre, although the interception and infiltration studies extended to much larger areas through the usual sampling methods.

Table 2, "Total Interception During Growing Season", indicates that representative canopies of alfalfa intercepted about 36 percent of the total rainfall, while corn, soybeans, and oats intercepted only 16, 15, and 7 percent, respectively, or total depths during the 1937 growing season of 3 87, 1.10, 0.91, and 0.47 inches These totals, each divided by the number of storms listed in their respective columns, represent average depths per storm of 0.08, 0.04, 0.04, and 0.01 inches,

---

[1] This Appendix was prepared by G W. Musgrave, in charge, Section of Soil and Water Conservation Experiment Stations, Department of Agriculture, C S. Jarvis, Hydraulic Engineer, and others (Not reviewed by editors.)

the first three being considerably in excess of estimates and observations of rainfall interception by vegetal cover, widely quoted heretofore.

Table 3a shows "Effect of Soil Type on Infiltration." The Ruston sandy loam absorbed 6.18 inches of water during a period of 3 hours, while the Iredell loam absorbed only 0.04 inch and the other three soil types ranged from 0.29 to 2.47 inches.

Table 3b illustrates "Effect of Erosion on Infiltration." The Cecil sandy loam, with a depth of 11 inches of A-horizon, absorbed 1.65 inches in 3 hours, while the 3-inch depth of Cecil sandy clay loam accounted for only 1.26 inches during the same period, and the 1-inch depth of Cecil clay loam, 0.28 inch. These infiltration rates, therefore, indicate the effectiveness of virgin soil intact as opposed to the residues from such soil after depletion through erosion, or to practically complete removal of top soil for this test.

Table 3c shows "Effect of Turbid Water on Infiltration," where the Ruston sandy loam absorbed 2.24 inches of clear water in 1 hour, and only 0.82 inch of turbid water, while the Davidson clay loam absorbed from 0.70 to 0.46 inch of clear and turbid water, respectively.

Table 3d illustrates "Effect of Organic Matter on Infiltration of Clarion Loam." For the untreated plots, the infiltration during 2 hours amounted to 1.71 inches. The application of 8 tons of manure per acre resulted in an increased infiltration up to 3.06 inches during an equal period, and doubling the tonnage of manure further increased it to 4.65 inches.

Table 4 shows the variation of "Soil and Water Losses Under: (a) Cultivation and (b) a Protective Vegetative Cover . . . ." The measurements were taken at Missouri Agricultural Experiment Station and the Spur Sub-station of the Texas Agricultural Experiment Station, and represent some 25,000 separate determinations, which were made along with an equal number covering related problems. This extensive tabulation shows remarkable disparities between the clean-tilled and dense cover-cropped areas; for example, the first item shows soil losses of 22.58 and 0.012 tons per acre with corresponding water losses of 10.21 against 0.33 percent of the total precipitation. The averages for all 13 of the soil types are 32.04 and 0.38 tons per acre respectively for clean-tilled and cover-cropped areas, while the corresponding water losses were 18.75 and 3.54 percent. Among approximately 25,000 separate determinations underlying this tabulation, there were only 101 reversals of trends from the averages shown therein, or 4 per 1,000.

Table 5 shows a comparison of "Soil and Water Losses from Clean Tilled Areas and Areas with Dense Cover of Vegetation by Seasons." Each of the 3 types of soil, Marshall and Shelby silt loams and Cecil sandy clay, showed almost negligible erosion under bluegrass cover, while under clean tillage it amounted to about 16 tons or more per acre during the summer. A pronounced disparity occurred likewise as to percent of precipitation represented by run-off where two-thirds to one-third or less flowed from the bluegrass areas as compared with the clean-tilled plots.

Table 6 shows "Effect of Direction of Rows on Run-off and Erosion for Several Soils." Unquestionably plowing up and down the slopes results in enormous increases of run-off percentage and soil loss. With the Marshall silt loam on 8 percent slope, under the rainfall observed during the tests, these factors increased respectively from 0.1 to 10.3 percent and from 0.0 to 11.75 tons per acre, as an extreme example; while Houston black clay on 4 percent slope showed corresponding increase from 5.2 to only 7.5 percent of run-off and 5.77 to 13.13 tons per acre.

Some data have been obtained by the Forest Service from investigation of small drainage areas of from 5 to 1,000 acres each. For example, at the Bent Creek Experimental Forest in North Carolina, measurement of storm run-off

TABLE 1.—*Analysis of 19 storms, July 1, 1933, to October 30, 1936, Bent Creek Experimental Forest, N. C.*

Watershed	Storm description				Intense rainfall				Run-off data					
	Average precipitation	Average duration		Average intensity per hour	Average duration		Rate per hour	Average maximum 20-minute period	Average peak flow (Q)[1]		Average duration	Intensity per hour (I)[2]	Area (A)[3]	Run-off coefficient (C)[4] $C=\dfrac{Q}{IA}$
	Inches	Hours	Min-utes	Inches	Hours	Min-utes	Inches	Inches	Cubic feet per second	Cubic feet per second —square mile	Minutes	Inches	Acres	
Forested	1 91	9	10	0.21	1	18	0 68	0 41	1.246	16 41	200	0 34	48 6	0 0753
Do.	1.76	9	15	.19	1	00	.73	.42	.455	16 83	160	.35	17.3	.0750
Pastured abandoned agricultural land	1 80	10	00	.18	1	30	.62	.48	5 328	316 2	60	.65	10 783	.7610
Forested	1 84	9	15	.20	1	32	.61	.47	30.74	25 42	220	.35	773 95	.1134
Do.	2 10	10	45	.20	2	00	.52	.34—	2 116	14 98	360	.25	90 4	0935
Do.	2 17	10	45	.20	2	00	.51	.34	2 167	19 40	255	.29	71 5	.1046
Abandoned agricultural land	1 97	10	40	.18	1	37	.56	.45	3 945	155 4	65	.61	16 25	3980

[1] Recorded peak flow in cubic feet per second
[2] Mean rate of precipitation during time of concentration
[3] Area of watershed in acres
[4] Watershed constant, indicating ratio of run-off to rainfall   It is an expression of the combined factors of slope, shade, cover, soil, geology, etc.

gives a comparison of two types of watershed cover, namely, forested and abandoned agricultural. These data are given in table 1, and show that peak run-off
rates from the abandoned agricultural land were from 10 to 20 times as great as
those from forested watersheds during the same storms.

Watershed conditions influence the amount and rate of run-off and erosion, and
these in turn affect both the life of the reservoir and the beneficial use of the water
Physical conditions in the watershed which can be evaluated readily are the soil
type, kind and degree of erosion, slope, and prevailing vegetal cover.

Soil types are differentiated on the basis of the texture, color, structure, and
other properties of the various horizons or layers in the soil profile. Soil properties
most significant in relation to run-off and erosion are the texture, structure, and
ease of dispersion of the surface soil; also the texture and degree of compaction
of the subsoil. Where full information regarding the area and distribution of soil
types cannot be obtained, general estimates of soil permeability and of the area of
highly erodible soils may be useful.

Erosion should be classified with respect to type and severity. In the erosion
legend used on the Soil Conservation Service maps, slight, moderate, severe, and
very severe sheet erosion are indicated by the numbers 2, 3, 4, and 5, respectively.
Occasional gullies are indicated by the symbol 7 in addition to the sheet erosion
symbol, and frequent gullies are similarly indicated by the symbol 8. These
gully sumbols are seldom used alone but usually with a sheet erosion symbol,
as 27, 38, or 47. An area that is highly dissected by gullies, where cultivation is
impossible without extensive reclamation, is indicated by the symbol 9.

Degree of slope affects the amount of run-off and the velocity, and hence the
erosivity of running water. In mapping done by the Soil Conservation Service,
4 slope classes are recognized. "A" slopes are those where little or no erosion
occurs if ordinary good tillage methods are followed. "B" slopes are those where
erosion-control practices are necessary, but on which clean-tilled crops can be
grown satisfactorily if the proper methods are used. "C" slopes are too steep
for clean-tilled crops, but may be safely used for close-growing crops or for pasture
"D" slopes are too steep for effective control of erosion unless they are maintained
in permanent vegetation. Where detailed mapping of slope classes cannot be
carried out, an estimate should be made of the relative proportions of each slope
class in the watershed.

Present cover is one of the most important watershed conditions to be considered
in the construction of reservoirs. Relative proportions of cropland, idle land,
pasture, and woodland should be ascertained as accurately as possible. Observations should be made also regarding the prevailing crops and the proportion of
the year during which cropland is left without vegetative cover adequate to provide
control of erosion. Further classification of the kind of crops grown, the type
and quality of pasture, and the type, size, and density of forest cover may be of
interest and importance.

Experimental Plots and Drainage Systems.—It should be borne in mind that
the results of experiments on small plots are not readily translated into terms of
either large fields or entire watersheds. Indications from fragmentary data
would lead us to expect some change from the comparative results derived from
small plots, when large areas are considered. Investigations now under way include among their principal objectives the determination of best practical means
for carrying the experimental results from small plots to small watersheds, and
thence to larger drainage systems.

Classification of Forest Cover.—Detailed classification of forest cover may be
desirable under some circumstances. The following suggestions are offered:

(a) Cover types: Wooded areas should be classified under one of the following
three general types:

1. Deciduous, when 80 percent or more of the entire stand by number of dominant individuals is of this type.

2. Coniferous, when 80 percent or more of the entire stand by number of dominant individuals is of this type.

3. Mixed, when both coniferous and deciduous are present but neither in sufficient number to qualify under (1) or (2).

The first letter may be used to designate type; that is, (D), (C), and (M), respectively.

(b) Size class: The size of the average dominant individuals should be classified under one of the following, the first letter being used as the designation where necessary:

1. Sapling (S) stands in which the dominant tree growth is 4 inches in diameter, or less, at breast height (4½ feet above ground, designated d. b. h.).

2. Pole (P) stands in which the dominant growth is from 4 to 12 inches d. b. h.

3. Veteran (V) stands in which the dominant growth is larger than 12 inches d. b. h.

(c) Density class: Cover types are further classified according to the extent to which the tree canopy covers the ground:

1. Dense (d), when the ground is from completely to three-fourths covered.

2. Moderate (m), when the ground is from three-fourths to one-half covered.

3. Thin (t), when the ground is less than one-half covered.

4. Sparse (s), when the ground is covered only by scattered groups of trees.

(d) In addition to the three classifications under "Cover types" there are three other classifications that may be used where applicable though they could not be considered wooded types. On the other hand, they do not fit into the treeless land types. They need not be classified as to size but they should be with regard to density.

1. Chaparral (chp), a permanent cover of shrubs and stunted trees occurring in southern California and adjacent regions.

2. Sagebrush (sgb), an area whose principal vegetation is sagebrush.

3. Brush (br), all other area the *present cover* of which is a stand of shrubs or stunted trees.

Symbols should be grouped to designate the type classifications; as for example, a stand predominantly deciduous of an average size of 8 inches and whose crown covers about one-half the ground would be designated (D.P·m).

A notation should be made as to whether erosion is increasing or whether it is decreasing as a result of either artificial or natural correction. Where terracing, strip cropping, contour tillage, or other run-off control methods are practiced, these should be designated.

Where noticeable erosion occurs, a continuing study of the silt load of the stream should be made to determine both the amount of the load and the turbidity. If the silt load is such that it will limit the life of the reservoir, or the resulting turbidity such that it will reduce the usefulness of the reservoir, an estimate should be prepared giving the cost, economic feasibility, and description of the necessary corrective measures.

TABLE 2.—*Total interception during growing season*

Crop	Inclusive dates	Number of storms	Precipitation	Interception	Interception
			Inches	*Inches*	*Percent*
Alfalfa	April 27–September 15	46	10 81	3 87	35 8
Corn	May 27–September 15	27	7 12	1 10	15 5
Soybeans	June 2–August 17	24	6. 25	91	14 6
Oats	April 15–June 29	35	6 77	47	6 8

TABLE 3a.—*Effect of soil type on infiltration* [1]

Soil	Depth of A-horizon	Total infiltration for 3 hours
	Inches	*Inches*
Davidson clay loam	6	2 47
Iredell loam	6	04
Ruston sandy loam	8	6 18
Greenville sandy clay loam	3	.60
Susquehanna clay loam	4	.29

TABLE 3b.—*Effect of erosion on infiltration* [1]

Soil	Depth of A-horizon	Total infiltration for 3 hours
	Inches	*Inches*
Cecil sandy loam	11	1 65
Cecil sandy clay loam	3	1 26
Cecil clay loam	1	.28

TABLE 3c.—*Effect of turbid water on infiltration* [1]

Soil	Treatment	Average per hour
		Inches
Ruston sandy loam	Uniformity-test, series A, clear water	2 12
	Series A, clear water, 1 hour later, as check against turbid water	2.06
	Uniformity-test, series B, clear water	2 24
	Series B, turbid water, 1 hour later	.82
Davidson clay-loam	Uniformity-test, series A, clear water	.62
	Series A, clear water, 1 hour later, as check against turbid water	.66
	Uniformity-test, series B, clear water	'.70
	Series B, turbid water, 1 hour later	46

[1] Musgrave, G. W., and Free, G. R · Preliminary report on a determination of comparative infiltration rates on some major soil-types, National Res. Coun , Trans Am Geophys U. 18th Annual Meeting.

TABLE 3d.—*Effect of organic matter on infiltration of Clarion loam* [1]

Treatment	Organic carbon	Total infiltration for 2 hours
	Percent	*Inches*
None	2 40	1 71
8 tons manure per acre	2. 38	3 06
16 tons manure per acre	2 62	4 65

[1] Smith, F. B , Brown, P. E., and Russell, J. A · "The effect of organic matter on the infiltration capacity of Clarion loam," J. Am. Soc. of Agron., Vol. 29, No. 7.

TABLE 4.—Soil and water losses under: (a) Cultivation and (b) a protective cover from 13 important agricultural soils representative of the principal types of erodible farm land within an area of approximately 250,000,000 acres. Measurements at the soil and water conservation experiment stations of the Soil Conservation Service, Missouri Experiment Station, and Spur substation, Texas Experiment Station

Soil, location, and years of measurement	Annual rainfall for period	Slope	Crop	Clean tilled soil loss annually	Water loss annually	Cover	Dense cover soil loss annually	Water loss annually	Approximate time to remove 7 inches of soil [1]		Approximate regional area represented
									Clean tilled	Dense cover	
	Inches	Percent		Tons per acre	Percent of precipitation		Tons per acre	Percent of precipitation	Years	Years	Million acres
Cecil clay loam, Statesville, N. C., 1931–35	45.22	10.0	Cotton	22.58	10.21	Mixed grasses	0.012	0.33	51	95,800	30
Nacogdoches fine sandy loam, Tyler, Tex., 1932–34	42.60	10.0	do	6.45	16.76	Bermuda grass	.007	.76	200	191,800	33
Kirvin fine sandy loam, Tyler, Tex., 1931–34	40.57	8.75	do	19.08	19.50	do	.17	1.34	72	8,070	} 19
Shelby silt loam, Columbia, Mo., 1918–31	40.37	3.7	Corn	19.72	29.4	Bluegrass	.36	12.0	58	3,150	
Shelby silt loam, Bethany, Mo., 1931–35	34.79	8.0	do	68.78	28.31	do	.29	9.30	16	3,900	} 25
Muskingum silt loam, Zanesville, Ohio, 1934–36	36.47	12.0	do	73.23	41.95	do	.04	6.5	16	28,900	
Clinton silt loam, La Crosse, Wis, 1933–35	34.12	16.0	do	88.66	20.84	do	.03	2.82	11	33,600	} 12
Dubuque silt loam, La Crosse, Wis, 1934–35	36.53	30.0	do	81.44	19.12	do	.22	6.83	11	4,110	
Vernon fine sandy loam, Guthrie, Okla, 1930–35	33.01	7	Cotton	24.29	14.22	Bermuda grass	.032	1.23	50	38,200	36
Houston black clay, Temple, Tex 1931–36	32.76	4.0	do	23.83	14.22	do	.03	.04	34	26,700	} 15
1933–36	34.90	2.0	Corn	10.62	13.38	do	.10	.95	75	8,010	
Marshall silt loam, Clarinda, Iowa, 1933–35	26.82	9.0	do	18.82	8.64	Bluegrass	.06	.97	48	15,200	30
Palouse silt loam, Pullman, Wash, 1932–35	21.74	30.0	Wheat, summer fallow.	8.52	9.40	Grass	1.45	3.71	130	740	15
Colby silty clay loam, Hays, Kans, 1930–35	21.74	30.0	Bare	27.82	25.03	Spring wheat	1.65	3.23	38	650	} 10
	20.36	5.0	Kafir	11.74	16.20	Native grass	.029	.30	91	36,900	
Miles clay loam, Spur, Tex, 1926–37	20.73	2.0	Cotton	7.03	15.53	Buffalo grass	1.29	4.34	150	830	24
Average [2]				32.04	18.75		.38	3.54			

[1] Based upon actual volume-weight determinations of these soils, except Miles clay loam estimated same as Colby silty clay loam.
[2] Because of differences in years, slopes, and soil types these averages are indicative of trends and not absolute differences.

TABLE 5.—*Soil and water losses from clean-tilled areas and areas with dense cover of vegetation by seasons*

[Averages of 1932–36]

Soil type and location	Average seasonal precipitation	Run-off		Erosion	
		Corn	Bluegrass	Corn	Bluegrass
	Inches	*Percent of precipitation*	*Percent of precipitation*	*Tons per acre*	*Tons per acre*
Marshall silt loam, 9 percent slope					
Winter	1 99	47 80	35 00	1 07	0 002
Spring	6 32	24 50	14 07	6 80	1.510
Summer	10 10	16 15	5 91	15 84	.260
Fall	7 85	12 85	1. 58	3 38	.004
Shelby silt loam, 8 percent slope					
Winter	2 94	25 15	16 05	.15	.003
Spring	8 05	21 05	1.11	15 68	.031
Summer	10 14	29 78	11 20	25 15	.020
Fall	10 30	31 30	55	13 35	.059
		Cotton mixed grasses			
Cecil sandy clay loam, 10-percent slope					
Winter	12 53	3 54	11	.34	.000
Spring	12 28	5 10	64	4 74	.024
Summer	12 80	15 92	18	17 87	.001
Fall	11 07	8 01	035	1 67	.001

TABLE 6.—*Effect of direction of rows on run-off and erosion for several soils*

Soil	Crop	Row direction	Rainfall	Runoff	Soil loss
			Inches	*Percent*	*T/A*
Marshall silt loam, 8 percent slope	Corn	Contour listed	26 94	0 1	0
Do	do	Up and down	26 94	10 3	10. 3
Houston black clay, 4 percent slope	Rotation of corn, oats, cotton.	Across	32 76	5 2	5 77
Do	do	Up and down	32 76	7 5	13 13
Vernon fine sandy loam,[1] 6 8-percent slope, 1932–35	Cotton (wheat cover)	Across	33 96	9 88	24 65
Do	do	Up and down	33. 96	11 11	55 19

[1] Soil loss determined from soil caught in silt box only

Causes Impounding, retarding, increase infiltration

APPENDIX E

WORKING STRESSES FOR STRUCTURAL LUMBER AND TIMBER [1]

The safe, economical use of lumber and timber in load-bearing structures requires, as in the case of other materials, experience and judgment as well as knowledge of the scientific facts about their physical and mechanical properties.

Working stresses shown herein for the various species and grades are those satisfactory as a general basis for engineering design of timber structures. The working stresses given are for long-time, permanent loads. Where the designer is thoroughly familiar with the possibilities of the material and is in a position to secure specified selection and condition, or where engineering service includes supervision of fabrication as well as design and there is reasonably frequent subsequent inspection and maintenance, higher working stresses may be used safely. Where those general conditions do not obtain, the working stresses shown herein will provide safe limits for the species and grade of lumber commonly used.

The statement on lumber and timber design, the allowable stresses, and the column load table (or the column formula if desired) herein given are recommended for inclusion in building codes and for use by designers. The stress-grades given correspond to standard commercial grades developed by lumber manufacturers' associations, both of which are in conformance with the procedure in the Guide to the Grading of Structural Timbers and Determination of Working Stresses, Miscellaneous Publication No. 185 of the United States Department of Agriculture. Further examples of this procedure are the basic provisions (which are not grading rules) given in the American Lumber Standards published by the Bureau of Standards, United States Department of Commerce, in Simplified Practice Recommendation R16–29—Lumber—Fourth Edition.

General.—(a) The quality and design of all wood used for load supporting members in buildings or other structures shall conform to the standards hereinafter specified.

(b) All members shall be so framed, anchored, tied, and braced together as to develop the strength and rigidity necessary for the purposes for which they are used.

(c) Workmanship in fabrication, preparation, installation, joining of wood members, and the connectors and mechanical devices for the fastening thereof shall conform throughout to good engineering practices.

Determination of Required Sizes.—(a) All wood structural members shall be of sufficient size to carry the dead and required live loads without exceeding the allowable working stresses as hereinafter specified.

(b) Minimum sizes of lumber members required by this code refer to nominal sizes. American lumber standard dressed sizes shall be accepted as the minimum net sizes conforming to nominal sizes. Computations to determine the required sizes of members shall be based on the net dimensions (actual size) and not the nominal sizes. If rough sizes or finish sizes exceeding American lumber standard dressed sizes are to be used, computations may be predicated upon such actual sizes, provided they are specified on the plans or in statement appended thereto. For convenience, nominal sizes may be shown on the plans.

[1] A National Lumber Manufacturers Association publication, 1936.

(c) The building commissioner may require the species and grade or the stress-grade of all wood used for load bearing purposes to be stated on the plans filed with the building department.

(d) "Grade," when used in connection with lumber for structural purposes, is a classification with respect to strength. Stress-grade is a grade of known strength and is so designated.

Allowable Unit Stresses.—(a) Allowable working stresses, in pounds per square inch of net cross-sectional area, for the respective species and grades conforming to the American lumber standards given in the following tables shall not be exceeded: *Provided, however,* That stresses exceeding those given for the lowest structural grades therein noted, in any species, shall be permitted only when the quality is identified by the official grade mark of the lumber manufacturers association under whose rules the lumber is graded, or by the official certificate of inspection of such association, or otherwise identified in an approved manner.

(b) For species and grades not given in the following tables the stresses therefor shall be established by the building commissioner in accordance with the principles in the Guide to the Grading of Structural Timbers and Determination of Working Stresses, Miscellaneous Publication No. 185 of the United States Department of Agriculture (Superintendent of Documents, Washington, D. C.).

(c) Stresses due to dead and live loads acting singly or in combination, but without wind or other lateral loads, shall not exceed the allowable stress permitted for the respective species and grade. For stresses produced by wind or other lateral loads only, or by combination of wind loads and dead and live loads, the allowable stresses herein permitted except modulus of elasticity may be increased 50 percent providing the resulting sections are not less than those required for dead and live loads alone.

(d) No allowance need be made for impact, using the stresses given in the following tables, when the impact stress produced by any load does not exceed the static live load stress.

(e) Stresses for joist and plank grades apply to material with the load applied to either the narrow or wide faces.

(f) For members in direct tension allowable stresses for the respective grades are the same as for extreme fiber stress in bending.

(g) In computing shear, the effect of uniform or concentrated static, or moving loads within a distance from the support in question equal to or less than the height of the beam may be neglected. All concentrated loads located at a distance from the support of one to three times the height of the beam may be considered as placed at three times the height of the beam from the support. All other loads shall be considered in the usual manner.

(h) Shearing stress for joint details may be taken as 50 percent greater than the horizontal shear values otherwise permitted.

(i) The stresses given for maximum horizontal shear are based on the maximum amount of checking (due to shakes or seasoning) permitted by the published association rules for the species and grade in question. Where checking is less than permitted for the grade the above shear values may be increased proportionately from 20 percent to 40 percent. When greater checking occurs than permitted in the published grade, the above shear values shall be proportionately reduced.

(j) In joists supported on a ribbon or ledger board and spiked to the studding, the allowable stress in compression perpendicular to the grain may be increased 50 percent.

(k) Beams notched upward in the face at their bearing on supports shall be limited to maximum end load V as determined by the formula

$$V = \frac{2bd^2H}{3h}$$

in which V is the maximum end load, H is the maximum permissible stress in shear, b is the breadth of joist, d is the height of joist above the notch, and h is the total depth of the joist.

(l) Allowable compression stresses perpendicular to grain may be increased in accordance with the following factors for bearings less than 6 inches in length and located 3 inches or more from the end of a timber:

Length of bearing (inches)	½	1	1½	2	3	4	6 or more
Factor	1 85	1 60	1 45	1 30	1 15	1 10	1.00

For stress under a washer or small plates the same factor may be taken as for a bearing, the length of which equals the diameter of the washer.

(m) Compression on surfaces inclined to grain shall be limited according to the following formula:

$$N = \frac{PQ}{P \sin^2 \theta + Q \cos^2 \theta}$$

in which

N = allowable unit stress on the inclined surface.

P = unit stress in compression parallel to the grain.

Q = unit stress in compression perpendicular to the grain.

θ = angle between the direction of the load and the direction of the grain.

(n) Compression parallel to the grain stresses, for ratios of length to least dimension intermediate of those given in the table, may be obtained by interpolation.

(o) The safe load on a column of round cross section shall not exceed that permitted for a square column of the same cross-sectional area.

(p) The diameter of tapered columns shall be measured at a point one-third the length from the small end, and in no case shall it be assumed as more than one and one-half times the least diameter of its small end. The compressive stress at the small end of the tapered column shall not exceed the allowable stress for a short column.

Stresses for beams and stringers, joists and planks

Stress-grade and species	Equivalent commercial grade	Rules under which graded	Allowable unit stresses in pounds per square inch [1] for joists or planks, beams and stringers			
			Extreme fiber in bending	Maximum horizontal shear	Compression perpendicular to grain	Modulus of elasticity
1800#f white ash	1800#f white ash	National Hardwood Lumber Association	1,800	120	500	1,500,000
1600#f white ash	1600#f white ash		1,600	120		
1400#f white ash	1400#f white ash		1,400	120		
1200#f white ash	1200#f white ash		1,200	100		
1800#f beech	1800#f beech		1,800	120	500	1,600,000
1600#f beech	1600#f beech	do	1,600	120		
1400#f beech	1400#f beech		1,400	120		
1200#f beech	1200#f beech		1,200	100		
1800#f birch	1800#f birch		1,800	120	500	1,600,000
1600#f birch	1600#f birch	do	1,600	120		
1400#f birch	1400#f birch		1,400	120		
1200#f birch	1200#f birch		1,200	100		
1600#f western red cedar	Structural	West Coast Lumbermen's Association	1,000	100	200	1,000,000
1200#f chestnut	1200#f chestnut	National Hardwood Lumber Association	1,200	100	300	1,000,000
1000#f chestnut	1000#f chestnut		1,000	100		
1400#f tidewater red cypress	1400#f tidewater red cypress	Southern Cypress Manufacturers Association	1,400	120	300	1,200,000
1100#f tidewater red cypress	1100#f tidewater red cypress		1,100	100		
1400#f southern cypress	1400#f southern cypress	National Hardwood Lumber Association	1,400	120	300	1,200,000
1100#f southern cypress	1100#f southern cypress		1,100	100		
1800#f rock elm	1800#f rock elm		1,800	120	500	1,300,000
1600#f rock elm	1600#f rock elm	do	1,600	120		
1400#f rock elm	1400#f rock elm		1,400	120		
1200#f rock elm	1200#f rock elm		1,200	100		
1400#f soft elm	1400#f soft elm		1,400	100	250	1,200,000
1200#f soft elm	1200#f soft elm	do	1,200	100		
1000#f soft elm	1000#f soft elm		1,000	100		
1800#f dense douglas fir (coast region)	Dense select structural	West Coast Lumbermen's Association	1,900	120	380	1,600,000
1600#f close-grained douglas fir (coast region)	Select structural		1,600	100	345	
1200#f douglas fir (coast region)	No 1 dimension		[2] 1,200	100	325	
1800#f dense douglas fir (inland empire)	Select structural	Western Pine Association	1,800	120	380	1,600,000
1600#f close-grained douglas fir (inland empire)	Structural		1,600	85	335	1,500,000
1200#f douglas fir (inland empire)	Common structural		1,200	80	315	
1400#f gum, black and red	1400#f gum, black and red	National Hardwood Lumber Association	1,400	100	300	1,200,000
1200#f gum, black and red	1200#f gum, black and red		1,200	100		
1000#f gum, black and red	1000#f gum, black and red		1,000	100		
1100#f eastern hemlock	Select structural	Northern Hemlock and Hardwood Manufacturers Association	1,100	70	300	1,100,000

Species	Grade	Association	Extreme fiber stress	¹	Compression ²	Modulus of elasticity
1040## west coast hemlock	No 1 dimension	West Coast Lumbermen's Association	² 1,040	100	300	1,400,000
1800## hickory	1800## hickory	National Hardwood Lumber Association	1,800	120	600	1,800,000
1600## hickory	1600## hickory		1,600	120		
1400## hickory	1400## hickory		1,400	120		
1800## dense larch	Select structural	Western Pine Association	1,800	120	380	1,300,000
1600## close-grained larch	Structural		1,600	100	345	
1200## larch	Common structural		1,200	100	325	
1800## hard maple	1800## hard maple	National Hardwood Lumber Association	1,800	120	500	1,600,000
1600## hard maple	1600## hard maple		1,600	120		
1400## hard maple	1400## hard maple		1,400	100		
1200## hard maple	1200## hard maple		1,200	100		
1800## oak, red and white	1800## oak, red and white		1,800	120	500	1,500,000
1600## oak, red and white	1600## oak, red and white	do	1,600	120		
1400## oak, red and white	1400## oak, red and white		1,400	100		
1200## oak, red and white	1200## oak, red and white		1,200	100		
1800## pecan	1800## pecan	do	1,800	120	600	1,800,000
1600## pecan	1600## pecan		1,600	120		
1400## pecan	1400## pecan		1,400	120		
2000## dense longleaf southern pine	Select structural	Southern Pine Association	2,000	120	380	1,600,000
1800## dense longleaf southern pine	Prime structural		1,800	120		
1600## dense longleaf southern pine	Merchantable structural		1,600	120		
1400## dense longleaf southern pine	Structural square edge and sound		1,400	100		
2000## dense shortleaf southern pine	No 1 structural		2,000	120	380	1,600,000
1800## dense shortleaf southern pine	Dense select structural		1,800	120		
1600## dense shortleaf southern pine	Dense structural	do	1,600	120		
1400## dense shortleaf southern pine	Dense structural square edge and sound		1,400	100		
1200## dense shortleaf southern pine	Dense No 1 structural		1,200	100		
1600## close-grained redwood	1600## close-grained redwood	California Redwood Association	1,600	80	267	1,200,000
1400## close-grained redwood	1400## close-grained redwood		1,400	80		
1200## close-grained redwood	1200## close-grained redwood		1,200	70		
1400## tupelo	1400## tupelo	National Hardwood Lumber Association	1,400	100	300	1,200,000
1200## tupelo	1200## tupelo		1,200	100		
1000## tupelo	1000## tupelo		1,000	100		

¹ For stresses in compression parallel to grain, see following table

² With slope of grain not more than 1 in 10.

Stresses for timber columns and compression members

Stress-grade and species	Equivalent commercial grade	Rules under which graded	Allowable unit stresses in compression parallel to grain (columns) (c) in pounds per square inch of net cross-sectional area for ratios of length-to-least dimension (l/d) equaling									
			Short columns l/d=11 or less	l/d=14	l/d=17	l/d=20	l/d=23	l/d=26	l/d=30	l/d=35	l/d=40	l/d=50
1200#c white ash	1200#c white ash	National Hardwood Lumber Association.	1,200	1,142	1,074	959	777	608	457	336	257	164
1000#c white ash	1000#c white ash		1,000	966	926	859	753					
900#c white ash	900#c white ash		900	876	847	798	722					
1300#c beech	1300#c beech	do	1,300	1,235	1,158	1,060	829	649	487	358	274	175
1200#c beech	1200#c beech		1,200	1,148	1,088	986	827					
1000#c beech	1000#c beech		1,000	970	935	876	783					
1300#c birch	1300#c birch	do	1,300	1,235	1,158	1,060	829	649	487	358	274	175
1200#c birch	1200#c birch		1,200	1,148	1,088	986	827					
1000#c birch	1000#c birch		1,000	970	935	876	783					
800#c western red cedar	Structural	West Coast Lumbermens Association.	800	762	716	638	519	405	304	223	170	109
900#c chestnut	900#c chestnut	National Hardwood Lumber Association.	900	845	780	671	466	365	274	201	154	99
1200#c tidewater red cypress	1200#c tidewater red cypress	Southern Cypress Manufacturers Association.	1,200	1,110	1,003	822	622	486	365	268	206	132
1000#c tidewater red cypress	1000#c tidewater red cy press		1,000	947	885	780						
1200#c southern cypress	1200#c southern cypress	National Hardwood Lumber Association.	1,200	1,110	1,003	822	622	486	365	268	206	132
1000#c southern cypress	1000#c southern cypress		1,000	947	885	780						
1300#c rock elm	1300#c rock elm	do	1,300	1,203	1,087	891	673	527	396	291	223	142
1200#c rock elm	1200#c rock elm		1,200	1,122	1,032	877	673					
1000#c rock elm	1000#c rock elm		1,000	955	902	813	672					
900#c soft elm	900#c soft elm	do	900	861	816	740	620	486	365	268	206	132
1300#c dense Douglas fir (coast region)	Dense select structural	West Coast Lumbermen's Association.	1,300	1,235	1,158	1,030	829	649	487	358	274	175
1200#c close-grained Douglas fir (coast region)	Select structural		1,200	1,148	1,088	986	827					
1100#c Douglas fir (coast region)	No 1 timbers		1,100	1,060	1,015	937	811	649	487	358	274	175
880#c Douglas fir (coast region)	No 1 dimension		880	860	837	796	705					
1300#c dense Douglas fir (inland empire)	Select structural	Western Pine Association.	1,300	1,235	1,158	1,030	829	608	457	336	257	164
1200#c close-grained Douglas fir (inland empire)	Structural		1,200	1,132	1,074	959	776					
1100#c Douglas fir (inland empire)	Common structural		1,100	1,055	1,013	914	774					
900#c gum, black and red	900#c gum, black and red	National Hardwood Lumber Association.	900	861	816	740	620	486	365	268	206	132

Species / grade	Grade	Association										
700#c eastern hemlock	Select structural	Northern Hemlock and Hardwood Manufacturers Association.	700	678	653	611	554	446	335	246	188	121
720#c west coast hemlock	No 1 dimension	West Coast Lumbermens Association.	720	706	688	660	615	549	448	313	240	153
1300#c hickory		National Hardwood Lumber Association.	1,300	1,249	1,190	1,087	928	730	548	403	308	197
1200#c hickory			1,200	1,159	1,111	1,031	904	730	548	403	308	197
1100#c hickory			1,100	1,069	1,033	971	876	730	548	403	308	197
1300#c dense larch	Select structural	Western Pine Association.	1,300	1,202	1,087	895	673	527	396	291	223	142
1200#c close-grained larch	Structural		1,200	1,122	1,032	877	673	527	396	291	223	142
1100#c larch	Common structural		1,100	1,041	970	851	673	527	396	291	223	142
1300#c hard maple		National Hardwood Lumber Association.	1,300	1,235	1,158	1,030	829	649	487	358	274	175
1200#c hard maple			1,200	1,148	1,088	986	827	649	487	358	274	175
1000#c hard maple			1,000	970	935	876	783	649	487	358	274	175
1100#c oak, red and white	do		1,100	1,055	1,003	914	774	608	457	336	257	164
1000#c oak, red and white			1,000	966	926	859	753	608	457	336	257	164
900#c oak, red and white			900	876	847	798	722	608	457	336	257	164
1300#c pecan	do		1,300	1,249	1,190	1,087	928	730	548	403	308	197
1200#c pecan			1,200	1,159	1,111	1,031	904	730	548	403	308	197
1100#c pecan			1,100	1,069	1,033	971	876	730	548	403	308	197
1400#c dense longleaf southern pine	Select structural		1,400	1,319	1,224	1,004	829	649	487	358	274	175
1300#c dense longleaf southern pine	Prime structural	Southern Pine Association.	1,300	1,235	1,158	1,030	829	649	487	358	274	175
1200#c dense longleaf southern pine	Merchantable structural		1,200	1,148	1,088	986	827	649	487	358	274	175
Do	Structural square edge and sound.		1,200	1,148	1,088	986	827	649	487	358	274	175
1000#c dense longleaf southern pine	No 1 structural		1,000	970	935	876	742	649	487	358	274	175
1400#c dense shortleaf southern pine	Dense select structural	Southern Pine Association.	1,400	1,319	1,224	1,064	829	649	487	358	274	175
1300#c dense shortleaf southern pine	Dense structural		1,300	1,235	1,158	1,030	827	649	487	358	274	175
1200#c dense shortleaf southern pine	Dense straight square edge and sound.		1,200	1,148	1,088	986	827	649	487	358	274	175
900#c dense shortleaf southern pine	Dense No. 1 structural		900	878	853	810	742	649	487	358	274	175
1200#c close-grained redwood		California Redwood Association.	1,200	1,110	1,003	822	622	486	365	288	206	132
1100#c close-grained redwood			1,100	1,031	948	810	620	486	365	288	206	132
1000#c close-grained redwood			1,000	947	885	780	620	486	365	288	206	132
900#c tupelo		National Hardwood Lumber Association.	900	861	816	740	620	486	365	288	206	132

For stresses other than compression parallel to grain, see preceding table.

NOTE.—Values for l/d ratios exceeding 11 were computed using the Forest Products Laboratory fourth-power-parabolic-Euler formula.

Timber Column Formula.—The following formulas apply to solid timber columns and other solid members stressed in compression parallel to grain:

(a) Short columns.—The safe load, in pounds per square inch of net cross-sectional area for solid columns, and other solid members stressed in compression parallel to the grain, with a ratio of unsupported length to least dimension (l/d) not exceeding eleven (11) (short columns) shall not exceed the allowable unit compression stress parallel to grain for short columns, i. e.,

$$P/A = c$$

(b) Intermediate columns.—For solid columns with a ratio of unsupported length to least dimension greater than eleven (11) (intermediate columns), the following formula shall be used until the reduction in allowable stress equals one-third (⅓) the stress permitted for short columns:

$$P/A = c\left[1 - \frac{1}{3}\left(\frac{l}{Kd}\right)^4\right] \text{ in which}$$

$$K = \frac{\pi}{2}\sqrt{\frac{E}{6c}} = 0.64\sqrt{\frac{E}{c}}; \text{ at which } \frac{P}{A} = \frac{2c}{3}$$

This means that the value of K is the minimum value of l/d at which the column will act as an Euler column. This maximum value is obtained when

$$\frac{P}{A} = \frac{2c}{3}$$

(c) Long columns.—For solid columns with a ratio of unsupported length to least dimension greater than K (long columns), the safe load shall be determined by the following formula:

$$P/A = \frac{\pi^2 E}{36\left(\frac{l}{d}\right)^2} = \frac{0.274E}{\left(\frac{l}{d}\right)^2}$$

(d) Notation:

P = total load in pounds

A = area in square inches of net cross-section

P/A = the working stress or maximum load per square inch

c = allowable unit stress in compression parallel to grain for short columns

l = unsupported length of column in inches

d = least dimension of column in inches

E = modulus of elasticity.

(e) The safe load on a column of round cross-section shall not exceed that permitted for a square column of the same cross-sectional area.

(f) Columns shall be limited in maximum length to $l/d = 50$.

APPENDIX F

SUGGESTED OUTLINE OF REPORT COVERING THE DESIGN OF A SMALL DAM

The report should contain a general description of the design, including the various factors involved, a copy of the detailed estimate, and a drawing showing the general plan and sections. Included on the drawing should be a vicinity map and curves showing hydraulic capacities.

The description of the design should incorporate as many of the items listed below as are pertinent to the project.

(Title) Project, _ _ _ _ _ _ _ _ _ _ _ _ _ _ _ _

Preliminary Design and Estimate Dam

Location and purpose—
1. Section, township, range, principal meridian, county, State, nearest city.
2. Location in respect to other features.
3. Accessibility.
4. Purpose (a) Amount of storage—live, dead.
        (b) Type of storage—irrigation, flood, power, domestic, etc.
        (c) Water surface elevations.
        (d) Place where water will be used.
5. Alternate designs, if any.

Summary of design:
1. Storage capacity_ _acre-ft_ _ _ _ _ _ _ _
2. Spillway capacity_ _cu. ft. per sec_ _ _ _ _ _ _ _
3. Outlet capacity_ _cu. ft. per sec_ _ _ _ _ _ _ _
4. Power outlet capacity_ _ _ _ _ _ _ _ _ _ _ _ _ _ _ _ _ _ _cu. ft. per sec_ _ _ _ _ _ _ _
5. Elevation top of dam_ _elevation_ _ _ _ _ _ _ _
6. Normal water surface_ _elevation_ _ _ _ _ _ _ _
7. Maximum water surface_ _ _ _ _ _ _ _ _ _ _ _ _ _ _ _ _ _ _elevation_ _ _ _ _ _ _ _
8. Minimum water surface_ _ _ _ _ _ _ _ _ _ _ _ _ _ _ _ _ _elevation_ _ _ _ _ _ _ _
9. Freeboard above maximum high water_ _ _ _ _ _ _ _ _ _ _ _ _feet_ _ _ _ _ _ _ _
10. Maximum height of dam_ _ _ _ _ _ _ _ _ _ _ _ _ _ _ _ _ _ _ _feet_ _ _ _ _ _ _ _
11. Estimated cost of dam (or dam and reservoir)_ _ _ _ _ _ _ _ _Dollars_ _ _ _ _ _ _ _
12. Estimated cost per acre-foot of storage_ _ _ _ _ _ _ _ _ _ _ _Dollars_ _ _ _ _ _ _ _ _
13. Total estimate, project_ _ _ _ _ _ _ _ _ _ _ _ _ _ _ _ _ _ _ _Dollars_ _ _ _ _ _ _ _
14. General plans and sections_ _ _ _ _ _ _ _ _ _ _ _ _ _ _ _ _ _Drawing No_ _ _ _ _ _ _ _

Design data—
1. Topography—
        (a) Scale.
        (b) Contour, interval.
        (c) Plane table sheet numbers.
        (d) Surveyed by.
        (e) Date of survey.
2. Geological report—author and title
3. Logs of test pits and drill holes.
4. Hydraulic data, capacities and requirements and by whom established—

(a) Storage, irrigation, flood, power.
(b) Spillway.
(c) Outlet.
(d) Diversion.
(e) Area-storage capacity curves for various elevations of water surface.
5. Hydrologic data—
    (a) Hydrographs.
    (b) Maximum recorded flood and maximum anticipated.
    (c) Mean annual runoff of drainage basin.
    (d) Tailwater curve.
    (e) Cross sections of stream bed.
    (f) Design values.
    (g) Climatic conditions.
6. Borrow facilities and aggregate deposits, location and transportation facilities available—
    (a) Laboratory tests.
7. Right-of-way information.
8. Photographs.

Reservoir—
1. Proposed capacities with water-surface elevations.
2. General dimensions.
3. Existing structures affected.
4. Nature of land flooded and clearing required.
5. Relocations: Railroad, highway, telephone lines, oil lines, power lines.
6. Limitations to maximum reservoir flow line.
7. Geology—
    (a) General formations.
    (b) Factors relating to reservoir losses.
    (c) Contributory springs.
    (d) Deleterious mineral and salt deposits.
8. Right-of-way.

Dam site—
1. Geological features, formations—
    (a) Nature of stream bed and abutments.
2. Interpretation of test pits and drill holes.
3. Percolation tests, ground water.

Dam—
1. Number and types of estimates prepared.
2. Features governing design.
3. Drawing number.
4. Water-surface elevations, storage capacities, freeboard.
5. General dimensions—
    (a) Top width.
    (b) Description of section—slopes.
    (c) Crest length; roadway.
    (d) Length of base at maximum section.
6. Percolation factor; sliding factor.
7. Cut-off trench and cut-off wall dimensions.
8. Grouting requirements.
9. Toe drains, drain holes.
10. Parapet and curbs.
11. Galleries.
12. Fishways, logways, etc.

Outlet works—
  1. Requirements—
        (a) Discharges and water surface elevations.
        (b) Diversion, capacities, water surface elevations.
  2. Factors affecting location.
  3. Tunnel dimension—material encountered. liner plates.
  4. Gate chamber—
        (a) Dimensions.
        (b) Location.
        (c) Accessibility.
  5. Gates, valves, and pipes—
        (a) Dimensions.
        (b) Elevations.
  6. Approaches, shafts, adits, plugs.
  7. Location of controls.
  8. Trashrack.
  9. Stilling basin.

Spillway—
  1. Requirements.
  2. Factors governing design and location.
  3. Type—
        (a) Controlled or uncontrolled.
        (b) Lining.
        (c) Dimensions.
        (d) Elevations.
  4. Gates, Gate structure—
        (a) Dimensions.
        (b) Operation.
  5. Stilling basin—
        (a) General description.
        (b) Dimensions.
  6. Approaches.

Construction facilities—
  1. Estimated time to complete.
  2. Power available.
  3. Construction railroad, shipping points, hauls.
  4. Construction camp.
  5. Local conditions.

Materials and unit prices—
  1. Location of borrow, hauls.
  2. Aggregate deposits, hauls.
  3. Cement, nearest mill, hauls.
  4. Railroad terminals.
  5. Basis for unit prices.

APPENDIX G

CONSTRUCTION METHODS AND SPECIFICATIONS

CONSTRUCTION METHODS

1. General.—The range of dam structures included within the scope of this manual lies between simple check dams, whose failure would have little probability of endangering human life, and more complex structures, the safety of which might be extremely important to permanent or transient residents below the structure. Good construction is important in all types since its cost is usually little more than that for structures built with indifference and carelessness. Close adherence to proper specifications and construction methods becomes more and more important as the possible dangers to life increase.

2. Construction Planning and Program.—Regardless of the size of the operation, all construction work for dams requires planning which may, depending upon the size and location, include consideration of some or all of the following items: (1) Required housing for the construction personnel; (2) selection of equipment; (3) facilities for the handling and storage of materials and equipment; (4) transportation for men, materials, and equipment; (5) diversion of the stream during construction; (6) clearing and preparation of the dam and reservoir site; (7) excavation and preparation of foundations for dam and abutments; (8) construction of the dam, abutments, outlet works, spillways and other appurtenances; (9) cleaning up and landscaping.

In many cases it is desirable to schedule the time of the various operations in a form such as is illustrated in figure 207 The organization of such a program

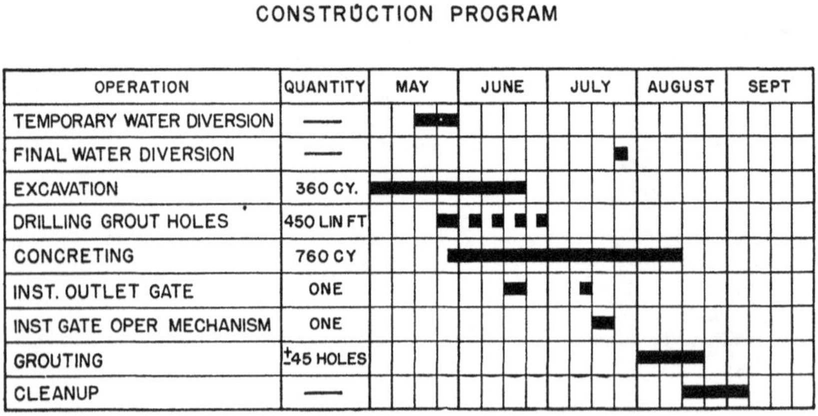

CONSTRUCTION PROGRAM

OPERATION	QUANTITY	MAY	JUNE	JULY	AUGUST	SEPT
TEMPORARY WATER DIVERSION	——					
FINAL WATER DIVERSION	——					
EXCAVATION	360 CY.					
DRILLING GROUT HOLES	450 LIN FT					
CONCRETING	760 CY					
INST. OUTLET GATE	ONE					
INST GATE OPER MECHANISM	ONE					
GROUTING	45 HOLES					
CLEANUP	——					

FIGURE 207

in advance serves to coordinate the operations with each other and with stream flow, climatic conditions, and other related factors. Such a program may also be useful in scheduling the ordering and delivery of materials so that unnecessary delays may be avoided. It also serves as the basis for a progress report during construction.

358

3. Housing.—Some type of field office is usually required regardless of the size of the job. If existing buildings are not available, the structure built for the purpose should be consistent with the size of thé project, the duration of the construction period, the climatic conditions, and other factors.

If housing for construction personnel is required, considerable attention should be given to provisions for the health and well-being of the occupants. Particular care should be exercised in providing and maintaining safe and potable water supplies and securing proper disposal of wastes both in the camp and on the job. Good health and contentment are usually important to the quality of work and the dispatch with which it is accomplished.

If animals are to be used, adequate housing, water supplies, and waste disposal must be provided for them.

4. The selection of equipment, with due regard to its adaptability, availibility and cost, is an important item in construction work. For example, the use of a concrete mixer which is too small or trucks in bad repair may cause delays which are far more expensive than the additional first cost of suitable equipment. On the other hand, limited funds may dictate the use of whatever equipment is readily available. The latter situation requires the exercise of ingenuity to make the best possible use of the equipment and labor at hand. On some work-relief projects the only immediate justification for the work may be employment for as many persons as it is possible to use. In such cases the use of equipment to secure better and more efficient mechanical operation may be entirely eliminated.

5. Handling and Storage of Materials and Equipment.—Saving of time, convenience, and reduced costs may be achieved through a careful consideration of ways and means for the handling of materials and equipment. If concrete materials can be placed near the mixer, much rehandling may be avoided. A slight downward inclination of a runway for concrete buggies may speed up the placing operations. Space limitations preclude the listing of more examples but those given indicate the importance of giving considerable thought to the relation of materials and equipment construction processes in the entire problem of plant and material storage layout.

Certain materials require protection from the weather. Cement storage is obviously quite important, as is also the care of mechanical equipment and other materials subject to weather deterioration.

6. Transportation of Men, Materials, and Equipment.—Isolated locations may require the building of roads or railways to facilitate the transportation of the necessary labor, materials, and construction equipment to the site. Each particular site will dictate the needs in this respect and studies will be necessary to a final decision.

Service roads should be adequate to satisfy the transportation needs of the project at all times and money spent on competent construction and proper maintenance usually will be less costly than the delays incident to bogged-down equipment. Wherever practicable service roads should be located on sites where permanent roads will be required.

7. Diversion of the Stream During Construction.—Considerable study is usually warranted where stream diversion is necessary. A well-developed plan for diversion will usually provide not only protection for work completed, construction materials, and equipment, but it may also serve to reduce the expense of pumping from excavations. In some cases, diversion works may be incorporated as a part of the finished structure; e. g., building an outlet conduit away from the stream channel and then diverting the stream into and through it while the other structures are being built. Tunnels may be employed as diversion channels and part of them used later as an outlet or spillway channel. A simple temporary wood or metal flume or pipe line may serve the purpose in other instances.   Again it is not possible to describe the many methods that may be used, but it is important to emphasize that diversion facilities ordinarily must be adequate to handle flood flows during the time they are in use and that their principal function is to protect the partially completed structure and prevent avoidable flow of water into excavations.   A study of stream hydrographs, if available, is of considerable assistance in designing diversion works and scheduling construction operations to avoid flood damage.   An example of specifications for river diversion is given in section **10.**

8. Clearing and Preparation of the Dam and Reservoir Sites.—For most purposes of water conservation it is necessary or desirable to clear the reservoir site of buildings, fences, logs, trees, and certain other vegetation.   The dam site requires clearing of trees, rocks, etc., in order to make room for the structure.   In some cases it may be desirable to remove other objectionable materials such as manure piles, straw stocks, and trash.   These operations may be carried on in the reservoir area while work is being done on the lower part of the dam.   They must be completed before the area is submerged.   All combustible material should be completely destroyed by fire.   Sample specifications are given in section **11.**

When borrow pits for earth-fill dams are located within the reservoir area the pit areas must be stripped of top soil containing roots or voids due to roots.   This stripping and that from other areas may be stored in a pile for use in landscaping purposes after the completion of the dam, if required.   Borrow pits not in the reservoir should be located so as to cause the least possible damage to the use or appearance of the area, but this objective must necessarily be subordinated to the requirement for soils of the proper type for dam construction.

This item also includes any necessary removal of highway or railway structures which remain after relocations have been provided.

9. Construction Methods.—Methods to be used for excavation and preparation of foundations for dam and abutments and for the construction of the dam, abutments, outlet works, spillways, and other appurtenances, depend largely upon the type or types of dam structure, outlet works, and spillway structures which has or have been selected for the particular site.   The types of structures included in this manual (chs. 7 to 12, inclusive) are: 1, Earth fill; 2, rock fill; 3, solid gravity; 4, arch; 5, buttress; and 6, timber.   Certain construction methods such as general excavation; composition, mixing, placing and curing of concrete; grouting operations, etc., are adaptable to all types of dams or structures for which they are required.

Suggestions relating to the construction of the various types of dams covered in chapters 7 to 12, inclusive, are given in the following subsections.

1. *Earth Dams.* The objective of the construction operations for earth dams of the rolled-fill type is to place properly selected earth materials in such a way that the resultant structure will offer adequate resistance to the passage of water.   Specific suggestions relating to good construction may be found in the sample specifications (see secs. **11, 12, 13, 14, 16,** and **17**).   Other references to construction will be found in chapter 7.

The importance of securing good compaction cannot be too strongly emphasized.   It may be obtained by a proper combination of the selection of good materials, with adequate compacting methods (with reference to equipment used and the thickness of layers) and proper control of moisture content.

For very small and relatively unimportant structures such as stock watering ponds sufficient compaction can be obtained by careful control of trucks and tractors used to haul and spread the material.   The material must be spread in thin layers, not exceeding 3 inches loose depth.   This is not in general a satisfactory method and should be used only where proper compacting equipment is not available at a cost commensurate with the cost of the structure. The definitely preferable method of compacting is by the use of sheepsfoot rollers and similar equipment. Each layer of earth should not be more than 6 inches in depth when compacted. The number of passages of the compacting equipment will vary usually between 8 and 16. The determining factor is the compactness of the material, control methods for which are given in appendix B.

Control of soil moisture content is very important. Too little moisture prevents good compaction and too much may induce shrinkage cracks or interfere with the placing and compacting of the materials. Irrigation of borrow pits is the best method of supplying the necessary added moisture. If this is not practicable it is necessary to sprinkle the fill during placement.

Section 20 should be consulted if there is any concrete work involved.

2. *Rock-fill Dams.* Sample specifications for river diversion are given in section 10; rock quarries in section 13; stripping and preparation of foundations in sections 14 and 15; and rock fill in section 18. These specifications will require modification to suit particular sites, but they indicate general requirements.

In selecting a quarry location, there is where possible, considerable advantage in having it at a higher elevation than the top of the dam to facilitate the hauling of the rock. The filling operation can be started at one or both ends of the dam by dumping from a stub trestle that is built out from the abutments. Dumping the rock from a considerable height has the advantage of breaking off thin edges and points of rock to give better compaction. When the fill has reached the height of the trestle, the loaded trucks or cars can be run over the completed fill and dumped over the edge. Filling should be started on the upstream face to give that part of the dam every opportunity to settle preparatory to placing the water seal.

If the rock contains a high percentage of fine material, compaction may be aided by the use of a water jet which is played onto the freshly dumped material. This tends to wash the fines from between the contact surfaces of the larger pieces, and deposit them in the larger void spaces. The result is a greater density and less danger of settlement after completion of the structure.

If a reverse filter construction (see ch. 8) is required, great care should be exercised to obtain the necessary gradation of the material so that the flow-retarding element will not be piped into and through the coarse material of the rock fill.

3. *Concrete Gravity Dams.* Details for the preparation of foundations for gravity dams may be found in section 15. The construction of the dam itself is principally concerned with the proper placement of good concrete, specifications for which may be found in section 20. Mass concrete should not contain aggregate larger than 6 inches and the mix should be carefully designed to insure concrete of adequate density and strength.

The dam is usually constructed in sections from 25 to 40 feet in length. Alternate sections are built and first allowed to set before the others are poured. This process minimizes the effect of shrinkage and tends to prevent the formation of dangerous vertical cracks. Keyways are formed in the ends of each section to develop resistance to shear. The exposed ends should be painted with a heavy coat of bituminous emulsion to prevent the adherence of the adjacent sections poured against them. Membrane sealing strips, preferably of copper, should be provided to prevent leakage through the vertical joints.

The top surface of each completed pour should be carefully cleaned and washed free of laitance and dirt before a fresh pour is started upon the old surface. To provide resistance against horizontal shear at construction joints, the surface of the joints is ordinarily finished off at two different elevations of approximately equal area, the downstream half being the higher, with a vertical step between them. This step should be not less than 0.5 foot high, and may be higher if the compressive strength on the vertical face exceeds conservative values.

Forms should be adequately braced to maintain the correct lines and thus insure good appearance of the finished structure.

4. *Single-arch Dams.* All soil overburden is stripped off for about 5 feet outside of the concrete limits to provide space for observation of possible seams and fissures in the foundation.   Within the limits of the concrete base, the excavation is carried to sound tight rock, avoiding the use of dynamite whenever possible. When dynamite is required, it should be used in small charges to avoid shattering seams unnecessarily.

Water should be excluded from the foundation by adequate pumping facilities during excavation, not only to provide good working conditions, but also to facilitate observation and inspection of the exposed rock.   Whenever possible, a sump that is lower than any point of the foundation should be provided.   The natural stream flow can be diverted a short distance upstream by whatever method is deemed satisfactory.

After all soft, loose, or seamy rock has been removed, holes should be drilled to test the soundness of the foundation and to provide for grouting.   The number of holes and the spacing will depend on the character of the rock.   A single row of holes 8 to 12 feet deep on 6 to 10 feet centers will meet the ordinary requirements.   For more severe cases, a double row of staggered holes can be used (see sec. **19**).   No grouting should be done until at least 10 feet of concrete have been placed above the foundation, but the best procedure is not to start grouting until the concreting of the dam is finished.

It is customary to divide the arch ring into sections from 25 to 40 feet long, separated by radial contraction joints.   The faces of the joints should be provided with shear boxes to lock the adjacent sections together and prevent movement.   The exposed ends of the sections poured first should be painted with a heavy coat of asphalt emulsion before the concrete in the adjacent sections is poured.   Membrane sealing strips are not ordinarily required to prevent leakage in small arch dams because, under load, the arch action keeps the joint closed.

Good concrete is extremely important to the success of an arch dam.   The design of the mix and mixing and placing must be carefully done to insure concrete of adequate strength and density (see sec. **20**).

There are two programs of concreting procedure, either of which may be followed.   In the first scheme, alternate sections are carried to the top of the dam before any concrete is poured in the other sections.   Thus, if a dam is divided into 25-foot sections numbered from 1 to 10, sections numbered 1, 3, 5, 7, and 9 would be concreted to the top before any concrete is poured in the others.   This scheme has the advantage of allowing the maximum amount of shrinkage and heat dissipation before the closure sections are concreted, and so tends to give a tighter structure.   Furthermore, leaving one or more sections low provides a passageway for streamflow during the first stage of concreting.   As a rule, one section is left low until the last to simplify the water problem.   There will usually be a permanent outlet built into one of the first sections, which, with the aid of the head built up in the rising pond, is depended on to keep the water level down below the tops of the low sections until they are finished.   It may be necessary to build a temporary timber bulkhead above the last of the low sections, to hold the water until the first lift or two of concrete has set.

The second scheme of concreting procedure consists of keeping the lowest sections within two or three lifts of the highest sections.   In this case, water is turned through the permanent outlet much sooner, and the advantage of one or more low sections to serve as an emergency construction spillway is lost.   The main advantage claimed for this scheme is that the range of concreting operations is confined to narrower limits. On small dams where the concreting is often done from buggies on scaffolding runways, the saving may be considerable.

Construction joints are permitted between the daily concreting pours. However, they should be laid out by the designer and his layout should be followed rigidly during construction. There are two guiding factors to be observed—(a) the capacity of the concrete plant, and (b) the strength of the forms. The fewer horizontal construction joints or limits of pour used the better, so that each lift should be as high as is practicable. In an arch dam, all horizontal joints should be continuous from end to end of the structure. The vertical spacing need not be uniform, and, in fact, is usually greater as the thickness decreases near the top. At the bottom, where the thickness is greatest, each lift should not contain a greater volume than can be turned out by the concreting plant within a reasonable working period. As the thickness decreases, the yardage per foot of height will also decrease, and the height of each lift can be increased up to the limit of the strength of the forms to hold the fresh concrete. Vertical lifts should not be less than 5 feet or greater than 10 feet, as a general rule.

Because of the thinness of the arch section, special care should be taken to assure good bond of freshly poured concrete to that which has set. Top surfaces of completed lifts should be chipped back to sound and dense faces in which the coarse aggregate is exposed, and brushed thoroughly with a neat cement grout before the succeeding concrete is placed. Keyways should be built in all horizontal joint surfaces to give mechanical bond. The difficulty of cleaning out depressions in concrete surfaces has led to the practice of making a raised keyway instead of a depressed one. This detail is no more difficult to build, and tends to give a better bond between the old and new concrete surfaces.

A construction program such as suggested in section 2 is particularly desirable for arch dams.

5. *Buttress Dams.* Rock foundations under buttress dams should be left rough and clean to secure a good bond with the concrete, and all water should be removed from depressions before concreting is begun. Blasting done in connection with excavation should be carefully controlled to prevent dangerous shattering or fracturing of foundations or abutments outside the excavation line (see section 15).

Concrete of adequate strength and density is particularly important in a buttress dam (see section 20) due to the relatively thin sections and the dependence placed upon the reinforcement and the concrete to carry the loads. Protection against horizontal seepage planes is afforded by proper cleaning and chipping of the top of each lift before new concrete is poured.

Special care should be taken to construct the buttresses and the face slabs or arches true to the lines, grades, and dimensions shown on the drawings. Irregularities in alinement can be avoided by the use of carefully built and adequately braced forms.

In the completed structure, all reinforcing bars should be in the positions indicated by the design drawings. Special care is therefore necessary to fix them firmly in their correct positions to avoid displacement during construction operations.

6. *Timber Dams.* Timber dams are usually considered as more or less temporary structures, nevertheless it is quite important that proper materials and methods be employed in their construction to secure as long a life as possible. Working stresses for timber will be found in appendix E.

If the timber in the dam is to be round logs, skilled axemen will be required to insure sound and economical work. For squared timber work, carpenters skilled in the manipulation of heavy timber are essential.

In building low crib dams with plank decks, cofferdams can often be avoided. The foundation can be stripped and prepared with a moderate amount of water flowing over it. By taking accurate soundings of the site, the crib can be built up in sections on shore, floated into place, and sunk into position by loading with rock. Cribs of this kind should be floored with timber in at least one-half of the cells so that the load of the rock-fill is transmitted to the cribwork. When the section which contains the outlet or sluiceway is in place, the gate is opened so that it will carry as much of the flow as possible, the balance passing through the cribwork. Then the placing of the deck plank is started at the sluiceway, working toward both ends until the deck is complete. The toe fill should be placed as soon as possible after the deck is completed to minimize the tendency toward the opening up of permanent leakage channels.

This same method of construction without unwatering the site may be used with the higher rock-filled crib type, although it is preferable that the foundation be exposed while being cleared of loose and unstable material. For the larger dams it is customary to drive the planks which compose the lower part of the deck into the foundation with a pile hammer. The depth to which they are driven will depend on the height of the dam and the nature of the material. If the foundation at the downstream toe of an overflow dam is erodable, protection against undermining of the structure can be afforded by the use of a row of sheet piling and heavy riprap.

SAMPLE SPECIFICATIONS

(Abstracted from specifications by the U. S. Bureau of Reclamation)

10. Diversion and Care of River During Construction and Unwatering Foundations.—The contractor shall construct and maintain all necessary cofferdams, channels, flumes, and/or other temporary diversion and protective works; shall furnish all materials required therefor; and shall furnish, install, maintain, and operate all necessary pumping and other equipment for unwatering the various parts of the work, and for maintaining the foundation, cut-off trenches, and other parts of the work free from water as required for constructing each part of the work. River discharge curves and diversion works capacity curves are shown on the drawings solely for the purpose of aiding the contractor to time his construction operations to prepare for such flood storage and/or to bypass such flow as may be necessary. The reliability or accuracy of any of these curves is not guaranteed. After having served their purpose, all cofferdams and other temporary protective works downstream from the dam shall be removed from the river channel or leveled to give a sightly appearance, so as not to interfere in any way with the operation or usefulness of the reservoir. All cofferdams or other temporary protective works constructed upstream from the dam shall be removed to the extent required to prevent obstruction in any degree whatever of the flow of water to the outlet works.

11. Clearing.—The area to be occupied by the dam, the surfaces of borrow pits, structure sites, and quarries shall be cleared of all trees, stumps, roots, brush, and rubbish; and all combustible materials shall be burned or otherwise disposed of in a satisfactory manner. All materials to be burned shall be neatly piled and when in a suitable condition shall be completely burned. Piling for burning shall be done in such manner and in such locations as to cause the least fire risk. All burning shall be so thorough that the materials are reduced to ashes. No logs, branches, or charred pieces shall be permitted to remain.

12. Stripping for Embankments.—The entire areas of the dam and dike sites, including the areas over cut-off trenches, shall be stripped or excavated to a sufficient depth to remove all materials not suitable, as determined by the contracting officer, for the foundations. The unsuitable materials to be removed shall include top soil, all rubbish, vegetable matter of every kind, roots, and all other perishable or objectionable materials which might interfere with the proper compacting of the materials in the embankments, or may be otherwise objectionable. The stripped materials shall be wasted or saved for landscaping.

13. Borrow Pits.—All materials required for the construction of the dam embankment and for backfill, which are not available from required excavations, shall be taken from borrow pits (and/or quarries).

These shall be operated so as not to mar the usefulness or appearance of any part of the work, and borrow pits (and/or quarries) and the surfaces of wasted material shall be left in a reasonably smooth and even condition. Should any borrow pits be located adjacent to the dam and below the level of the top of the dam, a berm of not less than 100 feet shall be left between the toe of the dam and edge of the borrow pit, with provision for a side slope of 4 to 1 to the bottom of the borrow pit. In order to avoid the formation of pools, drainage ditches from borrow pits (and/or quarries) to the nearest outlets shall be constructed where such drainage ditches are necessary. The sites of borrow pits (and/or quarries) or so much thereof as may be required shall be carefully stripped of top soil, vegetation, roots, brush, sod, loam, and other objectionable matter.

NOTE.—References to quarries will be applicable for rock-fill dams, and may also apply in other cases where concrete aggregates cr riprap materials are obtained and processed near the site. The proper wording will be determined by local conditions.

14. Preparation of Earth Foundations.—After all necessary stripping and excavation has been completed, the foundation area shall be unwatered and the foundation for the earth fill shall be prepared by scraping and rolling, such that the surface materials of the foundation will be as compact and well bonded with the first layer of the fill as is specified for the subsequent layers of the earth fill.

NOTE.—Specifications for the preparation of foundations for rockfill dams and for concrete structures on formations other than rock are determined by local conditions.

15. Preparation of Rock Foundations for Concrete.—The surfaces of all rock foundations upon or against which concrete is to be placed shall be prepared to provide adequate bond between the rock and the concrete by roughening and cleaning the rock surfaces. All loose rock fragments, spalls, dirt, gravel, grout, and other objectionable materials shall be removed from the rock surfaces. Immediately before placing concrete upon or against any rock surface, the surface shall be thoroughly cleaned by the use of stiff brooms, hammers, picks, jets of water and air applied at high velocity, wet sandblasting and other effective means satisfactory to the contracting officer. After cleaning and before concrete is placed, all water shall be removed from depressions so as to permit thorough inspection and proper bond of concrete with the foundation rock. The cost of all work described in this paragraph shall be included in the unit prices per cubic yard in the schedule for excavation.

16. Embankment Construction, General.—The term "embankment" includes the earth-fill portion of the dam, and the riprap on the upstream face of the dam. The embankment shall be constructed to the lines and grades shown on the drawings (increased by such heights and widths as may be determined to be necessary to allow for settlement). No brush, roots, sod, or other perishable or unsuitable materials shall be placed in the embankment. The suitability of each part of the foundation for placing embankment materials thereon and of all materials for use in the embankment construction will be determined by the engineer. No material shall be placed in the embankment when either the material or the foundation or embankment on which it would be placed is frozen.

17. Earth Fill in Embankment.—The earth-fill portions of the dam, including the fill in the cut-off trench under the upstream portion of the dam, shall consist of a mixture of the clay, sand, and gravel available from borrow pits in the vicinity of the work and from excavations required for other parts of the work. No material shall be placed in the earthfill portion of the dam until after the diversion of the river has been accomplished. No earth-fill material shall be placed until the foundation therefor has been unwatered and suitably prepared. The distribution and gradation of materials throughout the earth-fill portions of the dam shall be such that the earth embankment will be free from lenses, pockets, streaks, or layers of material differing materially in texture or gradation from the surrounding material.

The combined borrow-pit excavation and embankment-placing operations shall be such that the materials when compacted in the embankments will be blended sufficiently to secure the best practicable degree of compaction, impermeability, and stability. Successive loads of material shall be so dumped on the embankments as to produce the best practicable distribution of the material to the end that the finer material shall be placed in the central upstream portion of the earth fill, including the cut-off trench, and the sand and gravel content in the earth fill will be gradually increased toward the upstream and downstream slopes of the earth fill. No stones having maximum dimensions of more than 5 inches shall be placed in the earth-fill portions of the embankments.

The mixture of clay, sand, and gravel shall be placed in the earth embankment in continuous, approximately horizontal, layers not more than 6 inches in thickness after rolling. Tamping rollers having staggered, uniformly spaced knobs and equipped with suitable cleaners shall be used for compacting the earth fill. The projected face area of each knob and the number and spacing of the knobs shall be such that the total weight in pounds of the roller and ballast, if distributed over the equivalent area of one row of knobs parallel to the axis, shall not be less than 250 pounds per square inch.

The material in each layer while being compacted by rolling shall contain the optimum amount of water for compacting purposes within practicable limits, and this optimum water content shall be uniformly distributed throughout the layer. The application of water to material for this purpose shall be done at the site of excavation insofar as practicable and shall be supplemented as required by sprinkling on the embankment. Harrowing or other working of the material may be required to produce the required uniformity of water content. While in the above-described condition, each layer of material shall be compacted by passing the specified roller over the entire surface the number of times required to obtain 50 percent coverage as determined by the size and spacing of the roller feet or knobs and assuming that no part of the layer being compacted is covered by a roller knob more than once.

All portions of test-pit and cut-off trench excavation within the area to be covered by the embankment and below the required stripping lines for the embankment foundation shall be filled with compacted embankment material as herein specified for the earth fill. The earth fill on each side of the cut-off walls shall be kept at approximately the same level as the placing of the earth fill progresses, and the walls shall be carefully protected against displacement or other damage. Portions of the earth fill between projections on the dam abutments, near the cut-off walls, about the outlet conduit and spillway and other concrete structures, and elsewhere which cannot be properly compacted by the use of rolling equipment shall be thoroughly compacted by the use of mechanical tampers. The degree of compaction for such portions of the earth fill shall be equivalent to that obtained by moistening and rolling as specified for other portions of the earth fill. The upstream face of the earth fill shall be reasonably true to line and grade and all projections of more than 6 inches outside the neat lines of the earth fill shall be removed before the rock riprap is placed. The upper 12 inches of the crest of the dam embankment shall be constructed of selected gravelly material or selected fine-rock material.

18. Rock Fill.—The rock-fill portion of the dam embankment shall be constructed to the lines and grades shown on the drawings. The rock fill shall consist of a suitable free-draining mixture of rock fragments, boulders, and cobbles from a quarry or from required excavation. The largest rock in the rock fill shall be not more than 1 cubic yard in volume. The inclusion of gravel or rock spalls in the mass in an amount not in excess of that required to fill the voids in the coarser material will be permissible. Successive loads of material shall be so dumped as to secure the best practicable distribution of the materials. In general, the larger boulders and rock fragments shall be placed on the outer slope and the smaller cobbles and fragments next to the earth fill. The rock fill shall be placed in approximately horizontal layers not exceeding 3 feet in thickness, and during the placing of each layer the fine material shall be sluiced into the voids in the rock by a stream of water having sufficient force to move the material in place. The materials need not be hand placed or especially compacted but shall be dumped and roughly leveled so as to maintain a reasonably uniform surface and insure that the completed fill will be stable and that there will be no large unfilled spaces within the fill.

19. Pressure Grouting.—As the work progresses it may be found desirable to grout, under pressure, rock foundations of the dam or elsewhere.

Grout holes shall be drilled into the rock foundations of the cut-off walls and/or elsewhere as required, as shown on the drawing, and depth of each hole shall be as directed. It is expected that the depth of holes required will not exceed —[1] feet. The minimum diameter of each hole shall be not less than —[1] inches. Each grout hole shall be protected from becoming clogged or obstructed by being capped or otherwise protected until the hole is grouted.

Metal pipes for grout connections shall be set on the concrete cut-off walls, floors, or other parts of structures and foundations at such points as may be designated. Grout pipes set in concrete shall end not less than 1 inch inside of the finished inside surfaces of the concrete, and recesses shall be provided in the concrete to be filled with mortar after the grouting is completed. The size of grout pipe for each hole will be determined to meet the requirements of the drilling and grouting equipment used. The spaces between grout pipes and the rock or concrete into which they are inserted shall be carefully calked with oakum or other suitable material to prevent entry of concrete or other materials prior to grouting.

Grout shall be forced into each drilled grout hole and grout connection under pressure of up to 100 pounds per square inch as required. For ordinary grouting work the grout shall be composed of cement and water in proportions suitable for the particular requirements. Where grout of cement and water is found to be unsuitable or inadequate, sand shall be used in the grout mixture as directed. Sand, if required, for pressure grouting shall be of the fineness suitable for the work and shall be screened.[2] Before pressure grouting is begun, all grout holes shall be thoroughly washed out with clean water by inserting a pipe into the hole and introducing the wash water at the bottom of the hole. No grout hole or grout connection shall be grouted, except with permission, until all concrete required within a radius of 50 feet is placed and has set. The apparatus for mixing and placing grout shall be of a type approved by the contracting officer and shall be capable of effectively mixing and stirring the grout and forcing it into the

---

[1] To be supplied by the designer

[2] Admixtures such as diatomaceous earth, hydrated lime, calcium chloride, or aluminum sulphate are sometimes added to produce freer flow or to retard setting. A natural hydrous silicate of alumina known as bentonite has been used for grouting by the Forest Service with apparently good results. The permanence of its sealing qualities has not been determined.

holes at any desired pressure up to a maximum of 100 pounds per square inch. All grout shall be pumped with a duplex piston-type pump. The grouting equipment shall be maintained in a satisfactory manner so as to insure continuous and efficient performance during any grouting operation. Grouting shall not be stopped in any hole or grout connection until the hole or connection maintains a back pressure of at least two-thirds of the maximum pressure used in the grouting. After the grouting of the hole is completed, the pressure shall be maintained by means of a stopcock or other suitable device until the grout has set sufficiently so that it will be retained in the hole. If during the grouting of any hole grout is found to flow from adjacent grout connections in sufficient quantity to seriously interfere with the grouting operation or to cause appreciable loss of grout, such connections may be temporarily capped. Where such capping is not essential, ungrouted holes shall be left open to facilitate the escape of air and water. If during the grouting of any hole grout is found to flow from joints in the geologic formations at the site, such leaks shall be plugged or calked. As soon as possible after the concrete has cooled the desired amount, all contraction joints shall be pressure grouted with cement grout. The grouting of each joint shall be completed in the shortest practicable time in order to insure that the grout does not set in any part of the joint before the grouting of the joint is completed.

20. Concrete.—1. *Composition.*—Concrete shall be composed of cement, sand, broken rock or gravel, and water all well mixed and brought to the proper consistency. If required, powdered admixture shall be added. The exact proportions in which these materials are to be used for different parts of the work shall be as determined from time to time during the progress of the work and as analyses and tests are made of samples of the aggregates and the resulting concrete. In general, it is contemplated that 1 part, by weight, of cement shall be used with — parts, by weight, of sand and — parts, by weight, of broken rock or grave having a maximum size of — inches.[2] These proportions may be modified to suit the work or the nature of the materials used or to comply with the water-cement ratio limitation hereinafter specified. The individual mixes will be based upon securing concrete having suitable workability, density, impermeability, and required strength, without the use of an excessive amount of cement. Such means and equipment as are required shall be provided to accurately determine and control the relative amounts of the various materials, including water and each individual size of aggregate entering the concrete. All batches of concrete shall be proportioned on the basis of integral sacks of cement, unless the cement is weighed, and the amount of each individual size of aggregate entering each batch of concrete shall be determined by direct weighing. The amount of water shall be determined by direct weighing or volumetric measurement. The measuring and weighing equipment shall conform to the requirements of the United States Bureau of Standards for such equipment.

The amount of water used shall be changed as required to secure concrete of proper consistency and to adjust for any variation in moisture content of the aggregate as it enters the mixer, provided that a water-cement ratio of 0.60, by weight, shall not at any time be exceeded. The quantity of water entering any batch of concrete shall be just sufficient, with a normal mixing period, to produce concrete of the required consistency. Excessive overmixing, requiring additions of water to preserve the required concrete consistency, will not be permitted. Uniformity in concrete consistency from batch to batch will be required. Slump

---

[2] Proportional parts may be inserted to suit the characteristics of the material most likely to be used. These proportions will be fixed to produce the design working strength with the aggregates available for use. The maximum size of aggregates will be governed by the type of work. For reinforced concrete, sizes up to 1½ inches may be used. For mass concrete in the range of structures included in this manual sizes up to 6 inches may be used.

tests will be made in accordance with the Tentative Method of Test for Consistency of Portland-Cement Concrete (A. S. T. M. Designation: D138–32T) of the American Society for Testing Materials. For ordinary reinforced walls, beams, and slabs the slump shall not exceed 4½ inches at point of placement. A greater slump than 4½ inches at point of placement, but not exceeding a maximum of 6 inches, will be permitted only in exceptional cases where internal vibration of the concrete is not practicable and where specifically authorized for concrete in positions especially difficult to place, as in thin and/or heavily reinforced sections.

2. *Cement.*—Cement for concrete shall comply with the Government Specifications for Portland Cement (Designation C–33) or with the Standard Specifications for Portland Cement of the American Society for Testing Materials (Designation C–9). Note.—One specification or the other should be indicated, not both. Government specifications are usually required for Federal work.

3. *Admixtures.*—It may be required that diatomaceous earth or other admixtures be used in the concrete to secure increased uniformity, workability, impermeability, or to otherwise improve it. The materials to be used and the amount thereof shall be as directed. Diatomaceous earth, if used, shall not be in excess of 3 percent by weight of the amount of cement used. Other admixtures, if used, shall be in like moderate proportions. Not more than one admixture shall be used at one time.

4. *Aggregates.*—Sand.—Sand for concrete and grout may be obtained from natural deposits or may be made by crushing suitable rock. The sand particles shall be hard, dense, durable, uncoated, nonorganic rock fragments that will pass a ¼-inch square or a ⁵⁄₁₆-inch round opening. It must be free from injurious amounts of dust, lumps, soft or flaky particles, shale, alkali, organic matter, loam, mica, or other deleterious substances. The sand as it is used in the concrete must be so graded that concrete of the required workability, density, and strength can be made without the use of an excess of water or cement. The sand for concrete shall have a fineness modulus of not less than 2.75 nor more than 3.25, unless approval is given to use sand not meeting this requirement. The fineness modulus will be determined by dividing by 100 the sum of the accumulative percentages retained on Tyler standard sieves Nos. 4, 8, 14, 28, 48, and 100. The suitability of the sand will be determined with the aid of tests made in accordance with the standard practice of the United States Bureau of Standards. The sand shall be washed unless specific written authority is given to use unwashed sand. The sand shall be such that tests of briquets made in proportion of 3 parts sand to 1 part cement shall develop a tensile strength not less than the strength developed by such tests with standard Ottawa sand. Any crushing, rolling, blending, screening, washing, or other operation on the sand required to meet these specifications shall be done by the contractor, and the cost thereof shall be included in the unit prices bid in the schedule for the items of work in which the sand is used.

Broken Rock and Gravel.—The broken rock or gravel for concrete must be hard, dense, durable, uncoated rock fragments free from injurious amounts of soft friable, thin, elongated, or laminated pieces, alkali, organic, or other deleterious matter. It shall be so graded that concrete of the required workability, density, and strength can be made without the use of an excess of sand, water, or cement. The suitability of the broken rock or gravel will be determined with the aid of tests made in accordance with the standard practices of the United States Bureau of Standards. Any crushing, blending, screening, washing, or other operation on the broken rock or gravel required to meet these specifications shall be done by the contractor, and the cost thereof shall be included in the unit prices bid in the schedule for the items of work in which the broken rock or gravel is used. The broken rock or gravel shall be washed unless specific authority is given to use unwashed broken rock or gravel. The broken rock or gravel shall all pass through a screen having 2¾-inch square or 3-inch round openings and shall be retained on a screen having ¼-inch square or 5⁄16-inch round openings. It shall also be separated into three intermediate sizes by screens having ¾-inch square or ⅞-inch round openings and 1½-inch square or 1¾-inch round openings. Screens having openings of other sizes or shapes may be used, provided that equivalent results, as determined by tests, are obtained. The relative amounts of each size of broken rock or gravel to be used in each mix of concrete and in all parts of the work will be based on securing concrete having the required workability, density, impermeability, strength, and economy, without the use of an excess of sand, water, or cement, and using, insofar as practicable, the entire yield of suitable material from the natural deposits from which the broken rock or gravel is obtained. For very thin or heavily reinforced parts the maximum size will be that determined by the screen having ¾-inch square of ⅞-inch round openings; for somewhat heavier portions of the work the maximum size will be that determined by the screen having 1½-inch square or 1¾-inch round openings; and for the more massive portions the maximum size will be that determined by the screen having 2¾-inch square or 3-inch round openings.

Cobble Rock.—Cobble rock may be included in the concrete mixtures for mass concrete. Such cobble rock shall be sound, hard, clean gravel or broken rock of such size as will pass through a screen having 6-inch square openings and be retained on a screen having 2¾-inch square or 3-inch round openings. The suitability of cobble rock will be determined by the contracting officer. The amount of such cobble rock to be used shall be based on producing the most economical concrete of the required strength, and, insofar as practicable, utilizing the entire yield of the natural deposit or quarry from which the gravel or broken rock is obtained. The use of cobble rock will not be required or permitted where the concrete is reinforced, or in any part of the structure where the least dimension is less than 30 inches.

5. *Water.*—The water used in concrete shall be reasonably clean and free from objectionable quantities of silt, organic matter, alkali, salts, and other impurities.

6. *Mixing.*—The cement, admixture (if used), sand, and broken rock or gravel shall be so mixed and the quantity of water added shall be such as to produce a homogeneous mass of uniform consistency. Dirt and other undesirable substances shall be carefully excluded. All concrete shall be thoroughly mixed in a batch mixer of approved type and size, and one so designed as to positively insure a uniform distribution of all the component materials throughout the mass during the mixing operation.

In general, only sufficient water shall be used in mixing to produce a workable mix. Water shall be added prior to, during, and following the mixer-charging operation. The mixing of each batch shall continue not less than 1½ minutes after all materials except water are in the mixer, during which time the mixer shall rotate at the speed for which it was designed or at such speed as will produce a mass of uniform consistency before the expiration of the mixing period. Overloading of mixers will not be permitted.

7. *Temperature of concrete.*—Concrete when deposited shall have a temperature of not more than 90° F., and not less than 40° F. in moderate weather or 50° F. in freezing weather.

8. *Forms.*—Forms to confine the concrete and shape it to the required lines shall be used wherever necessary. All exposed concrete surfaces having slopes of 1 to 1 or steeper shall be formed. Where the character of the material cut into, to receive a concrete structure, is such that it can be trimmed to the prescribed lines, the use of forms will not be required. The forms shall be of sufficient strength and rigidity to hold the concrete and to withstand the necessary pressure,

ramming, and vibration without deflection from the prescribed lines. The surfaces of all forms in contact with the concrete shall be rigid, tight, and smooth Suitable devices shall be used to hold adjacent ends and edges of panels or other forms together in accurate alinement. The lagging of wooden forms for concrete surfaces that will be exposed to view, and for all other concrete surfaces that are to be finished smooth, shall be bevel-edged or matched. Wooden forms to be used more than once shall be maintained in serviceable condition and thoroughly cleaned before being re-used. Where metal sheets are used for lining forms, the sheets shall be placed and maintained on the forms with the minimum amount of wrinkles, bumps, or other imperfections. Before placing concrete, the surfaces of all forms shall be oiled with suitable nonstaining oil.

9. *Placing.*—Concrete shall be placed in the work before it has stiffened sufficiently to require re-tempering. If any concrete before placement has become so stiff that proper placement cannot be assured, the concrete shall be wasted. All surfaces upon or against which concrete is to be placed shall be thoroughly moistened immediately before the concrete is placed, so that moisture will not be drawn from the concrete. The surfaces of all rock foundations or previously placed concrete prepared as herein specified, upon or against which fresh concrete is to be placed shall be roughened to provide adequate bond between the rock or set concrete and the fresh concrete. Immediately before placing concrete upon or against any rock surface, the surface shall be thoroughly cleaned by the use of stiff brooms, picks, jets of water and air applied at high velocity, wet sand-blasting, or other effective means, and all water shall be removed from depressions before the concrete is placed, so as to permit thorough inspection and proper bond of concrete with the foundation rock. After cleaning, and immediately before placing concrete, all approximately horizontal rock surfaces shall be covered with a layer of mortar three quarters of an inch thick, consisting of the regular concrete mixture without the coarse aggregates. The concrete mortar shall be thoroughly worked with brooms or otherwise into all irregularities of the surface. Concrete shall then be placed immediately upon the fresh mortar. When the placing of concrete is to be interrupted long enough for the concrete to take its final set, the working face shall be given a shape as shown on the drawings by the use of forms or other means that will secure proper union with subsequent work. All concrete surfaces which have set for 12 hours or more, upon or against which concrete is to be placed and to which new concrete is to adhere, shall be cleaned of all laitance and loose or defective surface concrete, by means of jets of air and water applied at high velocity, by wet sand-blasting, and/or chipping. The surface of each lift or layer shall be washed immediately prior to the placing of the succeeding lift or layer of concrete and all water shall be removed from depressions before the concrete is placed.

Only methods of transporting and placing which will deliver concrete of the required consistency into the work, without segregation and without objectionable porosity, shall be used. Methods of conveying concrete to any of the structures, by which the mixed batch or combination of batches is progressively loaded into chutes, belt conveyors, or other similar equipment and carried in a thin continuous exposed flow to the forms will not be permitted. Concrete shall be deposited in all cases as nearly as practicable directly in its final position and shall not be caused to flow in the mass in a manner to permit or cause segregation. Dropping the concrete vertically, a distance such as to result in undesirable segregation, or depositing a large quantity at any point and running or working it along the forms, will not be permitted. No concrete shall be placed in water except with written permission, and the method of depositing the concrete shall be subject to approval. No concrete shall be placed in running water. Concrete which is not placed in accordance with these specifications and is of inferior quality, shall be removed and replaced with satisfactory material.

10. *Finishing.*—The surface of concrete finished against forms shall be smooth, free from projections, and thoroughly filled with mortar.   Immediately upon the removal of forms, all holes left by the removal of rods and all voids shall be neatly filled with cement mortar of the same proportions as the mortar in the concrete. All unsightly ridges or lips shall be removed, and undesirable local bulging on exposed surfaces shall be remedied by tooling and rubbing.   All patching required shall be done as directed by skilled workmen.

11. *Curing and protection.*—Exposed surfaces of concrete shall be protected from the direct rays of the sun for at least three days.   All concrete shall be kept continuously moist for at least two weeks after the concrete has been placed.   In freezing weather suitable means shall be provided for maintaining the concrete at a temperature of at least 50° F. for not less than 72 hours after placing or until the concrete has thoroughly hardened.   All concrete shall be protected from freezing for not less than two weeks by approved means.   Where artificial heat is employed special care shall be taken to keep the concrete from drying out. The method of keeping concrete moist shall be by sprinkling or spraying with water, or by other suitable and approved methods.

21. Reinforcement Bars.—Steel bars shall be placed in the concrete wherever shown on the drawings, or otherwise prescribed.   The steel shall be of _ _ _ _ _ _ _ _ _ _ grade.   The exact position, size, and shape of reinforcement bars are not shown in all cases on the drawings, and where not shown they shall be in all respects as specified by the officer in charge and, where necessary, the contractor will be furnished with supplemental detailed drawings or lists which will give the information necessary for cutting, bending, and placing the bars.   The steel used for concrete reinforcement shall be so secured in position that it will not be displaced during the depositing of the concrete, and special care shall be exercised to prevent any disturbance of the steel in concrete that has already been placed.   The cost of furnishing and attaching wire ties, of unloading, hauling, storing, cutting, bending, placing, and securing in position reinforcement bars, shall be included in the unit price bid for placing reinforcement bars.

22. Installing Metal Work.—The outlet gates, frames and hoists, trash racks, ladders, rungs, floor plates, and pipe handrail materials and other metal work required as parts of the completed structure will be furnished by the contractor. The contractor shall attach to or build into the dam or appurtenant works all such metal work and shall install the gates and hoists in a workmanlike manner, as shown on the drawings or as directed.   All moving parts and control mechanisms shall be carefully installed, tested for operation, and adjusted so that all parts move freely as intended and properly function to secure satisfactory operation.   Any changes or adjustments required shall be made to secure satisfactory operation.   A part of the metal work as furnished will have been given one or more shop coats of paint.   This paint coating shall be protected as much as practicable during the handling and storing and installing of metal work, and after installation all unfurnished surfaces not to be in contact with concrete shall be painted.

Trash-rack frames, bars, and I-beam supports furnished will be of the dimensions required for the structures.   Frames and I-beam supports shall be securely anchored to the concrete structures with anchor bolts or bars provided for this purpose.   The trash-rack bars are merely set into place and no other work is required for installation.   Dipping of trash-rack bars will be permitted in lieu of brush painting.

Pipe and fittings for handrailings will be furnished, cut to length, threaded or drilled, bent to shape, and complete with bolts, pins, rivets, flanges, and any other accessories required for installation. The contractor shall assemble, install, and paint this material as shown on the drawings. Railings to be set in concrete shall either be completely assembled and placed when concrete is poured, or recesses shall be left or holes drilled in the concrete to receive the railing posts and the railings completely assembled and grouted in position with cement grout at some later time.

23. Painting.—All exposed, unmachined metal surfaces shall be painted. Where metal parts have been painted before delivery, care shall be taken in unloading, handling, hauling, and installing such parts to preserve the shop paint in the best practicable condition. After installation, painted surfaces shall be thoroughly cleaned and all damaged places in the original paint film shall be repainted. After these repaired areas have thoroughly set, one or more field coats of paint shall be applied to the exposed surfaces. The leaves, frames, and other submerged parts of outlet gates and trash-rack bars and castings not in contact with concrete shall be given one coat of water-gas tar, if not painted in the shop, followed by two coats of coal-gas tar applied hot, or, painted with one or more coats of a metal preservative paint applied hot or cold, according to the specifications of the paint manufacturer. Machined surfaces and surfaces of metal work to be in contact with concrete shall not be painted except as directed. Pipe handrails shall be given one or more coats of oil paint or enamel of a color to be specified. Pipe handrails shall be carefully cleaned of rust, scale, and oil and given one priming coat of red lead and two additional coats of oil paint or enamel of a color to be specified by the contracting officer. All painting shall be performed in a skillful and workmanlike manner, and each coat of paint shall be permitted to properly dry before the succeeding coat is applied.

APPENDIX H

SUMMARY OF STATE LAWS AFFECTING DESIGN AND CONSTRUCTION OF LOW DAMS [1]

Cover statement.—No two States have precisely the same laws and regulations affecting the construction of dams. The chief restrictions in each State are summarized in the following tabulation, which is based, in part, on replies to questionnaires circulated to State officials during the fall of 1936.

In most States the State board of health has general supervision over all public water supplies, and the regulations which apply to municipal water supply primarily are not summarized here. Likewise the regulations affecting the appropriation of water in arid and semiarid States are not summarized.

[1] Prepared by William J. Ponorow

374

State	Design and construction		Operation	
	Supervising agency	Laws or regulations (reference shown in parentheses)	Supervising agency	Laws or regulations (reference shown in parentheses)
Alabama	State board of health	An application for a permit to impound water or raise the level of an existing pond must be filed with the State board of health and a permit obtained from this department. This regulation does not apply to ponds of less than one-tenth of an acre or if the project is more than 1 mile from a human habitation other than that of the owner and is not used by the public. All growth, brush, flotage, and similar material in the area to be flooded and within 15 feet of the water edge must be removed, burned, or otherwise satisfactorily disposed of prior to the impounding of the water. All trees and underbrush between the normal water level and 1 foot below the low water level must be cut off reasonably close to the ground. (Secs 1, 3a, 3b, and 3c State board of health regulations governing impounding of waters.)	State board of health	During the mosquito breeding season the permit holder is required to regularly and frequently remove all flotage and debris which produce mosquitoes in the reservoir. Prompt and proper measures must be taken by the owner to prevent the growth of cattails, bulrushes, and other aquatic or semiaquatic vegetation which offers protection to mosquito larvae from their fish enemies. A drain must be installed in the dam for the purpose of dewatering it. After the impounding of the water, representatives of the State board of health make occasional inspections, and any conditions found to be detrimental to public health must be remedied by the permit holder. (Secs 5a, 5b, and 5d State board of health regulations governing the impounding of waters)
Arizona	State highway engineer.	An application for a permit, together with plans and specifications must be filed to construct a dam 15 feet or more in height from ground level to the spillway crest or the impounding capacity of which will be more than 10 acre-feet, except dams or reservoirs used exclusively for watering livestock. (Ch 102, Session Laws of 1929, ch. 47, Session Laws of 1931.)	State highway engineer.	During large floods owners are required to secure an accurate record of inflow, by stream gaging or by taking measurements of reservoir elevations at frequent intervals. Stream flow measurements should be taken continuously at all dams which impound water for irrigation or power, regardless of size. Spillway gates are tested by State engineer at frequent intervals to insure satisfactory operation in case of emergency. Spillway gates are not permitted to be installed unless an operator is in continuous attendance. Spillway gates, except small ones, must have 2 separate methods of operation. Growths and debris must be cleared from spillway channel. Leakage under and through the dam shall receive periodic attention; shall be recorded, tabulated and submitted to office of State engineer. Water required to be tested to determine what materials are being carried in suspension or solution. (Circular No 2, State engineer office, rules and regulations governing maintenance and operation of dams.)
Arkansas	Corporation commission Secretary of State, county clerk	An application must be filed with the corporation commission for construction of power dams. A survey and estimate of expenses are required to be filed with the secretary of State and with the county clerk in the county where dam is to be located. (Secs 10460, 10461, Digest of Statutes of Arkansas, Crawford & Moses, 1921.)	State game and fish commission.	Persons owning or controlling dams are required to keep an opening for free passage of fish from Mar 1 to June 1. (Sec 4789, Digest of Statutes of Arkansas, Crawford & Moses, 1921)

State	Design and construction		Operation	
	Supervising agency	Laws or regulations (reference shown in parentheses)	Supervising agency	Laws or regulations (reference shown in parentheses)
Arkansas	Corporation commission. Secretary of State, county clerk	The permit is voided if construction is not begun within 2 years and completed within 4 years from date of permit. (Sec 10474, Digest of Statutes of Arkansas, Crawford & Moses, 1921.)		
	County court	A petition to and permit from county court is required for construction of a mill dam on private property on a nonnavigable river. (Secs 3943-3968, Digest of Statutes of Arkansas, Crawford & Moses, 1921.)		
California	State engineer	An application, with plans and approval secured from the State engineer for the construction or repair of any dam before construction is begun. A graduated filing fee based upon the estimated cost of the project must accompany the application for construction or enlargement, otherwise it will not be considered. Construction must be started within 1 year after date of approval of application, otherwise the application is voided, unless extended for good cause shown. This act applies to all dams except dams 6 feet or less in height regardless of storage capacity, dams storing 15 acrefeet or less regardless of height; and dams less than 25 feet in height which have a storage capacity of less than 50 acre-feet, and dams federally owned. (Secs. 2, 6, 7, ch. 766, Statutes of California of 1929, as amended by ch 808 of 1933; pamphlet of department of public works governing supervision of dams 1935.)	State engineer (department of public works).	After the completion of a dam it is not to be used until a certificate of approval, or authorization for use pending issuance of a certificate, has been issued by the State engineer. The department of public works makes inspections of all dams. Operation reports when necessary are required to be submitted by the owners. During times of emergency, the department has authority to lower the water level of a reservoir or may empty it entirely. The cost of any remedial work carried out by the department is recoverable from the owner of the dam. (Secs. 9 and 12, ch. 766, Statutes of California, 1929, as amended by ch 808 of 1933; pamphlet of department of public works governing supervision of dams, 1935.)
	Fish and game commission	A copy of the application that is filed with the State engineer for the construction or enlargement of a dam must also be filed with the fish and game commission, which has authority to order fishways built. (Sec. 520 5, 1933 Supplement to California Codes.)	Fish and game commission.	An application with maps and drawings is required to be filed with the State engineer for the right to appropriate waters of the State. (Sec 16, ch 586, Statute of 1913 of California, pamphlet of Department of Public Works of 1935 "Water Commission Act.")
				The commission has authority to examine all dams on rivers and streams in the State naturally frequented by salmon, trout, and other fish. (Sec. 521, 1933 Supplement to California Codes.)
				Owners of dams are required to keep fishways open and keep them in repair. The commission has authority to order an owner of a dam to establish a fish hatchery where the height of a dam makes it impractical to construct a fishway. (Secs 523, 526, 1933 Supplement to California Codes.)
				Owners of dams must accord the public right of access to water impounded by the dam during open season for taking fish in such river or stream where dam is located. (Sec. 531, 1933 Supplement to California Codes.)

State	Official	Requirement	Official	Requirement
Colorado	State engineer	An application, together with plans and specifications must be filed with and approval secured from the State engineer for the construction and repair of a dam or reservoir of a capacity of more than 1,000 acre-feet or having an embankment in excess of 10 feet in vertical height, or having a surface area at high water line in excess of 20 acres. (Secs. 3731, 3847, Courtright's Mills Annotated Statutes of Colorado, 1930.)	State engineer and water commissioner	The commission has authority to order screens installed on conduits to prevent fish from entering therein (Sec. 535, 1933 Supplement to California Codes) The water commissioner in the district where the reservoir is located is authorized to draw off the excess amount of water which is stored in the reservoir over the amount authorized by the State engineer to be safe. (Sec. 3734, Courtright's Mills Annotated Statutes of Colorado, 1930)
	County surveyor	Plans and specifications for dams to be built across a normally dry watercourse for domestic or irrigation purposes which are 10 feet or less in height, or which form a pool of 20 acres or less, must be submitted to and be approved by the county surveyor before construction is begun. (Ch. 185, Laws of 1937.)	State engineer	The State engineer annually determines the amount of water which it is safe to impound in all reservoirs of the State and the owners are prohibited from storing more water than determined by the State engineer (Sec. 3733, Courtright's Mills Annotated Statutes of Colorado, 1930). Upon complaint of three or more persons that a reservoir is unsafe the State engineer is authorized to make an inspection of such reservoir and if it is found unsafe, he is authorized to draw off a sufficient amount of water to make the reservoir safe. (Sec. 3735, Courtright's Mills Annotated Statutes of Colorado, 1930) An owner of an irrigation ditch, canal, flume or reservoir taking water from any stream must erect and maintain in good order at the point of intake a proper measuring flume, or weir, and headgate and a suitable wastegate. If the owner fails to construct a headgate or measuring device, the State engineer can refuse to deliver water to such owner. (Sec. 3772, Courtright's Mills Annotated Statutes of Colorado, 1930.) An owner of a reservoir is required to maintain a gage rod marked in feet, tenths, and one-hundredths of a foot at the outlet of the reservoir, otherwise the State engineer can refuse water to such reservoir. (Sec 3775, Courtright's Mills Annotated Statutes of Colorado, 1930)
			Fish commissioner	Persons operating dams must build and keep in repair fishways for free passage of fish, except where entire flow of stream is diverted. (Secs. 3231, 3232, 3233, Courtright's Mills Annotated Statutes of Colorado, 1930.) Persons owning or controlling reservoirs or other bodies of water into which public waters flow and which furnish water to any stream containing fish must not divert or lessen the flow to an extent detrimental to fish. (Sec. 3188, Courtright's Mills Annotated Statutes of Colorado, 1930.)

State	Design and construction		Operation	
	Supervising agency	Laws or regulations (reference shown in parentheses)	Supervising agency	Laws or regulations (reference shown in parentheses)
Connecticut	Board of civil engineers.	An application for a permit must be filed, together with the plans and specifications, for the construction, alteration or addition to all dams and reservoirs in the State. (Ch. 180, sec. 3004, General Statutes of Connecticut, Revision of 1930.)	Board of civil engineers.	The board has supervisory powers over all dams and reservoirs in the State, which by breakage or overflowing would cause injury to life or property. (Ch 180, sec. 3001, General Statutes of Connecticut, Revision of 1930.) Upon application in writing of two or more persons or corporations who would suffer by the breaking away of any dam or reservoir, the board makes inspection of such dam or reservoir and if same is found to be defective, can order the owner of such dam or reservoir to repair same. (Ch 180, sec 3003, General Statutes of Connecticut, Revision of 1930.)
Delaware	Court of general sessions	An application to the court of general sessions must be filed for the erection or the raising of a dam upon or across any nonnavigable waters of the State. (Ch 106, Revised Code of Delaware 1935, par 4221, sec 4)		
Florida	State board of health	An application, together with a plat of the area to be affected, must be filed with the State board of health to impound water or raise the level of existing impounded waters. All growths, brush, etc which are favorable to mosquito larvae production must be removed before impounding of the water. All trees and undergrowths must be cut off at least 1 foot below the water level Flashboards or other means must be provided for controlled fluctuation of the water This regulation does not apply to areas of less than 1 acre of water surface used exclusively for watering livestock or for other domestic purposes. (Secs. 1, 2, 3, 4, Florida State Board of Health, rule 106 of July 23, 1936)	State board of health.	The reservoir must not be filled during the malaria-carrying mosquito breeding season, April to September, unless authorized specifically in writing by the board of health. Owners are required to remove drift, floatage, and all other growths; apply larvacides, screening or any other measures effective in the control of malaria mosquitoes Owners are required to furnish to the board of health such reports of operation, maintenance, or condition of the impounded water, as may be specified by said board. After the impounding of the water the State board of health makes occasional inspections, and any conditions found to be detrimental to public health, must be remedied by the permit holder. (Secs 6, 7, 8, 9, Florida State Board of Health, rule 106 of July 23, 1936)
Georgia	State board of health	An application for a permit must be filed to impound water or raise the level of an existing pond. This regulation does not apply to ponds of less than 1/10 acre for watering stock, or if water is more than 1 mile from human habitation other than that of owner. All growths, brush, etc which are favorable to mosquito larvae production must be removed before impounding of the water. All trees and undergrowths must be cut off at least 1 foot below the water level (Secs. 1, 3a, 3b, Georgia Laws of 1918, p 275)	State board of health.	During the mosquito breeding period the permit holder is required to remove all floatage and debris which produce mosquitoes in reservoirs located within 1 mile of human habitation. After the impounding of the water the State board of health makes occasional inspections, and any conditions found to be detrimental to public health, must be remedied by the permit holder. (Secs. 5a, 5c, Georgia Laws of 1918, p 275)
Idaho	State department of reclamation.	Plans and specifications for dams more than 10 feet in height must be submitted to the department of reclamation for approval before beginning construction. (Ch. 15, Idaho Code of 1932, sec. 41-1507, p. 84, Eighth Biennial Report of Department of Reclamation, 1933-31.)	State department of reclamation.	Regulations provide for occasional inspection of dams of more than 20 feet in height If dam is found unsafe department may order owner to repair same. (Ch. 15, Idaho Code of 1932, Sec 41-1508; p. 84, Eighth Biennial Report of Department of Reclamation, 1933-34)

State	Agency	Requirements
Illinois	Department of public works and buildings	An application, together with plans, profiles, and specifications, must be filed with the department of public works and buildings, to build any breakwater, bulkhead or other structure or to do any work of any kind whatsoever in any of the public waters of the State. (Ch 19, sec 76, Cahill's Illinois Revised Statutes, 1935.)
	Department of Conservation	Durable and efficient fishways are required to be erected and kept in repair over dams in order to permit the passage of fish through and over same (Ch 56, sec 33, Cahill's Illinois Revised Statutes, 1935, Laws of 1937, p 629, sec 22.)
Indiana	Department of conservation	Persons controlling dams are required to keep same open at all times to permit a flow sufficient to maintain fish in the stream below, and also to keep a sufficient head of water in the pond above dam to maintain fish. This provision does not apply during low water period to municipal dams impounding water for municipal use, or used to impound water for fire protection. (Ch 59, Acts of 1935.)
	Township trustee	Owners of dams 4 feet high or over are required to construct, maintain, and keep in repair fish ladders. If owner fails to construct or repair fish ladders, the township trustee may construct or repair the fish ladders and charge the cost thereof to the owner of the dam. (Ch 21, secs 106 and 107, Acts of 1937.)
Iowa	Executive council	An application with map showing the lands to be affected must be filed with the executive council to construct, operate, or maintain a dam. Where water is to be used as a source of public or domestic water supply, plans and specifications must be submitted to the department of health for approval. The method of construction, operation, maintenance, and equipment of any and all dams in such waters is subject to the approval of the executive council. All dams must be provided with fishways, and pumping plants must provide screens for fish protection. Fishways and screens are required to be used and constructed in accordance with specifications of director of conservation. Owners must begin construction of project within 1 year, and if project is not completed within 3 years the permit is forfeited, unless extended by executive council. (Chs 86 and 363, Code of Iowa, 1935.)
	State department of health, Director of conservation	An annual inspection and license fee must be paid by all dam owners except municipalities. Department of health investigates all waters in the State used for public water supplies or for domestic use or which is deleterious to fish. No existing dam can be removed or destroyed so as to lower the water level without giving 10 days' written notice thereof to the director of conservation. (Chs. 86 and 363, Code of Iowa, 1935.)
Kansas	Division of water resources, State board of agriculture	An application, together with maps and specifications, must be filed for the construction, change, or addition to any dam or other water obstruction. These provisions do not apply to a dam on a purely private stream that is not more than 10 feet high and does not impound more than 15 acre-feet of water, these are under jurisdiction of the county engineer when built

State	Design and construction		Operation	
	Supervising agency	Laws or regulations (reference shown in parentheses)	Supervising agency	Laws or regulations (reference shown in parentheses)
Kansas	County engineer	across a dry watercourse and the landowner desires to secure a reduction in the assessed valuation of his land (Ch 263, Laws of 1929, secs. 1, 2, Chs. 330 and 332, sec 1, Laws of 1933.) Screens, flashboards, or other obstructions are not to be placed in spillways. All dams which are exposed to more than about 5 acres of water surface should be protected against the action of waves by suitable revetment on the upstream slope. All sod, brush, trees, roots, and stumps are required to be removed from the area to be occupied by the dam. (Bulletin No. 213-b, Regulations and Information of the Division of Water Resources, quarter ending March 1935, pp 11, 14, 20.)		
Kentucky	Game and fish commission.	Dams constructed across any streams where the annual tides are not sufficiently high to admit the passage of fish must maintain fish ladders during the months of April, May, and June. Fish ladders need not be built on dams constructed on large rivers to facilitate navigation or on dams constructed on streams for the purpose of generating water power. (Secs 1932a-1, 1932a-2, Carroll's Statutes, 1936.)	County court	Roads running over dams must be kept in good order and must be at least 14 feet wide, and the owner is required to keep in good order a bridge of like width over the pierhead or floodgates, and a rail must be erected on both sides of such bridge or dam. (See 4322, Carroll's Statutes of 1936.)
	County court	A petition must be filed in the county court for the erection of mill dams. Dams cannot be constructed below 10 miles from the head of a stream which is navigable for the running of push boats and floating of logs. This act does not apply to dams constructed for the purpose of generating water power (Sec 1392b, Carroll's Statutes of 1936.)		
Louisiana	State board of health, conservation commission, State land office, board of State engineers	An application with a plat of area to be affected, showing maximum and minimum water level, is required to impound water or for raising the level of an existing pond by the elevation of point of overflow of a dam. This section does not apply to ponds of less than 1/16 acre used for watering stock or other domestic purposes. All growths, brush, etc, which are favorable to mosquito larvae production must be removed before impounding of the water. All trees and undergrowths must be cut off at least 1 foot below water level. (Louisiana State Sanitary Code, Ch. XII, Art 195, Secs. I, II, III; letter of Dec 11, 1936, from chief engineer, board of State engineers of Louisiana.)	State board of health	During the mosquito breeding period the permit holder is required to remove all floatage and debris which produce mosquitoes in reservoirs located within 1 mile of human habitation. After the impounding of the water the State board of health makes occasional inspections, and any conditions found to be detrimental to the public health must be remedied by the permit holder. (Louisiana State Sanitary Code, Ch XII, Art. 195, Secs V, VII.) Persons taking water from the fresh waters of the State must provide suitable screens on intake pipes in order to prevent fish being removed from the streams. This act does not apply to intake pipes on the Mississippi and Red Rivers. (Dart's Louisiana General Statutes of 1932, Sec 3012.)

State	Agency	Construction / permit provisions	Inspection / enforcement provisions
Maine	Board of State engineers	A permit from the board of State engineers is required before anyone shall make, on the bank of any river or navigable stream of the State, any work tending to alter the course of the water or increase its rapidity, or make its navigation more difficult. (Act 185 of the State legislature of 1928)	
	Public utilities commission	Plans and a statement giving information relating to proposed structures must be filed with the public utilities commission before construction of any water storage basin or reservoir (Ch 62, sec 11, Revised Statutes of Maine, 1930)	
	Inspector		Upon complaint of county commissioners or taxpayers, dams are subject to inspection by an inspector appointed by the Governor. (Ch 106, Secs 45 and 46, Revised Statutes of Maine, 1930)
	State bureau of health		The State bureau of health exercises general supervisory powers over all bodies of waters of State used for domestic purposes Commissioner of health is authorized to enter upon any project in State to carry out control measures for abatement of mosquitoes and eradication of same. (Ch 22, secs 143 and 151, Revised Statutes of Maine, 1930)
Maryland	Commissioner of Inland Fisheries and Game		Commissioner may order owner of dam to provide a fishway or to make repairs to an existing fishway. (Ch. 331, sec. 5, Laws of 1929)
	Water resources commission	An application, together with maps, drawings, and specifications, must be filed and permit obtained from the water resources commission for the construction, addition, or repair of any dam or reservoir. This act does not apply to a dam which is 10 feet or less in height or to a reservoir with a storage capacity of less than 1 million gallons (Art 96B, secs. 5 and 6, 1935 Cumulative Supplement to 1924 Maryland Code) Construction must begin within 2 years and be completed within 5 years, unless commission for good cause shown extends the time (Art 96B, sec 10, 1935 Cumulative Supplement to 1924 Maryland Code)	Upon complaint or on its own initiative the water resources commission has power to investigate and examine any dam, reservoir, or other waterway obstruction, and if found unsafe the commission can order the owner to either repair or remove same (Art. 96B, sec. 9, 1935 Cumulative Supplement to 1924 Maryland Code)
	Conservation commission		Owners of dams are required to construct and keep in repair fishways and fish ladders. Where this is impracticable, the commission has authority to enter into an agreement with the owner of dam to pay a certain amount to the Conservation Commission of Maryland, which amount is to be used for stocking the pools with fish above and below the dam. (Art 39, secs 89 and 90, Maryland 1924 Code; Art. 39, sec 13, 1935 Cumulative Supplement to 1924 Code of Maryland)
Massachusetts	County commissioners	Plans and specifications must be submitted to the county commissioners for approval before the construction or material alteration of any dam or mill dam, and the commissioners shall inspect the work during its progress This act does not apply to small dams constructed for irrigation or for other purposes, the breaking of which would not cause loss of life or property, nor to dams where the area drained into the pond formed thereby does not exceed 1 square mile unless the dam is more than 10 feet in height above the natural bed of the stream at any point or unless the dam impounds more than 1 million gallons of water. (Ch 253, sec 44, General Laws of Massachusetts, Tercentenary Edition of 1932)	County commissioners shall make an examination at least once in 2 years of every dam and reservoir which by breakage would cause loss of life or property. They can also make examinations upon application of any mayor and alderman of a city or selectmen of a town or an individual whose property would be affected by the breakage. If dam is found defective, commissioners can order owner to make repairs, and upon his refusal they can make the repairs at owner's expense The county commissioners are also authorized to remove any part of the structure or they can order the water drawn off (Ch. 253, secs 45, 46, 47 and 48, General Laws of Massachusetts, Tercentenary Edition, 1932)

State	Design and construction		Operation	
	Supervising agency	Laws or regulations (reference shown in parentheses)	Supervising agency	Laws or regulations (reference shown in parentheses)
Massachusetts	Department of public works	Department may license and prescribe the terms for construction or extension of a dam, or certain other structures, in, over, or upon the waters below high water mark of the Connecticut River or certain portions of the Westfield and Merrimack Rivers (Ch. 91, sec 12, General Laws of Massachusetts, Tercentenary Edition of 1932)	Director of fisheries and game	The director of fisheries and game has authority to make examinations of all dams and structures and may order fishways built and prescribe the time when gates are to be opened (Ch. 131, sec. 31, General Laws of Massachusetts, Tercentenary Edition of 1932)
Michigan	County supervisors	Permission must be secured from the board of county supervisors to dam a navigable stream. (Sec 14, art 8, State Constitution of 1908)	Conservation commission	Conservation commission possesses power to erect and maintain proper facilities for the free passage of fish through and over dams (Act 123, Public Acts 1929)
Minnesota	Commissioner of conservation	An application, together with maps, drawings, and specifications, must be filed and a permit obtained from the commissioner of conservation for the construction, addition, or repair of any dam or reservoir This act does not apply to dams which have less than 1 square mile of drainage area or a normal flow of less than 2 cubic feet per second, or to dams to be used for the production of water power (Ch 468, secs 4, 5 and 6, Laws of 1937)	Commissioner of conservation	Upon complaint or on his own initiative the commissioner has power to investigate and examine any dam, reservoir, or other obstruction, and if found unsafe he can order the owner to either repair or remove same (Ch 468, sec 9, Laws of 1937)
	Department of conservation	Construction must begin within 2 years and be completed within 5 years, unless the commissioner for good cause shown extends the time (Ch. 468, sec. 10, Laws of 1937)		The department of conservation is charged with the responsibility for the control of all water impounded by dams in the State with the exception of those used for power purposes. (Letter of W. S. Olson, dated July 8, 1937)
	Executive council	Dams for public recreational uses or dams essential for logging or logging reservoirs that do not exceed 100 acres in extent may be constructed to maintain temporary water levels not higher than the normal high-water mark upon written approval from the department of conservation and the executive council (Sec 6602-2, 1934 Supplement to Mason's Minnesota Statutes of 1927) This provision applies only to the area known as the Superior-Quetico region in the northeastern part of the State (Letter of W. S. Olson, dated July 8, 1937)		
Mississippi			County mosquito control commission	The county board of supervisors in any county of the State may, with the approval of the State health officer, appoint 3 persons who shall constitute a board of commissioners of said county to be known as the _____ County Mosquito Control Commission This commission, when appointed, is given power to eliminate all breeding and producing places of mosquitoes in the county. (Sec. 4946, Code of Mississippi of 1930)

State	Agency	Provisions	Agency	Provisions
Missouri	Circuit court	Persons desiring to erect a dam or increase its altitude for the purpose of mill, electric power and light works, or water supply for any city, town, or village, must file a petition in the circuit court of the county where the dam is to be located Construction must begin within 1 year and be completed within 3 years (Secs. 9157 to 9184 and sec 4965, Revised Statutes of Missouri, 1929, vol 7, pp 5115 to 5123, and vol 3, p 2274, Missouri Statutes Annotated, 1932 permanent edition)	Game and fish commissioner	Persons owning or controlling any dams are required to erect and maintain durable and efficient fishways for the free passage of fish Whenever the height or character of a dam makes the installation of such ladders impractical the commissioner is authorized to require the establishment by the owner of a fish hatchery for stocking the waters above and below the dam (Session Laws of Missouri of 1931, p 228. vol 6, p 4102, Missouri Statutes Annotated, 1932 permanent edition) Dams constructed not wholly upon one's cwn property must be provided with an apron or chute not less than 15 feet wide and low enough for the free passage of fish (Sec 9183, Revised Statutes of Missouri, 1929, vol 7, p 5122, Missouri Statutes Annotated, 1632 permanent edition)
Montana	State engineer	Persons constructing dams and reservoirs are required to construct them in a substantial manner The State engineer does not exercise original jurisdiction over the construction of dams Persons who may be damaged by the improper construction of a dam must file a complaint with the State engineer or the county commissioners, in which case the State engineer makes an inspection of such dam and if it is found to be unsafe he can order that the dam be constructed in a substantial manner. (Secs. 2668, 7117, 7118, Montana Revised Statutes, 1935)	State engineer	Upon complaint of 3 or more persons that a dam is in a dangerous condition the State engineer has authority to examine such dam, and if it is found to be in a dangerous condition, he may order it repaired and the water drawn off (Ch 243, secs 2658-2671, Montana Revised Statutes, 1935)
	County commissioners		Water commissioner	All persons using waters under a decree whereby a water commissioner is appointed are required to maintain suitable headgates and also maintain a proper measuring box or weir for measurement of water flowing in such ditch (Sec 7151, Montana Revised Statutes, 1935)
Nebraska	Department of roads and irrigation	An application, together with plans and specifications, must be submitted for approval before construction of any dam for reservoir purposes or across the channel of any running stream (Sec 81-C327, Compiled Statutes of Nebraska 1929, p 30, rules of 1935 of department of roads and irrigation)	Department of roads and irrigation	Owners of dams which have or will have an impounding capacity of 10 acre-feet or more, are required to keep the dam in a state of repair to be approved by the department of public works Inspections are to be made annually If dam is found defective, the owner must make repairs within 3 months (See 81-6332, 1931 Supplement to Compiled Statutes of 1929, p 31, rules of 1935 of department of roads and irrigation)
Nevada	State engineer	There is no law directly relating to the construction of dams in the State of Nevada However, there is a law which provides that any person desiring to appropriate the public waters of the State or change the diversion or manner of use of such waters already appropriated must make an application and secure a permit from the State engineer before performing any work in connection with such appropriation or diversion. (Sec 7944, Compiled Laws of Nevada of 1929)	State engineer	The State engineer possesses supervisory powers over all waters and water projects in the State (Water Laws of Nevada of 1933, pp 20-34)
New Hampshire	Water control commission	An application together with such information as commission requires must be filed before beginning construction or reconstruction of any dam. All dams are subject to inspection during construction (Ch 133, secs 15 to 27, Laws of 1937)	Water control commission.	The commission makes inspection from time to time of all dams in State which may be a menace to public safety If found defective, commission may order owner to make repairs under penalty of a fine (Ch 133, secs 23 to 28, Laws of 1937)

State	Design and construction		Operation	
	Supervising agency	Laws or regulations (reference shown in parentheses)	Supervising agency	Laws or regulations (reference shown in parentheses)
New Jersey	State water policy commission	An application, together with plans and specifications, must be submitted to the State water policy commission for the construction, alteration, or repair of any dam or reservoir. (Ch. 243, P. L. 1912 of New Jersey; Title 58, Ch. 4, Rev. Stat. N. J. 1937; p 4 of State water policy commission pamphlet on information for construction of dams, 1931) This act applies only to dams which raise the waters of such stream more than 5 feet above their usual mean low-water height; it does not affect dams where the drainage area above same shall be less than ½ square mile in extent. (Ch. 243, P. L. 1912, sec 1, Title 58, Ch. 4, Rev. Stat. N. J. 1937; pamphlet of department of conservation and development of 1928, p. 15.)	State water policy commission	The State water policy commission has power to inspect dams and reservoirs which may become a menace to life or property and order the water drawn off or the dam repaired. (Ch. 243, P. L. 1912 of New Jersey, State water policy commission pamphlet, p. 5, on information for construction of dams, 1931) Permission must be obtained from the Board of fish and game commissioners before drawing off the waters of any pond, lake or stream. (Sec. 34, Laws of 1903, as amended by Ch. 33, sec. 6, Laws of 1921; Ch. 64, sec. 3, Laws of 1937)
			Fish and game commission	The board may require provision of fishways for passage of fish (Ch. 64, P. L. 1937)
New Mexico	State engineer	Persons desiring to appropriate natural public waters must file an application, together with maps, field notes, plans and specifications, with the State engineer before construction of the works. (Sec. 151-120, New Mexico Statutes, 1929) The State engineer makes all necessary rules, regulations and codes governing design and construction of dams (Ch. 178, Laws of New Mexico, 1937) The State engineer can grant extensions of time for the completion of works, and in the case of irrigation and power projects he can extend the time up to 2 years for any one extension. (Ch. 66, Laws of New Mexico, 1933)	Game and fish commission.	Persons owning or controlling reservoirs or other bodies of waters into which public waters flow and which furnish waters to any stream containing fish, cannot divert or lessen such water flow to an extent detrimental to fish in such reservoir or other body of water (Sec. 57-307, New Mexico Statutes, 1929) Persons owning lakes or other bodies of water must provide screens to prevent fish in public waters from entering therein. (Sec. 57-315, New Mexico Statutes, 1929)
			State engineer.	Persons controlling ditches must construct and maintain headgates when ordered to do so by the State engineer, and they are also required to construct measuring devices for measuring water (Sec. 151-158, New Mexico Statutes, 1929) Upon request of any party or upon his own volition, the State engineer can make inspection of works and if same are found to be unsafe and a menace to life or property, he can order same to be repaired. (Sec. 151-141, New Mexico Statutes, 1929) After the construction of a project, a license must be obtained from the State engineer before appropriating the water (Sec. 151-143, New Mexico Statutes, 1929)

State	Authority	Provisions
New York	Superintendent of public works.	All structures for impounding water requires a permit from the superintendent of public works This act does not apply where the area drained into the pond formed thereby does not exceed 1 square mile, unless the dam is more than 10 feet in height above the natural bed of the stream, or unless the quantity of water which the dam impounds exceeds 1 million gallons. (Ch. 10, sec. 948, Cahill's Consolidated Laws of New York, 1930) On all rivers and streams in the State recognized by law or used as a public highway for the purpose of floating logs or other timber, no dam shall be erected unless there be built in such dam an apron at least 15 feet in width in the middle of the current of such river of a proper slope for safe passage of logs, timber, etc (Ch. 38, sec 70, Cahill's Consolidated Laws of New York, 1930)
	Department of conservation.	No person shall obstruct the passage of fish in any stream or river by a screen, or otherwise, except as permitted by the department. Flumes and raceways in streams stocked with fish by the State shall be screened as the department may direct. (Ch. 40, sec. 213-2, Laws of 1938) Persons, associations, or corporations impounding water by a dam for power purposes, hereafter constructed, must allow fishing therein by the public which is to be conducted in accordance with rules and regulations of the commissioner, in agreement with the person, association, or corporation proposing to construct the dam (C². 602, Laws of 1938.)
North Carolina	Water power and control commission.	Reservoirs in river regulating districts cannot be constructed of a greater capacity than is required to maintain the average flow (Ch 10, sec 471, Cahill's Consolidated Laws of New York, 1930)
	State board of health.	An application for a permit must be filed with the State board of health to impound water or raise the level of existing impounded waters All growths, brush, etc, which are favorable to mosquito larvae production must be removed before impounding of the water. All trees and undergrowths must be cut off at least 1 foot below the water level Flashboards or other means must be provided for controlled fluctuation of the water level. Bottom drains or other means must be provided for removal of the impounded water This regulation does not apply to areas of less than 1 acre of water surface, used exclusively for watering livestock or for other domestic purposes. (Secs. 1, 2, 3, 4, North Carolina State Board of Health Regulations governing the impounding and maintenance of impounded waters of May 5, 1937)	The reservoir must not be filled during the malaria-carrying mosquito-breeding season, April to September, unless authorized specifically in writing by the board of health. Owners are required to use control measures approved by the board of health, they are required to furnish the board of health such reports of operation, maintenance, or condition of the impounded water as may be specified by said board After the impounding of the water the board of health makes occasional inspections, and any conditions found to be detrimental to the public health, must be remedied by the permit holder (Secs 6, 7, 8, North Carolina State Board of Health Regulations governing the impounding and maintenance of impounded waters of May 5, 1937)
	State board of agriculture.		Operators of dams must provide fishways and sluceways for the passage of fish (Secs 1974, 1975, and 1976, North Carolina Code of 1935)
North Dakota	Water conservation commission	An application together with plans and specifications must be submitted for approval before construction of any dam more than 10 feet in height or which is capable of impounding more than 30 acre-feet of water. This section does not apply to any works constructed by or under the supervision of the United States or any of its officers or employees. (Ch. 255, sec. 9, Laws of 1937) All appropriators of water of the natural flow of streams are required to maintain headgates and measuring devices, at their respective points of diversion. (Ch. 255, sec. 18, Laws of 1937)
	Fish commissioner		Persons owning or operating dams or other obstructions across any river, creek or stream are required to construct, keep in repair and maintain durable and efficient fishways. If any owner fails to construct or repair a fishway after 10 days' notice from the fish commissioner to do so, the commissioner may construct or repair same at the owner's expense (Sec 10282, Compiled Laws of North Dakota, 1913)

State	Design and construction		Operation	
	Supervising agency	Laws or regulations (reference shown in parentheses)	Supervising agency	Laws or regulations (reference shown in parentheses)
Ohio	Superintendent of public works	A petition must be filed with the superintendent of public works for the construction of dams or reservoir projects used in connection with the construction of any highway, highway bridge or culvert (Secs 412-16, 412-17, 412-18, Throckmorton's 1934 Ohio Code)	County commissioners	Upon petition of 5 freeholders of a county, the county commissioners thereof must erect, maintain and keep open for the free passage of fish a sufficient passageway or chute over dams across rivers and creeks (Sec 2496, Throckmorton's 1934 Ohio Code)
	Board of public works.	Persons desiring to construct any dam or other obstruction connecting with any canal of the State which is navigable and used by boats and vessels must obtain a permit from the board of public works authorizing such construction (Sec. 12503, Page's Annotated Code of Ohio of 1926)	Secretary of agriculture	The secretary of agriculture has general supervision over the fish in the public lakes and reservoirs of the State (Sec. 1446, Page's Annotated Code of Ohio of 1926)
	State board of health	Plans must be submitted to and be approved by the State board of health before construction or alteration of any public water supply works (Sec 1240 Throckmorton's 1934 Ohio Code)		
Oklahoma.	Conservation commission	All dams more than 10 feet in height must be approved by the conservation commission (Letter Jan. 18, 1937 from chief engineer, Conservation Commission of Oklahoma)	Conservation commission	Conservation commission has power to prevent the contamination or pollution of all reservoirs, ponds, lakes, and other bodies of water in State. (Art 3, Ch 70, Sec. 5 (b), Session Laws of Oklahoma 1935)
	State board of health.	A permit must be obtained from the State board of health for construction of water-supply projects used for domestic purposes (Sec. 49, p 15, Engineering Bulletin No 2 of Oklahoma State Health Department, June 1932)	State game and fish commission	Owners of dams are required to provide suitable fish ladders or runways for fish to pass over the dam. (Sec. 49, p 14, Engineering Bulletin No 2 of Oklahoma State Health Department, June 1932; Title 29, sec. 274, Oklahoma Statutes Annotated)
Oregon	State engineer	An application, together with plans, must be submitted to the State engineer before the construction of works for the appropriation of waters of the State (Secs 47-502, 47-503, Oregon Code, 1930) The State engineer must approve all dams and hydraulic structures the failure of which would result in damage to life and property. (Sec 47-702, Oregon Code, 1930)	Master fish wardens	Master fish wardens make inspections of dams and have authority to order passageways to be constructed where none exist (Sec 40-215, Oregon Code Supplement, 1935)
	State board of health.	State board of health must approve all projects for public water supplies before beginning of construction. (Sec. 59-1102, Oregon Code, 1930) Actual construction work, except by municipal corporations, must begin within 1 year from date of approval of application (Sec 47-506, Oregon Code, 1930)	Game commission	Persons operating canals or millraces receiving water from a river or creek in which fish have been placed are required on order of game commission to place a screen or grating over the inlet to prevent fish from entering (Sec. 39-426, Oregon Code, 1930)
			Water masters	Water masters divide the waters in their districts among the several ditches and reservoirs taking waters therefrom and control the headgates and controlling works of reservoirs in time of scarcity of water. (Sec 47-310, Oregon Code, 1930)

State	Agency	Provisions
Pennsylvania	County courts	The county courts have jurisdiction to grant permits for the building of dikes and dams for the protection of lands from overflowing. (Sec 34-401, 402, Oregon Code, 1930)
	Fish commission	An application must be made to the fish commission to secure permission to use explosives, gas, or lime for removing any obstructions from the foundation for dams and other structures. (Sec. 40-213, Oregon Code Supplement, 1935)
	Water and power resources board (departmental administrative board in the department of forests and waters)	An application together with plans and specifications must be filed for a permit with the water and power resources board for the construction or repair of any dam or similar structure on a drainage area of more than ½ square mile, or if its construction will imperil life or property, except dams not exceeding 3 feet in height in streams not more than 50 feet in width, built to create pools for fish and fishing purposes (Acts of June 25, 1913, P. L. 555, amended by act of May 6, 1937, Act No. 137, and June 25, 1931, P L 1371) A filing fee of $5 is required and an annual charge for inspection and investigation is made for dams over 5 feet in height, except to Federal, State, county, or municipal authorities (Regulations of Water and Power Resources Board) A limited power permit, subject to an annual charge, must be secured from the water and power resources board for the construction of any power dam for the development of water power or the supply of water for steam purposes. (Act of June 14, 1923, P. L 704)
	Department of health	A reservoir site which is to be used for public water supply must be cleared of all stumps, brush, and debris. The sides of the reservoir site from the level of the spillway down to a distance equal to 5 feet vertical shall be stripped of topsoil containing any considerable proportion of vegetable matter. Grass and weeds on the remainder of the reservoir area shall be burned prior to filling the reservoir. (Regulations of State Department of Health)
	Board of fish commissioners.	The owner of a dam must install a fishway upon written order of the board of fish commissioners, if fishway is impracticable, the board has authority to enter into an agreement for annual payment to board to be used for stocking pool above dam (Fish law of 1925, sec. 185, as amended Apr 22, 1929)
	Water and power resources board (departmental administrative board in the department of forests and waters)	A permit holder must notify the board at least 1 week in advance when it is proposed to begin the storage of water in the reservoir The board exercises general powers as to storage, discharge, and level of reservoir The permit holder building a dam for power or industrial purposes is required to permit fishing in the reservoir by holders of fishing licenses (Regulations of Water and Power Resources Board)
	Department of forests and waters.	Upon complaint or upon own initiative, the department of forests and waters has power to investigate and examine dams and other obstructions to determine safety and order repairs or removal (Act of June 25, 1913, P L 555, Administrative Code of 1929, Sec 1804, amended by act of May 6, 1937, Act No 137)
	Board of fish commissioners.	No person responsible for a dam holding back waters inhabited by fish shall draw off water without written permission from the board of fish commissioners, nor obstruct the flow of water through such dam without allowing sufficient water to flow in the stream at all times to enable fish to live Dams, retards, or similar devices placed across streams with permission of owners of adjacent land shall not be destroyed or disturbed without written permission of the board of fish commissioners (Fish law of 1925, sec. 191, as amended by act of May 29, 1935)

State	Design and construction		Operation	
	Supervising agency	Laws or regulations (reference shown in parentheses)	Supervising agency	Laws or regulations (reference shown in parentheses)
Rhode Island	Division of harbors and rivers.	Plans and specifications for the construction or alteration of any dam or reservoir must be filed with and approved by the division of harbors and rivers before the work is begun (Sec 4, Ch 180, General Laws of Rhode Island of 1923, Ch 2250, sec. 64, Public Laws of Rhode Island of 1935.)	Division of harbors and rivers.	The division of harbors and rivers is authorized to make inspection of all dams and reservoirs in the State as often as it deems necessary to insure their safety (Sec. 2, Ch 180, General Laws of Rhode Island of 1923, Ch. 2250, Sec. 64, Public Laws of Rhode Island of 1935.) Upon written application of a property owner or the mayor or board of aldermen of a town that a dam is unsafe the division makes examination of such dam, and if it is found unsafe it may order the water drawn off in whole or in part and can also order the dam to be repaired. (Sec. 5, Ch. 180, General Laws of Rhode Island of 1923, Ch 2250, Sec. 64, Public Laws of Rhode Island of 1935.)
			Superior court	No person owning any (mill) dam on any river or stream can detain the natural stream thereof, at any one time more than 12 hours out of 24, except on Sundays, whenever he shall be requested by the owner of any dam within 1 mile below on the same stream to suffer the natural run of such river or stream to pass his said dam. (Sec. 16, Ch. 178, General Laws of Rhode Island of 1923.)
South Carolina	State board of health.	An application, together with a plat showing the maximum and minimum water level, must be filed with the State board of health to impound water for any purpose. All growths, brush, etc, which are favorable to mosquito larvae production must be removed before impounding of the water All trees and undergrowths must be cut off at least 1 foot below the water level These regulations do not apply to stock watering projects. (Secs 1, 3b, 3c of Regulations Governing the Control of Mosquito Production on Impounded Waters)	State board of health.	During the mosquito-breeding period the permit holder is required to remove all floatage and debris which produce mosquitoes, he is also required to apply such larvacides as are approved by the State board of health. After the impounding of water, the State board of health makes occasional inspections, and any conditions found to be detrimental to public health must be remedied by the permit holder. When practicable, the water level from Nov. 1 to Apr. 30 is required to be 3 feet above that level which is the normal high-water level for the period from May 1 to Oct. 30. (Secs. 3e, 3f, and 5b, Regulations Governing the Control of Mosquito Production on Impounded Waters)
			County commissioners.	The owners of dams constructed for the purpose of reservoirs which have an insufficient wasteway and are inadequate to sustain the weight of the water against the same are required to cause the same to be enlarged or strengthened on recommendation of county freeholders. (Sec. 6092, Code of 1932 of South Carolina)

State	Agency	Provisions	Agency	Provisions
South Dakota	State Engineer	There is no law directly relating to the construction of dams in the State of South Dakota However, there is a law which provides that before construction of any works for the appropriation of any waters of the State, an application must be filed with the State engineer therefor. (Ch 2, Secs. 8220–8271, South Dakota Compiled Laws of 1929)	State game warden	Owners of dams or other obstructions across any river or stream are required to construct fishways. (Sec 10525, South Dakota Compiled Laws, 1929) Owners of mill races receiving waters from any river are required to place screens over the inlet to keep fish from entering the same. (Sec. 10526, South Dakota Compiled Laws, 1929)
Tennessee	State board of health	An application for a permit must be filed with the State department of health before construction of any works to impound water or raise the level of existing impounded waters All growths, brush, etc., which are favorable to mosquito larvae production must be removed before impounding of the water. All trees and undergrowths must be cut off at least 1 foot below the water level Flashboards or other means are required to be provided for controlled fluctuation of the water level at any season of the year in the manner and amount as specified by the State department of health Bottom drains or other means must be provided in the construction of the project which will permit removal of the impounded waters These regulations do not cover waters of less than 1 acre of water surface used exclusively for watering live-stock or for other domestic purposes (Rules and Regulations of the State Department of Health Covering the Impounding and Maintenance of Impounded Waters)	State board of health	The State board of health makes occasional inspections and any conditions found to exist which are conducive to the production of malaria-carrying mosquitoes must be abated by the permit holder. The permit holder is required to furnish to the State Board of Health such reports of operation, maintenance, or condition of the impounded water as may be required by said board. (Rules and Regulations of the State Department of Health Covering the Impounding and Maintenance of Impounded Waters)
	Division of game and fish.	Persons owning dams are required to construct fish ladders for the passage of fish. This section does not apply to locks and dams constructed across large rivers to facilitate navigation, nor to dams constructed for the purpose of generating electricity for distribution and sale, where the height or type of the dam makes it impracticable for the passage of fish over same. (Sec 5193 (46), Michie's Code of 1938)		
Texas	Board of water engineers	An application for a permit must be filed with the board of water engineers before commencing the construction, enlargement, or extension of any dam, reservoir, or other works Plans must be submitted and approved before permit is granted No permit is necessary for a dam or reservoir that is built entirely on one's own property which impounds less than 500 acre-feet of water Board has authority to inspect any such construction The owner is required to commence construction at a time fixed by board, not to exceed 2 years from granting of the permit. (Arts. 7492, 7494, 7496 (last 5 lines), 7500, 7514, 7536, Vernon's Annotated Texas Statutes, 1925)	Game, fish, and oyster commissioner.	Owners of dams or other structures on regular flowing streams are required to construct and keep in repair fish ladders at such dams upon written order of game, fish, and oyster commissioner (Art. 951a, Revised Criminal Statutes of Texas, 1925)

State	Design and construction		Operation	
	Supervising agency	Laws or regulations (reference shown in parentheses)	Supervising agency	Laws or regulations (reference shown in parentheses)
Texas	State reclamation engineer	Approval of plan is required by State reclamation engineer for any levee or other such improvement on or along any river which is subject to floods or freshets, so as to control, regulate, or otherwise change the flood waters of such stream: *Provided,* That provisions of this section shall not apply to dams, canals, or other improvements made or to be made by irrigation, water improvements, or irrigation improvements. (Art 8028, Vernon's Annotated Texas Statutes, 1925)		
Utah	State engineer	Plans, drawings, and specifications in duplicate must be filed and approval thereof secured by the State engineer for the construction and repair of dams 10 feet or more in height, or for any dam of less than 10 feet which will impound more than 100 acre-feet of water While the bureau of reclamation does not have to secure the approval of the State engineer of its plans, drawings, and specifications, they are required, nevertheless, to file these plans, drawings, and specifications, in duplicate, with the State engineer (Sec 100-5-5, Water Laws of Utah of 1937, pp 44-45)	State engineer	The State engineer has power to make examination of any dam at any time Any person may apply to the State engineer requesting an examination of any dam Upon finding that such dam is in an unsatisfactory condition, he may limit the amount of water to be stored thereby; he may order the release of all or any part of the water impounded, and he may regulate future storage or forbid it entirely until the dam is repaired or reconstructed pursuant to the provisions of sec 100-5-5 To prevent waste, loss, pollution, or contamination of any waters, the State engineer may require the repair or construction of headgates or other devices (Sec 100-5-6 and sec 100-5-11, Water Laws of Utah of 1937, pp 45, 47)
Vermont	Public service commission.	An application, together with plans and specifications, must be filed with the public service commission to secure its approval to construct dams that impound more than 500,000 cubic feet of water in any stream or at the outlet of any body of water (Sec 6122, Public Laws of Vermont, 1933) All lands which may be overflowed by damming, storing, or diverting, or raising the level or increasing the area of the volume of the waters of any stream, river, lake, or pond, shall prior to such flowing be cleared of all trees and bushes. (Sec 6469, Public Laws of Vermont, 1933)	Public service commission	The public service commission has authority to make investigation upon petition of 10 taxpayers that a dam is unsafe or is a menace If the dam is found to be unsafe, the commission may require the owner to repair such dam. (Sec 6127, Public Laws of Vermont, 1933)
			Commissioner of fish and game	Authority is needed from the commissioner of fish and game to prevent the passing of fish in streams or the outlet or inlet of a natural or artifical pond by means of a rack, screen, weir, or other obstruction (Sec. 5667, Public Laws of Vermont, 1933)

State	Agency	Provision	Agency	Provision
Virginia	State corporation commission	An application, together with plans, maps, and other necessary data, must be filed with the State corporation commission to secure permission to construct all dams; a copy of said application must also be filed within 10 days thereafter with the State commission on conservation. (Sec. 3581 (2), (3), Virginia Code, 1936.)	Commissioner of game and fish.	Persons owning dams which may interfere with the passage of fish during the months of March, April, May, and June are required to keep and maintain fish ladders, no fish ladders are required for dams 20 feet or more in height or on any others which the commissioner deems unnecessary. (Sec. 3305 (42), Virginia Code, 1936.)
Washington	State commission on conservation			
	State supervisor of hydraulics	Plans and specifications must be submitted to the State supervisor of hydraulics for his examination and approval as to safety before any dam or controlling works are constructed for the storage of 10 acre-feet or more of water (Sec 7388, Remington's Revised Statutes of Washington, 1932) A permit must be obtained from the State supervisor of hydraulics for the storage of water (Sec 7390, Remington's Revised Statutes of Washington, 1932)		
	State board of health	Plans must be submitted to and approval thereof secured from the State board of health before any public water supply may be constructed or before any alterations may be made which will affect the quality of a public water supply (Book V, Pt 2 of the Rules and Regulations of the State Board of Health)		
	Fish Commissioner	Persons erecting or managing dams across any river, creek, or stream must construct, maintain, and keep in repair fishways and fish ladders (Secs 5730, 5731, 5963, Remington's Revised Statutes of Washington, 1932)		
West Virginia	Public service commission	Dams in excess of 10 feet constructed in any stream or watercourse must first be passed upon by the public service commission In all navigable and floatable streams, provision must be made for the passage of boats and other craft and logs Fish ladders must be provided for the passage of fish (Sec. 5968, Michie's Code of West Virginia, 1932) Dams located across streams which are floatable must be provided with chutes, booms, and sluices, upon plans to be approved by the public service commission Dams that are more than 30 feet high need not provide passage for fish (Sec 3234, Michie's Code of West Virginia, 1932)	Municipalities, public service corporations	Any municipality or public service corporation that is authorized to supply water to a municipality is authorized to purchase or condemn and take water from the reservoirs constructed by any licensee (See 3236, Michie's Code of West Virginia, 1932)

State	Design and construction		Operation	
	Supervising agency	Laws or regulations (reference shown in parentheses)	Supervising agency	Laws or regulations (reference shown in parentheses)
Wisconsin	Public service commission.	An application, together with flowage map, plans, and specifications, must be filed with the public service commission to secure permission to construct, operate, and maintain a dam on waters of the State navigable for any purpose whatsoever Plans and specifications must be approved by the commission for construction on nonnavigable streams also The commission has power to order any existing or future dam to be equipped with slides and chutes for the passage of logs and timber, to provide locks, boat hoists, or other devices to accommodate navigation, to provide fishways, spillways or floodgates Dams must be constructed within 5 years from the date of the permit unless the time is extended by the commission (Chs 30 01 (2) and 31 01-31 34, Wisconsin Statutes, 1933)	Public service commission Water regulatory board.	The public service commission has power to examine any dam about which a complaint has been made of its unsafety The commission makes examinations at least once each year of all dams having a theoretical capacity of 750 horsepower or more and which is maintained and operated in or across navigable waters. (Ch 31 19, Wisconsin Statutes, 1933) The board has power and supervision over the operation, repair and maintenance of dams and dykes, constructed across drainage ditches and streams in drainage districts. (Ch. 31 36, Wisconsin Statutes 1933, Ch 379, Laws of 1937)
Wyoming	State engineer	An application must be filed with the State engineer and a permit secured before commencement of construction of any reservoir intended for storing the unappropriated waters of the State Duplicate plans must be submitted to the State engineer for his approval before construction of a dam across the channel of a running stream, above 5 feet in height, or of any other dam intended to retain water above 10 feet in height (Ch 122, Secs 1401 and 1502, Wyoming Revised Statutes, 1931)	State game and fish commissioner Division water superintendents	The State game and fish commissioner has power to order the erection of fishways in obstructions on streams (Ch. 49, Sec 203, Wyoming Revised Statutes, 1931) Division water superintendents have power to regulate and control the storage and use of waters under all permits approved by the State engineer; they further have power to close the headgates of ditches and reservoirs In time of scarcity of water, they have power to shut and fasten the headgates and controlling works of reservoirs (Ch 122, Secs. 203 and 303, Wyoming Revised Statutes, 1931)

APPENDIX I

AVERAGE AND MAXIMUM RIVER DISCHARGES [1]

Various methods were outlined in chapter 2 for approaching the problem of maximum flood flows to be expected, corresponding to given periodicities or frequencies. Deficiencies in available rainfall data west of the one hundred and third meridian seemed to require recourse to empirical flood formulas as the basis of the comparison and estimate. Accordingly, a map of the United States was prepared, on which were recorded the maximum ratings attained on the Myers scale during periods of record or observation. It was deemed advisable to publish also the list of maximum flood flows represented by the percentage ratings on the above-described map (fig. 4 in ch. 2) in order to furnish essential information as to drainage areas, peak and average discharges, and dates of maximum flow.

The main criteria of eligibility for inclusion in this list were (1) length of stream-flow record, (2) permanency of the gage control, and dependability of record, and (3) outstanding maxima recorded or observed.

Special efforts were made to secure at least two gaging-station records on each river and on each main tributary in order to provide comparisons between head-water and lower valley stations both as to peak flow and time required for flood travel.

In general, the same sequence has been followed for listing gaging stations as prevails in the United States Geological Survey water supply papers; namely, from headwaters proceeding downstream along each main channel, thence returning for the principal lateral tributaries.

In many instances, maximum gage heights were recorded for which no official estimates of discharge had been published. It seemed incumbent in a number of such cases to derive appropriate values for discharge by extension of rating curves for the respective gaging stations, or else to omit the item. Figures representing such unofficial estimates are enclosed in parenthesis; likewise for drainage areas heretofore undetermined. A number of these, particularly along the Snake River system and the Rio Grande Valley, were derived from the best maps available.

In a number of instances the floods of the early months in 1938 superseded maxima previously recorded on the map. In such cases, the new values were added both on the map and in the tabulations to afford comparisons between the previous and the newly recorded maxima.

---

[1] By C. S. Jarvis, Soil Conservation Service.

78961°—39——27

Average and maximum river discharges [1]

Item No	River and station	Drainage area (square miles)	Years of record — Average	Years of record — Maximum	Years of record — Final	Discharge in cubic feet per second — Average	Discharge in cubic feet per second — Maximum	Discharge in cu ft. per sec. per sq. mi. — Average	Discharge in cu ft. per sec. per sq. mi. — Maximum	Date of maximum	Myers rating (percent)
1	St Croix, Baileyville, Maine	1,320	19	20	1935	2,098	23,300	1.59	17.65	May 1923	6.4
2	Machias, Whitneyville, Maine	465	21	24	1935	997	11,100	2.14	23.9	September 1909	5.2
3	West Branch Union, Amherst, Maine	139	16	16	1935	273	2,560	1.96	18.4	April 1934	2.2
4	West Branch Penobscot, Millinocket, Maine	1,910	34	35	1936	3,039	36,600	1.59	19.2	March 1936	8.4
5	West Branch Penobscot, near Medway, Maine	2,120	19	19	1935	3,669	24,100	1.73	11.4	May 1928	5.2
6	Penobscot, West Enfield, Maine	6,600	28	35	1936	11,650	153,000	1.76	23.2	May 1923	18.9
							125,000		18.9	March 1936	15.4
7	East Branch Penobscot, Grindstone, Maine	1,070	28	33	1935	1,863	35,100	1.74	32.4	April 1923	10.6
8	Piscataquis, near Dover-Foxcroft, Maine	286	32	33	1936	653	21,700	2.28	75.9	September 1909	12.8
							19,300		67.6	March 1936	11.4
9	Piscataquis, near Medford, Maine	1,170	11	12	1936	2,240	50,200	1.92	42.9	March 1936	14.7
							35,000		29.9	September 1932	10.2
10	Sebec River, Sebec, Maine	344	10	11	1935	608	11,400	1.77	33.0	March 1936	6.1
							4,780		13.9	April 1934	2.6
11	Pleasant River near Milo, Maine	322	15	16	1936	682	24,400	2.12	75.8	April 1923	13.6
							23,400		72.7	March 1936	13.1
12	Passadumkeag, Lowell, Maine	301	20	20	1935	495	5,680	1.64	18.9	May 1923	3.3
13	Kennebec, Moosehead, Maine	1,240	16	17	1936	1,810	13,600	1.46	11.0	May 1929	3.9
							10,200		8.2	April 1936	2.9
14	Kennebec, The Forks, Maine	1,570	30	30	1936	2,512	23,700	1.60	15.1	June 1917	6.0
15	Kennebec, Bingham, Maine	2,710	8	9	1936		55,200		20.4	March 1936	10.6
16	Kennebec, Waterville, Maine	4,270	42	43	1936	7,065	157,000	1.66	36.8	December 1901	24.1
							154,000		36.1	March 1936	23.6
17	Dead River, The Forks, Maine	878	25	26	1936	1,382	28,700	1.58	32.7	March 1936	9.7
							23,800		27.1	April 1923	8.1
18	Carrabassett, North Anson, Maine	351	10	10	1936	675	24,100	1.92	68.7	March 1936	12.9
							20,100		57.1	September 1932	10.7
19	Sebasticook, near Pittsfield, Maine	598		8	1936		14,400		24.0	March 1936	5.9
							9,400		15.8	April 1934	3.9
20	Cobbosseecontee, Gardiner, Maine	220	45	46	1936	319	4,320	1.45	19.6	May 1922	2.9
							4,250		19.3	March 1936	2.9
21	Androscoggin, Rumford, Maine	2,090	39	44	1936	3,483	74,000	1.66	35.4	April 1895	16.2
							55,200		26.4	March 1936	12.1
22	Androscoggin, near Auburn, Maine	3,257		8	1936		135,000		41.5	April 1933	23.7
							45,400		14.0	June 1917	8.0
23	Magalloway, Aziscohos Dam, Maine	233	23	24	1936	488	4,660	2.09	20.0	May 1936	3.1
							2,430		10.4	September 1932	1.6
24	Swift River, near Roxbury, Maine	95		7	1936		13,000		137	March 1936	13.4
							10,500		111		10.8

No.	Stream and locality	Drainage area			Year	Discharge				Date
25	Little Androscoggin, near South Paris, Maine	76 2	14	15		139	6,980	1 82	91 8	March 1936
26	Presumpscot, Outlet of Sebago Lake, Maine	436	48	49	1936	635	3,540	1 45	46 4	April 1920
27	Saco River near Conway, N H	386		15	1936		7,000; 3,790		16 0; 8 7	April 1902; April 1936
28	Saco River, Cornish, Maine	1,298	19	20	1936	2,540	40,600; 24,000	1 96	105; 62 2	November 1907; March 1936
29	Saco River, West Buxton, Maine	1,572	24	26	1936	2,959	51,300; 23,000	1 88	39 5; 17 7	March 1936; May 1923
30	Ossipee, Cornish, Maine	453	19	20	1936	847	58,200; 27,800	1 84	37 0; 17 0	March 1936; May 1923
31	East Branch Pemigewasset, near Lincoln, N H	104		8	1936		17,200; 7,440		38 0; 16 4	March 1936; April 1923
32	Pemigewasset, Plymouth, N H	622	32	33	1936	1,354	17,000; 8,000	2 17	164; 77 0	March 1936; May 1929
33	Pemigewasset, Bristol, N H	746		9	1936		65,400; 60,000		105; 96 0	March 1936; November 1927
34	Merrimack, Franklin Junction, N H	1,507	30	31	1936	2,732	71,400; 62,300; 83,000	1 81	96 0; 83 4; 55 0	March 1936; November 1927; November 1927
35	Merrimack, Garvins Falls, N H	2,427	11	9	1936		63,000; 122,000		41 8; 50 2	March 1936; November 1927
36	Merrimack, Manchester, N H	2,854		12	1936	4,366	64,600; 144,000	1 53	26 6; 50 6	March 1936; November 1927
37	Merrimack, below Concord Junction, Lowell, Mass	4,424	12	13	1936	6,780	70,300; 173,800; 76,800	1 53	24 7; 39 0; 17 3	March 1936; November 1927; November 1927
38	North Branch Contoocook near Antrim, N H	54 8	11	12	1936	93 6	108,000; 6,160	1 71	24 4; 113	April 1852; March 1936
39	Contoocook, Penacook, N H	766		8	1936		2,370; 46,800		43 5; 61 0	April 1933; March 1936
40	Blackwater near Webster, N H	129		18	1936		17,600; 17,000		23 0; 132	April 1933; March 1936
41	Suncook, North Chichester, N H	157	12	13	1936	222	2,929; 12,900	1 41	22 6; 82 0	April 1934; March 1936
42	Souhegan, Merrimack, N H	171	26	27	1936	279	12,900; 16,580	1 63	41 8; 98 8	April 1923; March 1936
43	North Nashua near Leominster, N H	107 8	39	40	1936	191	6,900; 16,300	1 78	54 0; 152	April 1924; March 1936
44	South Branch Nashua, Clinton, N H	108 8			1936		11,100; 9,550		101; 87 5	March 1936; February 1900
45	Lake Cochituate, Outlet Cochituate, Mass	17 58	72	73	1936	26 5	218; 240	1 51	14 7; 13 7	March 1936; March 1900
46	Blackstone, Worcester, Mass	31 3	12	13	1936	50	2,520; 1,020	1 60	80 7; 32 6	March 1936; January 1935
47	Connecticut near Pittsburgh, N H	83	18	19	1936	209	1,810; 1,320	2 52	21 8; 15 5	May 1930; March 1936
48	Connecticut, North Stratford, N. H	796		6	1936		28,400; 21,700		35 6; 27 2	March 1936; April 1934
49	Connecticut, South Newbury, Vt	2,825	16	18	1936	4,985	77,800; 65,900	1 76	27 5; 23 3	March 1936; November 1927

See footnotes at end of table

Average and maximum river discharges—Continued

Item No	River and station	Drainage area (square miles)	Years of record — Average	Years of record — Maximum	Years of record — Final	Discharge in cubic feet per second — Average	Discharge in cubic feet per second — Maximum	Discharge in cu ft per sec per sq. mi — Average	Discharge in cu ft per sec per sq. mi — Maximum	Date of maximum	Myers rating (percent)
50	Connecticut, White River Junction, Vt	4,098	24	25	1936	7,274	136,000	1 79	33 5	November 1927	21 3
							120,000		29 5	March 1936	18 9
51	Connecticut, Springfield, Mass	9,587		84	1936		281,000		29 4	March 1936	28 8
							188,000		19 7	November 1927	19 3
52	Connecticut, Thompsonville, Conn	9,637		8	1936		282,000		29 8	March 1936	28 8
							190,000		19 8	March 1936	19 4
53	Connecticut, Hartford, Conn	10,480		135	1936		313,000		29 9	May 1854	30 6
							208,000		19 9	November 1927	20 4
							200,000		19 1	November 1927	19 6
54	White, West Hartford, Vt	690	19	20	1936	1,136	120,000	1 65	174	March 1936	45 7
							45,400		65 6	March 1920	17 3
55	Ashuelot, Hinsdale, N. H	420	25	26	1936	634	18,001	1 51	43 0	March 1936	8 8
							16,600		39 5	November 1927	8 1
56	South Branch Ashuelot River, Webb, N H	36 6	14	15	1936	59 5	3,880	1 63	106	April 1934	6 4
							3,560		97 5	April 1933	5 9
57	Millers River near Winchendon, Mass	83 8	18	19	1936	138	5,530	1 65	66 0	July 1915	6 0
							1,610		19 2	March 1936	1 8
58	Millers River, Erving, Mass	370	21	22	1936	619	19,700	1 67	53 2	September 21, 1938	10 2
							6,001		16 2	March 1936	3 1
59	Deerfield, Charlemont, Mass	362	22	23	1936	872	38,200	2 41	105	September 1933	20 1
							32,200		89 0	March 1936	16 9
59a	Deerfield, Charlemont, Mass	362 6		25	1933		56,400		155	November 1927	29 4
60	Middle Branch Westfield, Goss Heights, Mass	52 6	25	26	1936	102	8,400	1 94	160	March 1936	11 6
							8,023		153	September 21, 1938	11 1
61	Westfield River, Knightville, Mass	162	26	27	1936	315	25,700	1 96	159	November 1927	20 2
							16,000		99 0	September 21, 1938	12 6
61a	Westfield, Knightville, Mass	162		29	1938		33,700		208	March 1936	26 5
62	Westfield River near Westfield, Mass	497	21	22	1936	900	48,200	1 81	97 0	March 1936	21 6
							42,590		85 5	November 1932	19 1
62a	Westfield, near Westfield, Mass	497		24	1938		55,500		112	March 1936	25 0
63	Farmington, near New Boston, Mass	92	22	23	1936	179	9,080	1 95	99 0	November 1932	9 5
							6,610		72 0	September 21, 1938	6 6
64	Farmington, Riverton, Conn	216		7	1936		19,900		92 0	November 1927	13 5
							9,720		45 0	March 1936	6 6
65	Farmington, Tariffville, Conn	578		8	1936		22,200		38 4	March 1936	9 2
							7,610		13 1	November 1932	3 2
65a	Housatonic, Coltsville, Mass	57 1			1938		8,293		145	September 21, 1938	12 0
66	Housatonic, Great Barrington, Mass	280	22	23	1936	525	8,990	1 83	32 0	March 1936	5 4
							7,910		23 4	November 1927	4 8

No.	Station	Drainage area (sq. mi.)	(c1)	(c2)	Year	Mean discharge	Max. discharge	Ratio	Per sq. mi.	Date	(b)
67	Housatonic, Falls Village, Conn.	632	23	24	1936	1,029	14,500	1.62	22.9	March 1936.	5.8
							11,700		18.4	November 1927.	8.6
68	Housatonic, Stevenson, Conn.	1,545		8	1936		69,500		45.0	March 1936.	17.0
							23,700	2.02	15.3	March 1934.	6.0
69	Hudson, near Indian Lake, N. Y.	419	19	19	1935	846	13,900	1.98	33.2	April 1922.	6.8
70	Hudson, North Creek, N. Y.	792	28	29	1936	1,568	27,400	1.73	34.5	March 1913.	9.7
							23,000		29.1	March 1936.	8.2
71	Hudson, Hadley, N. Y.	1,664	14	15	1936	2,880	41,300		24.8	March 1936.	10.1
							33,100	1.65	19.9	April 1922.	8.1
72	Hudson, Mechanicville, N. Y.	4,500	48	49	1936	7,423	120,000		26.7	March 1913.	17.9
							86,100		19.1	March 1936.	12.8
72a	Poesten Kill, near Troy, N. Y.	89	12	15	1938	135	16,000	1.52	180	September 22, 1938.	17.0
72b	Wappinger Creek, near Wappinger Falls, N. Y.	182		10	1938		13,200		72.5	September 22, 1938.	9.8
72c	Kinderhook Creek, Rossman, N. Y.	329	10	18	1933	436	27,500	1.32	83.5	September 22, 1938.	15.1
72d	Hoosic, near Eagle Bridge, N. Y.	510		27	1938		55,000		108	September 22, 1938.	24.4
							29,800	1.75	58.5	November 4, 1927.	13.2
							23,500		46.1	March 1936.	10.4
73	Mohawk, below Delta Dam, near Rome, N. Y.	151		25	1936		4,210		27.9	March 1921.	3.4
74	Mohawk, near Little Falls, N. Y.	1,348	17	15	1936	893	23,200	1.63	17.2	March 1936.	6.3
							21,300		15.8	March 1929.	5.8
75	Mohawk, Cohoes, N. Y.	3,456		18	1936	5,625	130,000		37.7	March 1936.	22.1
							72,000		20.8	March 1929.	12.3
76	Passaic, Paterson, N. J.	785	15	38	1935		28,000	1.55	35.7	October 1903.	10.0
77	South Branch Raritan, Stanton, N. J.	147	12	19	1935	228	7,070	1.48	48.0	July 1935.	5.8
78	North Branch Raritan, Milltown, N. J.	190	14	13	1936	282	17,800	1.44	93.8	September 1934.	8.9
79	Raritan, Manville, H. J.	490		19	1936	708	25,000		51.0	October 1903.	11.3
							20,600	1.80	42.0	January 1936.	9.3
80	West Branch Delaware, Hale Eddy, N. Y.	593	22	33	1936	1,068	46,000		77.5	October 1903.	18.9
							26,500		44.8	September 1924.	10.9
81	East Branch Delaware, Fishs Eddy, N. Y.	783	22	23	1936	1,677	53,300	2.14	68.0	August 1933.	19.0
82	Lehigh, Tannery, Pa.	322	16	22	1936	674	21,800	2.10	68.0	March 1936.	12.2
							21,000		65.0	July 1935.	11.7
83	Delaware, Port Jervis, N. Y.	3,076	31	32	1936	5,503	155,500	1.80	50.4	October 1903.	28.0
							92,700		30.3	March 1914.	18.6
							108,000		35.2	March 1936.	19.6
84	Delaware, Riegelsville, N. J.	6,344	29	34	1936	10,770	275,000	1.70	43.4	October 1903.	34.6
							210,000		33.1	March 1936.	26.4
							144,000		22.7	March 1913.	18.1
85	Delaware, Trenton, N. J.	6,796	22	23	1936	11,290	227,000	1.66	33.5	March 1936.	18.1
							160,000		23.0	March 1936.	17.6
86	Schuylkill, Philadelphia, Pa.	1,893	13	67	1936	2,505	130,000		69.0	March 1913.	19.4
							82,000	1.33	43.0	October 1869.	29.9
87	Brandywine, Chadds Ford, Pa.	287	24	25	1936	372	30,500	1.30	106.0	March 1920.	18.0
88	Susquehanna, Conklin, N. Y.	2,240	22	23	1936	3,654	61,600	1.63	27.5	March 1936.	13.0
							52,000		23.2	March 1913.	11.0
							41,900		18.7	July 1935.	8.9
89	Susquehanna, Towanda, Pa.	7,797	17	18	1936	9,937	188,900	1.28	24.1	March 1865.	21.3
							188,000		24.1	March 1936.	21.3
							182,000		23.3	March 1902.	20.6

See footnotes at end of table.

Average and maximum river discharges—Continued

Item No	River and station	Drainage area (square miles)	Years of record — Average	Maximum	Final	Discharge in cubic feet per second — Average	Maximum	Discharge in cu. ft per sec. per sq. mi. — Average	Maximum	Date of maximum	Myers rating (percent)
90	Susquehanna, Wilkes-Barre, Pa	9,960	36	152	1936	13,510	232,000	1.36	23.3	March 1865	23.4
							232,000		23.3	March 1936	23.3
91	Susquehanna, Harrisburg, Pa	24,100	45	196	1936	34,630	213,000	1.43	21.0	March 1902	21.5
							740,000		30.6	March 1936	47.6
							699,000		29.0	June 1889	45.0
							613,000		25.4	May 1894	39.4
92	Susquehanna, West Branch, Bower, Pa	315	22	47	1936	557	32,600	1.76	104	June 1889	18.4
							31,500		100	March 1936	17.7
93	Susquehanna, West Branch, Renovo, Pa	2,975	23	90	1936	4,709	236,000	1.59	79.5	June 1889	43.3
							211,000		71.3	March 1936	38.8
94	Susquehanna, West Branch, Williamsport, Pa	5,682	40	90	1936	8,852	264,000	1.56	46.5	June 1889	35.1
							252,000		44.5	June 1880	33.6
95	Raystown Branch Juniata, Saxton, Pa	756	24	47	1936	925	80,500	1.22	106	March 1936	29.2
							71,300		94.0	June 1889	27.9
96	Frankstown Branch Juniata, Williamsburg, Pa	291	16	126	1936	375	47,600	1.29	164	March 1936	27.5
							35,500		122	June 1889	20.6
97	Frankstown Branch Juniata, Huntingdon, Pa	849		47	1936		80,000		94.3	March 1936	41.0
							60,000		70.5	June 1889	37.2
98	Juniata, Newport, Pa	3,354	34	134	1936	4,408	237,000	1.32	70.6	June 1889	21.1
							215,000		64.2	August 1933	10.7
99	West Conewago Creek, near Manchester, Pa	510	7	8	1936		47,600		93.5	March 1936	25.9
							24,100		47.2	May 1889	30.1
100	Great Cacapon, near Great Cacapon, W Va	670	12	12	1936	554	67,000	.83	100	March 1924	19.1
101	North Branch Potomac, near Cumberland, Md	875		11	1936		89,000		102	October 1896	34.6
102	Shenandoah, Millville, W Va	3,040		21	1936		105,000		34.6	May 1889	49.1
103	Potomac, Hancock, Md	4,070		46	1936		220,000		54.2	March 1936	47.1
104	Potomac, Point of Rocks, Md	9,651	39	47	1936	9,380	480,000	.97	50.2	June 1889	45.0
							460,000		48.0	March 1936	25.4
105	Potomac, near Washington, D C	11,560		7	1937		484,000		41.8	June 1889	9.1
106	Rock Creek, Sherrill Drive, Washington, D C	62.2	10	44	1937		20,000		322	October 1885	7.3
107	Rappahannock, Kellys Ford, Va	641		50	1935	596	23,000	.93	35.9	October 1929	16.6
							18,660		29.0	May 1924	16.2
108	Rappahannock, near Fredericksburg, Va	1,599	28	28	1935	1,640	66,000	1.03	41.3	September 1877	14.3
109	James River, Lick Run, Va	1,369	11	58	1935	1,570	60,000	1.10	43.8	March 1913	14.0
							53,000		38.6	March 1936	10.8
							51,600		47.1	January 1935	20.2
							40,000		29.2	March 1913	16.9
110	James River, Buchanan, Va	2,084	37	41	1936	2,530	92,200	1.21	44.2	March 1936	13.9
							76,400		37.4	January 1935	
							63,400		30.5		

No.	Stream and place	Drainage area (sq. mi.)	Rec.	Yrs.	Year	Mean disch.	Maximum discharge (sec.-ft.)	Gage	Disch. per sq. mi.	Date	(per sq. mi.)
111	James River, Cartersville, Va.	6,242	36	37	1936	7,250	149,000	1 15	23 9	March 1936	18 9
							134,000		21 5	September 1935	17 0
112	Roanoke, Roanoke, Va.	388	36	40	1936		16,900	1 04	41 3	August 1901	8 6
113	Roanoke, Brookneal, Va.	2,420	13	14	1936	402	68,300	94	28 3	August 1928	13 9
							45,700		18 9	January 1936	9 3
114	Roanoke, Roanoke Rapids, N C	8,410		25	1936	2,270	190,000		22 6	March 1912	20 7
							110,000	1 51	13 1	January 1936	12 0
115	Dan River near Francisco, N C	119	11	13	1935	180	8,700	1 05	73 1	December 1924	8 0
116	Dan River, So Boston, Va.	2,730	18	36	1935	2,860	58,200		21 4	January 1936	11 1
							52,600		19 3	December 1901	10 1
117	Cape Fear, Lillington, N C	3,530	12	12	1936	3,290	101,100	93	28 8	October 1929	17 1
118	Cape Fear, Fayetteville, N C	4,290	21	47	1936	4,775	133,000	1 11	31 0	August 1908	20 3
							110,000		25 7	October 1929	16 8
119	Deep River, Ramseur, N C	343	13	13	1936	353	21,100	1 03	61 5	September 1928	11 4
120	Yadkin, Wilkesboro, N C	480	16	16	1936		23,000	1 56	48 0	January 1934	10 5
121	Yadkin, Yadkin College, N C	2,250		8	1936	749	67,900		30 1	October 1929	14 3
							47,900		21 3	January 1936	10 1
122	Pee Dee, near Rockingham, N C	6,910		9	1936		212,000		30 5	September 1928	25 5
							188,000		27 3	April 1936	22 6
123	Catawba, Catawba, N C	1,540		8	1936		81,500		53 0	May 1901	20 4
124	Wateree, near Camden, S C	5,010		13	1936		199,000		39 6	October 1929	28 1
							168,000		33 6	April 1936	23 7
125	Santee, Ferguson, S C	14,800	28	28	1936	18,900	368,000	1 28	24 8	July 1916	30 3
							245,000		16 6	April 1936	20 1
126	Linville, Branch, N C	65	14	14	1936	136	16,800	2 10	258	August 1928	20 8
127	Broad, near Chimney Rock, N C	97		11	1936		20,500		214	August 1928	20 8
128	Broad, near Boiling Springs, N C	815	11	11	1936	1,398	56,800	1 72	69 6	August 1928	9 9
							26,000		31 9	April 1936	31 9
129	Broad, Richtex, S C	4,800		11	1936		228,000		47 5	October 1929	32 9
							157,000		32 8	April 1936	32 8
130	Saluda, Chappells, S. C	1,290		28	1936		(90,000)		70 0	August 1908	25 1
							63,700		49 5	October 1929	17 8
131	Saluda, near Silverstreet, S C	1,570	11	9	1936		83,800		53 3	do	21 0
							63,000		40 3	April 1936	15 6
132	Saluda, near Columbia, S C	2,450		11	1936	3,201	67,000	1 31	27 3	October 1929	13 6
							61,600		25 3	April 1936	12 4
133	Seneca, near Anderson, S C	1,026		8	1936		77,000		75 0	August 1928	24 0

NOTE.—Figures in parentheses are unofficial estimates

No.	Stream and place	Drainage area (sq. mi.)	Rec.	Yrs.	Year	Mean disch.	Maximum discharge (sec.-ft.)	Gage	Disch. per sq. mi.	Date	(per sq. mi.)
134	Savannah, Augusta, Ga.	7,304		19	1932	2,877	350,000		47 9	October 1929	41 0
135	Ocmulgee, Macon, Ga.	2,290	23	28	1936		51,000	1 26	22 2	January 1925	10 7
							50,900		22 2	March 1902	10 7
136	Oconee, Dublin, Ga.	4,350	19	24	1936	5,282	96,700	1 21	22 2	April 1936	14 7
							88,600		20 4	January 1925	13 4
							57,200		19 2	March 1913	8 7
137	St Marys, near Macclenny, Fla.	859		10	1936		16,500		19 2	September 1928	5 6
138	St Johns, near Christmas, Fla.	1,320		10	1936		10,000		7 6	September 1926	2 8
139	North Fork of Black Creek, near Middleburg, Fla.	207		17	1936		18,000		87 0	June 1919	12 5
140	Kissimmee, near Okeechobee, Fla.	3,260		8	1936		20,000		6 1	August 1928	3 5

See footnotes at end of table.

Average and maximum river discharges—Continued

Item No.	River and station	Drainage area (square miles)	Years of record			Discharge in cubic feet per second		Discharge in cu. ft. per sec. per sq. mi.		Date of maximum	Myers rating (percent)
			Average	Maximum	Final	Average	Maximum	Average	Maximum		
141	Peace, Arcadia, Fla.	1,330		24	1936		43,000		32.4	1912	11.8
142	Suwannee, White Springs, Fla.	1,990		11	1936		36,200		27.3	September 1933	9.9
143	Suwannee, Ellaville, Fla.	6,580		9	1936		20,600		10.4	September 1928	4.6
144	Suwannee, Luraville, Fla.	6,900		9	1936		73,000		11.1	August 1928	9.0
145	Suwannee, near Bell, Fla.	9,260		8	1936		66,000		9.6	August 1928	8.0
146	Chattahoochee, West Point, Ga.	3,550	38	40	1936	5,859	134,000	1.65	37.8	December 1919	22.5
147	Chattahoochee, Columbus, Ala.	8,040		8	1936		203,000		25.3	March 1929	22.6
148	Flint, Montezuma, Ga.	2,920		10	1936		102,000		25.7	April 1936	13.9
149	Flint, Albany, Ga.	5,160	25	35	1936	6,451	101,000	1.25	24.0	January 1925	14.2
150	Flint, Bainbridge, Ga.	7,290		16	1936		83,200		11.4	January 1925	9.8
151	Choctawhatchee, near Bruce, Fla.	4,580		7	1936		220,000		48.0	March 1929	32.5
152	Escambia, near Century, Fla.	3,700		7	1936		315,000		85.0	March 1929	51.8
153	Coosa, Childersburg, Ala.	8,390	19	22	1936	14,350	130,000	1.71	15.5	January 1936	14.2
154	Coosa, lock 18, near Wetumpka, Ala.	10,200	10	13	1936	16,340	207,000	1.60	20.3	March 1929	20.5
155	Alabama, near Montgomery, Ala.	15,100		12	1936		209,000		13.8	March 1929	17.0
156	Alabama, Selma, Ala.	17,100	21	51	1936	26,730	196,000	1.56	11.5	February 1936	15.6
157	Alabama, Claiborne, Ala.	22,000		6	1936		177,000		8.3	February 1936	12.4
158	Tallapoosa, Wadley, Ala.	1,660	13	13	1936	2,488	52,800	1.50	31.8	January 1933	13.0
159	Tallapoosa, below Tallassee, Ala.	3,320		8	1936		115,000		34.7	May 1929	20.0
160	Cahaba, Centerville, Ala.	1,050		20	1936		76,200		72.5	February 1936	23.5
161	Tombigbee, Aberdeen, Miss.	2,210		44	1936		33,100		15.0	April 1892	7.1

No.	Station	Drainage area (sq mi)		Yrs	Year		Max flood discharge	Factor	per sq mi	Date	per sq mi
162	Tombigbee, Coatopa, Ala.	15,500	---	8	1936	---	179,000	---	11 5	March 1929	14 4
							145,500		9 3	February 1936	11 6
163	Tombigbee, near Leroy, Ala.	19,100	---	8	1936	---	123,000	---	7 9	March 1935	9 9
							190,000		9 9	April 1929	13 7
164	Black Warrior, lock 17, near Bessemer, Ala.	3,980	---	8	1936	---	134,000	---	7 0	February 1936	8 9
							123,000		6 4	November 1929	7 8
165	Black Warrior, Tuscaloosa, Ala.	4,830	16	23	1936	7,978	133,000	1.65	33 4	February 1936	21 1
							118,000		29.6	April 1900	18 7
166	Sipsey Fork of Mulberry Fork of Black Warrior River, near Sipsey, Ala.	1,020	---	8	1936	---	215,000	---	44.5	March 1935	31 0
							106,000		22 0	February 1936	15 2
167	Locust Fork of Black Warrior, Trafford, Ala.	622	---	6	1936	---	51,400	---	505 0	November 1929	16 1
							50,400		49 4	February 1936	15 8
168	Pearl, Edinburg, Miss.	898	---	8	1936	---	45,500	---	73 0	October 1934	18 2
							28,100		45 2	March 1902	11 3
169	Pearl, Jackson, Miss.	3,100	17	19	1936	3,689	(70,000)	1 19	78 0	March 1935	23 4
							31,400		34 8	April 1902	10 5
170	Strong River, Dlo, Miss.	361	---	8	1936	---	(100,000)	---	32 3	December 1932	18 0
							60,000		19 4	March 1935	10 8
171	Allegheny, Red House, N.Y.	1,690	31	33	1936	2,724	22,900	1 62	63 4	March 1910	12 1
							41,000		24 3	March 1913	10 0
172	Allegheny, Franklin, Pa.	5,982	18	72	1936	9,492	191,000	1 59	32 0	March 1865	24 7
							159,200		26 6	March 1865	20 6
173	Ohio, Pittsburgh, Pa.	19,106	---	174	1936	---	550,000	---	28 8	March 1936	39.8
							440,000		23 0	March 1936	31 8
174	Ohio, Sewickley, Pa.	19,500	---	103	1936	---	574,000	---	29 5	March 1936	41 1
							413,000		21 2	January 1937	29 6
175	Ohio, Cincinnati, Ohio	75,800	---	79	1937	---	950,000	---	12 5	January 1937	34 5
							660,000		8 7	March 1913	30 5
176	Ohio, Louisville, Ky.	90,600	---	104	1937	---	1,100,000	---	12 1	March 1913	36 5
							770,000		8 5	March 1936	25 6
177	Ohio, Paducah, Ky.	202,700	---	63	1937	---	622,000	---	6 9	February 1937	20 7
							1,850,000		9 2	April 1913	41 2
178	Ohio, Metropolis, Ill.	203,000	---	55	1937	(300,000)	1,600,000	1 48	8 0	April 1936	35 6
							1,100,000		5 4	February 1937	24 4
179	Kiskiminetas, Avonmore, Pa.	1,723	29	31	1937	2,999	1,850,000	1 74	9 1	March 1936	41 0
							200,000		116	March 1907	48 2
180	Youghiogheny, Connellsville, Pa.	1,326	27	29	1937	---	110,000	1 88	64 0	March 1908	26.5
							91,700		53 1	March 1936	22 0
181	Monongahela, Charleroi, Pa.	5,213	---	64	1937	2,482	92,500	---	68 0	March 1924	24 8
							86,000		65 0	July 1888	23 7
182	Tygart, Belington, W Va.	390	29	29	1936	824	156,000	2 11	29 9	March 1936	21 6
							138,000		26 5	March 1917	19 2
183	Muskingum, Dresden, Ohio	5,982	15	23	1936	5,881	24,400	.98	51 5	March 1913	10 2
							228,000		38 1	August 1935	29 5
184	Muskingum, McConnelsville, Ohio	7,411	15	23	1936	6,811	100,000	.92	16 8	March 1913	12 9
							270,000		36 4	August 1935	31 4
185	New River, Eggleston, Va.	2,941	22	22	1936	3,870	104,000	1 32	14 0	July 1916	12 1
							152,000		51 7	May 1901	28 0
186	New River near Hinton, W Va.	4,600	13	35	1936	5,531	(190,000)	1 20	41 3	October 1929	28.0
							93,600		20 4		13.8

See footnotes at end of table

Average and maximum river discharges—Continued

Item No	River and station	Drainage area (square miles)	Years of record — Average	Years of record — Maximum	Years of record — Final	Discharge in cubic feet per second — Average	Discharge in cubic feet per second — Maximum	Discharge in cu ft per sec per sq mi — Average	Discharge in cu ft per sec per sq mi — Maximum	Date of maximum	Myers rating (percent)
187	Elkhorn Creek, Keystone, W. Va.	44		35	1936		60,000		1,363	June 1901	90.5
188	Kanawha, Kanawha Falls, W. Va.	8,367	59	59	1936	13,040	270,000	1.56	32.3	September 1878	29.6
189	Greenbrier, Alderson, W. Va.	1,357	39	39	1936	2,088	159,000	1.54	19.0	January 1935	17.4
190	Scioto, near Dublin, Ohio	988	15	23	1936	739	77,500	.75	57.0	March 1918	21.0
191	Scioto, Columbus, Ohio	1,624	15	23	1936	1,326	(80,000)		80.8	March 1918	25.1
192	Scioto, Chillicothe, Ohio	3,847	15	23	1936	3,306	28,500	.82	28.8	May 1933	9.1
193	Miami, Sidney, Ohio	545	22	23	1936	485	138,000	.86	85.0	March 1913	34.3
194	Miami, Dayton, Ohio	2,513	23	23	1936	2,250	46,200		28.4	March 1927	11.5
195	Miami, Hamilton, Ohio	3,639	16	23	1936	3,381	260,000	.89	67.6	March 1913	42.0
196	Loramie Creek, Lockington, Ohio	261	21	23	1936	207	88,770		23.1	February 1929	14.3
197	Mad, near Springfield, Ohio	485	22	23	1936	498	44,000	.90	80.7	March 1927	18.9
198	Mad, near Dayton, Ohio	632	18	23	1936	645	20,700	.93	38.0	March 1913	8.0
199	Kentucky, near Winchester, Lock 10, Ky.	3,990	27	27	1936	5,038	250,000		98.4	March 1913	8.9
200	Kentucky, Lock 6, Warwick, Ky.	5,140	11	11	1936	5,996	(350,000)	.79	96.4	March 1913	50.0
201	Little Wabash, Wilcox, Ill.	1,130	22	22	1936	784	70,300	1.03	19.4	February 1929	58.1
202	Wabash, Bluffton, Ind.	500	13	13	1936	469	25,600	1.02	98.0	March 1913	11.7
203	Wabash, Logansport, Ind.	3,830	13	16	1936	3,326	55,400	1.27	114	March 1913	15.8
204	Wabash, Lafayette, Ind.	7,200	12	15	1936	6,372	75,700	1.17	119	March 1913	25.2
205	Wabash, Mount Carmel, Ill.	28,600		23	1936		68,500		17.2	March 1913	30.1
206	Embarrass, Ste Marie, Ill.	1,540	22	22	1936	1,181	71,400	.69	13.9	January 1930	10.8
207	West Fork White, Muncie, Ind.	233	11	13	1936	260	16,300	.94	14.5	January 1930	10.0
208	West Fork White, near Noblesville, Ind.	800	17	17	1936	799	13,200	.87	26.4	March 1913	4.9
209	West Fork White, Spencer, Ind.	2,910	11	11	1936	3,155	116,000	.87	30.3	January 1930	5.9
210	White, Hazleton, Ind.	11,200		12	1936		61,400		16.0	January 1930	18.7
211	East Fork White, Seymour, Ind.	2,380	13	13	1936	2,088	144,000	.89	20.0	March 1913	9.9
212	East Fork White, Shoals, Ind.	4,900	13	24	1936	5,174	74,600	1.06	10.4	May 1927	17.0
213	South Fork Cumberland, Nevelsville, Ky.	1,260	20	20	1936	2,340	428,000	1.86	15.0	May 1933	8.8
214	Cumberland, Burnside, Ky.	2,010	24	25	1936	3,345	39,000	1.67	25.4	March 1927	25.3
215	Cumberland, Celina, Tenn.	4,890	22	22	1936	7,830	10,500	1.60	45.0	May 1933	10.1
216	Cumberland, Carthage, Tenn.	7,320	14	14	1936	12,060	25,000	1.65	31.4	January 1930	6.9
217	Cumberland, Carthage, Tenn.	10,700	14	14	1936	17,740	56,000	1.66	19.2	December 30, 1926	17.1
218	Cumberland, Clarksville, Tenn.	16,000	12	13	1936	25,510	139,000	1.60	13.5	January 2, 1927	17.7

No.	Stream and place of gaging station	Drainage area (sq. mi.)	Yrs.	Yrs.	Year	Mean discharge	Per sq. mi.	Maximum discharge	Per sq. mi.	Date of maximum	Max. per sq. mi.
219	New River, near New River, Tenn.	312	11	11	1934	623	2.00	70,000	22.4	March 1929	39.6
220	Caney Fork, near Rock Island, Tenn.	1,640	17	24	1936	3,453	2.11	210,000	128.0	March 23, 1929	51.9
221	Caney Fork, near Silver Point, Tenn.	2,100	13	14	1936	3,698	1.76	220,000	105	March 23, 1929	48.0
222	Obey River, near Byrdstown, Tenn.	452	17	17	1936	850	1.88	35,000	77.5	June 1928	16.5
223	French Broad, Calvert, N C	104	12	12	1936	331	3.19	16,100	155	August 15, 1928	15.4
224	French Broad, Blantyre, N. C	296	15	15	1936	923	3.12	26,500	89.5	August 16, 1928	15.4
225	French Broad, Asheville, N. C	949	35	37	1936	2,189	2.30	110,000	33.5	July 19, 1916	35.7
226	French Broad, near Newport, Tenn.	1,860	15	18	1936	2,832	1.53	62,200	33.7	April 8, 1903	14.4
227	French Broad, near Dandridge, Tenn.	4,450	18	35	1936	6,813	1.52	(150,000)		May 21, 1901	22.5
228	Tennessee, Knoxville, Tenn.	8,990	37	37	1936	13,370	1.49	84,500	19.0	April 2, 1920	12.7
229	Tennessee, Loudon, Tenn.	12,300	14	14	1936	19,540	1.58	195,000	21.7	March 1, 1902	20.6
230	Tennessee, Chattanooga, Tenn.	21,400	62	69	1936	38,410	1.79	459,000	13.7	March 28, 1936	15.3
231	Tennessee, Decatur, Ala.	26,300	12	12	1936	43,700	1.66	410,000	21.5	March 11, 1867	31.4
232	Tennessee, Florence, Ala.	30,800	42	42	1936	52,000	1.69	444,000	19.2	March 1, 1875	28.0
233	Tennessee, near Johnsonville, Tenn.	38,500	47	47	1936	63,640	1.65	410,000	19.7	March 29, 1936	16.6
234	Tennessee, near Buchanan, Tenn.	39,700		2	1936			339,000	10.7	January 1, 1927	17.4
235	Little Pigeon, Sevierville, Tenn.	352	15	16	1936	578	1.64	32,000	14.4	April 3, 1936	17.4
236	South Fork Holston, Bluff City, Tenn.	828	36	36	1936	1,237	1.50	(28,000)	10.7	March 19, 1897	25.3
237	South Fork Holston, Kingsport, Tenn.	1,960	11	11	1936	2,763	1.41	45,000	10.7	April 6, 1936	17.8
238	Holston, near Rogersville, Tenn.	3,060	34	34	1936	4,214	1.38	70,900	8.8	March 24, 1897	20.9
239	Little Tennessee, Judson, N C	668		68	1936	1,810	2.71	(160,000)	8.4	April 10, 1936	17.3
240	Little Tennessee, Calderwood, Tenn.	1,870	40	40	1936	4,352	2.34	40,800	91.1	April 13, 1936	16.8
241	Little Tennessee, McGhee, Tenn.	2,470	20	20	1936	6,004	2.44	70,300	33.8	June 1928	17.1
242	Tuckasegee, Bryson City, N. C	673	31	31	1936	1,588	2.36	92,000	23.1	May 23, 1901	9.7
243	Clinch, Cleveland, Va	539	37	37	1936	730	1.46	40,300	23.1	March 26, 1935	10.2
244	Clinch, Speer Ferry, Va	1,131	14	16	1936	1,650		25,000	52.3	January 29, 1918	12.8
245	Clinch, near Coal Creek, Tenn.	2,960	16	16	1936			37,200	61.0	March 10, 1867	28.9
246	Powell, near Arthur, Tenn.	685		9	1936	1,225	1.79	63,400	37.6	February 28, 1902	28.9
247	Hiwassee, Murphy, N C	410	17	17	1936	964	2.35	(37,000)	37.2	April 6, 1936	15.8
248	Hiwassee, near Reliance, Tenn.	1,220	38	40	1936	2,593	2.13	27,800	59.8	April 2, 1920	16.3
249	Hiwassee, Charleston, Tenn.	2,300	10	10	1936	4,763	2.07	23,100	46.4	November 19, 1906	18.5
250	Toccoa, near Dial, Ga.	175	17	17	1936	509	2.90	9,200	52.4	July 9, 1916	15.5
251	Toccoa, near Blue Ridge, Ga.	232	23	50	1936	613	2.65	13,900	60.0	July 9, 1916	10.8
252	Ocoee, Eml, Tenn.	525	23	23	1936	1,266	2.41	29,400	56.0	July 10, 1916	10.1
253	Ocoee, Parksville, Tenn.	600	23	23	1936	1,301	2.17	19,000	31.6	April 6, 1936	11.7
254	Niagara, Buffalo, N. Y.	263,452	20	23	1936	191,000	.74	297,000	1.13		14.2
255	St. Lawrence, Ogdensburg, N. Y.	301,200	31	20	1936	219,000	.73	320,000	1.07		10.6
256	Menominee, Twin Falls, Mich.	1,790	17	31	1936	1,723	.96	16,700	9.34	April 1916	11.4
257	Menominee, below Koss, Mich.	3,790	22	17	1936	3,121	.82	23,200	6.12	April 1916	13.0
258	Pike, Amberg, Wis.	250	23	22	1936	231	.92	2,730	10.9	April 1922	11.6
259	Oconto, near Gillett, Wis.	678	22	24	1936	600	.89	6,470	9.54	April 1922	16.2

See footnotes at end of table.

Average and maximum river discharges—Continued

Item No	River and station	Drainage area (square miles)	Years of record			Discharge in cubic feet per second		Discharge in cu ft per sec per sq mi		Date of maximum	Myers rating (percent)
			Average	Maximum	Final	Average	Maximum	Average	Maximum		
260	Fox, Berlin, Wis.	1,430	38	38	1936	1,105	6,620	0 77	4 64	March 1929	1 8
261	Fox, near Wrightstown, Wis.	6,150	40	40	1936	4,290	20,600	70	3 35	April 1929	2 6
262	Wolf, Keshena Falls, Wis.	812	25	25	1936	782	4,390	96	5 40	April 1922	1 5
263	Wolf, New London, Wis.	2,240	23	23	1936	1,785	15,500	80	6 92	April 1922	3 3
264	Embarrass, near Embarrass, Wis.	395	17	17	1936	289	6,760	73	17 1	April 1922	3 4
265	Milwaukee, Milwaukee, Wis.	661	22	22	1936	414	15,100	.63	22 8	March 1918	3 5
266	Grand, Grand Rapids, Mich.	4,900			1936		53,000		10 8	March 1904	5 9
267	Tittabawassee, Freeland, Mich	2,530	24	30	1936	1,766	24,500	70	9 70	May 1933	7 6
268	Huron, Barton, Mich.	723	22	22	1936	380	5,840	53	8 07	March 1918	4 9
269	Maumee, Antwerp, Ohio.	2,049	14	14	1935	1,568	22,000	77	10 8	January 1930	2 2
270	Maumee, near Defiance, Ohio.	5,530	10	10	1935	3,704	87,000	67	15 8	January 1930	4 8
271	Maumee, Waterville, Ohio.	6,314	14	17	1935	4,320	94,700	.68	15 0	January 1930	11 0
272	Auglaize, near Defiance, Ohio.	2,329	20	20	1935	1,671	38,700	70	16 7	January 1930	8 0
273	Sandusky, near Upper Sandusky, Ohio.	299	14	14	1935	257	6,750	86	22 5	December 1927	8 9
274	Sandusky, near Fremont, Ohio.	1,248	12	12	1935	887	21,000	71	16 8	March 1933	6 0
275	Cuyahoga, Old Portage, Ohio.	405	14	14	1935	434	3,820	1 07	9 40	January 1929	1 9
276	Cuyahoga, Independence, Ohio.	709	13	13	1936		9,780		13 8	January 1929	3 7
277	Genesee, Scio, N. Y.	309	20	20	1936	372	10,600	1 20	34 3	May 1919	6 0
278	Genesee, St. Helena, N. Y.	1,017	28	28	1936	1,198	44,400	1 18	43 7	May 1916	13 9
279	Genesee, near Mount Morris, N. Y.	1,419	26	29	1936	1,564	55,100	1 11	39 0	May 1916	14 6
280	Black, near Boonville, N. Y.	295	25	25	1936	659	10,000	2 23	33 8	March 1913	15 8
281	Black, Watertown, N. Y.	1,876	16	16	1936	3,898	33,900	2 09	18 0	April 1928	8 8
282	Oswegatchie, near Heuvelton, N Y	973	20	20	1936	1,710	15,600	1 76	18 0	January 1930.	5 0
283	West Branch Oswegatchie, near Harrisville, N. Y.	258	20	20	1936	521	6,920	2 02	26 8	January 1930	4 3
284	Raquette, Piercefield, N. Y.	722	28	28	1936	1,285	7,580	1 78	10 5	April 1922	2 8
285	St. Regis, Brasher Center, N. Y.	616	23	24	1936	1,087	16,200	1 76	26 3	March 1914.	6 5
286	West Branch Ausable, near Newman, N. Y.	116	17	17	1936	219	6,200	89	53 5	October 1932	5 8
287	East Branch Ausable, Ausable Forks, N. Y.	198	12	12	1936	309	11,000	1 56	55 5	October 1924	7 8
288	Ausable, near Ausable, Forks, N Y.	448	12	12	1936	687	19,100	1 54	42 6	October 1924	9 0
288a	Ausable River, near Ausable Forks, N. Y.	198			1938		20,800		105	September 22, 1938	14 8
289	Otter Creek, Middlebury, Vt.	628	20	22	1936	954	13,600	1 52	21 6	November 1927	5 4
290	Winooski, Montpelier, Vt.	433	17	22	1936	583	57,000	1 35	131	November 1927	27 4
291	Winooski, near Essex Junction, Vt.	1,079		67	1936		113,000		105	November 1927	34 4
292	Missisquoi near Richford, Vt.	479	16	21	1936	920	45,000	1 92	94 0	November 1927.	20 6
293	Clyde, Newport, Vt.	140	16	23	1936	244	3,900	1 74	27 9	March 1936.	3 3

Hudson Bay and Upper Mississippi River Basins

No.	Stream and place of determination	Drainage area			Year		Max. discharge			Date	
294	St. Mary, near Kimball, Alberta, Canada	497	34	34	1936	845	18,000	1.69	36.2	June 1908	8
295	Red, Fargo, N. Dak.	6,420	34	35	1936	444	7,740	.069	1.21	July 1916	1
296	Red, Grand Forks, N. Dak.	25,500	54	54	1936	2,199	43,000	.086	1.69	April 1897	0
297	Red, Emerson, Manitoba, Canada	34,600	23	23	1936	2,052	46,200	.059	1.33	April 1916	2
298	Red Lake River, Crookston, Minn.	5,320	35	35	1936	936	14,700	.176	2.75	July 1919	7
299	Thief, near Thief Falls, Minn.	1,010	18	18	1936	73.3	4,080	.073	4.05	April 1916	5
300	Pembina, Neche, N. Dak.	2,950	29	17	1936	118	3,870	.040	1.31	May 1904	2
301	Souris, above Minot, N. Dak.	10,270	33	23	1936	133	12,000	.013	1.17	April 1904	0
302	Namakan, outlet of Lac LaCroix, Ontario, Canada	5,165	15	14	1936	2,829	16,700	.55	3.23	May 1927	1
303	Kawishwi near Winton, Minn.	1,300	14	19	1936	777	7,210	.60	5.55	May 1934	3
304	Little Fork, Little Fork, Minn.	1,620	19	14	1936	772	19,300	.48	1.9	April 1916	7
305	Mississippi, Elk River, Minn.	14,500	14	21	1936	4,115	27,000	.284	1.86	April 1916	1
306	Mississippi, St. Paul, Minn.	33,800	21	44	1936	8,997	107,000	.232	2.76	April 1881	2
307	Mississippi, Le Claire, Iowa	83,600	55	63	1936	47,960	250,000	.540	2.82	June 1880	2
308	Mississippi, Keokuk, Iowa	119,000	63	58	1936	60,800	314,000	.510	2.64	May 1888	0
309	Mississippi, Louisiana, Mo.	140,700	53		1935		197,000		1.40	June 1935	8
310	Mississippi, Alton, Ill.	171,500	84 / 92		1936		(300,000) / 255,000 / (218,000) / (350,000)		2.14 / 1.55 / 2.05 / 1.27	—1851 / June 1844 / March 1936	4 / 6 / 5 / 8
311	Minnesota, near Montevideo, Minn.	6,300	27	27	1936	483	22,000	.077	3.50	June 1919	3
312	Minnesota, Mankato, Minn.	14,600	55	33	1936	2,233	65,000	.153	4.45	April 1881	2
313	St. Croix, near Danbury, Wis.	1,550	22	22	1936	1,147	8,480	.74	5.48	April 1916	4
314	St. Croix, near St. Croix Falls, Wis.	5,930	31	31	1936	3,432	35,800	.580	6.05	March 1920	2
315	Apple, near Somerset, Wis.	550	35	35	1925	297	2,280	.540	4.15	June 1905	4
316	Chippewa, near Bruce, Wis.	1,600	22	22	1935	1,292	15,300	.810	9.60	March 1935	7
317	Chippewa, Chippewa Falls, Wis.	5,600	48	26	1936	4,679	96,000	.837	17.2	September 1884	0
318	Red Cedar, near Colfax, Wis.	1,100	22	22	1936	737	78,000	.670	14.0	March 1920	3
319	Red Cedar, near Menomonie, Wis.	1,760	21	21	1935	1,182	21,900	.673	19.9	April 1934	8
320	Black, Neillsville, Wis.	756	31	31	1936	562	40,100	.730	22.8	April 1934	9
321	La Crosse, near West Salem, Wis.	412	22	22	1936	303	37,100	.735	49.0	June 1905	6
322	Root, near Houston, Minn.	1,280	15	15	1936	622	4,780	.487	11.6	September 1928	5
323	Wisconsin, Merrill, Wis.	2,780	34	28	1936	2,589	26,600	.93	20.8	March 1933	4
324	Wisconsin, Knowlton, Wis.	4,520	15	15	1936	3,957	45,000	.88	16.2	July 1912	7
325	Wisconsin, near Nekoosa, Wis.	5,500	22	22	1936	4,866	49,800	.89	11.0	April 1922	9
326	Wisconsin, Muscoda, Wis.	10,300	22	22	1936	8,607	68,500	.83	12.5	March 1935	2
327	Maquoketa, near Maquoketa, Iowa	1,550	23	23	1936	899	72,100	.58	7.0	April 1922	1
328	Rock, Afton, Wis.	3,300	22	22	1935	1,804	21,800	.55	14.1	March 1929	5
329	Rock, Como, Ill.	8,700	22	22	1936	5,319	13,000	.61	3.94	March 1929	3
330	Iowa, Marshalltown, Iowa	1,500	15	15	1936	656	38,100	.44	4.38	March 1916	1
331	Iowa, Iowa City, Iowa	3,230	26	26	1936	1,484	42,000	.46	28.0	June 4, 1918	0
332	Iowa, Wapello, Iowa	12,480	21	21	1936	5,833	36,200	.47	11.2	June 7, 1918	9
333	Cedar, Janesville, Iowa	1,660	17	17	1936	629	67,500	.379	5.4	March 1929	6
334	Cedar, Cedar Rapids, Iowa	6,640	33	33	1936	2,996	27,700	.451	16.7	April 1, 1933	8
335	Skunk, Augusta, Iowa	4,290	21	21	1936	2,094	72,000	.490	10.9	March 19, 1929	8
336	Devils Creek, near Viele, Iowa	143	31	31	1936		44,500 / 50,880		10.4 / 600	June 17, 1930 / June 1905	6 / 7

See footnotes at end of table.

Average and maximum river discharges—Continued

Item No	River and station	Drainage area (square miles)	Years of record — Average	Years of record — Maximum	Years of record — Final	Discharge in cubic feet per second — Average	Discharge in cubic feet per second — Maximum	Discharge in cu ft per sec per sq mi — Average	Discharge in cu ft per sec per sq mi — Maximum	Date of maximum	Myers rating (percent)
	Hudson Bay and Upper Mississippi River Basins—Continued										
337	Des Moines, Des Moines, Iowa	6,180	16	18	1936	1,879	(90,000)	0 305	14 6	May 1903	11 4
338	Des Moines, Keosauqua, Iowa	13,900	23	29	1936	4,785	41,500	.343	6 72	June 1918	5.3
							97,000		6 96	June 1903	8 2
							80,000		5 74	June 1851	6 7
339	Salt, near New London, Mo	2,480	14	14	1936	1,588	58,700	64	23 6	June 1928	11 8
340	Des Plaines, Lemont, Ill	687	21	21	1936	415	5,520	.60	8 0	March 1919	2 1
341	Des Plaines, Riverside, Ill	645	45	45	1936	413	11,620	.64	18 1	May 1, 1909	4 6
342	Illinois, Peoria, Ill	13,480	26	29	1936	16,510	58,300	1 22	4 34	October 1926	5 02
343	Illinois, Beardstown, Ill	(22,500)	16	32	1936	23,680	115,000	1 05	5 1	April 4, 1904	7.7
344	Illinois, at mouth	27,914		31	1936		125,000		4 5	April 1904	7.5
345	Kankakee, Momence, Ill	2,340	21	21	1936	1,729	14,000	74	6 0	January 22, 1916	2 9
346	Sangamon, Monticello, Ill	550	22	26	1936	402	15,400	.73	28 0	October 4, 1926	6 6
347	Sangamon, Riverton, Ill	2,560	22	26	1936	1,705	30,200	67	11 8	October 4, 1926	6 0
348	Kaskaskia, Vandalia, Ill	1,980	22	26	1936	1,372	20,000	69	10 1	October 4, 1926	4 5
349	Kaskaskia, New Athens, Ill	5,220	14	14	1936	4,053	63,100	78	12 1	August 26, 1915	8 7
350	Big Muddy, Plumfield, Ill	753	22	22	1936	721	16,300	96	21 6	February 1, 1916	5 9
	Missouri River Basin										
351	Jefferson, near Silverstar, Mont	7,840	22	22	1936	1,790	19,800	229	2 53	June 1927	2 24
352	Missouri, below Hauser Lake Dam, near Helena, Mont.	16,600	13	13	1936	4,190	33,300	.252	2 00	June 1927	2 58
353	Missouri, Fort Benton, Mont	24,600		55	1936		140,000		5 70	June 1908	8 9
354	Missouri, near Williston, N. Dak	164,500		8	1936		231,000		1 40	April 1930	5 7
355	Missouri, Bismarck, N. Dak	186,400		10	1936		201,000		1 08	March 1928	4 7
356	Missouri, near Mobridge, S. Dak	208,700		8	1936		164,000		.785	March 1929	3 59
357	Missouri, Pierre, S Dak	243,500		7	1936		131,000		.538	July 14, 1935	2 46
358	Missouri, Yankton, S. Dak	279,500		6	1936		(150,000)		.465	July 16, 1935	2 84
359	Missouri, Omaha, Nebr	322,800		8	1936		198,000		616	June 7, 1929	3 51
							(400,000)		1 24	April 5, 1881	7.1
360	Missouri, St. Joseph, Mo	424,300		8	1936		196,000		.465	April 25, 1881	3.01
							(350,000)		.828	April 29, 1881	7.2
361	Missouri, Kansas City, Mo	489,200		93	1937		196,000		1 02	June 16, 1844	3.64
							(500,000)		.520	June 5, 1929	5.4
362	Missouri, Boonville, Mo	505,700	11	11	1936	55,400	254,000	110	755	April 23, 1927	4.3
							381,000		608	June 4, 1935	
							306,000				

No.	Station	Date	Year	(1)	(2)	(3)	(4)
363	Missouri, near Bonnots Mill, Mo.	June 6, 1935	1936	523,400		8	
364	Missouri, Hermann, Mo.	June 1844	1937	528,200		93	
365	Madison, near West Yellowstone, Mont.	June 7, 1935	1936	419	480	23	20
366	Gallatin, near Gallatin Gateway, Mont.	June 1917	1936	810		11	
367	Milk River, Milk River, Alberta, Canada.	June 1932	1936	1,104	264	27	23
368	Marias, Shelby, Mont.	May 1927	1936	2,610		31	
369	Yellowstone, Corwin Springs, Mont.	June 24, 1907	1936	2,630		26	26
370	Yellowstone, near Sidney, Mont.	June 1918	1938	69,450	3,023	2	
370a	Custer Creek, northeast of Miles City, Mont.	June 18, 1935	1933	155			
370b	Beaver Creek, Wibaux, Mont. (Little Missouri Basin).	June 19, 1938		311			
371	Wind, Riverton, Wyo.	June 7, 1929	1936	2,320	1,188	26	26
372	Big Horn, Thermopolis, Wyo.	June 14, 1906	1936	8,080	2,015	31	31
373	Powder, Arvada, Wyo.	July 1923	1936	6,050	500	21	20
374	Cheyenne, near Eagle Butte, S. Dak.	September 1923	1936	24,500		8	
375	Belle Fourche, near Belle Fourche, S. Dak.	May 24, 1933	1936	4,310	407	25	25
376	James, near Scotland, S. Dak.	April 1924	1936	21,550		8	
377	North Platte, Saratoga, Wyo.	March 31, 1929	1936	2,880	1,299	29	29
378	North Platte, Mitchell, Nebr.	June 8, 1909	1936	24,300	1,398	32	16
379	North Platte, North Platte, Nebr.	June 17, 1921	1936	32,000	2,464	41	41
380	South Platte, South Platte, Colo.	June 6, 1909	1936	2,550	398	34	34
381	South Platte, North Platte, Nebr.	June 3, 1935	1936	24,300		21	
382	Thompson, near Drake, Colo.	July 1919	1936	277	188	17	17
383	Cache la Poudre, near Fort Collins, Colo.	June 9, 1923	1936	1,048	421	52	52
384	Republican, Scandia, Kans.	June 1935	1936	23,000	793	14	13
385	Kansas, Topeka, Kans.	June 1935	1936	56,400	3,940	19	19
386	Saline, Tescott, Kans.	June 3, 1935	1936	2,820	157	17	17
387	Smoky Hill, Ellsworth, Kans.	August 1927	1936	7,580	199	26	23
388	Solomon, Niles, Kans.	July 5, 1895	1936	6,770	471	23	23
389	Little Blue, Waterville, Kans.	June 3, 1903	1936	3,440	416	11	10
390	Big Blue, Hull, Kans.	June 2, 1935	1936	4,510	519	14	13
391	Big Blue, Randolph, Kans.	May 1903	1936	9,100	1,191	18	18
392	Delaware, Valley Falls, Kans.	May 31, 1903	1936	922	326	14	14
393	South Grand, near Brownington, Mo.	June 16, 1925	1936	1,660	879	15	15
394	Grand, near Gallatin, Mo.	November 19, 1928	1936	2,250	1,119	15	15
395	Grand, near Sumner, Mo.	July 1909	1936	6,880	4,018	12	12
396	Lamine, Clifton City, Mo.	June 2, 1929	1936	598	456	14	14
397	Marmaton, near Fort Scott, Kans.	June 4, 1929	1936	411	237	11	10
398	Osage, near Quenemo, Kans.	September 18, 1905	1936	1,030	326	14	14
399	Osage, near Ottawa, Kans.	May 19, 1929	1936	1,260	532	21	20
400	Osage, Trading Post, Kans.	May 28, 1935	1936	2,910	1,264	10	10
401	Osage, Osceola, Mo.	April 1927	1936	8,220	5,205	12	12
402	Osage*, near Bagnell, Mo.	June 1844	1936	14,000	9,161	11	11

See footnotes at end of table.

Average and maximum river discharges—Continued

Item No.	River and station	Drainage area (square miles)	Years of record — Average	Maximum	Final	Discharge in cubic feet per second — Average	Maximum	Discharge in cu ft per sec. per sq. mi. — Average	Maximum	Date of maximum	Myers rating (percent)
	Missouri River Basin—Continued										
403	Pomme de Terre, Hermitage, Mo.	655	15	15	1936	632	70,000	0 962	107 0	August 1927	27 3
404	Gasconade, near Waynesville, Mo.	1,680	15	15	1936	1,371	69,000	815	41 0	March 13, 1935	16 8
				21			(85,000)		50 6	August 1915	20 4
405	Gasconade, Jerome, Mo.	2,840	13	16	1936	2,486	76,800	877	27 0	March 13, 1935	14 4
				39			100,000		35 2	January 6, 1897	18 8
406	Gasconade, near Rich Fountain, Mo.	3,180	15	15	1936	2,818	86,000	883	27 0	March 14, 1935	15 2
	Lower Mississippi										
407	Mississippi, St Louis, Mo.	701,000	6	93	1937	156,200	1,300,000	223	1 85	June 28, 1844	15 5
					1933		1,146,000		1 63	May 21, 1892	13 7
408	Mississippi, Columbus, Ky.	921,900	6	56	1937		2,020,000		2 19	April 2, 1912	21 0
							2,010,000		2 18	April 9, 1913	20 9
409	Mississippi, Memphis, Tenn.	932,800	6	66	1933	485,800	1,780,000	528	1 93	February 4, 1937	18 5
							1,630,000		1 77	April 6, 1927	17 0
410	Mississippi, Helena, Ark	941,800	6	66	1937	505,500	2,020,000	536	2 16	February 7, 1937	20 9
					1933		2,040,000		2 09	April 23, 1912	21 0
411	Mississippi, Arkansas City, Ark	1,130,700	6	58	1937	543,300	1,970,000	480	1 91	February 12, 1937	20 5
							1,800,000		1 87	April 22, 1913	18 5
							1,760,000		1 92	April 29, 1927	18 1
412	Mississippi, Vicksburg, Miss.	1,144,500	6	67	1933	569,800	2,160,000	497	1 78	February 16, 1937	19 2
							2,010,000		1 80	April 16, 1912	20 3
							2,060,000		1 60	April 30, 1922	18 9
							1,830,000		1 55	May 1, 1927	19 3
							1,780,000			May 2, 1913	17 1
										April 12, 1912	16 6
413	Mississippi, Natchez, Miss.	1,149,400	6	126	1937		2,050,000		1 78	February 19, 1937	19 1
414	Mississippi, Red River Landing, La.[3]	(1,202,700) / 1,050,000	6	57	1937	497,300	2,210,000	472	1 77	February 18, 1937	19 8
					1933		1,630,000		1 31	June 7, 1929	14 6
							1,460,000		1 17	May 17, 1927	13 0
							1,400,000		1 13	May 15, 1922	12 6
415	Atchafalaya, Simmesport, La[3]	(40,000–) / 193,000	6	53	1937	127,200	592,000	660	3 07	May 25, 1927	13 5
					1933		471,000		2 45	February 27, 1937	10 7
							395,000		2 05	May 9, 1913	9 0
416	Mississippi at Red River Landing and Atchafalaya at Simmesport, combined[3]	1,242,900	42		1933	(857,700)	(2,000,000)	528	1 61	May 1927	17 9
			6		1933	662,900		532			

No.	Stream and place	Drainage area (sq. mi.)	Yrs.	Yrs.	Year	Mean annual discharge (sec.-ft.)	Per sq. mi.	Maximum discharge (sec.-ft.)	Per sq. mi.	Date	Gage height (feet)
417	Old River, Angola, La.	(100,000)			1937	[4]75,800	.758	514,000	5.14	February 22, 1937	16.3
								447,000	4.47	April 25, 1922	14.2
418	Mississippi, Carrollton, La.[5]	1,243,600	6	66	1933	479,300	.472	1,560,000	1.48	May 18, 1927	15.2
		[6](1,050,600)	6	65				1,340,000	1.28	March 1, 1916	13.1
419	Meramec, near Eureka, Mo.	3,800	15	21	1936	2,942	.772	175,000	46.0	February–March 1937	28.4
								64,000	16.8	August 22, 1915	10.4
420	Bourbeuse, Union, Mo.	767	15	21	1936	643	.847	50,000	65.0	April 3, 1927	18.0
								22,500	29.4	August 22, 1915	8.1
421	South fork Obion, near Greenfield, Tenn.	431		7	1936			21,100	48.9	April 3, 1927	10.2
422	North fork Obion, near Union City, Tenn.	490		7	1936			23,600	48.2	January 21, 1935	10.6
423	Obion, Obion, Tenn.	1,880		7	1936			47,000	25.0	January 10, 1930	10.8
424	South fork of Forked Deer, Jackson, Tenn.	574		7	1936			35,800	62.4	January 11, 1930	15.0
425	St. Francis, near Patterson, Mo.	956	15	15	1936	1,100	1.15	(98,000)	103	January 21, 1935	31.6
								79,200	83.0	August 1915	25.6
426	North fork White River, Tecumseh, Mo.	1,180	14	31	1936	1,230	1.04	(80,000)	68.0	March 11, 1935	23.3
								53,000	45.0	July 1905	13.1
427	White, Beaver, Ark.	1,270	13	14	1936	1,654	1.30	65,000	51.2	June 13, 1928	18.6
428	White, Forsyth, Mo.	4,610		9	1936			160,000	34.8	April 16, 1927	18.2
								127,000	27.5	April 11, 1935	23.6
429	White, Clarendon, Ark.	25,750	53	53	1937	34,300	1.33	440,000	17.1	February 7, 1916	18.7
								322,000	12.5	February 7, 1916	20.1
429A	White, Clarendon, Ark.	25,750	6		1933			215,000	8.34	January 23, 1927	13.4
								230,000	8.90	January 27, 1937	14.8
430	Black, Leeper, Mo.	957	15	32	1936	941	.98	(120,000)	125	March 1904	38.7
								78,400	82.0	June 30, 1928	25.4
431	Current, Van Buren, Mo.	1,640	15	32	1936	1,829	1.12	(150,000)	91.5	May 14, 1933	37.1
								86,600	52.8	March 26, 1904	21.4
432	Current, Doniphan, Mo.	2,030	15	32	1936	2,753	1.36	130,000	64.0	March 11, 1935	28.8
								94,400	46.5	March 1904	21.0
433	Jacks Fork, Emmence, Mo.	376	15	15	1936	443	1.18	40,000	106	March 12, 1935	20.6
434	Arkansas, Salida, Colo.	1,210	26	26	1936	632	.524	5,100	4.23	June 13, 1928	1.47
435	Arkansas, Canon City, Colo.	3,090	49	49	1936	735	.238	19,000	6.15	June 16, 1924	3.42
436	Arkansas, near Pueblo, Colo.	4,730	32	44	1934	775	.164	103,000	21.8	August 2, 1921	15.0
								12,400	2.62	June 3, 1921	1.8
437	Arkansas, Nepesta, Colo.	9,130	9	31	1936	624	.132	180,000	19.7	July 22, 1927	18.9
438	Arkansas, La Junta, Colo.	12,200	24	28	1936	268	.022	200,000	16.4	June 4, 1921	18.1
439	Arkansas, Lamar, Colo.	19,800	23	24	1936	298	.015	165,000	8.30	June 4, 1921	11.7
440	Arkansas, Holly, Colo.	25,000	29	29	1936	360	.014	136,000	5.45	June 5, 1921	8.6
441	Arkansas, Syracuse, Kans.	25,500	15	19	1936	341	.013	(50,000)	1.96	October 20, 1908	3.13
442	Arkansas, Garden City, Kans.	28,800	14	14	1936	242	.008	21,200	.734	June 6, 1921	1.25
443	Arkansas, Larned, Kans.	34,900	14	14	1936	240	.007	14,300	.410	August 9, 1929	.77
444	Arkansas, Arkansas City, Kans.	44,700	15	15	1936	1,214	.027	(70,000)	1.57	August 25, 1923	3.32
445	Arkansas, near Muskogee, Okla.	96,800	2	2	1936			243,000	2.61	June 11, 1923	7.8
446	Arkansas, Van Buren, Ark.	150,300	9	9	1936			418,000	2.77	June 9, 1935	10.8
447	Arkansas, Little Rock, Ark.	157,900	7	7	1936			422,000	2.68	June 19, 1935	10.6
448	Cameron Arroyo, near Pueblo, Colo.	7.3			1936			13,900	1,900	June 22, 1935	51.4
449	Rock Creek, near Pueblo, Colo.	59.0			1936			53,800	913	June 1921	70.0
450	Purgatoire, Trinidad, Colo.	742	24	15	1936	87	.117	45,400	61.1	September 30, 1904	16.7
451	Purgatoire, near Higbee, Colo.	2,900	12	40	1936	99.1	.034	64,500	22.3	September 15, 1934	12.0

See footnotes at end of table.

Average and maximum river discharges—Continued

Lower Mississippi—Continued

Item No.	River and station	Drainage area (square miles)	Years of record: Average	Years of record: Maximum	Years of record: Final	Discharge in cubic feet per second: Average	Discharge in cubic feet per second: Maximum	Discharge in cu. ft. per sec per sq. mi.: Average	Discharge in cu. ft. per sec per sq. mi.: Maximum	Date of maximum	Myers rating (percent)
452	Verdigris, Independence, Kans	2,952	15	15	1936	1,497	124,000 / 80,600	0.508	42.0 / 27.3	October 3, 1927 / April 10, 1922	22.8 / 14.8
453	Neosho, near Iola, Kans.	3,795	19	32	1936	1,147	74,000	.303	19.5	July 1904	12.0
454	Neosho, near Parsons, Kans	4,828	15	15	1936	2,023	48,100	.420	10.0	November 24, 1928	6.9
455	Canadian, near Bell Ranch, N Mex	6,400		11	1936		49,300		7.70	June 27, 1935	6.2
456	Canadian, Logan, N Mex	11,200		32	1936		141,000 / 102,000		12.6 / 9.10	October 1904 / October 11, 1930	13.3 / 9.6
457	East Quartermaster Creek, Okla	41.5					(54,800)		1,320	April 1934	85.2
458	West Quartermaster Creek, Okla	108					(69,000)		640	April 1934	66.6
459	Red, near Colbert, Okla	38,700	13	13	1936	5,065	201,000	.131	5.20	May 21, 1935	10.2
460	Red, Garland, Ark	51,500		9	1938	15,060	(180,000) / 143,000	.292	3.49 / 2.77	February 1938 / June 25, 1935	7.9 / 6.3
461	Sulphur, near Darden, Tex	2,751	7	13	1936	1,832	67,200	.665	24.4	May 19, 1930	12.8
462	Cypress Creek, near Jefferson, Tex	848	13	12	1936	560	22,600	.660	26.7	May 20, 1930	7.8
463	Ouachita at Remmel Dam, near Malvern, Ark	1,540	14	14	1936	2,050	140,000 / 138,000	1.33	91.0 / 89.7	May 16, 1923 / April 21, 1927	35.7 / 35.2
464	Ouachita, Monroe, La	15,700	6	53	1933	(14,200)	95,400	.904	6.08	March 3, 1932	7.6
464A	Ouachita, Monroe, La	15,700	6	66	1937		95,400 / 72,600		6.08 / 4.62	March 3, 1932 / February 7, 1937	7.6 / 5.8
465	Red, Alexandria, La	65,900		6	1933	29,200	210,000 / 186,000	.444	3.20 / 2.83	July 2, 1908 / February 4, 1932	8.2 / 7.3
466	Tallahatchie, near Sardis, Miss	1,680	3	6	1933	1,677	11,900	1.00	7.10	March 10, 1930	2.9
467	Yocona, near Enid, Miss	560	3	3	1931	572	10,100	1.02	18.0	March 20, 1930	4.3
468	Coldwater, near Coldwater, Miss	617	3	3	1931	655	41,800	1.06	67.6	January 9, 1930	16.8
469	Yalobusha, near Grenada, Miss	1,550	3	3	1931	1,207	33,800	.78	21.8	May 20, 1930	8.6
470	Yazoo, Greenwood, Miss	7,450	6	33	1937	10,412	73,000	1.40	9.80	January 19, 1932	8.5
471	Yazoo, near Yazoo City, Miss	(9,000)		55	1933		139,000		15.5	April 14, 1933	14.7
472	Sabine, Logansport, La	4,858	13	16	1937	2,917	41,100 / (70,000)	.600	8.34 / 14.6	March 18, 1882 / February 1932	5.9 / 10.1
473	Sabine, near Ruliff, Tex	9,448	12	12	1936	7,636	76,600	.808	8.10	May 1884	7.9
473a	Flat Fork near Center, Tex	58			1936		42,200		728	May 1935	55.3
473b	Tenaha Creek, near San Joaquin, Tex	374	13	13	1936		117,000		313	July 24, 1933	60.5
474	Neches, near Rockland, Tex	3,539			1936	2,427	(90,000) / 48,500	.687	25.4 / 13.7	July 24, 1933 / May 1884	15.1 / 8.2
475	Neches, Evadale, Tex	7,908	13	15	1936	6,338	175,000 / 83,800	.801	22.1 / 10.6	May 1935 / June 1929	19.7 / 9.4
476	Angelina, near Lufkin, Tex	1,575		28	1936		65,000		41.3	May 1908	16.4

The table on this page is printed sideways and is extremely dense. The column headings are not present on this page. The following is a best-effort transcription of the principal legible columns; minor columns of record data could not be reliably aligned.

No.	Station	Drainage area (sq. mi.)	Yrs.	Yr.	Max. discharge	Date of maximum
477	Angelina, Horger, Tex	3,435			82,000	August 1915
478	West Fork Trinity, Fort Worth, Tex	2,431	15		85,000	April 1922
479	Trinity, Dallas, Tex	6,001	15		184,000	May 26, 1908
480	Trinity, Oakwood, Tex	12,840	13	1936	76,700	May 1935
481	Trinity, Riverside, Tex	15,500	13	1936	84,400	May 1930
482	Trinity, Romayor, Tex	17,190	12	1936	86,600	June 1908
483	Clear fork Trinity, Fort Worth, Tex	522	12	1936	81,100	May 1929
484	Elm fork Trinity, near Carrollton, Tex	2,535	13	1936	74,300	April 1922
485	East Fork of Trinity, near Rockwall, Tex	831		1936	82,100	May 1935
486	San Jacinto, near Humble, Tex	1,811	12	1936	64,800	June 1935
487	Brazos, Seymour, Tex?	5,250		1936	111,000	May 1929
488	Brazos, near Mineral Wells, Tex?	13,860		1936	95,400	October 1926
489	Brazos, near Waco, Tex?	19,260	35	1936	270,000	1876
490	Brazos, near Bryan, Tex?	29,190	15	1936	246,000; (350,000); (200,000); 134,000	September 27, 1936; December 1913; May 1908; May 1930
491	North Bosque, near Clifton, Tex	974	13	1936	139,000	May 1935
492	Little River, Cameron, Tex	7,034	18	1936	38,300	May 1935
493	Salado Creek, near Salado, Tex	148		1936	647,000	September 10, 1921
494	Colorado, Ballinger, Tex	16,840	21	1936	143,000	September 10, 1921
495	Colorado, Austin, Tex	38,150	38	1936	75,400	September 1936
496	Concho, near San Angelo, Tex	4,492	21	1936	500,000; 481,000	May 1935; July 1869
497	North Concho, near Carlsbad, Tex	1,529	12	1936	246,000; 230,000	June 15, 1935; August 6, 1906
498	Pecan Bayou, Brownwood, Tex	1,614	11	1936	139,000	September 17, 1936
499	San Saba, Menard, Tex	1,151	21	1936	94,600	April 26, 1922
500	San Saba, San Saba, Tex	3,046	21	1936	35,000; 235,000	September 17, 1936; April 28, 1922
500a	East Fork of James, Old Noxville, Tex	60.8		1936	(90,000); 68,600	May 1925; June 5, 1899
500b	Onion Creek, near Buda, Tex	151		1936	60,000	July 3, 1932
500c	Onion Creek, near Delvalle, Tex	137		1936	57,000	June 5, 1899
500d	Copperas Creek, near Roosevelt, Tex	118		1936	105,000	September 16, 1936
500e	San Gabriel, near Georgetown, Tex	431		1936	53,200	
501	North Llano, near Junction, Tex	914		1936	139,000	June 1899
502	Llano, near Junction, Tex	1,762	21	1936	98,900	April 28, 1922
503	Llano, near Castell, Tex	3,514	21	1936	160,000	July 1, 1932
503a	Llano, near Castell, Tex	110	12	1936	94,800	May 28, 1929
503b	North Fork of Guadalupe, near Hunt, Tex	65.3		1936	319,000	September 10, 1921
503c	South Fork of Guadalupe, near Hunt, Tex	336		1936	388,000	September 16, 1936
503d	Johnson Creek, near Ingram, Tex	111		1936	108,000	September 10, 1921
504	Guadalupe, Kerrville, Tex	570	66	1936	84,300; 206,000; 138,000	June 14, 1935; July 1, 1932
505	Guadalupe, near Comfort, Tex	916	67	1936	196,000; (220,000); 182,000	July 1869; July 1, 1932

See footnotes at end of table

Average and maximum river discharges—Continued

Item No.	River and station	Drainage area (square miles)	Years of record — Average	Years of record — Maximum	Years of record — Final	Discharge in cubic feet per second — Average	Discharge in cubic feet per second — Maximum	Discharge in cu ft per sec per sq mi — Average	Discharge in cu ft per sec per sq mi — Maximum	Date of maximum	Myer rating (percent)
	Lower Mississippi—Continued										
506	Guadalupe, near Spring Branch, Tex	1,432	14	36	1936	279	121,000	0.195	84.5	July 3, 1932	32.0
507	Guadalupe, New Braunfels, Tex	1,666		67	1936		167,000		100	July 1869	40.9
							101,000		60.6	December 1913	24.8
508	Guadalupe, below Cuero, Tex	5,073	17	17	1935	1,365	101,000	0.269	19.6	June 15, 1935	14.2
509	Guadalupe, Victoria, Tex	5,676		32	1936		179,000		31.6	May 1929	23.8
509a	Bunton Branch, near Kyle, Tex	4.1			1936		13,800		3,370	July 3, 1936	68.1
509b	O'Neal Creek, near Leesville, Tex	30			1936		30,000		1,000	June 30, 1936	54.8
510	Blanco, Wimberley, Tex	378	10	57	1936	101	113,000	0.266	298	July 1, 1936	58.0
511	Blanco, near San Marcos, Tex	429		7	1936		139,000		324	May 1929	67.2
512	Sandies Creek near Dewitt, Tex	95			1936		54,300		574	May 1929	55.7
513	Sandies Creek near Westhoff, Tex	493		38	1936		92,700		188	July 1, 1936	41.8
513a	Olmas Creek, near San Antonio, Tex	26.4			1936		28,000		1,061	July 2, 1936	54.5
513b	San Pedro Creek, near San Antonio, Tex	46.5			1936		32,443		698	September 9, 1921	47.6
514	San Antonio, near San Antonio, Tex	85		117	1936		42,400		499	September 9, 1921	46.0
							(80,000)		38.8	July 1819	17.6
							16,200		7.9	September 1921	3.6
							(290,000)		380	October 1913	105
515	San Antonio, near Falls City, Tex	2,067	11	61	1936	277	226,000	0.134	296	July 3, 1936	81.8
516	Nueces, Laguna, Tex	764	13	33	1936	173	213,000	0.226	279	June 1913	77.1
517	Nueces, near Uvalde, Tex	1,930		9	1936		616,000		319	September 21, 1923	140
518	Nueces, near Three Rivers, Tex	15,600	19	21	1936	802	85,000	0.052	5.45	June 14, 1935	6.8
519	West Nueces, near Brackettville, Tex	402			1936		580,000		1,440	September 18, 1919	289
520	West Nueces, near Cline, Tex	880			1936		536,000		609	June 14, 1935	181
520a	East Fork of Frio, near Leakey, Tex	75.0			1936		89,500		1,190	June 14, 1935	103
520b	Frio, Rio Frio, Tex	71			1936		128,000		345	July 1, 1932	66.6
520c	Sabinal, Vanderpool, Tex	45.7			1936		52,300		1,140	July 1, 1932	77.4
520d	Sabinal, Sabinal, Tex	258			1936		71,700		278	July 2, 1932	44.6
521	Frio, Conran, Tex	485	11	13	1936	130	162,000	0.258	334	July 1, 1932	73.5
522	Frio, near Derby, Tex	3,493	21	22	1936	196	230,000	0.056	65.8	July 1, 1932	38.9
523	Dry Frio, near Reagan Wells, Tex	120			1936		64,700		539	July 4, 1932	59.0
524	Seco Creek, near D'Hanis, Tex	153		64	1936		230,000		1,500	May 31, 1935	186.1
525	Atascosa, near Benton City, Tex	21.3		6	1936		25,900		1,220	June 22, 1924	56.1
526	Atascosa, Whitsett, Tex	1,171			1936		38,300		32.8	June 14, 1935	11.2
527	Conejos, near Mogote, Colo	282	34	34	1936	379	6,000	1.36	21.2	October 5, 1911	21.2
528	Conejos, near La Sauses, Colo	887	15	15	1936	244	[2]3,660	0.275	4.10	May 24, 1932	1.2
529	Rio Grande, near Creede, Colo	163	23	27	1936	232	7,500	1.42	46.0	June 28, 1927	5.9

No.	Stream and place	Drainage area			Year					Date	
530	Rio Grande, below Creede, Colo.	705	29	29	1936	651	9,750	925	13 8	June 28, 1927	3 6
531	Rio Grande, near Del Norte, Colo.	1,320	47	47	1936	970	15,000	.731	11 4	June 29, 1927	4.1
532	Rio Grande, Alamosa, Colo.	1,840	24	24	1936	353	18,000	182	13 6	October 5, 1911	5 0
533	Rio Grande, near Lobatos, Colo [8]	4,760	37	37	1936	773	14,000	163	7 60	July 1, 1927	3 3
534	Rio Grande, Embudo, N. Mex [8]	7,360	23	23	1936	1,001	13,100	136	2 75	June 8, 1905	1 0
535	Rio Chama, Park View, N. Mex.	405		10	1936		8,360		1 13	May 25, 1932	2 9
536	Rio Grande, near San Ildefonso, N. Mex [8]	11,260	19	37	1936		5,840		14 4	April 22, 1936	2 1
537	Santa Fe Creek, near Santa Fe, N. Mex.	22		19	1936	10 5	21,900		94	August 20, 1935	1 4
538	Rio Grande, San Felipe, N. Mex [8]	13,060		12	1936		²655	477	4	August —, 1921	3 7
539	Rio Puerco, Rio Puerco, N. Mex.	(5,500)		15	1936		42,100		29 8	August 21, 1935	3 4
540	Rio Grande, San Marcial, N. Mex [9]	24,200	39	39	1936	1,541	40,000	064	3 22	September 23, 1929	3 0
541	Rio Grande, below Elephant Butte Dam, N. Mex [9]	(26,000)	20	20	1936	1,176	47,000	045	1 94	September 24, 1929	20
542	Rio Grande, El Paso, Tex [9]	32,819	48	108	1936	1,010	3,200	031	123	July —, 1917	1 35
543	Rio Grande, Tornillo Bridge, Tex.	(33,600)	13	13	1936	287	24,000	008	728	June 12, 1905	14
544	Rio Grande, Fort Quitman, Tex.	34,450	14	14	1936	304	6,500	009	193	September 5, 1925	39
545	Rio Grande, La Nutria, Tex.	36,292	2	2	1936	224	²2,600	006	076	September 11, 1925	78
546	Rio Grande, Upper Presidio, Tex.	37,488	13	13	1936	336	7,480	009	206	August 31, 1935	6 9
547	Rio Grande, Lower Presidio, Tex.	60,109	13	36	1936	1,780	15,000	030	40	June 12, 1912	3 6
548	Rio Grande, Boquillas, Tex.	69,373	13	13	1936	1,913	168,000	028	2 79	September —, 1904	7 2
549	Rio Grande, Langtry, Tex.	77,518	13	36	1936	2,350	95,030	030	1 37	October 4, 1932	17 0
550	Rio Grande, near Del Rio, Tex.	123,164	13	36	1936	4,050	200,000	033	2 58	June 18, 1922	16 0
551	Rio Grande, Eagle Pass, Tex.	126,962	13	36	1936	4,350	605,000	036	4 93	September 1, 1932	11 3
552	Rio Grande, Laredo, Tex.	132,915	13	5	1936	4,785	569,000	036	4 48	September 2, 1932	6 6
553	Rio Grande, Zapata, Tex.	156,714	13	36	1936	6,600	402,000	042	3 02	September 3, 1932	5 1
554	Rio Grande, Roma, Tex.	160,014	13	27	1936	5,920	261,160	037	1 67	September 4, 1932	4 8
555	Rio Grande, near Rio Grande City, Tex.	174,268	13	32	1936	7,350	203,000	042	1 14	September 5, 1932	11 3
556	Rio Conchos, near Ojinaga, Chihuahua, Mexico	22,450	5	5	1936	580	198,750	064	7 50	September 5, 1932	10 7
557	Terlingua Creek, near Terlingua, Tex.	1,070	4	31	1936	934	170,000	054	32 7	September 11, 1904	47 4
558	Lozier Creek, near Langtry, Tex.	1,728	6	16	1936	117	34,900	054	114	May 24, 1935	3 9
559	Pecos, near Anton Chico, N. Mex.	1,080	6	23	1936	106	197,000	108	12 0	September 4, 1935	2 8
560	Pecos, Santa Rosa, N. Mex.	2,880	30	30	1935	205	12,900	037	5 20	July 16, 1933	4 0
561	Pecos, Guadalupe, N. Mex.	4,470	24	24	1935	355	15,000	042	6 05	July 12, 1936	0 6
562	Pecos, near Dayton, N. Mex.	(20 000)	16	16	1936	294	27,000	018	3 51	October 11, 1930	5 7
563	Pecos, Carlsbad, N. Mex.	(22,500)	21	22	1936	272	50,300	013	3 80	July 1905	3 7
564	Pecos, near Malaga, N. Mex.	(23,500)	9	9	1936	353	85,000	012	3 40	September 7, 1916	3 1
565	Pecos, Angeles, Tex.	(26,000)	38	38	1936	257	80,000	014	2 31	September 1919	1 5
566	Pecos, Comstock, Tex.	38,283	38	38	1936	470	60,000	010	30 3	August 8, 1916	59 3
567	Goodenough Spring, near Comstock, Tex.		13	13	1936	632	23,500	012	28 0	June 13, 1935	54 7
568	Devils, near Juno, Tex.	2,733	54	54	1936	161	116,000	017	136	September 1, 1932	70 7
569	Devils, near Del Rio, Tex.	4,060	36	36	1936	223	107,000	.082	147	April 6, 1900	93 6
570	Dry Devils, near mouth, Texas	748	6	6	1936	723	742	.178	172	November 5, 1932	6 7
571	San Felipe Creek, near Del Rio, Tex [10]	62	3	3	1936	78	370,000		725	September 1, 1932	47 0
572	Syeamore Creek, near Del Rio, Tex.	524	8	8	1935	39	597,000	26	410	September 1, 1932	57.2
573	Pinto Creek, near Del Rio, Tex.	229	9	12	1936	36	129,000	074	239	September 1, 1932	94 0
574	San Diego, Jimenez, Coahuila, Mexico	840	9	14	1936	213	45,000	.157	38 2	June 14, 1935	36 1
575	San Rodrigo, near El Moral, Mexico	750	8	14	1936	305	215,000	.253	108	August 31, 1932	11.1
576	Escondido, Villa Fuente, Mexico	1,170			1936	122	54,650	.406	15 1	June 14, 1935	29.6

See footnotes at end of table.

Average and maximum river discharges—Continued

Item No.	River and station	Drainage area (square miles)	Years of record — Average	Years of record — Maximum	Years of record — Final	Discharge in cubic feet per second — Average	Discharge in cubic feet per second — Maximum	Discharge in cu ft per sec per sq mi. — Average	Discharge in cu ft per sec per sq mi. — Maximum	Date of maximum	Myers rating (percent)
	Lower Mississippi—Continued										
577	Salado, Cd Guerrero, Tamaulipas, Mexico	21,830	13	24	1936	710	43,800	0 325	2 00	September 7, 1933	3 0
578	Alamo, Cd Mier, Tamaulipas, Mexico	1,840	13	13	1936	200	76,600	109	41 6	September 7, 1933	17 9
579	San Juan, Santa Rosalia, Tamaulipas, Mexico	13,000	12	26	1936	1,120	353,000	086	27 2	August 30, 1909	31 0
580	Rio Grande, Penitas Pumps, near Hidalgo, Tex	(174,700)		28	1937		187,000		14 4	September 29, 1932	16 4
581	Rio Grande, Matamoras, Tamaulipas, Mexico [11]	175,138	13	30	1936	5,405	360,000	.031	2 06	August 30, 1909	8 6
							[2]38,300		22	July 20, 1906	.9
	Colorado River Basin										
582	Fraser, West Portal, Colo	28	26	26	1936	44.4	820	1 59	29 3	June 13, 1918	1 6
583	Colorado, near Grand Lake, Colo	101	15	15	1936	138	1,840	1 36	18 2	June 15, 1918	1 8
584	Williams, near Parshall, Colo	184	23	23	1936	170	2,750	92	15 0	June 16, 1918	2 0
585	Colorado, Hot Sulphur Springs, Colo	782	29	30	1936	749	10,300	96	13 2	June 15, 1921	3 7
586	Roaring Fork, Aspen, Colo	109	14	14	1936	172	3,170	1 58	29 0	June 18, 1917	3 0
587	Roaring Fork, Glenwood Springs, Colo	1,460	29	29	1936	1,541	17,600	1 06	12 1	June 14, 1918	4 5
588	Colorado, Glenwood Springs, Colo	4,560	37	37	1936	3,048	30,100	67	6 60	June 19, 1918	4 9
589	Colorado, near Cisco, Utah	24,100	18	18	1936	8,544	76,800	354	3 18	June 19, 1917	5 8
590	Colorado, Lees Ferry, Ariz	107,900	14	14	1936	17,300	190,000	161	1 76	June 18, 1921	3 4
591	Colorado, Bright Angel Creek, Ariz	138,700	14	14	1936	17,910	127,000	129	92	July 2, 1927	4 2
592	Colorado, near Topock, Ariz	174,300	17	17	1934	20,300	174,000	116	1 00	June 22, 1921	.5
593	Colorado, near Topock, Ariz (regulated by Lake Mead)	174,300	2	2	1936	7,760	18,600	045	.11	June 24, 1935	
594	Colorado, Yuma, Ariz [12]	244,800	28	58	1936	22,800	240,000	093	97	January 22, 1916	4 8
595	Taylor, Almont, Colo	440	26	26	1936	374	3,760	85	8 56	June 9, 1920	4 8
596	Gunnison, near Grand Junction, Colo	8,020	16	18	1936	2,906	35,700	363	4 45	May 23, 1920	4 0
597	Dolores, Dolores, Colo	508	15	24	1936	477	10,000	94	19 6	October 5, 1911	4 4
598	Green River, Warren Bridge, near Daniel, Wyo	468		5	1936		3,260		6 96	June 26, 1932	4 5
599	Green River, Wyo	7,670	21	23	1936	1,902	22,200	248	2 90	June 19, 1918	2 4
600	Green River, Utah	40,600	36	36	1936	7,253	68,800	178	1 69	May 29, 1897	3 4
601	Yampa, Steamboat Springs, Colo	604	26	28	1936	507	6,820	836	11 3	June 14, 1921	2 8
602	Yampa, near Maybell, Colo	3,410		20	1936		17,900		5 27	May 19, 1917	3 1
603	Ashley, near Vernal, Utah	101	20	22	1936	105	2,050	1 04	20 2	May 29, 1921	2 0
604	Duchesne, Duchesne, Utah	660	19	19	1936	386	4,420	585	6 70	June 10, 1922	1 7
605	Duchesne, Myton, Utah	2,750	28	36	1936	660	12,800	240	4 66	June 10, 1922	2 4
606	Strawberry, Duchesne, Utah	1,040	22	22	1936	171	3,230	164	3 10	May 27, 1922	1 0
607	Lake Fork, near Myton, Utah	468	27	32	1936	153	5,600	327	12 0	November 24, 1927	2 6
608	Whiterocks, near Whiterocks, Utah	115	12	22	1936	119	2,750	1 03	23 8	June 21, 1922	2 2
609	White, near Meeker, Colo	762	31	31	1936	657	6,070	96	7 94	June 16, 1921	

No.	Station				Year		Max. discharge	Per sq. mi.	Gage height	Date	
610	White, near Watson, Utah	4,020		13	1936	144	8,160	272	2 00	July 15, 1929	1.3
611	Price, near Helper, Utah	530	28	30	1934	101	10,000	536	18 8	September 1927	4.3
612	Huntington Creek, near Huntington, Utah	188	21	25	1936	109	2,500	545	13 5	August 1930	3.8
613	Cottonwood Creek, near Orangeville, Utah	200	20	22	1936		2,500		12 5	August 1922	1.8
614	San Juan, near Blanco, N. Mex.	3,320		11	1935		17,700		5 34	August 21, 1932	3.1
615	San Juan, near Farmington, N. Mex.	6,580		24	1936		32,800		5 00	September 28, 1935	3.0
616	San Juan, near Shiprock, N. Mex.	(11,600)	5	11	1936	2,035	(80,000)	175	6 90	August 11, 1929	4.7
617	San Juan, near Bluff, Utah	24,000	11	25	1936	2,848	(120,000), (90,000), (70,000)	118	10 3, 3 75	October 6, 1911; October 1911	11.1, 10.3
618	Animas, Durango, Colo.	692	35	36	1936	923	11,000	1 34	2 82	September 10, 1927	5.8
619	Animas, Farmington, N. Mex.	1,360		24	1936		9,350		20 2	June 29, 1927	4.5
620	Florida, near Durango, Colo.	96	18	25	1936	129	(6,000)	1 34	6 88	June 16, 1935	5.3
621	La Plata, Hesperus, Colo.	37	19	19	1936	47	4,640	1 30	62 5	October 5, 1911	2.5
622	La Plata, at Colorado-New Mexico State line	331	16	16	1936	35	1,460	105	48 3	June 28, 1927	6.1
623	Mancos, near Towaoc, Colo.	558	15	15	1936	55	4,750	099	39 5	June 28, 1927	4.7
624	Para, Lees Ferry, Ariz.	1,520	13	13	1936	35	4,900	023	14 3	August 24, 1927	2.4
625	Little Colorado, near Woodruff, Ariz.	9,060		17	1936		16,100		8 78	August 26, 1934	2.6
626	Little Colorado, Grand Falls, Ariz.	22,100	11	13	1936	323	25,000	015	10 6	October 5, 1925	2.2
627	Chevelon Fork, near Winslow, Ariz.	1,010	6	13	1936	49	120,000	049	2 76	December 5, 1919	2.1
628	Clear Creek, near Winslow, Ariz.	607	6	9	1936	88	16,100	145	5 42	September 19, 1923	4.1
629	Moenkopi Wash, near Tuba, Ariz.	2,270	9	10	1936	25	39,000	011	15 9	April 4, 1929	2.6
630	Bright Angel Creek, near Grand Canyon, Ariz.	100	13	13	1936	35	15,100	350	64 0	April 4, 1929	8.1
631	North Fork of Virgin River, near Springdale, Utah	(450)	10	10	1936	89	4,400, 4,480	198	6 65, 41 0, 10 0	August 4, 1929; August 19, 1936; September 2, 1936	5.1
632	Virgin River, Virgin, Utah	990	17	17	1936	216	12,000	218	12 2	October 27, 1912	15.8
633	Virgin River, Littlefield, Ariz.	4,400	7	7	1936	241	25,000	055	5 69	August 27, 1932	3.2
634	Santa Clara Creek, near Central, Utah	84	22	22	1930	22	1,450	262	17 3	October 6, 1916	2.4
635	Williams, Planet, Ariz.	5,140	7	12	1936	122	(100,000)	024	19 5	January 19, 1916	2.1
636	Gila, Red Rock, N. Mex.	2,840	8	28	1936	141	55,500	050	10 8	August 5, 1931	3.8
637	Gila, near Clifton, Ariz.	4,040	13	14	1936	279	12,600	069	4 45	July 3, 1931	3.8
638	Gila, near Solomonsville, Ariz.	7,900	19	22	1936	488	15,000	062	3 70	December 1914	1.6
639	Gila, Coolidge Dam, Ariz (Regulation since November 1928)	12,880	42	42	1936	480	11,500	037	12 7	August 26, 1934	14.0
640	Gila, Kelvin, Ariz.	18,260	22	22	1936	508	100,000, 130,000, 130,000	039, 093	10 1, 10 1, 7 22	January 19, 1916; January 20, 1916; do	7.8
641	Gila, Gillespie Dam, Ariz.	49,700	25	25	1936	642	132,000	035	1 41	December 28, 1923	4.4
642	Gila, near Dome, Ariz.	58,100	15	21	1936	456	70,000	009	3 43	January 22, 1923	2.2
643	San Francisco, Clifton, Ariz.	2,790	10	12	1936	1,136	200,000	020	5 23	December 1914	4.1
644	San Simon Creek, near Solomonsville, Ariz.	2,280	2	5	1936	258	14,600	009	3 08	August 26, 1934	11.3
645	San Carlos, near Peridot, Ariz.	1,070	7	21	1936	20	8,600	038	12 1	August 9, 1931	11.4
646	San Pedro, Charleston, Ariz.	1,480	7	23	1936	41	27,500	045	23 4	January 18, 1916	4.9
647	San Pedro, near Mammoth, Ariz.	3,850	5	10	1936	66	25,000	014	13 5	February 17, 1936	9.8
648	Aravaipa Creek, near Feldman, Ariz.	535	7	17	1936	52	14,400	056	66 3	September 28, 1926	3.3
649	Santa Cruz, near Nogales, Ariz.	473	10	17	1936	30	98,000	057	23 4	September 28, 1926	8.1
650	Santa Cruz, Tucson, Ariz.	2,100	26	31	1936	27	90,000	011	37 6	August 2, 1919	2.6
651	Sonoita Creek, near Patagonia, Ariz.	210	4	6	1936	24	20,000	.048	5 44	September 28, 1926	5.8

See footnotes at end of table

Average and maximum river discharges—Continued

Item No.	River and station	Drainage area (square miles)	Years of record			Discharge in cubic feet per second		Discharge in cu. ft per sec. per sq. mi.		Date of maximum	Myers rating (percent)
			Average	Maximum	Final	Average	Maximum	Average	Maximum		
	Colorado River Basin—Continued										
652	Rillito Creek, near Tucson, Ariz.	903	23	26	1936	20	24,000	0 022	26 6	September 23, 1929	8 0
653	Salt, near Chrysotile, Ariz.	2,830	12	12	1936	628	36,000	222	12 7	February 10, 1932	6 8
654	Salt, near Roosevelt, Ariz.	4,310	29	23	1936	1,160	100,000	270	23 2	January 19, 1916	15 8
654a	Salt River, Granite Reef Dam, Ariz.	(12,500)			1937		300,000		24	February 1891	26 8
655	Tonto Creek, near Roosevelt, Ariz.	813	22	23	1936	153	2 20,000	188	24 6	December 28, 1923	7 0
656	Verde, near Camp Verde, Ariz.	5,010	9	10	1936	383	32,000	076	6 39	February 22, 1920	4 5
657	Verde, above Camp Creek, near McDowell, Ariz.	6,230	32	39	1936	849	96,000	136	15 4	November 27, 1905	12 2
658	Whitewater Draw, near Douglas, Ariz.	1,023	7	25	1936	10	4,050	010	3 95	July 28, 1919	1 26
658A	Cave Creek, near Phoenix, Ariz.	200					25,000		125	August 1921	17 7
	GREAT BASIN										
659	Bear, near Evanston, Wyo.	645	23	23	1936	244	3,690	379	5 71	June 14, 1921	1 46
660	Bear, Harer, Idaho.	2,780	20	20	1936	543	3,860	196	1 39	June 2, 1920	73
661	Bear, Alexander, Idaho.	3,840	21	21	1936	844	4,590	220	1 20	June 7, 1909	74
662	Bear, near Collinston, Utah.	6,000	46	47	1936	1,826	11,600	305	1 94	March 21, 1916	50
663	Logan, above State dam, near Logan, Utah.	218	23	23	1936	125	2,000	572	9 15	May 15, 1917	1 35
664	Blacksmith Fork, near Hyrum, Utah.	260	22	22	1936	129	1,620	496	6 22	July 6, 1907	1 01
665	Weber, near Oakley, Utah.	163	30	30	1936	244	4,000	1 49	24 5	July 6, 1907	3 13
666	Weber, Devils Slide, Utah.	1,100	31	31	1936	482	6,000	440	5 45	May 22, 1920	1 81
667	Weber, Gateway, Utah.	1,610	16	16	1936	624	7,980	388	4 96	May 31, 1896	96
668	Weber, near Plain City, Utah.	2,060	29	29	1936	834	7,580	405	3 68	June 6, 1909	1 67
669	South Fork of Ogden River, near Huntsville, Utah.	148	15	15	1936	110	1,780	742	12 0	May 4, 1936	1 46
670	Ogden, near Ogden, Utah.	360	12	12	1936	265	3,700	0 735	1 03	April 24, 1936	1 95
671	Farmington Canyon, Farmington, Utah.	7					2,450		350	August 1923	9 3
672	Jordan River at Narrows, near Lehi, Utah.	2,610	23	23	1936	392	1,370	150	525	June 8, 1923	27
673	Provo, Washington Lake, Utah.	3 4					1,020		300	June 1911	6 00
674	Provo, Forks, Utah.	600	24	24	1936	361	3,180	602	5 30	June 11, 1921	1 30
675	Provo, near Provo, Utah.	640			1936		4,100		6 40	May 1907	1 63
676	South Fork Provo, Forks, Utah.	30			1936	30 6	123	1 02	4 10	May 27, 1922	22
677	Sevier, near Kingston, Utah.	1,110	24	24	1936	149	1,460	135	1 32	May 21, 1922	.44
678	Sevier, below Piute Dam, near Marysvale, Utah.	2,440	22	24	1936	256	3,000	105	1 23	September 3, 1909	.61
							2,600		1 06	May 23, 1922	.53
679	Sevier, below San Pitch River, near Gunnison, Utah.	4,880	19	19	1936	225	2,620	046	54	June 1, 1922	.38

No.	Station										
680	Sevier River, near Juab, Utah (below Sevier Bridge Dam).	5,120	25	25	1936	267	2,140	.052	.42	June 2, 1922	.30
681	East Fork Sevier River, near Kingston, Utah	1,260	23	23	1936	93 5	2,000	074	1 59	August 27, 1929	56
682	Salt Creek, near Nephi, Utah	95	11	11	1936	24 2	800	255	8 40	July 17, 1932	82
683	Diamond Fork, near Thistle, Utah	137	10	10	1917	34 8	735	276	5 36	May 1909	63
684	Spanish Fork, Thistle, Utah	490	18	18	1925	95	1,250	194	2 55	May 26, 1922	55
685	Spanish Fork, Castilla, Utah	670	17	17	1925	228	1,440	340	2 15	May 7, 1922	55
686	American Fork, near American Fork, Utah	43	3	15	1914	53	860	1 23	20 0	May 1914	1 31
687	Big Cottonwood Creek, near Salt Lake City, Utah.	48 5	15	15	1913	83	835	1 71	17 2	June 1909	1 20
688	Mill Creek, near Salt Lake City, Utah	21 3	15	15	1913	18 0	112	845	5 27	June 1909	24
689	Parleys Creek, near Salt Lake City, Utah.	50 1	15	15	1913	28 5	274	570	5 47	June 1909	39
690	Emigration Creek, near Salt Lake City. Utah	29	13	13	1913	4 4	174	152	6 0	April 1913	32
691	City Creek, near Salt Lake City, Utah	19 2	15	15	1913	17 9	154	90	8 0	May 1907	35
692	Beaver, near Beaver, Utah	82	22	23	1936	54 9	1,080	670	13 2	July 22, 1936	1 19
693	Beaver, Adamsville, Utah	272	22	23	1936	37 0	989	136	3 64	September 1, 1936	60
694	Beaver, near Minersville, Utah.	512	22	22	1936	38 3	727	075	1 42	June 10, 1921	31
694a	Las Vegas Wash near Las Vegas, Nev.	(6)					(1,500)		(250)	July 10, 1932	6 1
695	Deep Creek, near Hesperia, Calif	137	7	7	1936	37 7	7,900	275	57 8	February 9, 1932	6 75
696	Mojave, Victorville, Calif.	211	33	33	1936	50.3	13,600	238	64 0	March 1903	9 35
697	Mojave, below Victorville, Calif.	400	7	7	1936	18 0	12,500		59 0	February 9, 1932	8 60
698	West Fork Mojave, near Hesperia, Calif.	74 8	13	13	1937	11 4	13,200	242	33 0	March 1903	6 60
699	Rock Creek, near Valyermo, Calif	23 0	29	29	1935	225	6,000	496	80 4	February 8, 1932	6 95
700	Owens, near Round Valley, Calif.	450	30	30	1935	330	510	500	21 9	February 16, 1927	1 05
701	Owens, near Big Pine, Calif.	1,930	13	13	1935	20 2	1,190	172	2 64	June 30, 1907	66
702	Rock Creek, near Bishop, Calif	51 7	26	26	1935	38 1	3,220	392	1 67	January 26, 1914	73
703	Rock Creek, near Round Valley, Calif.	96	14	14	1936	36 3	162	398	3 15	June 17, 1927	63
704	Pine Creek, near Bishop, Calif.	37 9	26	26	1935	21 3	360	960	3 65	January 25, 1914	37
705	Pine Creek, near Round Valley, Calif.	58	13	13	1936	97 5	315	367	8 30	June 20, 1922	51
706	East Walker, near Bridgeport, Calif.	362	26	26	1936	277	370	260	6 40	January 22, 1911	49
707	West Walker, near Coleville, Calif.	245	15	31	1936		1,050	1 13	2 90	June 29, 1922	55
708	East Forks Carson, Stateline, Nev.	298	25	25	1936	346	4,200		17 2	July 3, 1907	2 70
709	Carson, near Fort Churchill, Nev.	1,450	25	25	1936	331	2,710	240	11 1	June 12, 1921	1 73
710	Humboldt, Palisade, Nev.	5,010	28	78	1936	217	3,460	066	11 6	June 16, 1911	2 00
711	Humboldt, near Rye Patch, Nev.	13,700	28	24	1936	247	6,150	016	4 25	January 25, 1914	1 62
712	Truckee, Tahoe, Calif.	519	24	36	1936	630	4,300	476	86	March 3, 1921	61
713	Truckee, Iceland, Calif.	937	36	24	1936	56 6	3,050	670	22	May 12, 1897	26
714	Donner Creek, near Truckee, Calif.	30	24	21	1936	130	1,340	1 89	2 59	July 1907	59
715	Chewaucan, near Paisley, Oreg.	266	8	26	1936	26 5	15,300	488	16 3	March 18, 1907	5 00
716	Silver Creek, near Silver Lake, Oreg.	22 1	26	26	1936	13 2	980	119	32 7	March 18, 1907	1 79
717	Silvies, near Burns, Oreg.	940	23	31	1936		4,000	014	15 0	November 23, 1904	2 45
	Snake River Basin										
718	Snake, near Moran, Wyo	820	33	33	1936	1,430	15,100	1.74	8 10	June 12, 1918	5 3
719	Snake, near Heise, Idaho.	5,740	16	16	1936	6,680	60,000	1.16	5 01	May 19, 1927	7 9
720	Snake, near Shelley, Idaho.	(9,000)	21	21	1936		47,200		18 4	June 17, 1918	5.0
721	Snake, near Blackfoot, Idaho.	11,700	26	26	1936		46,200		10 5	June 18, 1918	4 3
722	Snake, Neeley, Idaho.	(16,300)	30	30	1936		48,400		2 96	June 20, 1918	3.8

See footnotes at end of table.

Average and maximum river discharges—Continued

Item No.	River and station	Drainage area (square miles)	Years of record — Average	Years of record — Maximum	Years of record — Final	Discharge in cubic feet per second — Average	Discharge in cubic feet per second — Maximum	Discharge in cu ft per sec per sq mi — Average	Discharge in cu ft per sec per sq mi — Maximum	Date of maximum	Myers rating (percent)
	Snake River Basin—Continued										
723	Snake, near Minidoka, Idaho	(19,000)		26	1936		45,900		2 41	June 21, 1918	3 3
724	Snake, Milner, Idaho	(22,000)		27	1936		44,400		2 02	June 12, 1909	3 0
725	Snake, near Kimberly, Idaho	(24,000)		13	1936	2,430	27,200	105	1 13	July 4, 1927	1 8
726	Snake, near Twin Falls, Idaho	(24,500)	13	22	1936	4,160	32,200	170	1 31	June 10, 1914	2 1
727	Snake, near Hagerman, Idaho	(28,200)	20	22	1936	8,270	35,100	293	1 24	June 10, 1914	2 1
728	Snake, King Hill, Idaho	(33,000)	27	27	1936	10,920	47,200	330	1 43	June 22, 1918	2 6
729	Snake, near Murphy, Idaho	41,900	23	23	1936	10,930	47,300	260	1 13	June 22, 1918	2 3
730	Snake, Weiser, Idaho	(60,000)	25	42	1936	17,710	(120,000)	295	2 00	June 1894	4 9
							(100,000)		1 67	March 3, 1910	4 1
							83,100		1 38	May 23, 1921	3 4
731	Snake, Oxbow, Oreg	(65,000)	13	13	1936	15,690	63,100	241	97	April 25, 1936	2 5
732	Snake, near Clarkston, Wash	103,200	27	42	1936	48,860	409,000	473	3 96	February 6, 1925	12 7
							270,000		2 62	June 5, 1894	8 4
							219,000		2 12		6 8
733	Henrys Fork, Warm River, Idaho	660	22	22	1936	1,010	3,540	1 53	5 35	May 20, 1921	1 4
734	Henrys Fork, near Rexburg, Idaho	3,010		27	1936		9,490		3 15	May 16, 1936	1 7
735	Portneuf, Pocatello, Idaho	(900)	23	23	1936	257	(2,200)	285	2 44	May 1917	.7
736	Big Wood, Hailey, Idaho	640	21	21	1936	280	3,560	437	5 56	June 12, 1921	1 4
737	Big Wood, below Magic Dam, near Richfield, Idaho	(1,600)	24	24	1936	376	5,070	235	3 16	May 18, 1911	1 3
738	Little Wood, near Carey, Idaho	312	13	32	1936	124	(1,500)	396	4 80	May 22, 1904	9
739	Owyhee, below Owyhee Dam, Oreg	(10,900)	7	7	1936	430	14,600	039	1 34	March 21, 1932	1 4
740	Boise, near Twin Springs, Idaho	830	25	25	1936	1,120	10,300	1 35	12 4	May 11, 1928	3 6
741	Boise, near Arrowrock, Idaho	2,230	25	25	1936	2,230	17,600	1 00	7 88	May 19, 1921	3 7
742	Boise, Notus, Idaho	(3,500)	14	16	1936	1,100	14,500	314	4 15	May 15, 1917	2 5
743	South Fork of Boise River, near Lenox, Idaho	1,090	25	25	1936	961	9,200	880	8 45	July 1913	2 7
743a	Hull's Gulch near Boise, Idaho	5			1936		5,000		1,000		22 4
744	Malheur, near Drewsey, Oreg	1,010	10	12	1936	118	3,800	117	3 75	March 19, 1932	2 2
745	Malheur, near Riverside, Oreg	1,100	24	26	1936	166	5,490	151	5 00	March 2, 1910	1 7
746	Malheur, near Hope, Oreg	3,030		17	1936		8,100		2 67	February 5, 1925	1 5
747	North Fork of Payette River, Lardo, Idaho	131	25	25	1936	345	4,260	2 63	32 5	June 10, 1933	3 7
748	South Fork of Payette River, near Garden Valley, Idaho	779	12	12	1936	1,160	10,600	1 49	13 6	May 26, 1928	3 7
749	South Fork of Payette River, near Banks, Idaho	1,200	15	15	1936	1,530	13,800	1 28	11 5	May 17, 1927	4 0
750	Payette, near Horseshoe Bend, Idaho	2,230	25	27	1936	2,990	22,100	1 34	9 90	June 9, 1921	4 7

No.	Stream and place of measurement	Drainage area			Year	Min. discharge	Max. discharge (cfs)		Per sq. mi.	Date of maximum	
751	Weiser, near Weiser, Idaho	1,160	15	16	1936	772	14,000	667	12.1	March 19, 1932	4 1
752	Powder, Salisbury, Oreg	230	18	18	1936	109	1,820	473	7.90	March 20, 1910	1 2
753	Powder, near Robinette, Oreg	1,700	8	8	1936	360	4,180	212	2.45	June 15, 1933	1 0
754	Salmon, near Clayton, Idaho	841	13	15	1936	834	8,000	990	9.50	June 27, 1927	2 8
755	Middle Fork of Salmon, near Meyers Cove, Idaho.	2,020	5	5	1936	1,867	17,000	930	8.44	June 10, 1933	3 8
756	Salmon, Salmon, Idaho	3,600	19	19	1936	1,770	16,400	492	4.55	June 12, 1921	2 7
757	Salmon, Whitebird, Idaho	13,400	24	42	1936	10,400	120,000	775	8.95	June 1894	10 4
758	South Fork of Salmon, near Warren, Idaho	1,160	5	5	1936	1,604	20,000	1 38	6.62	June 9, 1921	7 7
759	Yankee Fork of Salmon, near Clayton, Idaho	195	13	15	1936	177	3,360	910	17.3	June 9, 1933	7 5
760	Grande Ronde, La Grande, Oreg	(900)	22	28	1936	353	8,880	292	17.3	June 12, 1921	9 4
761	Grande Ronde, Rondowa, Oreg	2,555	10	10	1936	1,890	22,400	740	9.87	March 18, 1932	3 0
762	Selway, near Lowell, Idaho	1,510	7	8	1936	3,373	33,800	2 23	8.76	do	4 7
763	South Fork of Clearwater, near Grangeville, Idaho.	865	17	19	1936	833	9,830	960	22.4	June 14, 1933	8 3
764	Clearwater, Kamiah, Idaho	4,850	26	26	1936	8,310	81,400	1 72	16.8	May 30, 1912	11 7
765	Clearwater, Spalding, Idaho	9,570	10	10	1936	15,270	172,000	1 60	18.0	June 10, 1933	17 6
766	North Fork of Clearwater, near Ahsahka, Idaho.	2,440	10	10	1936	585	100,000	249	41.0	December 23, 1933	20 2
	Pacific Slope Basins									do	
767	Wynoochee, Oxbow, Wash	65	11	11	1936	780	18,000	12 0	277	January 22, 1935	22 3
768	Quinault, Quinault Lake, Wash	294	25	25	1936	2,728	37,000	10 3	140	December 12, 1921	22 8
769	Queets, near Clearwater, Wash	454	6	6	1936		100,000		220	January 22, 1935	47 0
770	Hoh, near Spruce, Wash	193	10	10	1936	2,019	40,000	10 5	207	November 5, 1934	28 5
771	Elwha, near Port Angeles, Wash	262	22	22	1936	1,484	26,700	5 65	102	December 21, 1933	16 5
772	North Fork of Skokomish, near Hoodsport, Wash	60	12	12	1936	464	23,300	7 74	339	November 5, 1934	30 1
773	South Fork of Skokomish, near Union, Wash	81	5	5	1925	754	17,000	9 34	210	January 22, 1935	18 9
774	Nisqually, near Alder, Wash	250	5	5	1936	1,393	25,000	5 57	100	December 22, 1933	15 8
775	Puyallup, Puyallup, Wash	911	22	22	1936	2,302	57,000	3 60	62	December 10, 1933	18 0
776	Cedar, Cedar Falls, Wash	83	22	22	1936	308	6,290	3 72	76	December 19, 1917	6 9
777	Cedar, near Landsberg, Wash	136	38	38	1936	708	13,600	5 20	100	November 19, 1911	11 7
778	North Fork of Skykomish, Index, Wash	149	19	19	1936	1,226	(26,500)	8 20	17	December 21, 1933	21 7
779	South Fork of Skykomish, near Index, Wash	355	28	28	1936	2,378	57,000	6 70	160	December 18, 1917	30 3
780	Skykomish, near Gold Bar, Wash	535	8	8	1936	3,902	79,000	7 28	148	December 21, 1933	34 1
781	North Fork Snoqualmie, near North Bend, Wash.	105	26	26	1936	698	11,500	6 63	109	October 25, 1934	11 2
782	South Fork Snoqualmie, North Bend, Wash	84	26	26	1936	541	7,620	6 45	91 0	do	8 3
783	South Fork Stillaguamish, near Granite Falls, Wash.	119	8	8	1936	1,060	26,700	8 88	223	February 26, 1932	24 4
784	South Fork Stillaguamish, near Arlington, Wash.	254	8	8	1936	1,781	35,000	7.00	138	do	22 0
785	Skagit, near Newhalem, Wash	1,160	28	28	1936	4,431	60,000	3 83	51 8	December 12, 1921	17 6
786	Skagit, near Concrete, Wash	2,700	12	12	1936	14,590	117,000	3 40	54 5	February 27, 1932	28 3
787	Sauk, near Darrington, Wash	152	13	13	1936	1,147	23,000	7 54	151	December 12, 1921	18 7
788	Sauk, near Sauk, Wash	714	8	8	1936	4,270	68,500	5 97	96	February 26, 1932	25 6
789	Columbia, Trail, British Columbia	34,000	23	23	1936	71,740	312,000	2 11	9 17	June 14, 1913	16 8
790	Columbia, Kettle Falls, Wash	64,500	23	42	1936	100,700	700,000 / 468,000	1 55	10 85 / 7.25	June 14, 1913	27 6 / 18

See footnotes at end of table

Average and maximum river discharges—Continued

Item No.	River and station	Drainage area (square miles)	Years of record			Discharge in cubic feet per second		Discharge in cu. ft. per sec. per sq mi		Date of maximum	Myers rating (percent)
			Average	Maximum	Final	Average	Maximum	Average	Maximum		
791	Columbia, Grand Coulee, Wash	74,100	23	42	1936	109,200	725,000	1.47	9.78	June 1894	26.6
							492,000		6.64	June 15, 1913	18.1
792	Columbia, Trinidad, Wash	89,700	23	42	1936	120,000	740,000	1.34	8.25	June 7, 1894	24.7
							528,000		5.88	June 15, 1913	17.6
793	Columbia, The Dalles, Oreg	237,000	58	58	1936	199,800	1,170,000	.845	4.94	June 6, 1894	24.1
							766,000		3.23	May 29, 1925	15.7
794	Kootenai, Libby, Mont	11,000	26	26	1936	11,700	130,000	1.06	11.8	June 21, 1916	12.4
795	Clark Fork, St. Regis, Mont	10,500	20	20	1936	7,705	62,800	.733	5.97	May 30, 1913	6.1
796	Clark Fork, near Plains, Mont	19,900	26	26	1936	19,760	126,000	.99	6.33	May 28, 1928	8.9
797	Clark Fork, near Heron, Mont	21,800	8	42	1936	20,106	(300,000)	.920	13.7	June 1894	20.3
							137,000			June 17, 1933	9.3
798	Clark Fork, Priest River, Idaho	24,200	33	42	1936	25,850	217,000	1.07	8.97	June 1894	14.0
							136,000		5.62	June 16, 1913	8.7
799	Clark Fork, near Metaline Falls, Wash	25,200	24	26	1936	26,570	139,000	1.05	5.51	June 19, 1916	8.8
800	South Fork Flathead, near Columbia Falls, Mont	1,640	13	26	1936	3,612	46,200	2.20	28.2		11.4
801	Flathead, Columbia Falls, Mont	4,440	8	14	1936	9,333	102,000	2.10	23.0	June 1, 1923	15.3
802	Flathead, near Polson, Mont	7,010	29	29	1936	11,670	82,100	1.67	11.7	May 30, 1928	9.8
803	Priest, outlet of Priest Lake, near Coolin, Idaho	572	22	24	1936	1,090	7,290	1.90	12.7	May 30, 1917	3.0
804	Coeur d'Alene, near Cataldo, Idaho	1,220	17	17	1936	2,490	55,300	2.04	45.5	December 23, 1933	15.8
805	Spokane, Post Falls, Idaho	3,880	23	23	1936	6,110	50,100	1.58	12.9	December 25, 1933	8.0
806	Spokane, Spokane, Wash	4,350	45	45	1936	6,963	49,000	1.60	11.3	May 31, 1894	8.4
807	Spokane, below Little Falls, near Long Lake, Wash	6,380	24	24	1936	7,777	48,000	1.22	7.52	December 26, 1933	6.0
808	St Joe, Calder, Idaho	1,080	17	17	1936	2,370	53,000	2.20	49.1	December 23, 1933	16.1
809	St Maries, Lotus, Idaho	420	16	17	1936	509	23,800	1.21	56.8	December 23, 1933	11.6
810	Chelan, Chelan, Wash	950	33	33	1936	2,068	12,800	2.17	13.5	June 3, 1936	4.2
811	Wenatchee, Plain, Wash	591	32	32	1936	2,239	20,800	3.79	35.2	December 13, 1921	8.6
812	Yakima, near Martin, Wash	55	32	32	1936	332	7,370	6.04	134	March 26, 1915	10.0
813	Yakima, Cle Elum, Wash	500	30	30	1936	1,997	25,600	4.00	50.5	November 14, 1906	11.4
814	Yakima, Umtanum, Wash	1,620	1	21	1936	2,476	41,100	1.53	25.3	November 15, 1906	10.2
815	Yakima, Kiona, Wash	5,520	21	21	1936	4,577	71,100	.83	12.9	December 23, 1933	9.6
816	Kachess, near Easton, Wash	64	33	33	1936	290	2,240	4.53	34.0	August 27, 1920	2.8
817	Cle Elum, near Roslyn, Wash	202	33	33	1936	920	18,700	4.55	92.5	November 15, 1906	13.2
818	Naches, near Naches, Wash	942	18	27	1936	1,678	33,200	1.78	35.3	December 23, 1933	10.8
819	Bumping, near Nile, Wash	68	27	27	1936	297	5,180	4.38	76.1	December 29, 1917	6.3
820	Teton, at Teton Dam, near Naches, Wash	187	17	22	1936	479	8,450	2.56	45.2	December 22, 1933	6.2

821	Tieton, at Tieton Canal Headworks, near Naches, Wash	240	27	28	1936	558	8, 910	2 32	37 0	December 22, 1933	5 8
822	South Fork of Walla Walla, near Milton, Oreg	67	13	13	1936	170	(3, 000)	2 54	44 7	May 30, 1906	3 7
823	Umatilla, Pendleton, Oreg	(800)	13	30	1936	443	(15, 000)	554	18 7	May 30, 1906	5 3
							13, 500		16 9	April 1, 1931	4 8
824	Umatilla, Yoakum, Oreg	1, 280	33	33	1936	672	20, 000	526	15 6	May 30, 1906	5 6
825	Umatilla, near Umatilla, Oreg	2, 290	32	32	1936	508	19, 600	222	8 55	May 31, 1906	4 1
826	Willow Creek, near Heppner, Oreg	20		33			36, 000		1, 800	June 1903	80 5
827	John Day, McDonald Ferry, Oreg	7, 580	31	42	1936	1, 912	33, 000	253	4 35	May 1894	3 8
							24, 900		3 29	March 20, 1932	2 9
828	Deschutes, Benham Falls, Oreg	(1, 300)	20	22	1936	1, 381	5, 000	1 06	3 85	November 27, 1909	1 4
829	Deschutes, below Bend, Oreg	(1, 500)	22	31	1936	704	4, 820	.470	3 22	November 27, 1909	1 2
830	Deschutes, Moody, Oreg	10, 500	31	32	1936	5, 879	43, 600	559	4 15	January 7, 1923	4 3
831	White, below Tygh Valley, Oreg	393	19	19	1936	428	13, 300	1 09	33 9	January 6, 1923	6 7
832	Klickitat, near Glenwood, Wash	356	19	27	1936	850	9, 870	2 38	27 6	December 22, 1933	5 2
833	Klickitat, near Pitt, Wash	1, 170		11	1936		21, 000		18 0	December 22, 1933	6 1
834	White Salmon, Husum, Wash	300	18	18	1936	977	10, 800	3 25	36 0	December 22, 1933	6 2
835	Hood, near Hood River, Oreg	329	23	23	1936	1, 086	34, 000	3 30	10 3	January 6, 1923	6 0
836	Sandy, near Marmot, Oreg	262	25	25	1936	1, 341	29, 200	5 12	112	January 6, 1923	18 1
837	Sandy, near Bull Run, Oreg	440	10	11	1936	2, 246	58, 000	5 10	132	March 31, 1931	27 7
838	Salmon, Welches, Oreg	100	13	13	1936	438	13, 000	4 38	130	March 31, 1931	13 0
839	Middle Fork Willamette, Eula, Oreg	941	12	13	1936	2, 337	55, 100	2 48	58 6	February 21, 1927	18 0
840	Willamette, Springfield, Oreg	2, 030	18	19	1936	4, 942	(150, 000)	2 43	74 0	January 25, 1903	33.3
										December 1861	
										February 1890	
							73, 300		36 2	February 21, 1927	16 3
841	Willamette, Albany, Oreg	4, 840	41	75	1936	13, 740	274, 000	2 84	56 6	December 4, 1861	39 4
							229, 000		47 4	January 14, 1881	33 0
							[2] 208, 000		43 0	January 26, 1903	29 9
842	Willamette, Salem, Oreg	7, 280	16	75	1936	22, 010	500, 000	3 03	68 7	December 4, 1861	38 6
							(440, 000)		60 4	February 5, 1890	51 5
							315, 000		43 3	November 25, 1909	36 9
843	McKenzie, McKensie Bridge, Oreg	345	17	26	1936	1, 636	18, 000	4.74	52 1	January 6, 1923	9 7
844	McKenzie, near Vida, Oreg	930	12	14	1936	3, 633	60, 000	3 90	64 4	January 6, 1923	19 7
							47, 200		50 7	February 20, 1927	15 5
845	North Santiam, Mehama, Oreg	665	19	19	1936	3, 272	62, 900	4 92	94 5	November 20, 1921	24 4
846	South Santiam, Waterloo, Oreg	640	14	16	1936	2, 728	70, 000	4 25	109	March 31, 1931	27 6
847	Clackamas, Big Bottom, Oreg	132	16	16	1936	451	6, 750	3 42	51 0	March 31, 1931	5 9
848	Clackamas, above Three Lynx Creek, Oreg	488	17	17	1936	1, 849	34, 800	3 79	71 2	March 31, 1931	15 8
849	Clackamas, near Cazadero, Oreg	665	27	27	1936	2, 607	60, 800	3 92	91 2	March 31, 1931	23 6
850	Canyon Creek, near Amboy, Wash	62	11	11	1933	410	10, 000	6 62	161	November 25, 1927	12 7
851	East Fork of Lewis, near Heisson, Wash	124	7	7	1936	735	15, 600	5 90	126	December 22, 1933	14 0
852	Lewis, near Cougar, Wash	483	12	15	1936	2, 819	54, 400	5 83	112	December 21, 1933	24 7
853	Lewis, Ariel, Wash	733	13	15	1936	4, 544	129, 000	6 20	176	December 22, 1933	47 7
854	Cowlitz, Packwood, Wash	287	15	15	1936	1, 645	36, 600	5 74	128	December 21, 1933	21 6
855	Cowlitz, Castle Rock, Wash	2, 210	8	10	1936	8, 858	139, 000	4 00	63.0	December 23, 1933	29.6
856	Clear Fork of Cowlitz, near Packwood, Wash	56	11	11	1936	255	8, 030	4 55	144	December 22, 1933	10.7
857	Siletz, Siletz, Oreg	204	16	19	1936	1, 707	40, 800	8 38	200	November 20, 1921	28 6
858	North Umqua, above Rock Creek, near Glide, Oreg.	866	12	12	1936	2, 221	55, 000	2 56	63.5	February 20, 1927	18.7

See footnotes at end of table

Average and maximum river discharges—Continued

Item No.	River and station	Drainage area (square miles)	Years of record			Discharge in cubic feet per second		Discharge in cu ft per sec per sq mi		Date of maximum	Myers rating (percent)
			Average	Maximum	Final	Average	Maximum	Average	Maximum		
	Pacific Slope Basins—Continued										
859	North Umpqua, near Glide, Oreg	1,210	25	31	1936	3,200	90,000	2 65	74 5	November 22, 1909	25 9
							59,500		49 4	March 19, 1932	17 1
860	Umpqua, near Elkton, Oreg	3,680	31	75	1936	7,070	(230,000)	1 92	62 5	1861	37 9
861	South Fork of Coquille, Powers, Oreg	169	17	18	1936	685	25,300	4 05	150	February 21, 1927	19 5
862	Middle Fork of Coquille, near Myrtle Point, Oreg	305	8	12	1936	616	(30,000)	2 02	98 5	October 31, 1924	17 2
							22,600		74 0	October 31, 1924	13 0
863	North Fork of Coquille, near Myrtle Point, Oreg	276	8	27	1936	810	(20,000)	2 94	72 5	January 2, 1933	12 0
							10,400		37 6	November 1909	6 3
864	Rogue, above Prospect, Oreg	332	14	28	1936	681	9,300	2 05	28 0	January 3, 1933	5 1
865	Rogue, Raygold, Oreg	2,020	31	31	1936	2,713	91,500	1 34	45 5	November 22, 1909	20 4
866	Illinois, Kerby, Oreg	367	8	10	1936	904	50,000	2 46	136	February 21, 1927	26 1
867	Drew Creek, near Lakeview, Oreg	211	21	21	1930	61	3,000	289	14 2	February 20, 1927	2 1
868	Williamson, near Chiloquin, Oreg	3,000	19	19	1936	825	7,000	275	2 33	March 1, 1910	1 3
869	Link, Klamath Falls, Oreg	3,800	32	32	1936	1,623	9,400	426	2 47	April 27, 1917	1 5
870	Klamath, Keno, Oreg	3,920	15	32	1936	1,610	9,250	411	2 36	May 12, 1904	1 5
871	Klamath, below Fall Creek, near Copco, Calif	4,370	13	13	1936	1,324	6,950	304	1 59	May 10, 1904	1 1
872	Klamath, Somesbar, Calif	8,480	13	9	1936	5,359	60,300	.631	7 12	March 26, 1928	6 6
							(160,000)		18 9	February 21, 1927	17 4
873	Salmon, Somesbar, Calif	737	10	10	1936	1,335	21,600	1 81	28 4	January 14, 1936	8 0
874	Trinity, Lewiston, Calif	724	25	25	1936	1,395	31,900	1 93	44 0	November 30, 1926	11 8
875	Trinity, near Hoopa, Calif	2,820	5	10	1936	3,982	89,000	1 42	31 6	December 31, 1913	16 8
876	Smith, near Crescent City, Calif	613	5	5	1936	3,271	61,700	5 33	101	March 18, 1932	24 9
877	Eel, Hullville, Calif	(270)	13	13	1936	360	32,600	1 34	121	March 26, 1928	19 8
878	Eel, near Potter Valley, Calif	(350)	13	13	1936	[1]4,396	40,000	12 6	114	March 26, 1928	21 4
879	Eel, Scotia, Calif	3,070	23	25	1936	5,626	290,000	1 84	94 7	February 2, 1915	52 4
							196,000		64 0	January 16, 1936	35 4
880	Russian, Geyserville, Calif	662	17	17	1936	907	[2]16,500	1 37	24 9	March 1911	6 4
881	Sacramento, Antler, Calif	461	10	11	1936	5,762	34,900	874	73 5	March 26, 1928	15 8
882	Sacramento, Kennett, Calif	6,600	41	41	1936	11,220	94,900	1 21	14 4	March 26, 1928	11 7
883	Sacramento, near Red Bluff, Calif	9,300	40	41	1929	34,300	278,000	1 25	29 9	February 3, 1909	28 9
884	Sacramento, Rio Vista, Calif	27,260	20	67	1929	28,400	(254,000)	1 04	9 30	February 1909	15 4
							(600,000)		22 0	January 1862	36 3
885	Pit, near Canby, Calif	1,500	5	9	1936	136	13,600	091	9 30	March 8, 1904	3 6
886	Pit, Big Bend, Calif	4,290	25	26	1936	2,769	16,100	645	3 75	July 9, 1925	2 5
887	Pit, near Ydalpom, Calif	5,350	25	26	1936	3,829	47,000	716	8 80	December 31, 1913	6 7
888	McCloud, Baird, Calif	668	25	26	1936	1,637	27,600	2 45	41 3	February 2, 1917	10 7
889	Thomas Creek, Paskenta, Calif	188	15	15	1936	209	16,600	1 11	88 0	March 26, 1928	12 1

No.	Stream and place	Drainage area (sq. mi.)	Year	Date of maximum	Maximum discharge (sec.-ft.)	Per sq. mi.	Mean (sec.-ft.)
890	North Fork of Feather, near Prattville, Calif	507	1936	March 19, 1907	10,000	19.7	864
891	North Fork of Feather, near Big Bar, Calif	1,934	1936	February 21, 1936	38,400	19.9	2,412
892	Feather, near Oroville, Calif	3,611	1936	March 26, 1928	211,000	58.4	5,917
893	Middle Fork of Feather, near Clio, Calif	699	1936	March 26, 1928	11,000	15.8	189
894	Middle Fork of Feather, near Bidwell Bar, Calif	1,353	1936	March 26, 1928	100,000	73.9	1,655
895	South Fork of Feather, near Enterprise, Calif	134	1936	March 26, 1932	16,200	113	278
896	Middle Fork of Yuba, near North San Juan, Calif	207	1936	March 25, 1928	26,000	125	417
897	Yuba, Smartville, Calif	1,201	1936	March 26, 1928	120,000	100	2,927
898	North Fork of Yuba, near Sierra City, Calif	91 [3]	1936	March 25, 1928	5,920	65.0	198
899	Bear, near Wheatland, Calif	295	1936	January 14, 1909	29,600	100	315
900	North Fork of American, near Colfax, Calif	308	1936	April 8, 1935	55,000	178	611
901	Middle Fork of American, near Auburn, Calif	619	1936	March 25, 1928	100,000	161	1,313
902	South Fork of American, near Camino, Calif	497	1936	March 25, 1928	31,500	63.4	664
903	American, Fair Oaks, Calif	1,921	1936	March 25, 1928	182,000	94.8	3,629
904	Cache Creek, Yolo, Calif	1,230	1936	March 19, 1907	119,000	62.0	488
905	Putah Creek, near Guenoc, Calif	112	1936	February 2, 1915	21,100	17.2	123
906	Putah Creek, near Winters, Calif	(650)	1936	March 10, 1904	24,600	220	270
907	North Fork of Cosumnes, near El Dorado, Calif	202	1936	December 31, 1913	60,000	92.4	188
908	Cosumnes, Michigan Bar, Calif	537	1936	February 6, 1925	7,600	37.5	445
909	Middle Fork of Mokelumne, West Point, Calif	67 [2]	1936	January 23, 1914	23,800	44.3	49
910	North Fork of Mokelumne, below Salt Springs Dam, Calif	160	1936	March 25, 1928	2,550	38.0	412
911	Mokelumne, near Clements, Calif	630	1928	March 25, 1928	8,740	54.5	1,110
912	Calaveras, Jenny Lind, Calif	395	1936	January 31, 1911	25,600	40.6	255
913	North Fork of Stanislaus, near Avery, Calif	163	1936	May 11, 1915	69,600	176	392
914	Middle Fork of Stanislaus, near Avery, Calif	329	1936	March 19, 1907	5,250	32.1	661
915	Stanislaus, below Melones power house, Calif	898	1936	February 22, 1936	9,769	29.7	1,294
916	Eleanor Creek, near Hetch Hetchy, Calif	79	1936	March 25, 1928	17,200	19.2	211
917	Tuolumne, near Hetch Hetchy, Calif	462	1936	June 26, 1929	6,400	81.0	902
918	Tuolumne, near Buck Meadows, Calif	934	1936	January 14, 1909	12,000	25.9	501
919	Tuolumne, near La Grange, Calif	1,540	1936	March 25, 1928	27,200	29.1	2,083 [15]
920	Merced, near Yosemite, Calif	181	1936	January 31, 1911	38,100	24.8	2,519
921	Merced, Exchequer, Calif	1,035	1936	May 28, 1919	60,300	39.2	311
922	Merced, at Pohono Bridge, near Yosemite, Calif	321	1936	June 17, 1916	3,800	21.0	1,086
923	Fresno, near Knowles, Calif	132	1936	February 21, 1917	22,000	21.2	529
924	Mono Creek, near Vermilion Valley, Calif	92	1936	June 16, 1927	6,370	19.8	67
925	South Fork of San Joaquin. near Florence Lake, Calif	171	1936	June 4, 1922	4,500	34.2	126
926	San Joaquin, above Big Creek, Calif	1,042	1936	June 5, 1922	1,420	15.4	242
927	San Joaquin, near Friant, Calif	1,632	1936	January 25, 1914	3,460	20.2	1,067
928	San Joaquin, near Newman, Calif	(12,500)	1929	January 27, 1914	18,000	17.3	2,220
929	San Joaquin, near mouth, distributed among various channels	18,178	1929		(200,000)	28.3	2,141

See footnotes at end of table.

Average and maximum river discharges—Continued

Item No	River and station	Drainage area (square miles)	Years of record — Average	Years of record — Maximum	Years of record — Final	Discharge in cubic feet per second — Average	Discharge in cubic feet per second — Maximum	Discharge in cu. ft per sq mi — Average	Discharge in cu. ft per sq mi — Maximum	Date of maximum	Myers rating (percent)
	Public Slope Basins—Continued										
930	North Fork of Kings, near Cliff Camp, Calif.	174	13	14	1936	306	6,030	1 76	34 6	June 4, 1922	4 6
931	Kings, near Hume, Calif.	838	12	15	1936	1,052	11,700	1 26	14 0	June 4, 1922	4 0
932	Kings, Piedra, Calif.	1,694	41	41	1936	2,254	59,700	1 37	35 3	January 25, 1914	14 5
933	North Fork of Kaweah, Kaweah, Calif.	128	25	25	1936	85 7	7,400	670	57 9	January 25, 1914	6 5
934	Kaweah, near Three Rivers, Calif.	520	33	33	1936	531	14,780	1 02	28 2	January 17, 1916	6 4
935	Tule, near Porterville, Calif.	266	35	35	1936	132	6,780	.495	25 4	January 17, 1916	4 2
936	South Fork of Kern, near Onyx, Calif.	531	15	21	1936	75 7	2,360	.142	4 43	January 25, 1914	1 02
937	Kern, near Kernville, Calif.	845	24	24	1936	642	9,690	.760	11 4	January 17, 1916	3 3
938	Kern, near Bakersfield, Calif.	2,345	41	43	1936	922	18,287	393	7 80	January 26, 1914	3 8
939	Tehachapi Creek, below Tehachapi, Calif.	29 1			1938		20,500		704	September 1932	38 0
940	Tehachapi Creek, above Woodford-Keene, Calif.	69 5			1938		41,100		592	September 1932	49 4
941	Alameda Creek, above Sunol, Calif.	33 1	18	18	1929	18 4	[2] 1,660	.558	50 2	February 21, 1917	2 9
942	Alameda Creek, near Sunol, Calif.	(626)	10	10	1900	190	12,200	.304	19 5	December 3, 1892	4 9
943	Alameda Creek, near Niles, Calif.	633	19	19	1936	58 3	13,900	.092	22 0	February 10, 1922	5 5
944	San Antonio Creek, near Sunol, Calif.	39 7	18	18	1929	10 9	[2] 1,460	.275	36 8	January 3, 1916	2 3
945	Arroyo de la Laguna, near Pleasanton, Calif.	412	17	17	1929	42 1	[2] 9,810	.102	23 8	January 25, 1914	4 8
946	Coyote Creek, near Madrone, Calif.	193	30	30	1936	70 7	25,000	.365	130	March 7, 1911	8 6
947	Coyote Creek, near Edenvale, Calif.	229	20	20	1936	26 3	10,000	.115	43 6	February 10, 1922	6 6
948	Los Gatos Creek, Los Gatos, Calif.	40	7	7	1936	22 8	3,340	.570	83 5	December 27, 1931	5 3
949	Guadalupe Creek, San Jose, Calif.	131	7	7	1936	13 9	6,700	.106	51 0	December 27, 1931	5 8
950	San Francasquito Creek, Stanford University, Calif.	37 7	6	6	1936	6 9	1,660	.183	44 0	February 21, 1936	2 7
951	Uvas Creek, near Morgan Hill, Calif.	30 2	6	6	1936	21 0	4,340	695	144	December 27, 1931	7 9
952	Arroyo Seco, near Soledad, Calif.	238	35	35	1936	166	22,000	695	92 2	February 21, 1917	14 2
953	Salinas, near Santa Margarita, Calif.	150	4	4	1936	28 1	4,050	187	27 1	April 8, 1935	3 3
954	Salinas, near Spreckels, Calif.	4,180	5	8	1936	278	42,100	067	10 1	December 29, 1931	6 5
955	Santa Ynez, Gibraltar Dam, near Santa Barbara, Calif.	219	22	16	1936	45 7	7,245	208	33 0	April 1926	4 9
956	Santa Ynez. near Lompoc, Calif.	790	20	23	1936	213	41,800	270	53 0	January 25, 1914	14 9
957	Ventura, near Ventura, Calif.	187	6	9	1936	38 9	23,000	.208	123	December 31, 1933	16 8
958	Santa Paula Creek, near Santa Paula, Calif.	39 8	9	11	1938	21 8	(50 000)	505	267	March 2, 1938	36 5
959	Sespe Creek, near Fillmore, Calif.	257	9	10	1936	54 2	34,000 (66,000)	211	132 256 5	December 31, 1933	5 8
960	Piru Creek, near Piru, Calif.	432	9	13	1936	25 9	15,800 65,000	060	36 5 150 9	February 9, 1932	21 1
961	Santa Clara, near Saugus, Calif.	355	6	6	1935	2 32	3,870	007	10 9	January 1, 1934	2 1

No.	Station	
962	Santa Clara, near Montalvo, Calif	1,610
963	Arroyo Seco, near Pasadena, Calif	16 4
964	Eaton Creek, near Pasadena, Calif	6 5
965	Eaton Creek, Huntington Drive, Calif	12 9
966	Little Santa Anita Creek, near Sierra Madre, Calif	1 9
967	Little Santa Anita Creek, near Sierra Madre, Calif	10 5
968	Sawpit Creek, near Monrovia, Calif	5 3
969	Haines Creek, near Tujunga, Calif	1 2
970	Big Tujunga, Tujunga Dam, Calif	81 4
971	Los Angeles, Los Angeles, Calif	510
972	Los Angeles, near Downey, Calif	614
973	Los Angeles, Long Beach, Calif	1,060
974	Rio Hondo, Stewart and Grey Road, Calif	370
975	Ballona Creek, Centinela Blvd Bridge, near Culver City, Calif	112
976	San Jose Creek, near Whittier, Calif	85 2
977	Fish Creek, near Duarte, Calif	6 5
978	Rogers Creek, near Azusa, Calif	6 4
979	West Fork San Gabriel, Dam No. 2, Calif	40 4
980	West Fork San Gabriel, Camp Rincon, Calif	102
981	San Gabriel, Dam No 1, Calif	202
982	San Gabriel, near Azusa, Calif	214
983	Dalton Creek, near Glendora, Calif	7 5
984	San Dimas Creek, near San Dimas, Calif	18 3
985	San Antonio Creek, near Claremont, Calif	16 9
986	Santiago Creek, near Villa Park, Calif	83 8
987	San Jacinto, near San Jacinto, Calif	140 9
988	Cajon Creek, near Keenbrook, Calif	40 9
989	Lytle Creek, near Fontana, Calif	47 9
990	Lytle Creek, San Bernardino, Calif	60
991	Devil Canyon Creek, near San Bernardino, Calif	6 16
992	Plunge Creek, near East Highlands, Calif	16 9
993	Mill Creek, near Craftonville, Calif	42 9

See footnotes at end of table

Average and maximum river discharges—Continued

Item No.	River and station	Drainage area (square miles)	Years of record			Discharge in cubic feet per second		Discharge in cu. ft per sq. mi.		Date of maximum	Myers rating (percent)
			Average	Maximum	Final	Average	Maximum	Average	Maximum		
	Pacific Slope Basins—Continued										
994	City Creek, near Highland, Calif	19.8	17	17	1936	7.43	2,360	0 375	119	April 5, 1926	5 3
995	Waterman Canyon, near Arrowhead Springs, Calif	4.55	18	20	1936	2 46	182	540	40 0	February 2, 1936	9
996	Strawberry Creek, near Arrowhead Springs, Calif	8.6	16	17	1936	3.64	408	423	47 5	January 2, 1922	1 4
997	Warm Creek, near Colton, Calif	50	16	18	1938	58.7	(30,000)	1 17	600	March 2, 1938	42 5
							2,780		55 7	December 21, 1921	3 94
							3,000		25 4	February 16, 1927	2 8
998	San Timoteo Creek, near Redlands, Calif	118	10	10	1936	1.51	3,000	.013	50 8	March 2, 1938	2 6
							6,000		238	March 2, 1938	5 6
999	Santa Ana, near Mentone, Calif	189	38	42	1938	36.6	45,000	194	154	January 27, 1916	32 7
			38	40	1936	18 87.8	29,000	465	46 5	March 3, 1938	21 2
1000	Santa Ana, near Prado, Calif	2,288	17	19	1938	141	(106,500)	615		January 27, 1916	22 3
1001	San Luis Rey, near Mesa Grande, Calif	209	25	25	1938	40 6	58,600	194	280	January 27, 1916	40 5
1002	San Luis Rey, near Bonsall, Calif	514	7	45	1936	16 9	128,100	033	249	February 1891	56 5
1003	Temecula Creek, Railroad Canyon, near Temecula, Calif	592	12	12	1935	19 1	27,600	033	46 6	February 16, 1927	11 3
1004	Santa Margarita, near Fall Brook, Calif	645	10	11	1935	24.6	33,000	038	51 0	February 16, 1927	13 0
			10	10	1935	59 4	2 37,200	198	124	January 27, 1916	21 5
1005	San Dieguito, Lake Hodges, Calif	299		20	1936		38,000		201	January 27, 1916	27 6
1006	San Diego, Lakeside, Calif	189	20	20	1935	38.6	70,200	103	187	...do...	36 2
1007	San Diego, Santee, Calif	375		20	1935		75,000		173	...do...	36 0
1008	San Diego, San Diego, Calif	434	22	22	1927	11.3	9,900	258	226	January 1916	15 0
1009	Sweetwater, near Descanso, Calif	43 7	40	40	1927	(26.4)	43,000	153	250	...do...	32 8
1010	Sweetwater, Jamacha, Calif	172	9	49	1936	8 47	45,500	047	251	...do...	33 8
1011	Sweetwater, Sweetwater Dam, Calif	181								...do...	

1 Compiled mainly from U. S. Geological Survey water supply papers, but supplemented from other Federal and State publications as well as technical journals, and occasionally from unofficial sources, by C. S. Jarvis, hydraulic engineer　　2 Daily mean.　　3 Influenced by both direct and reversed flow between Red River and the Mississippi

4 Net contribution to Atchafalaya from Mississippi River.　　5 Reduced by Bonnett Carre Spillway in 1937, by crevass in 1927, and by discharge into Old River during high flood periods.　　8 Excluding closed basins, 2,940 square miles　　SK

6 Ordinarily contributing; area of about 193,000 square mile drains via Atchafalaya.　　7 9,240 square miles probably noncontributing.　　11 Flood crests are largely diverted to floodways and low flows to irrigation canals

9 Excluding closed basins, 4,200 square miles.　　10 Largely spring-fed

12 Including Yuma Canal, diversion of 1,840 cubic feet per second average, about two-thirds of which returns through wasteway to river channel below gaging station

13 Including regulated run-off from Strawberry Valley, draining 175 square miles, average yield about 100 cubic feet per second, or .57 cubic square miles.

14 Including the diversion of 198 cubic feet per second to Russian River Basin for power and irrigation purposes.

16 Reflecting increased consumptive use, mainly for irrigation.

15 River and canal flow.　　17 Combined creek and canal flow.

18 Combined river and canal flow.

NOTE.—Figures in parentheses are unofficial estimates

INDEX

O